T0141896

Advances in Computer Vision and Pattern Recognition

More information about this series at http://www.springer.com/series/4205

Saurabh Prasad · Jocelyn Chanussot
Editors

Hyperspectral Image Analysis

Advances in Machine Learning and Signal Processing

 Springer

Editors
Saurabh Prasad
Department of Electrical
and Computer Engineering
University of Houston
Houston, TX, USA

Jocelyn Chanussot
CNRS, Grenoble INP, GIPSA-lab
Université Grenoble Alpes
Grenoble, France

ISSN 2191-6586   ISSN 2191-6594   (electronic)
Advances in Computer Vision and Pattern Recognition
ISBN 978-3-030-38619-1   ISBN 978-3-030-38617-7   (eBook)
https://doi.org/10.1007/978-3-030-38617-7

This Springer imprint is published by the registered company Springer Nature Switzerland AG
The registered company address is: Gewerbestrasse 11, 6330 Cham, Switzerland

Contents

Chapter 1
Introduction

Saurabh Prasad and Jocelyn Chanussot

Hyperspectral imaging entails acquiring a large number of images over hundreds (to thousands) of narrowband contiguous channels, spanning the visible and infrared regimes of the electromagnetic spectrum. The underlying premise of such imaging is that it captures the underlying processes (e.g., chemical characteristics, biophysical properties, etc.) at the pixel level. Recent advances in optical sensing technology (miniaturization and low-cost architectures for spectral imaging) and sensing platforms from which such imagers can be deployed (e.g., handheld devices, unmanned aerial vehicles) have the potential to enable ubiquitous multispectral and hyperspectral imaging on demand to support a variety of applications, such as biomedicine and sensing of our environment. In many applications, it is possible to leverage data acquired by other modalities (e.g., Synthetic Aperture Radar, SAR, and Light Detection and Ranging (LiDAR)) in conjunction with hyperspectral imagery to paint a complete picture—for example, hyperspectral imagery and LiDAR data when used together provide information about the underlying chemistry (e.g., as provided by hyperspectral data) and the underlying topography (as provided by LiDAR data) and can facilitate robust land-cover classification. Although this increase in the quality and quantity of diverse multi-channel optical data can potentially facilitate improved understanding of fundamental scientific questions, there is a strong need for robust image analysis methods that can address the challenges posed by these imaging paradigms. While machine learning approaches for image analysis have evolved

S. Prasad (✉)
Department of Electrical and Computer Engineering, University of Houston,
Houston, TX 77578, USA
e-mail: saurabh.prasad@ieee.org

J. Chanussot
University of Grenoble Alpes, CNRS, Grenoble INP, GIPSA-lab,
38000 Grenoble, France
e-mail: jocelyn.chanussot@grenoble-inp.fr

© Springer Nature Switzerland AG 2020
S. Prasad and J. Chanussot (eds.), *Hyperspectral Image Analysis*,
Advances in Computer Vision and Pattern Recognition,
https://doi.org/10.1007/978-3-030-38617-7_1

1

to exploit the rich information provided by hyperspectral imagery and other high-dimensional imagery data, key challenges remain for effective utilization in an operational environment, including the following:

– Representation and effective feature extraction from such high-dimensional datasets,
– Design of effective learning strategies that are robust to a limited quantity of training samples (*in situ data*), missing or noisy labels, and spatial–temporal non-stationary environments,
– Design and optimization of analysis algorithms that can effectively handle nonlinear, complex decision boundaries separating classes (objects of interest on ground) in the feature space,
– The need to address variability in light source $\rightarrow$ sensor $\rightarrow$ object geometry and variation in orientation and scale of objects in ubiquitous sensing environments from a multitude of sensors and sensing platforms, and
– Effective utilization of the rich and vast quantity of unlabeled data available in geospatial imagery in conjunction with limited ground truth for robust analysis.

This book focuses on advances in machine learning and signal processing for hyperspectral image analysis and presents recent algorithmic developments toward robust image analysis that address challenges posed by the unique nature of such imagery. We note that although a majority of the chapters in this book focus on hyperspectral imagery, these ideas extend to data obtained from other modalities, such as microwave remote sensing, multiplexed immunofluorescence imaging, etc. Chapters in this book are grouped in the challenges they address based on the following broad thematic areas.

Challenges in Supervised, Semi-Supervised, and Unsupervised Learning: The high-dimensional nature of hyperspectral imagery implies that many learning algorithms that seek to leverage the underlying spatial–spectral information are associated with a large number of degrees of freedom, necessitating a rich (in both quality and quantity) representative ground reference data. Leveraging the limited quantity and varying quality of labeled data associated with remote sensing and biomedicine applications is a critical requirement of successful learning algorithms, and a vast number of recent developments address this aspect of learning, under the umbrella of supervised, semi-supervised, and unsupervised learning. Further, the end goal of learning may not always be discrete classification. Numerous applications with hyperspectral imagery entail mapping spectral observations to prediction (e.g., posed as a regression problem) of continuous-valued quantities (such as biophysical parameters)—although there exist commonalities between learning algorithms that are carrying out discrete classification and regression, care must be taken to understand the needs and constraints of each application.

In Chap. 2, Moreno-Martinez et al. survey recent developments in machine learning for estimating spatial and temporal parameters from multi-channel earth-observation images (both microwave imaging and passive optical imaging). Chapters 3 and 4 are a two-part series introducing the foundations of deep learning as applied to hyperspectral image analysis. In Chap. 3 (Part I), Berisha et al. review the foundations of convolutional and recurrent neural networks as they can be applied for spatial–spectral analysis of hyperspectral imagery. In Chap. 4 (Part II), Shahraki et al. present practical architectures and design strategies to successfully deploy such networks for hyperspectral image analysis tasks. Results with hyperspectral imagery in the areas of remote sensing and biomedicine are presented, along with a detailed discussion of the "successful" network configurations relative to the data characteristics. In Chap. 5, Zhou and Prasad review recent developments in deep learning that address the label scarcity problem—including semi-supervised, transfer, and active learning. In Chap. 6, Jiao et al. present multiple instance learning as a mechanism to address imprecise ground reference data that is commonly encountered in hyperspectral remote sensing. In Chap. 7, Rise et al. survey supervised, semi-supervised, and unsupervised learning for hyperspectral regression tasks. In Chap. 8, Wu et al. survey sparse-representation-based methods for hyperspectral image classification. In Chap. 9, Gu et al. review multiple kernel learning for hyperspectral image classification.

Subspace Learning and Feature Selection: Given the high dimensionality of spectral features and the inherent inter-channel correlations due to the dense, contiguous spectral sampling, algorithms that learn effective subspaces (e.g., subspaces where much of the discriminative information is retained) and that learn the most relevant spectral channels are often a crucial pre-processing to image analysis. In Chap. 10, Zhu et al. present a low-dimensional manifold model for hyperspectral image reconstruction. In Chap. 11, Taherkhani et al. present a deep sparse band selection for hyperspectral face recognition.

Change and Anomaly Detection: In many applications, the ability to reliably detect changes between sets of hyperspectral imagery is highly desirable. In Chap. 12, Ziemann and Matteoli present recent developments toward robust detection of large-scale and anomalous changes.

Spectral Unmixing: The spatial resolution of hyperspectral imagery acquired from airborne or spaceborne sensors often is not fine enough relative to the size of objects of interest in the scene, resulting in mixed pixels. Over recent years, numerous advances have been made in the area of spectral unmixing—the process of estimating the relative abundance of the endmembers (e.g., objects in these mixed pixels) in each mixed pixel. In Chap. 13, Zhang et al. review recent advances in spectral unmixing using sparse techniques and deep learning.

Image Superresolution: In many remote sensing applications, a common imaging scenario entails simultaneous acquisition of very high spatial resolution color/multispectral/monochromatic(pan) images and lower spatial resolution hyperspectral images. One can leverage this by extracting spatial information available in the higher resolution imagery, which can then be fused in the lower resolution hyperspectral imagery. In Chap. 14, Yang et al. present a deep-learning-based approach to fuse high spatial resolution multispectral imagery with hyperspectral imagery.

Target Detection: An important application of hyperspectral imagery has been identification of targets of interest in a scene. In Chap. 15, Bitar et al. present an automatic target detection approach for sparse hyperspectral images.

Chapter 2
Machine Learning Methods for Spatial and Temporal Parameter Estimation

Álvaro Moreno-Martínez, María Piles, Jordi Muñoz-Marí, Manuel Campos-Taberner, Jose E. Adsuara, Anna Mateo, Adrián Perez-Suay, Francisco Javier García-Haro and Gustau Camps-Valls

Abstract Monitoring vegetation with satellite remote sensing is of paramount relevance to understand the status and health of our planet. Accurate and constant monitoring of the biosphere has large societal, economical, and environmental implications, given the increasing demand of biofuels and food by the world population. The current democratization of machine learning, big data, and high processing capabilities allow us to take such endeavor in a decisive manner. This chapter proposes three novel machine learning approaches to exploit spatial, temporal, multi-sensor, and large-scale data characteristics. We show (1) the application of multi-output Gaussian processes for gap-filling time series of soil moisture retrievals from three spaceborne sensors; (2) a new kernel distribution regression model that exploits multiple observations and higher order relations to estimate county-level crop yield from time series of vegetation optical depth; and finally (3) we show the combination of radiative transfer models with random forests to estimate leaf area index, fraction of absorbed photosynthetically active radiation, fraction vegetation cover, and canopy water content at global scale from long-term time series of multispectral data exploiting the Google Earth Engine cloud processing capabilities. The approaches demonstrate that machine learning algorithms can ingest and process multi-sensor data and provide accurate estimates of key parameters for vegetation monitoring.

Á. Moreno-Martínez, M. Piles, J. Muñoz-Marí, M. Campos-Taberner, J. E. Adsuara, G. Camps-Valls—Authors contributed equally.

Á. Moreno-Martínez (✉) · M. Piles · J. Muñoz-Marí · J. E. Adsuara · A. Mateo · A. Perez-Suay · G. Camps-Valls
Image Processing Laboratory (IPL), Universitat de València, Valencia, Spain
e-mail: alvaro.moreno@uv.es

G. Camps-Valls
e-mail: gustau.camps@uv.es

M. Campos-Taberner · F. Javier García-Haro
Department of Earth Physics and Thermodynamics, Universitat de València, Valencia, Spain
e-mail: manuel.campos@uv.es

© Springer Nature Switzerland AG 2020
S. Prasad and J. Chanussot (eds.), *Hyperspectral Image Analysis*,
Advances in Computer Vision and Pattern Recognition,
https://doi.org/10.1007/978-3-030-38617-7_2

2.1 Introduction

2.1.1 Remote Sensing as a Diagnostic Tool

The Earth is a complex, dynamic, and networked system, and this system is under pressure and in continuous change. Population is increasingly demanding more food and biofuels, at a faster pace, worldwide. Consequently, monitoring the planet in a spatially explicit and timely resolved manner is an urgent need to address important societal, environmental, and economical questions. This is exactly the main goal of Earth Observation (EO) from space, and current satellite sensors operating in different bands of the electromagnetic spectrum help in this challenge as accurate diagnostic tools.

The analysis of the acquired sensor data can be done either at local or global scales by looking at biogeochemical cycles, atmospheric situations, and vegetation dynamics [1–5]. All these complex interactions are studied through the definition of bio-geophysical parameters, either representing different properties for land (e.g., surface temperature, soil moisture, crop yield, defoliation, biomass, leaf area coverage), water (e.g., yellow substance, ocean color, suspended matter, or chlorophyll concentration), or the atmosphere (e.g., temperature, moisture, or trace gases). Every single application considers the specific knowledge about the physical, chemical, and biological processes involved, such as energy balance, evapotranspiration, or photosynthesis.

However, remotely sensed observations only sample the energy reflected or emitted by the surface and thus, an intermediate modeling step is necessary to transform the measurements into estimations of the biophysical parameters [6]. From a pure statistics standpoint, this is considered to be as an *inverse modeling* problem, because we have access to observations generated by the system and we are interested in the unknown parameters that generated those. A series of international study projections, such as the International Geosphere-Biosphere Programme (IGBP), the World Climate Research Programme (WCRP), and the National Aeronautics and Space Administration (NASA) Earth Observing System (EOS), established remote sensing model inversion as one of the most important problems to be solved with EO imagery in the near future.

2.1.2 Data and Model Challenges

Current EO, however, faces two very important challenges that we hereby define as the *data problem* and the *model problem*:

- *The data problem*: The data involved in EO applications is *big, diverse, and unstructured*. We often deal with remote sensing data acquired by many satellite sensors working with different and ever-increasing spatial, temporal, and vertical resolutions. Not to mention that data may also come from high-resolution simulations and re-analysis. At the same time, data is heterogeneous and covers space and time with uneven resolutions, different footprints, signal and noise levels, and feature characteristics. EO applications on land monitoring have mainly considered optical sensors, like the **NASA A-Train** (http://atrain.nasa. gov/) satellite constellations including MODIS and Landsat, and recently the European Space Agency (ESA) Sentinels 2–3 sensors. More recently, sensors operating in the microwave range of the spectrum were introduced. Unlike optically based technologies, microwaves are not affected by atmospheric conditions, and a total coverage of the Earth's surface is obtained every 2–3 days. Microwave radiometry is optimal for sensing the water content in soils and vegetation, but the passive measurement is presently limited in spatial resolution by the size of the instrument antenna aperture to ~25 km (e.g., ESA's SMOS, NASA's SMAP). Active microwave remote sensing can overcome this limitation but often it is accompanied by constraints on spatial coverage and temporal data refresh rate and require complex scattering models for inversion of geophysical parameters (e.g., ESA's Sentinel 1). Optical sensing technology, in turn, is at a maturity level today that allows providing very fine spatial resolution on a weekly basis (e.g., ESA's Sentinel 2). Undoubtedly, the combination of satellite-based microwave and optical sensory data offers an unprecedented opportunity to obtain a unique view of the Earth system processes.

- *The model problem*: Dealing with such data characteristics and big data influx requires (semi)automatic processing techniques that should be accurate, robust, reliable, and fast. Over the last few decades, a wide diversity of bio-geophysical retrieval methods have been developed, but only a few of them made it into operational processing chains. Lately, machine learning has attained outstanding results in the estimation of climate variables and related bio-geophysical parameters at local and global scales [1]: leaf area index (LAI) [7] and Gross Primary Production (GPP) [8–11] are currently derived with neural networks, kernel methods, and random forests, while multiple regression is used for retrieving biomass [12], support vector methods were also proposed to derive vegetation parameters [13, 14], and kernel methods and Gaussian processes (GPs) [15] have been paid wide attention in the last years in deriving vegetation properties [16]. However, it is important to observe here that, very often, these methods are applied blindly, without being adapted to the data specificities. On the one hand, data exhibits clear spatial and temporal structures that could be useful to design new kernel functions in GPs [17] or rely on convolutional networks [18]. On the other hand, data from different sensors should be synergistically combined in the model, but this is often done via *ad hoc* data re-sampling or statistics summarization, as a convenient way to data preparation for the algorithm. These practices are far from being optimal, and a lot is yet to be done in the algorithm development arena to improve algorithms that respect data characteristics, learn structures from data, fuse heterogeneous multi-sensor and multi-resolution data naturally, and scale well to big data volumes.

Fig. 2.1 Normalized worldwide interest (i.e., popularity) of terms "remote sensing", "machine learning", "artificial intelligence", and "big data" in the last decade, as measured in Google trends©

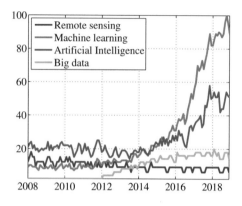

Tackling the two sides of the EO challenge is nowadays possible. The current popularization of machine learning, big data, and high processing capabilities allows us to take such an endeavor in a decisive manner, cf. Fig. 2.1.

Nowadays, both data and algorithms are mostly freely available, while large-scale data processing platforms, clusters, and infrastructures are accessible to everyone:

- Machine learning code is now ready to (re)use in different forms: from excellent packages and frameworks like scikit-learn or TensorFlow, to open accessible repositories and developer's platforms like GitHub.
- Earth observation data is also currently accessible through the main space agencies hubs: for example, ESA provides Sentinels data through the ESA open access hub, and NASA grants access via its NASA open data portal.

This unprecedented situation has sowed the seed for the development of applications and the creation of EO-centered companies. Google allows not only accessing but also processing data through the Google Earth Engine, which will be subject of study in this chapter (cf. Sect. 2.4), Descartes Labs offers an EO data processing facility in the cloud, and an increasing number of SMEs has grown around and created what is called the "EO exploitation ecosystem". Altogether, they have allowed tackling problems that were unthinkable just a decade ago.

> Earth observation through remote sensing offers great opportunities to monitor our planet by the estimation of key parameters of the land, ocean, and atmosphere. The combined action of machine learning, big data, and high-performance computing platforms, like the Google Earth Engine (GEE), is currently paving the way toward this goal.

2.1.3 Goals and Outline

In this chapter, we will focus on modern machine learning methods for deriving land parameters (e.g., about the vegetation status and crop production) from remote sensing data: we will introduce three recent ML developments that can deal with multisensor and multi-resolution data, that exploit nonlinear feature relations and higher order moments of data (observational) distributions, and that can be implemented in the Google cloud platform to derive global maps of parameters of interest. We will mainly focus on new kernel methods, Gaussian processes, and random forests, which fulfill the needs of the field: mathematical tractability and big data scalability, respectively.

We will treat three main problems with different particular data characteristics:

- *Non-uniform temporal sampling and sensor fusion*: First, we will focus on problems of interpolating remote sensing parameters when several variables are available and heavy non-uniform sampling is present. This is a common problem when trying to fuse information from different sensors or in optical remote sensing due to the presence of clouds. Microwave remote sensing is not affected by clouds, but measurements can also be limited in some regions due to combined effects of Radio Frequency Interferences (RFIs), presence of snow, dense vegetation canopies, and high topography [19]; since these effects are sensor- and frequency-dependent, the optimal blend of available microwave-based soil moisture products holds great promise, particularly for observational climate data records [20]. In Sect. 2.2, we will show the exploitation of multi-output Gaussian processes to fill in the temporal gaps in satellite-based estimates of soil moisture from SMOS (L-band passive), AMSR2 (C-band passive), and ASCAT (C-band active) [21, 22]. The method will allow to treat non-uniform sampling and "transfer information across sensors" when samples are missing.

- *Non-uniform spatial sampling*: In remote sensing and geospatial applications, we often encounter problems where one aims to spatialize a variable of interest from a sparse set of measurements, while having access to a finer grid of observations. This is the case of non-uniform spatial sampling. This mismatch in quantity and location is typically resolved by summarizing (e.g., averaging) the observations and co-locating them with the measure. This procedure is ad hoc and suboptimal. In Sect. 2.3, we introduce a new kernel distribution regression model that exploits multiple observations to estimate county-level yield of major crops (wheat, corn, and soybean) from SMAP-based vegetation optical depth (VOD) time series [23]. The method exploits all the available observations and their feature relations.

- *Uniform spatial–temporal data spatialization*: Finally, we deal in Sect. 2.4 with the exploitation of big data in the cloud by spatializing vegetation parameters of interest when long time series of data are available. We will show the combination of radiative transfer models (RTMs) with random forests to estimate various vegetation parameters, namely, LAI, Fraction of Absorbed Photosynthetically Active Radiation (FAPAR), Fraction Vegetation Cover (FVC), and Canopy water content (CWC), globally from long-term time series of MODIS data exploiting the GEE.

The platform will allow us to generate products of almost any variable of interest modeled in an RTM [24, 25].

We conclude in Sect. 2.5 with some remarks and an outline of future work. The approaches demonstrate that machine learning algorithms can ingest and process multi-sensor data and provide accurate estimates of key parameters for vegetation monitoring.

2.2 Gap Filling and Multi-sensor Fusion

Measurements of soil moisture (SM) are needed for a better global understanding of the land surface-climate feedbacks at both local and global scales. Satellite sensors operating in the low-frequency microwave spectrum (from 1 to 10 GHz) have proven to be suitable for soil moisture retrievals. These sensors now cover nearly 4 decades, thus allowing for global multi-mission climate data records. The ESA Climate Change Initiative (CCI) soil moisture product combines various single-sensor active and passive microwave soil moisture products into three harmonized products: an only-active, an only-passive, and a combined active–passive microwave product [26]. In its current version, the presence of data gaps in time and space has been acknowledged as a major shortcoming which makes it difficult for users to integrate the data in their applications [20]. From a scientific perspective, the presence of "intermittent" data gaps in satellite-based soil moisture estimates impacts the analysis of spatiotemporal dynamics and trends, which may be limited to certain regions [27]. Also, the presence of missing data in time series prevents a robust computation of temporal autocorrelation and e-folding times, as a measure of soil moisture persistence [22]. In this regard, recent studies on the use of Gaussian process regression techniques to mitigate the effect of missing information in Earth observation data are very promising (e.g., [17, 21]).

> The presence of gaps in EO data limits their applicability in a number of applications. In contrast with the standard temporal interpolation techniques, the LMC multi-output GP-based gap-filling regression allows taking into account information from other collocated sensors measuring the exact same variable. The method learns the relationships among the different sensors and builds a cross-domain kernel function able to transfer information across the time series and do predictions and associated confidence intervals on regions where no data are available.

In this section, a subset of 6 years of SMOS L-band passive, ASCAT C-band active, and AMSR2 C-band passive soil moisture measurements, starting in June 2010, have been used. SMOS and ASCAT estimates are available for the whole period, whereas AMSR2 estimates start on May 18, 2012 (its launch date). Each product presents different observational gaps due to the presence of RFI at their operating frequency or a too high uncertainty in their inversion algorithm (e.g., due to the presence of snow masking observations, dense vegetation, or high topography). The problem we face here is that we need a gap-filling methodology able to handle several outputs together and force a "sharp" reconstruction of the time series so that fast dry-down and wetting-up dynamics are preserved (avoid smoothing). We show how we can efficiently deal with our problem by employing a multi-output Gaussian Process model based on the Linear Model of Coregionalization (LMC) [28]. This model implicitly exploits the relationships among the three microwave sensors and predicts an output for each of them. The reconstructed time series are provided with an estimate of its uncertainty and are shown to preserve the statistics from comparison to in situ data over a selection of catchments from the International Soil Moisture Network.

2.2.1 Proposed Approach

The presence of temporal data gaps in satellite-based estimates of soil moisture limits their applicability in a number of applications that need continuous estimates. Standard techniques for gap-filling temporal series such as linear or cubic interpolation, or auto-regressive functions fail to reconstruct sharp transitions or long data gaps and do not take into account information from other collocated sensors measuring exactly the same biophysical variable. Given that we have three different soil moisture products presenting no data in different time and space locations, we employ here an LMC multi-output GP regression (LMC-GP) to maximize the spatiotemporal coverage of the datasets. We illustrate the procedure at three in situ soil moisture networks where the SMOS satellite presents good, average, and poor temporal coverage, see Fig. 2.2. We will show how LMC-GP exploits the relationships among SMOS, ASCAT, and AMSR2 soil moisture time series to do inferences on regions where no data (gaps) are available, and provides a reconstructed prediction with and associated uncertainty for each dataset. Statistical scores from comparison with in situ data at the selected sites of the original and reconstructed time series will be shown.

2.2.2 LMC-GP

First, we will start introducing the formulation of standard GP models. Then, we will extend it to the LMC-GP model.

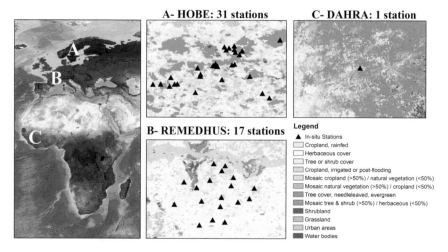

Fig. 2.2 Location and land use map of the three International Soil Moisture Network (ISMN) validation sites used in the study: **a** HOBE in Denmark (31 stations), **b** REMEDHUS in Spain (17 stations), and **C** DAHRA in Senegal (1 station)

2.2.2.1 Gaussian Processes

GPs [15] are state-of-the-art statistical methods for regression and function approximation, and have been used with great success in biophysical variable retrieval by following statistical and hybrid approaches [29]. We start assuming that we are given a set of n pairs of measurements, $\{\mathbf{x}_i, y_i\}_{i=1}^{n}$, where $\mathbf{x}_i$ is the feature/measurement space and y_i is the biophysical parameter from field data or other sources, perturbed by an additive independent noise e_i. We consider the following model:

$$y_i = f(\mathbf{x}_i) + e_i, \quad e_i \sim \mathcal{N}(0, \sigma_n^2), \tag{2.1}$$

where $f(\mathbf{x})$ is an unknown latent function, $\mathbf{x} \in \mathbb{R}^d$, and σ_n^2 represents the noise variance. Defining $\mathbf{y} = [y_1, \ldots, y_n]^\mathsf{T}$ and $\mathbf{f} = [f(\mathbf{x}_1), \ldots, f(\mathbf{x}_n)]^\mathsf{T}$, the conditional distribution of $\mathbf{y}$ given $\mathbf{f}$ becomes $p(\mathbf{y}|\mathbf{f}) = \mathcal{N}(\mathbf{f}, \sigma_n^2 \mathbf{I}_n)$, where $\mathbf{I}_n$ is the $n \times n$ identity matrix. It is assumed that $\mathbf{f}$ follows a n-dimensional Gaussian distribution $\mathbf{f} \sim \mathcal{N}(\mathbf{0}, \mathbf{K})$. The covariance matrix $\mathbf{K}$ of this distribution is determined by a squared exponential (SE) kernel function with entries $\mathbf{K}_{ij} = k(\mathbf{x}_i, \mathbf{x}_j) = \exp(-\|\mathbf{x}_i - \mathbf{x}_j\|^2/(2\sigma^2))$, encoding the similarity between input points. In order to make a new prediction y_* given an input $\mathbf{x}_*$, we obtain the joint distribution over the training and test points,

$$\begin{bmatrix} \mathbf{y} \\ y_* \end{bmatrix} \sim \mathcal{N}\left(\mathbf{0}, \begin{bmatrix} \mathbf{C}_n & \mathbf{k}_*^\mathsf{T} \\ \mathbf{k}_* & c_* \end{bmatrix}\right),$$

where $\mathbf{C}_n = \mathbf{K} + \sigma_n^2 \mathbf{I}_n$, $\mathbf{k}_* = [k(\mathbf{x}_*, \mathbf{x}_1), \ldots, k(\mathbf{x}_*, \mathbf{x}_n)]^\mathsf{T}$ is an $n \times 1$ vector and $c_* = k(\mathbf{x}_*, \mathbf{x}_*) + \sigma_n^2$. Using the standard Bayesian framework, we obtain the distribution over y_* conditioned on the training data, which is a normal distribution with predictive mean and variance given by

$$
\begin{aligned}
\mu_{\text{GP}}(\mathbf{x}_*) &= \mathbf{k}_*^\mathsf{T}(\mathbf{K} + \sigma_n^2 \mathbf{I}_n)^{-1}\mathbf{y}, \\
\sigma_{\text{GP}}^2(\mathbf{x}_*) &= c_* - \mathbf{k}_*^\mathsf{T}(\mathbf{K} + \sigma_n^2 \mathbf{I}_n)^{-1}\mathbf{k}_*.
\end{aligned}
\tag{2.2}
$$

One of the most interesting things about GPs is that they yield not only predictions $\mu_{\text{GP}*}$ for test data, but also the uncertainty of the mean prediction, $\sigma_{\text{GP}*}$. Model hyperparameters $\boldsymbol{\theta} = (\sigma, \sigma_n)$ determine, respectively, the width of the SE kernel function and the noise on the observations, and they are usually obtained by maximizing the log-marginal likelihood.

2.2.2.2   Linear Model of Corregionalization for GPs

LMC-GPs [28] extend standard GPs so it is possible to both handle several outputs at the same time (i.e., it is a multi-output model) and to deal with missing data in the considered outputs. This model is well known in the field of geostatistics as *co-kriging* [30].

In the LMC-GP model, we have a vector function, $\mathbf{f} : \mathcal{X} \to \mathbb{R}^D$, where D is the number of outputs. Given a reproducing kernel, defined as a positive definite symmetric function $\mathbf{K} : \mathcal{X} \times \mathcal{X} \to \mathbb{R}^{n \times n}$, where n is the number of samples of each output, we can express $\mathbf{f}(\mathbf{x})$ as

$$
\mathbf{f}(\mathbf{x}) = \sum_{i=1}^{n} \mathbf{K}(\mathbf{x}_i, \mathbf{x})\mathbf{c}_i,
\tag{2.3}
$$

for some coefficients $\mathbf{c}_i \in \mathbb{R}^n$. The coefficients $\mathbf{c}_i$ can be obtained by solving the linear system, obtaining

$$
\bar{\mathbf{c}} = (\mathcal{K}(\mathbf{X}, \mathbf{X}) + \lambda n \mathbf{I})^{-1}\bar{\mathbf{y}},
\tag{2.4}
$$

where $\bar{\mathbf{c}}, \bar{\mathbf{y}}$ are nD vectors obtained by concatenating the coefficients and outputs, respectively, and $\mathcal{K}(\mathbf{X}, \mathbf{X})$ is an $nD \times nD$ matrix with entries $(\mathbf{K}(\mathbf{x}_i, \mathbf{x}_j))_{d,d'}$ for $i, j = 1, \ldots, n$ and $d, d' = 1, \ldots, D$. The blocks of this matrix are $(\mathbf{K}(\mathbf{X}_i, \mathbf{X}_j))_{i,j}$ $n \times n$ matrices. Predictions are given by

$$
\mathbf{f}(\mathbf{x}_*) = \mathcal{K}_{\mathbf{x}_*}^\mathsf{T} \bar{\mathbf{c}},
\tag{2.5}
$$

with $\mathcal{K}_{\mathbf{x}_*} \in \mathbb{R}^{D \times nD}$ composed of blocks $(\mathbf{K}(\mathbf{x}_*, \mathbf{x}_j))_{d,d'}$. When the training kernel matrix $\mathcal{K}(\mathbf{X}, \mathbf{X})$ is block diagonal, that is, $(\mathbf{K}(\mathbf{X}_i, \mathbf{X}_j))_{i,j} = \mathbf{0}$ for all $i \neq j$, then each output is considered to be independent of the others, and we thus have individual GP models. The non-diagonal matrices establish the relationships between the outputs.

In the LMC-GP model, each output is expressed as a linear combination of independent latent functions,

$$f_d(\mathbf{x}) = \sum_{q=1}^{Q} a_{d,q} u_q(\mathbf{x}), \tag{2.6}$$

where $a_{d,q}$ are scalar coefficients, and $u_q(\mathbf{x})$ are latent functions with zero mean and covariance $k_q(\mathbf{x}, \mathbf{x}')$. It can be shown [28] that the full covariance matrix of this model can be expressed as

$$\mathcal{K}(\mathbf{X}, \mathbf{X}) = \sum_{q=1}^{Q} \mathbf{B}_q \otimes k_q(\mathbf{X}, \mathbf{X}), \tag{2.7}$$

where $\otimes$ is the Kronecker product. Here, each $\mathbf{B}_q \in R^{D \times D}$ is a positive definite matrix known as a *co-regionalization matrix*, and it encodes the relationships between the outputs.

2.2.3 Data and Setup

The temporal period of study is 6 years, starting in June 2010. Three global satellite soil moisture products have been extracted for the study period: SMOS BEC L3 (1.4 GHz, L3 SM v3.0), Metop A/B ASCAT (5.3 GHz, Eumetsat H-SAF), and GCOM W1 AMSR2 L3 (6.9 GHz, LPRM v05 retrieval algorithm, NASA). ASCAT and AMSR2 products have been resampled from their 0.25° grid to the SMOS EASE2 25-km grid using bilinear interpolation. These products have been widely validated under different biomes and climate conditions by comparison with ground-based observations (e.g., [26, 31, 32]) and outputs of land surface models (e.g., [33–35]).

We show the robustness of the multi-sensor gap-filling approach at three in situ soil moisture networks: REMEDHUS in Spain (17 stations [36]), HOBE in Denmark (31 stations [37]), and DAHRA in Senegal (1 station [38]). In terms of temporal coverage, they are representative of best-case (REMEDHUS), average-case (HOBE), and wort-case (DAHRA) scenarios, with SMOS providing a coverage during the study period of 96, 65, and 45%, respectively. The locations and land use maps of the in situ networks used for this study are presented in Fig. 2.2.

2.2.4 Results

Let us start with an illustrative example of method's performance. Figure 2.3 shows with a real example how the LMC-GP transfers information across SMOS, ASCAT, and AMSR2 satellite time series for the predictions when no data is available and provides associated confidence intervals.

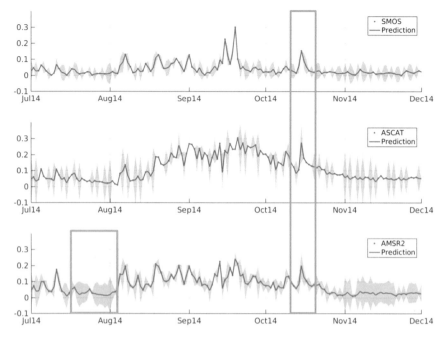

Fig. 2.3 Time series of original (orange dots) and reconstructed (blue lines) SMOS, ASCAT, and AMSR2 time series using the LMC-GP gap-filling technique. The uncertainty on the predictions is shown in shaded gray. The orange square points out a specific rainfall event that was captured only by SMOS and is accounted for in the reconstruction of ASCAT and AMSR2 time series. The green square exemplifies how the method reconstructs long data gaps in AMSR2 based on no-rain information from the other two sensors, assigning a higher uncertainty when no original data is available

A more thorough experimental analysis follows. Results of the application of the proposed LMC-GP over REMEDHUS, HOBE, and DAHRA networks are shown in Fig. 2.4, together with the original satellite time series and the in situ data as a benchmark. It can be seen that the reconstructed soil moisture time series follow closely the original time series, capturing the wetting-up and drying-down events and filling the missing information (e.g., see in HOBE the dry-down in February 2014 which was captured only by AMSR2 during consecutive days and is reproduced by the three reconstructed time series). In DAHRA, the limited temporal coverage of AMSR2 in the dry seasons is completed in the reconstructed time series using information from the other two sensors. It is worth to remark that for AMSR2 the reconstructed time series back-propagate to dates where the satellite was not yet launched (shown here for illustration purposes), yet they look very consistent with the real satellite data. Also importantly, we fixed the kernel lengthscale parameter in LMC-GP model to force a sharp reconstruction, to prevent the predictions being smoothed with respect to the original time series.

Á. Moreno-Martínez et al.

Fig. 2.4 Time series of in situ (black lines) and satellite-based soil moisture estimates from SMOS, ASCAT, and AMSR2 (orange dots denote the original time series and blue lines the predicted using the LMC-GP gap-filling technique) over **a** REMEDHUS, **b** HOBE, and **c** DAHRA networks

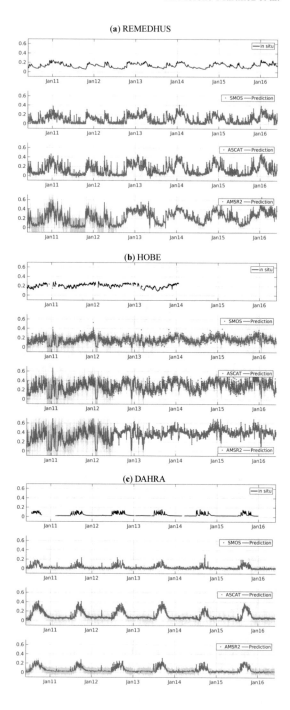

A statistical analysis of the original and reconstructed satellite time series has been undertaken following the recommended performance metrics in [39]. Table 2.1 shows that Pearson's correlation coefficient R, bias (as estimated by the mean error, ME) and root-mean-squared error (RMSE) with respect to in situ data in the three networks are not affected to a high degree by the reconstruction, and they remain within reasonable bounds. For SMOS, the reconstructed time series preserve the statistical scores of original time series in REMEDHUS and DAHRA and improve the R in HOBE from 0.62 to 0.68 (note the other sensors in HOBE have higher correlations of 0.66 and 0.73). The increase in coverage is notable, with an improvement of 37% for HOBE and of 54% for DAHRA. SMOS has the largest coverage over REMEDHUS, and the improvement of coverage is therefore limited (of 8%). For ASCAT, the statistical scores are preserved in the reconstructed time series, and the increase in coverage is also remarkable: 23% for REMEHDUS, 31% for HOBE, and 36% for DAHRA. For AMSR2, the validation is limited to four annual cycles (from its launch date in May 18, 2012, onward). Over REMEDHUS, AMSR2 presents a wet bias with respect to the in situ data that is reduced in the reconstructed time series; its correlation is reduced from 0.86 to 0.81, probably due to the lower correlations of the other two sensors, and the increase in coverage is of 27%. Similar results are obtained for reconstructed AMSR2 over HOBE, but with a lower number of collocated observations due to the lack of in situ data in early January 2014. Over DAHRA, correlation is improved from 0.73 to 0.79, with a 66% improvement of coverage. These results provide confidence in the proposed technique and show how it exploits the complementary spatiotemporal coverage of the three microwave sensors.

2.3 Distribution Regression for Multiscale Estimation

Non-uniform spatial sampling is a common problem in geostatistics and spatialization problems. When the variable of interest is available at the same resolution that the remote sensing observations, standard algorithms such as random forests, Gaussian processes, or neural networks are available to establish the relationship between the two. Nevertheless, we often deal with situations where the target variable is only available at the group level, collectively associated to a number of remotely sensed observations. This kind of problem is known in statistics and machine learning as *multiple instance learning* (MIL) or *distribution regression* (DR). Chapter 6 introduces the MIL framework and methodology, and reviews different approaches to address the particular issue of imprecision in hyperspectral images analysis. We here present a nonlinear method based on kernels for distribution regression that solves the previous problems without making any assumption on the statistics of the grouped data. The presented formulation considers distribution embeddings in reproducing kernel Hilbert spaces and performs standard least squares regression with the empirical means therein. A flexible version to deal with multisource data of different dimensionality and sample sizes is also introduced. The potential of the

Table 2.1 Mean error (ME) ($m^3 m^{-3}$), unbiased RMSE (ubRMSE) ($m^3 m^{-3}$) and Pearson correlation (R) for the original and reconstructed satellite time series against in situ measurements from REMEDHUS, HOBE, and DAHRA networks. Variable "days" reports the number of collocated satellite and in situ data available to compute the statistics

	REMEDHUS				HOBE				DAHRA			
	ME	ubRMSE	R	Days	ME	ubRMSE	R	Days	ME	ubRMSE	R	Days
SMOS	−0.032	0.003	0.81	2004	−0.042	0.002	0.62	741	−0.0143	0.001	0.79	841
SMOS rec	−0.033	0.003	0.81	2192	−0.048	0.002	0.68	1185	−0.014	0.001	0.78	1834
ASCAT	0.002	0.004	0.79	1673	0.086	0.003	0.73	816	0.071	0.002	0.70	1171
ASCAT rec	−0.001	0.004	0.78	2192	0.071	0.003	0.70	1185	0.064	0.002	0.70	1834
AMSR2	0.118	0.005	0.86	1047	0.199	0.002	0.66	486	0.026	0.002	0.73	417
AMSR2 rec	0.084	0.005	0.81	1428	0.153	0.003	0.62	513	0.019	0.001	0.79	1242

presented approach is illustrated by using SMAP VOD time series for the estimation of crop production in the US Corn Belt.

2.3.1 Kernel Distribution Regression

In distribution regression problems, we are given several sets of observations each of them with a single output target variable to be estimated. The training dataset $\mathcal{D}$ is formed by a collection of B bags (or sets) $\mathcal{D} = \{(\mathbf{X}_b \in \mathbb{R}^{n_b \times d}, y_b \in \mathbb{R}) | b = 1, \ldots, B\}$. A training set from a particular bag b is formed by n_b examples, here denoted as $\mathbf{X}_b = [\mathbf{x}_1, \ldots, \mathbf{x}_{n_b}]^\mathsf{T} \in \mathbb{R}^{n_b \times d}$, where $\mathbf{x}_i \in \mathbb{R}^{d \times 1}$. Let us denote all the available data collectively grouped in matrix $\mathbf{X} \in \mathbb{R}^{n \times d}$, where $n = \sum_{b=1}^{B} n_b$, and $\mathbf{y} = [y_1, \ldots, y_B]^\mathsf{T} \in \mathbb{R}^{B \times 1}$. In this setting, the direct application of regression algorithms is not possible because not just a single input point $\mathbf{x}_b$ but a set of points $\mathbf{X}_b$ is available for each target output, and latter for prediction we may have test points or sets from each bag, $\mathbf{x}_b^* \in \mathbb{R}^{d \times 1}$ or $\mathbf{X}_b^* \in \mathbb{R}^{m_b \times d}$, which we denote with a star superscript. The problem boils down to finding a function f that learns the mapping from $\mathbf{x}$ to y exploiting the many-to-one dataset. To solve the problem, two main approaches are typically followed: (1) *output expansion*, that is, replicating the label y_b for all points in bag b; or (2) *input summary* most notably with the empirical average $\bar{\mathbf{x}}_b = \frac{1}{n_b} \sum_i \mathbf{x}_i$, or a set of centroids $\mathbf{c}_b$, $b = 1, \ldots, B$. What makes DR distinctive is that it instead exploits the rich structure in $\mathcal{D}$ by performing regression with the group distributions directly. Statistically, this consists of considering all higher order statistical relationships between the groups, not just the first- or second-order moments. The method we are going to introduce here works by embedding the bag distribution in a Hilbert space and performing linear regression therein. We essentially need the definition of a mean embedding, its induced kernel function, and how the regression is done with it.

Distribution regression problems rely very often on using non-uniformly spatial sampled datasets, where the variables of interest are associated with sets of observations instead of single observations. While some approaches summarize the sets of observations using some kind of aggregation, such as the mean of the standard deviation, kernel distribution regression uses all higher moments by computing mean map embeddings in high-dimensional Hilbert spaces, and hence improved ability for function approximation.

2.3.1.1 Mean Map Embeddings

We frame the problem in the theory of mean map embeddings of distributions [40–42]. The kernel mean map from the set of all probability distributions $\mathcal{B}_X$ into $\mathcal{H}$ is defined as

$$\boldsymbol{\mu} : \mathcal{B}_X \rightarrow \mathcal{H}, \quad \mathbb{P} \rightarrow \int_X k(\cdot, \mathbf{x}) d\mathbb{P}(\mathbf{x}) \in \mathcal{H}.$$

Assuming that $k(\cdot, \mathbf{x})$ is bounded for any $\mathbf{x} \in X$, it can be shown that for any $\mathbb{P}$, letting $\boldsymbol{\mu}_{\mathbb{P}} = \boldsymbol{\mu}(\mathbb{P})$, the $\mathbb{E}_P[f] = \langle \boldsymbol{\mu}_{\mathbb{P}}, f \rangle_{\mathcal{H}}$, for all $f \in \mathcal{H}$. Here $\boldsymbol{\mu}$ represents the expectation function on $\mathcal{H}$. Every probability measure has a unique embedding and the $\boldsymbol{\mu}$ fully determines the corresponding probability measure [41]. Here, we show how to estimate the mean map embeddings from empirical samples. For one particular bag, $\mathbf{X}_b$, drawn i.i.d. from a particular $\mathbb{P}_b$, the empirical mean estimator of $\boldsymbol{\mu}_b$ can be computed as

$$\widehat{\boldsymbol{\mu}}_b = \boldsymbol{\mu}_{\mathbb{P}_b} = \int k(\cdot, \mathbf{x}) \hat{\mathbb{P}}(d\mathbf{x}) \approx \frac{1}{n_b} \sum_{i=1}^{n_b} k(\cdot, \mathbf{x}_i). \tag{2.8}$$

This is an empirical mean map estimator whose dot product can be computed via kernels:

$$\langle \widehat{\boldsymbol{\mu}}_{\mathbb{P}_b}, \widehat{\boldsymbol{\mu}}_{\mathbb{P}_{b'}} \rangle_{\mathcal{H}} = \frac{1}{n_b n_{b'}} \sum_{i=1}^{n_b} \sum_{j=1}^{n_{b'}} k(\mathbf{x}_i^b, \mathbf{x}_j^{b'}), \tag{2.9}$$

which is the base of a useful kernel algorithm for hypothesis testing named maximum mean discrepancy (MMD) [41, 42] and estimates the distance between two sample means in a reproducing kernel Hilbert space $\mathcal{H}$ where data are embedded

$$\text{MMD}(\mathbb{P}_b, \mathbb{P}_{b'}) := \|\boldsymbol{\mu}_{\mathbb{P}_b} - \boldsymbol{\mu}_{\mathbb{P}_{b'}}\|_{\mathcal{H}}^2.$$

This can be computed using kernel functions in Eq. (2.9). Figure 2.5 shows how MMD and mean map embeddings can detect differences between distributions in higher order moments.

2.3.1.2 Distribution Regression with Kernels

The distribution regression task is carried out by standard least squares regression using the mean embedded data in Hilbert spaces. The solution leads to the kernel ridge regression (KRR) algorithm [43] working with mean map embeddings. We need to minimize a loss function composed of two terms: the least square errors of the approximation of the mean embedding, and a regularization term that acts over the class of functions to be learned in Hilbert space $f \in \mathcal{H}$:

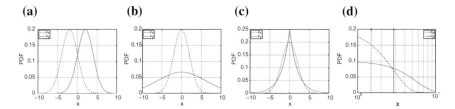

Fig. 2.5 The two-sample problem consists of detecting whether two distributions $\mathbb{P}_x$ and $\mathbb{P}_y$ are different or not. When they have different means (**a**), a simple t-test can differentiate them. When they have the same first moments (mean in **b**, mean and variance in **c**) but different higher order moments, mapping the data to higher dimensional spaces allows to distinguish them (**d**). Kernel mean embeddings are able to do so without having to map the data explicitly

$$f^* = \arg\min_{f \in \mathcal{H}} \left\{ \frac{1}{n} \sum_{i=1}^{n} \| y_i - f(\boldsymbol{\mu}_i) \|^2 + \lambda \| f \|_{\mathcal{H}}^2 \right\},$$

where $\lambda > 0$ is the regularization term. The ridge regression has an analytical solution for a test set given a set of training examples:

$$\hat{f}_{\boldsymbol{\mu}_t} = \mathbf{k}(\mathbf{K} + n\lambda \mathbf{I})^{-1} \mathbf{y}, \tag{2.10}$$

where $\boldsymbol{\mu}_t$ is the mean embedding of the test set $\mathbf{X}_t$, $\mathbf{k} = [k(\boldsymbol{\mu}_1, \boldsymbol{\mu}_t), \dots, k(\boldsymbol{\mu}_n, \boldsymbol{\mu}_t)]^{\mathsf{T}} \in \mathbb{R}^{n \times 1}$, $\mathbf{K} = [k(\boldsymbol{\mu}_i, \boldsymbol{\mu}_j)] \in \mathbb{R}^{n \times n}$ and $\mathbf{y} = [y_1, \dots, y_n]^{\mathsf{T}}$ represents all outputs. Now, for a set of B bags each one containing n_b samples, and exploiting (2.9), one can readily compute the kernel entries of $\mathbf{K}$ as follows:

$$[\mathbf{K}]_{b,b'} = \boldsymbol{\mu}_b^{\mathsf{T}} \boldsymbol{\mu}_{b'} = \frac{1}{n_b n_{b'}} \mathbf{1}_{n_b}^{\mathsf{T}} \mathbf{K}_{bb'} \mathbf{1}_{n_{b'}},$$

where the matrix $\mathbf{K}_{bb'} \in \mathbb{R}^{n_b \times n_{b'}}$. Therefore, we have an analytic solution of the problem in (2.10):

$$\hat{y}_b^* = \frac{1}{m_b n} \mathbf{1}_{m_b}^{\mathsf{T}} \mathbf{K}_{bb'} \mathbf{1}_{n_{b'}} \boldsymbol{\alpha}, \tag{2.11}$$

where $\mathbf{K}_{bb'} \in \mathbb{R}^{m_b \times n_{b'}}$ which is computed given a valid Mercer kernel function k.

Kernel methods also allow to combine multisource (also known as multimodal) information in each bag, as was previously done with standard paired settings in either remote sensing or signal processing applications [42, 44, 45]. This is the case when bags have different numbers of both features and sizes, e.g., we aim to combine different spatial, spectral, or temporal resolutions. Notationally, now we have access to different matrices $\mathbf{X}_f^b \in \mathbb{R}^{n_b^f \times f}$, $f = 1, \dots, F$. The multimodal kernel distribution method summarizes each dataset into a mean and then exploits the direct sum of Hilbert spaces in the mean embedding space. Therefore, we define F Hilbert spaces $\mathcal{H}_f$, $f = 1, \dots, F$, and the direct sum of all of them, $\mathcal{H} = \oplus_{f=1}^{F} \mathcal{H}_f$. We

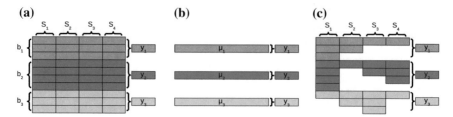

Fig. 2.6 Distribution regression approaches presented in this chapter. The DR problem is illustrated in **a** for $B = 3$ bags different numbers of samples per bag ($n_1 = 3$, $n_2 = 4$, $n_3 = 3$), three corresponding target labels, y_b, $b = 1, 2, 3$, and columns represent different features (sources, sensors) S_i, $i = 1, 2, 3, 4$. The standard approach **b** summarizes the distributions $\mathbb{P}_b$ with the mean vectors μ_b and then applies standard regression methods. Alternatively, this can be done in Hilbert spaces too with the advantage of considering all moments of the distributions. In **c**, we show the case of multisource distribution regression (MDR) in which some features are missing for particular bags and samples, which is often the case when different sensors are combined

summarize the bag feature vectors with a set of mean map embeddings of samples in bag b, which we denoted as μ_b^f. The collection of all mean embeddings in $\mathcal{H}$ is defined as $\mu_b = [\mu_b^1, \ldots, \mu_b^F] \in \mathcal{H}$, and then we define the mean map embedding as $\mathbf{M} = [\mu_1 | \cdots | \mu_B]^\mathsf{T} \in \mathbb{R}^{B \times H}$. The multimodal kernel matrix is computed as follows:

$$[\widetilde{\mathbf{K}}]_{b,b'} = \mu_b^\mathsf{T} \mu_{b'} = \sum_{f=1}^{F} \frac{1}{n_b^f n_{b'}^f} \mathbf{1}_{n_b^f}^{f\mathsf{T}} \mathbf{K}_{bb'} \mathbf{1}_{n_{b'}^f}, \tag{2.12}$$

Fig. 2.6 graphically illustrates the DR approaches used in this chapter. The algorithm reduces to the application of a standard kernel ridge regression with the kernel function Eq. (2.11) for the standard case or Eq. (2.12) for the multisource case. We provide source code of our methods in http://isp.uv.es/code/dr.html.

2.3.2   Data and Setup

We show results for crop yield estimation, which is a particular problem of distribution regression in the context of remote sensing. We show results for our KDR (kernel distribution regression) and several baseline standard approaches like least squares regularized linear regression model (RLR) and its nonlinear (kernel) counterpart, the kernel ridge regression (KRR) method, both working on the empirical means of each bag as input feature vectors. We use as evaluation criteria the standard mean error (ME) to account for bias, the root-mean-square-error (RMSE) to assess accuracy, and the coefficient of determination or explained variance (R^2) to account for the goodness-of-fit.

Specifically, for the crop yield estimation, satellite-based retrievals of vegetation optical depth (VOD) from SMAP [46] is related to crop production data from the

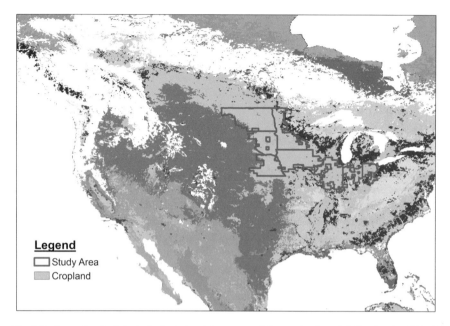

Fig. 2.7   Area of study. It includes both the eight states and the cropland mask following the MODIS IGBP land cover classification

2015 US agricultural survey (total yield and yield per crop type), and then the proposed methods are evaluated. VOD is a measure of the attenuation of soil microwave emissions when they pass through the vegetation canopy, being sensitive not only to the amount of living biomass, but also to the amount of water stress experienced by the vegetation [47]. SMAP VOD has been shown to carry information about crop growth and yield in a variety of agro-ecosystems [48, 49].

We focus on eight states within the so-called Corn Belt of the Midwestern United States: Illinois, Indiana, Iowa, Minnesota, Nebraska, North Dakota, Ohio, and South Dakota (Fig. 2.7). Also, the United States Department of Agriculture, in particular, the National Agricultural Statistics Service (USDA-NASS), publish reports and survey of agricultural information every year at the country, state and county levels. There is a total of 385 counties with yield and satellite data for prediction of total yield. We also predict per crop type. In particular, the three main crops in the region, i.e., corn, soybean, and wheat, are predicted. All the 363, 361, and 204 counties reporting corn, soybean, and wheat yields, independent of their relative importance at the county level, are included in the corresponding crop-specific experiments.

2.3.3 Results

The methodology for evaluating the algorithms is as follows. A 66% of the counties (bags) are used to train/validate and the remaining 33% are used for testing. With the first ones, we perform a fivefold cross-validation also at a bag level, i.e., we split the data into five subsets, one reserved for validation and the rest used for training the regression model. After this, we only apply the best model found to the test data. Finally, all this process is repeated ten times, and the average over all test results is computed. Only test errors are reported.

Table 2.2 shows the crop yield predictions for all the approaches. Notably, these results outperform those obtained in previous literature for corn–soy croplands ([48] and references therein), even with the simplest models like RLR and KRR. Results of the best regression model between VOD and official corn yields at county level are illustrated in Fig. 2.8. Except in few counties, corn predictions are reasonably good, with relative errors below 3%. The proposed DR approaches will be particularly useful for regional crop forecasting in areas covering different agro-climatic conditions and fragmented agricultural landscapes (e.g., Europe), where scale effects need to be properly addressed for adequate analysis and predictions [50].

Table 2.2 Results for prediction of total yield and crop yield prediction using VOD (Kg m^{-2})

Total crop yield	ME×1000	RMSE ×100	R^2
RLR	1.19 ± 7.36	9.67 ± 0.74	0.80 ± 0.02
KRR	2.22 ± 10.77	9.34 ± 0.73	0.81 ± 0.02
KDR	2.27 ± 10.95	9.35 ± 0.71	0.81 ± 0.02
Corn yield	ME×1000	RMSE ×100	R^2
RLR	−1.20 ± 5.89	7.54 ± 0.50	0.85 ± 0.02
KRR	1.68 ± 8.52	6.54 ± 0.72	0.88 ± 0.02
KDR	1.59 ± 7.88	6.47 ± 0.74	0.89 ± 0.02
Soybean yield	ME×1000	RMSE ×100	R^2
RLR	-1.99 ± 1.85	2.45 ± 0.13	0.85 ± 0.03
KRR	-0.70 ± 2.92	2.47 ± 0.21	0.85 ± 0.04
KDR	-0.64 ± 2.43	2.40 ± 0.21	0.86 ± 0.03
Wheat yield	ME×1000	RMSE ×100	R^2
RLR	2.72 ± 6.65	5.46 ± 0.48	0.64 ± 0.08
KRR	2.42 ± 8.47	5.07 ± 0.38	0.69 ± 0.05
KDR	2.91 ± 7.31	5.10 ± 0.40	0.69 ± 0.05

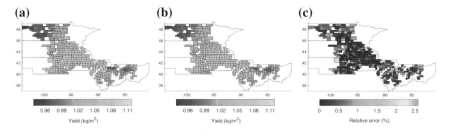

Fig. 2.8 a Map of official corn yield for year 2015 from USDA-NASS survey given in (Kg/m^2). **b** KDR predicted corn yield and **c** KDR relative error prediction per county (%)

2.4 Global Parameter Estimation in the Cloud

From an operational point of view, the implementation of biophysical parameter retrieval chains on ongoing basis demands high storage capability and efficient computational power, mainly when dealing with long time series of remote sensing data at global scales. There exist a wide variety of free available remote sensing data which could be potentially ingested in these processing chains. Among them, one can find remote sensing data disseminated by NASA (e.g., MODIS), the United States Geological Survey (USGS) (e.g., Landsat), and ESA (e.g., data from the Sentinel constellation). To deal with this huge amount of data, Google developed the Google Earth Engine [51], a cloud computing platform specifically designed for geospatial analysis at the petabyte scale. Due to its unique features, GEE is the state of the art in remote sensing big data processing. The GEE data catalog is composed by widely used geospatial datasets. The catalog is continuously updated and data are ingested from different government-supported archives such as the Land Process Distributed Active Archive Center (LP DAAC), the USGS, and the ESA Copernicus Open Access Hub. The GEE data repository embrace a wide variety of remote sensing datasets including meteorological records, atmospheric estimates, vegetation, and land properties and also surface reflectance data. Data processing is performed in parallel on Google's computational infrastructure, dramatically improving processing efficiency and speed. These features, among others, make GEE an extremely valuable tool for multitemporal and global studies which include vegetation, temperature, carbon exchange, and hydrological processes [24, 52, 53].

Here, we present an example of biophysical parameter estimation in the GEE cloud computing platform. The developed processing chain includes the joint estimation of LAI, FAPAR, FVC, and CWC parameters at global scale from long-term time series (15 years) of MODIS data exploiting the GEE cloud processing capabilities. The retrieval approach is based on a hybrid method, which combines the physically based PROSAIL radiative transfer model with random forests (RFs) regression. The implementation on GEE platform allowed us to use global and climate data records (CDR) of both MODIS surface reflectance and LAI/FAPAR datasets which provided

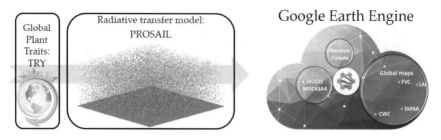

Fig. 2.9 Schema of the developed biophysical retrieval chain in the cloud

us with global biophysical variable maps at unprecedented timeliness. Figure 2.9 shows an schema summarizing the developed retrieval chain.

> Cloud-based geospatial computing platforms such as Google Earth Engine offer opportunities to create a broad range of applications with precision and accuracy over unprecedented large areas with medium and high spatial resolutions. In this section, we illustrate the advantages of using algorithms implemented in a cloud computing infrastructure dealing with a common problem in remote sensing science, the retrieval of land biophysical parameters.

2.4.1 Data and Setup

As shown in Fig. 2.9, to model the spectral response of the vegetation we chose the PROSAIL radiative transfer model. This model results from the PROSPECT leaf optical reflectance model [54] coupled with the SAIL canopy model [55]. PROSAIL has been widely used in many remote sensing studies [56] and successfully applied for local and global parameter estimation [24, 57–59]. PROSAIL assumes the canopy as a turbid medium and simulates vegetation reflectance along the optical spectrum (from 400 to 2500 nm) depending on the leaf biochemistry, structure of the canopy, as well as the background soil reflectance and the sun–satellite geometry. At leaf level, the parametrization was based on the distributions derived from a massive global leaf trait measurements (TRY) [60] in order to account for a realistic representation of global leaf trait variability to optimize PROSAIL at global scale, whereas distributions of the canopy variables were similar to those adopted in other global studies [59]. The TRY database embrace 6.9 million trait records for 148,000 plant taxa at unprecedented spatial and climatological coverage [60]. Although the database is recent, due to the TRY unique properties, these data have been widely used and hundreds of top publications (TRY database) have been presented covering topics ranging from ecology and plant geography to vegetation modeling and

Table 2.3 General information about leaf traits measurements used in this work

Trait	No. samples	No. of species
C_{ab}	19,222	941
C_{dm}	69,783	11,908
C_w	32,020	4802

Table 2.4 Spectral specifications of the MODIS MCD43A4 product

MODIS band	Wavelength (nm)
Band 1 (red)	620–670
Band 2 (NIR)	841–876
Band 3 (blue)	459–479
Band 4 (green)	545–565
Band 5 (SWIR-1)	1230–1250
Band 6 (SWIR-2)	1628–1652
Band 7 (MWIR)	2105–2155

remote sensing [25, 61]. In this section, instead of using the usual lookup tables available in the literature, we use the TRY to parametrize PROSAIL. Our objective is to exploit the TRY database to infer more realistic distributions and correlations among some key leaf traits such as leaf chlorophyll (C_{ab}), leaf dry matter (C_{dm}), and water (C_w) contents. Table 2.3 shows some basic information about the considered traits extracted from the TRY.

The reflectance simulations obtained with PROSAIL were set up to mimic the MCD43A4 product bands which are available in GEE. The MCD43A4 MODIS product is generated combining data from Terra and Aqua spacecrafts, being disseminated as a level-3 gridded dataset. This product provides a bidirectional reflectance distribution function (BRDF) from a nadir view in the seven land MODIS bands (see Table 2.4 for more details), thus offering global surface reflectance data at 500 m spatial resolution with 8-day temporal frequency.

PROSAIL's forward mode provides a reflectance spectrum given a set of input parameters (leaf chemical components/traits, structural parameters of the vegetation canopy, etc.). After running PROSAIL in forward mode, its inversion was undertaken using RFs. This inversion allows, in turn, to retrieve the selected biophysical parameters (LAI, FAPAR, FVC, and CWC). RFs have been applied both for classification and regression in multitude of remote sensing studies [62] including forest ecology [63, 64], land cover classification [65], and feature selection [66]. We chose RFs to invert the PROSAIL model mainly because they can cope with high-dimensional problems due to their optimal pruning strategy and efficiency. RF is an ensemble method that builds up a stack of decision trees. This approach has been proven to be very beneficial to alleviate over-fitting problems in single decision tree models. On the ensemble, every tree is trained with different subsets of features and examples (selected randomly) yielding an individual prediction. The combined prediction (usually the mean value) of the considered trees composing the RFs is the final prediction of the model [67]. The computed simulations obtained with PROSAIL were split into two groups: (1) a training dataset to optimize the models, and (2) an independent test set which was only used to assess the models (RFs). After our models were trained and validated, we predicted the chosen biophysical variables using real MODIS spectral information (land bands, see Table 2.4). In addition, RFs, once trained, are easily parallelized to cope with large-scale problems routinely encountered in global remote sensing applications. This is specifically the case of the problem described here, where we exploit large datasets and run predictions covering many years within the Google Earth Engine platform. A toy example of the code is available at https://code.earthengine.google.com/e3a2d589395e4118d97bae3e85d09106.

2.4.2   Results

The PROSAIL simulations were uploaded to GEE and randomly split into train (2/3 of the simulations) and test (the remaining 1/3 of the samples never used in the RFs training) datasets. The RFs theoretical performance evaluated in the GEE platform (assessed over the test dataset) revealed high correlations (R^2 = 0.84, 0.89, 0.88, and 0.80 for LAI, FAPAR, FVC, and CWC, respectively), low errors (RMSE = 0.91 m^2/m^2, 0.08, 0.06, and 0.27 kg/m^2 for LAI, FAPAR, FVC, and CWC, respectively), and practically no biases in all cases. Subsequently, the RFs retrieval model was executed over the computing cloud to obtain 15 years of global biophysical parameters from the MCD43A4 product available on GEE. Figure 2.10 shows the global mean values of LAI, FAPAR, FVC, and CWC derived from 2010 to 2015. The computation of the mean biophysical maps implied processing 230 (46 yearly images × 5 years) FAPAR images at 500 m spatial resolution (~440 million cells), and compute their annual mean, which took around 6 h.

Validation of the estimates was achieved by means of intercomparison over a network of sites named BELMANIP-2.1 (Benchmark Land Multisite Analysis and Intercomparison of Products) especially selected for representing the global variability of Earth vegetation. Over this network, we compared the LAI and FAPAR estimates against the official LAI/FAPAR MODIS product (MCD15A3H) on GEE. We selected the MODIS pixels for every BELMANIP-2.1 location, and then we computed the mean value of the MODIS valid pixels within a 1 km surrounding area. In addition, since the MCD15A3H and MCD43A4 differ in temporal frequency, only the coincident dates between them were selected for comparison. For validation, we selected only high-quality MODIS pixels which resulted in ~60000 valid pixels from 2002–2017 accounting for vegetation biomes: evergreen broadleaf forests (EBF), broadleaf deciduous forest (BDF), needle leaf forest (NLF), cultivated (C), shrublands (SH), herbaceous (H), and bare areas (BA). For FAPAR, very good agreement (R^2 ranging from 0.89 to 0.92) and low errors (RMSE ranging from 0.06 to 0.08) were found between retrievals and the MODIS FAPAR product over bare areas, shrub-

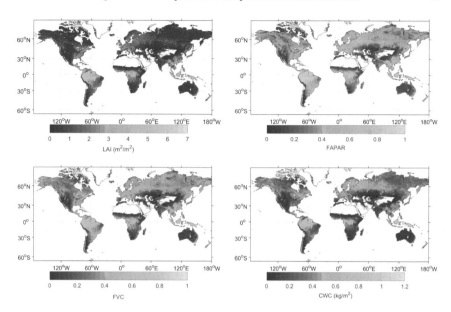

Fig. 2.10 LAI, FAPAR, FVC, CWC global maps corresponding to the mean values estimated by the proposed retrieval chain for the period 2010–2015

lands, herbaceous, cultivated, and broadleaf deciduous forest biomes. For needle-leaf and evergreen broadleaf forests, lower correlations ($R^2 = 0.57$ and 0.41) and higher errors (RMSE = 0.18 and 0.09) were obtained. It is worth mentioning that over bare areas, the MODIS FAPAR presents an unrealistic minimum value (~ 0.05) through the entire period. In the case of LAI, goodness-of-fit ranging from 0.70 to 0.86 and low errors (RMSE ranging from 0.23 to 0.57 m^2/m^2) were found between estimates in all biomes except for evergreen broadleaf forest, where $R^2 = 0.42$ and RMSE = 1.13 m^2/m^2 are reported.

Figure 2.11 shows the LAI and FAPAR difference maps calculated using the mean outcomes (2010–2015) of our processing chain and the mean reference MODIS LAI/FAPAR product for the same period. The mean difference LAI map shows that the discrepancies among both products range within the $\pm 0.5 \, m^2/m^2$ range, indicating that both products are consistent. However, our high LAI values present a significant underestimation over heavily vegetated areas (dense canopies) that reaches values up to 1.4 m^2/m^2. When comparing both FAPAR products, a constant negative bias of $\approx 0.05 \, m^2/m^2$ our estimates is observed. These differences could be related with a documented systematic overestimation of operational MODIS FAPAR [68], meaning that our approach is partly correcting some of the flaws of the official MODIS product.

State-of-the-art cloud computing platforms like GEE provides routinely time series of global land surface variables related with vegetation status and an unprecedented computational power. Despite the variety of regression and classification methods implemented in GEE, the user could be limited by the number of state-of-

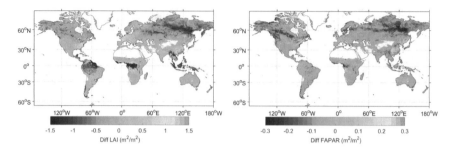

Fig. 2.11 LAI and FAPAR global maps corresponding to the difference of mean values between derived estimates by the proposed retrieval chain and the GEE MODIS reference product for the period 2010–2015

the-art algorithms which are currently implemented. However, GEE is being updated at a fast pace due to an increasing number of users developing new approaches and methods that may be potentially implemented in GEE for a wide range of geoscience applications. Here, we have illustrated an application that takes advantage of GEE capabilities to retrieve standard biophysical variables at a global scale. The validation of our estimates revealed, in general, good spatial consistency. However, differences in mean LAI values over dense forests are still noticeable and could be attributed mostly to differences in retrieval approaches. Other possible source for discrepancies shown could be associated to (i) product definition, such as those related with considering or not vegetation clumping [69], (ii) embedded algorithm assumptions (RTM, optical properties, canopy architecture), and (iii) satellite input data and processing. In relation with the FAPAR, as mentioned above, an overall negative bias is found for all biomes, which is not an issue since different studies have pointed out a systematic overestimation of MODIS retrievals in both C5 and C6 at low FAPAR values. Finally, it is worth mentioning that neither the FVC nor the CWC products are available on GEE. Moreover, there is no global and reliable CWC product with which compare the CWC estimates derived by the proposed retrieval chain. Regarding FVC, there are only a few global products that differ in retrieval approaches and spatiotemporal features.

2.5 Conclusions

This chapter focused on the problem of parameter estimation from remotely sensed optical sensor data. We identified two main challenges related to the data and the used models. To satisfy the urgent needs of fast and accurate data processing and product generation, we relied on three main building blocks: advanced machine learning, big and heterogeneous EO data, and large-scale processing platforms. In this scenario, machine learning has to be redesigned to accommodate data characteristics (spatiotemporal and multi-sensor structures, higher order, and nonlinear feature relations), to be accurate and flexible, and to scale well to millions of observations.

To deal with these challenges, we introduced three machine learning approaches to exploit different spatial, multi-sensor, temporal, and large-scale data characteristics. In particular, we exploited multi-output Gaussian processes for gap-filling time series, kernel distribution regression models that exploits multiple observations and avoid working with arbitrary summarizing statistics, and random forests trained on RTM simulations and implemented in the GEE computation cloud. The approaches allow us to estimate key land parameters from optical and microwave EO data synergistically: SM, LAI, FAPAR, FVC, CWC, and crop yield.

Synergistic benefits of machine learning, big data, and scalable cloud computing are here to stay, and we envision many exciting developments in the near future. EO data allows to monitor continuously in space and time the Earth and can be used to "spatialize" almost any arbitrary quantity measured on the ground or simulated with appropriate transfer codes. Plant, vegetation, and land parameters will readily benefit from ML-based approaches in the cloud to make reliable and accurate products accessible to everyone.

References

1. Camps-Valls G, Tuia D, Gómez-Chova L, Jiménez S, Malo J (eds) (2011) Remote Sens Image Process. Morgan & Claypool Publishers, LaPorte, CO, USA
2. Liang S (2004) Quantitative Remote Sensing of Land Surfaces. Wiley, New York
3. Liang S (2008) Advances in land remote sensing: system, modeling. inversion and applications. Springer, Germany
4. Lillesand TM, Kiefer RW, Chipman J (2008) Remote sensing and image interpretation. Wiley, New York
5. Rodgers CD (2000) Inverse methods for atmospheric sounding: theory and practice. World Scientific Publishing Co., Ltd
6. Baret F, Buis S (2008) Estimating canopy characteristics from remote sensing observations: review of methods and associated problems. In: Advances in land remote sensing: system, modeling, inversion and applications. Springer, Germany
7. Baret F, Weiss M, Lacaze R, Camacho F, Makhmara H, Pacholcyzk P, Smets B (2013) GEOV1: LAI and FAPAR essential climate variables and FCOVER global time series capitalizing over existing products. part1: principles of development and production. Remote Sens Environ 137(0):299–309
8. Beer C, Reichstein M, Tomelleri E, Ciais P, Jung M, Carvalhais N, Rödenbeck C, Arain MA, Baldocchi D, Bonan GB, Bondeau A, Cescatti A, Lasslop G, Lindroth A, Lomas M, Luyssaert S, Margolis H, Oleson KW, Roupsard O, Veenendaal E, Viovy N, Williams C, Woodward FI, Papale D (2010) Terrestrial gross carbon dioxide uptake: global distribution and covariation with climate. Science 329(5993):834–838
9. Jung M, Reichstein M, Margolis HA, Cescatti A, Richardson AD, Arain MA, Arneth A, Bernhofer C, Bonal D, Chen J, Gianelle D, Gobron N, Kiely G, Kutsch, W, Lasslop G, Law BE, Lindroth A, Merbold L, Montagnani L, Moors EJ, Papale D, Sottocornola M, Vaccari F, Williams C (2011) Global patterns of land-atmosphere fluxes of carbon dioxide, latent heat, and sensible heat derived from eddy covariance, satellite, and meteorological observations. J Geophys Res: Biogeosci 116(G3)

10. Jung M, Reichstein M, Schwalm CR, Huntingford C, Sitch S, Ahlström A, Arneth A, Camps-Valls G, Ciais P, Friedlingstein P, Gans F, Ichii K, Jain AK, Kato E, Papale D, Poulter B, Raduly B, Rödenbeck C, Tramontana G, Viovy N, Wang YP, Weber U, Zaehle S, Zeng N (2017) Compensatory water effects link yearly global land CO_2 sink changes to temperature. Nature 541(7638):516–520

11. Tramontana G, Jung M, Camps-Valls G, Ichii K, Raduly B, Reichstein M, Schwalm CR, Arain MA, Cescatti A, Kiely G, Merbold L, Serrano-Ortiz P, Sickert S, Wolf S, Papale D (2016) Predicting carbon dioxide and energy fluxes across global FLUXNET sites with regression algorithms. Biogeosci Discuss 2016:1–33. https://doi.org/10.5194/bg-2015-661

12. Sarker LR, Nichol JE (2011) Improved forest biomass estimates using ALOS AVNIR-2 texture indices. Remote Sens Env 115(4):968–977

13. Durbha SS, King RL, Younan NH (2007) Support vector machines regression for retrieval of leaf area index from multiangle imaging spectroradiometer. Rem Sens Env 107(1–2):348–361

14. Yang F, White MA, Michaelis AR, Ichii K, Hashimoto H, Votava P, Zhu AX, Nemani RR (2006) Prediction of continental-scale evapotranspiration by combining MODIS and ameriflux data through support vector machine. IEEE Trans Geosci Remote Sens 44(11):3452–3461

15. Rasmussen CE, Williams CKI (2006) Gaussian processes for machine learning. The MIT Press, Cambridge, MA

16. Verrelst J, Muñoz J, Alonso L, Delegido J, Rivera JP, Moreno J, Camps-Valls G (2012) Machine learning regression algorithms for biophysical parameter retrieval: opportunities for Sentinel-2 and -3. Remote Sens Environ 118:127–139

17. Camps-Valls G, Verrelst J, Muñoz-Marí J, Laparra V, Mateo-Jimenez F, Gómez-Dans J (2016) A survey on gaussian processes for earth-observation data analysis: a comprehensive investigation. IEEE Geosci Remote Sens Mag 4(2):58–78

18. Reichstein M, Camps-Valls G, Stevens B, Denzler J, Carvalhais N, Jung M (2019) Prabhat: deep learning and process understanding for data-driven Earth system science. Nature

19. Ulaby FT, Long D, Blackwell W, Elachi C, Fung A, Ruf C, Sarabandi K, van Zyl J, Zebker H (2014) Microwave radar and radiometric remote sensing. University of Michigan Press

20. Dorigo W, Wagner W, Albergel C, Albrecht F, Balsamo G, Brocca L, Chung D, Ertl M, Forkel M, Gruber A, Haas E, Hamer PD, Hirschi M, Ikonen J, de Jeu R, Kidd R, Lahoz W, Liu YY, Miralles D, Mistelbauer T, Nicolai-Shaw N, Parinussa R, Pratola C, Reimer C, van der Schalie R, Seneviratne SI, Smolander T, Lecomte P (2017) ESA CCI soil moisture for improved earth system understanding: state-of-the art and future directions. Remote Sens Environ 203:185–215. Earth Observation of Essential Climate Variables

21. Mateo-Sanchis A, Muñoz-Marí J, Campos-Taberner M, García-Haro J, Camps-Valls G (2018) Gap filling of biophysical parameter time series with multi-output gaussian processes. In: IGARSS 2018—2018 IEEE international geoscience and remote sensing symposium, pp 4039–4042

22. Piles M, van der Schalie R, Gruber A, Muñoz-Marí J, Camps-Valls G, Mateo-Sanchis A, Dorigo W, de Jeu R (2018) Global estimation of soil moisture persistence with L and C-band microwave sensors. In: IGARSS 2018—2018 IEEE international geoscience and remote sensing symposium, pp 8259–8262

23. Adsuara JE, Pérez-Suay A, Muñoz-Marí J, Mateo-Sanchis A, Piles M, Camps-Valls G (2019) Nonlinear distribution regression for remote sensing applications. IEEE Trans Geosci Remote Sens (2019) (Submitted)

24. Campos-Taberner M, Moreno-Martínez A, García-Haro FJ, Camps-Valls G, Robinson NP, Kattge J, Running SW (2018) Global estimation of biophysical variables from google earth engine platform. Remote Sens 10:1167

25. Moreno A, Camps G, Kattge J, Robinson N, Reichstein M, van Bodegom P, Kramer K, Cornelissen J, Reich P, Bahn M et al (2018) A methodology to derive global maps of leaf traits using remote sensing and climate data. Remote Sens Environ 218:69–88

26. Dorigo WA, Gruber A, Jeu RAMD, Wagner W, Stacke T, Loew A, Albergel C, Brocca L, Chung D, Parinussa RM, Kidd R (2015) Evaluation of the ESA CCI soil moisture product using ground-based observations. Remote Sens Environ 162:380–395

27. Piles M, Ballabrera-Poy J, Muñoz-Sabater J (2019) Dominant features of global surface soil moisture variability observed by the SMOS satellite. Remote Sens 11(1):95
28. Alvarez MA, Rosasco L, Lawrence ND (2011) Kernels for vector-valued functions: a review. arXiv:1106.6251 [cs, math, stat]. ArXiv: 1106.6251
29. Verrelst J, Alonso L, Camps-Valls G, Delegido J, Moreno J (2012) Retrieval of vegetation biophysical parameters using gaussian process techniques. IEEE Trans Geosci Remote Sens 50(5/P2):1832–1843
30. Journel A, Huijbregts C (1978) Mining geostatistics. Academic Press
31. Albergel C, de Rosnay P, Gruhier C, Muñoz-Sabater J, Hasenauer S, Isaksen L, Kerr Y, Wagner W (2012) Evaluation of remotely sensed and modelled soil moisture products using global ground-based in situ observations. Remote Sens Environ 118:215–226
32. González-Zamora Á, Sánchez N, Martínez-Fernández J, Gumuzzio Á, Piles M, Olmedo E Long-term SMOS soil moisture products: a comprehensive evaluation across scales and methods in the duero basin (spain)
33. Al-Yaari A, Wigneron JP, Ducharne A, Kerr YH, Wagner W, Lannoy GD, Reichle R, Bitar AA, Dorigo W, Richaume P, Mialon A (2014) Global-scale comparison of passive (SMOS) and active (ASCAT) satellite based microwave soil moisture retrievals with soil moisture simulations (MERRA-land). Remote Sens Environ 152:614–626
34. Albergel C, Dorigo W, Balsamo G, noz Sabater JM, de Rosnay P, Isaksen L, Brocca L, de Jeu R, Wagner W (2013) Monitoring multi-decadal satellite earth observation of soil moisture products through land surface reanalyses. Remote Sens Environ 138:77–89
35. Polcher J, Piles M, Gelati E, Barella-Ortiz A, Tello M (2016) Comparing surface-soil moisture from the SMOS mission and the ORCHIDEE land-surface model over the iberian peninsula. Remote Sens Environ 174:69–81
36. Sanchez N, Martinez-Fernandez J, Scaini A, Perez-Gutierrez C (2012) Validation of the SMOS L2 soil moisture data in the REMEDHUS network (Spain). IEEE Trans Geosci Remote Sens 50(5):1602–1611
37. Bircher S, Skou N, Jensen KH, Walker JP, Rasmussen L (2012) A soil moisture and temperature network for SMOS validation in western denmark. Hydrol Earth Syst Sci 16(5):1445–1463
38. Torbern T, Rasmus F, Idrissa G, Olander RM, Silvia H, Cheikh M, Monica G, Stéphanie H, Inge S, Bo HR, Marc-Etienne R, Niklas O, Jørgen LO, Andrea E, Mathias M, Jonas A (2014) Ecosystem properties of semiarid savanna grassland in west africa and its relationship with environmental variability. Global Change Biol 21(1):250–264
39. Entekhabi D, Reichle RH, Koster RD, Crow WT (2010) Performance metrics for soil moisture retrievals and applications requirements. J Hydrometeorol 11:832–840
40. Harchaoui Z, Bach F, Cappe O, Moulines E (2013) Kernel-based methods for hypothesis testing: a unified view. IEEE Signal Proc Mag 30(4):87–97
41. Muandet K, Fukumizu K, Sriperumbudur B, Schölkopf B (2016) Kernel mean embedding of distributions: a review and beyond. now foundations and trends
42. Rojo-Álvarez JL, Martínez-Ramón M, Muñoz-Marí J, Camps-Valls G (2017) Digital signal processing with Kernel methods. Wiley, UK
43. Shawe-Taylor J, Cristianini N (2004) Kernel methods for pattern analysis. Cambridge University Press, Cambridge, MA, USA
44. Camps-Valls G, Bruzzone L (2009) Kernel methods for remote sensing data analysis. Wiley
45. Camps-Valls G, Gómez-Chova L, Muñoz-Marí J, Vila-Francés J, Calpe-Maravilla J (2006) Composite kernels for hyperspectral image classification. IEEE Geosci Remote Sens Lett 3(1):93–97
46. Konings AG, Piles M, Das N, Entekhabi D (2017) L-band vegetation optical depth and effective scattering albedo estimation from SMAP. Remote Sens Environ 198:460–470
47. Jackson RD, Huete AR (1991) Interpreting vegetation indices. Prev Vet Med 11(3–4):185–200
48. Chaparro D, Piles M, Vall-llossera M, Camps A, Konings AG, Entekhabi D (2018) L-band vegetation optical depth seasonal metrics for crop yield assessment. Remote Sens Environ 212:249–259

49. Piles M, Camps-Valls G, Chaparro D, Entekhabi D, Konings AG, Jagdhuber T (2017) Remote sensing of vegetation dynamics in agro-ecosystems using smap vegetation optical depth and optical vegetation indices. In: IGARSS17, pp 4346–4349

50. López-Lozano R, Duveiller G, Seguini L, Meroni M, García-Condado S, Hooker J, Leo O, Baruth B (2015) Towards regional grain yield forecasting with 1km-resolution EO biophysical products: strengths and limitations at pan-european level. Agric For Meteorol 206:12–32

51. Gorelick N, Hancher M, Dixon M, Ilyushchenko S, Thau D, Moore R (2017) Google earth engine: planetary-scale geospatial analysis for everyone. Remote Sens Environ 202:18–27

52. He M, Kimball JS, Maneta MP, Maxwell BD, Moreno A, Beguería S, Wu X (2018) Regional crop gross primary productivity and yield estimation using fused landsat-MODIS data. Remote Sens 10:372

53. Robinson NP, Allred B, Jones MO, Moreno A, Kimball JS, Naugle D, Erickson TA, Richardson AD (2017) A dynamic landsat derived normalized difference vegetation index (NDVI) product for the conterminous united states. Remote Sens 9:823

54. Jacquemoud S, Baret F (1990) PROSPECT: a model of leaf optical properties spectra. Remote Sens Environ 43:75–91

55. Verhoef W (1984) Light scattering by leaf layers with application to canopy reflectance modeling: the SAIL model. Remote Sens Environ 16:125–141

56. Berger K, Atzberger C, Danner M, D'Urso G, Mauser W, Vuolo F, Hank T (2018) Evaluation of the PROSAIL model capabilities for future hyperspectral model environments: a review study. Remote Sens 10:85

57. Campos-Taberner M, García-Haro FJ, Camps-Valls G, Grau-Muedra G, Nutini F, Busetto L, Katsantonis D, Stavrakoudis D, Minakou C, Gatti L, Barbieri M, Holecz F, Stroppiana D, Boschetti M (2017) Exploitation of SAR and optical sentinel data to detect rice crop and estimate seasonal dynamics of leaf area index. Remote Sens 9:248

58. Campos-Taberner M, García-Haro FJ, Camps-Valls G, Grau-Muedra G, Nutini F, Crema A, Boschetti M (2016) Multitemporal and multiresolution leaf area index retrieval for operational local rice crop monitoring. Remote Sens Environ 187:102–118

59. García-Haro FJ, Campos-Taberner M, noz Marí JM, Laparra V, Camacho F, Sánchez-Zapero J, Camps-Valls G (2018) Derivation of global vegetation biophysical parameters from EUMET-SAT polar system. ISPRS J Photogramm Remote Sens 139:57–75

60. Kattge J, Díaz S, Lavorel S, Prentice I, Leadley P, Bönisch G et al (2011) TRY-a global database of plant traits. Glob Change Biol 17:2905–2935

61. Madani N, Kimball J, Ballantyne A, Affleck D, van Bodegom P, Reich P, Kattge J, Sala A et al (2018) Future global productivity will be affected by plant trait response to climate. Sci Rep 8(2870)

62. Belgiu M, Lucian D (2016) Random forest in remote sensing: a review of applications and future directions. ISPRS J Photogramm Remote Sens 114:24–31

63. De'ath G, Fabricius K (2000) Classification and regression trees: a powerful yet simple technique for ecological data analysis. Ecology 81:3178–3192

64. Evans J, Cushman S (2009) Gradient modeling of conifer species using random forests. Landsc Ecol 24:673–683

65. Cutler D, Edwards J, Thomas C, Beard K, Cutler A, Hess K, Gibson J, Lawler J (2007) Random forests for classification in ecology. Ecology 88:2783–2792

66. Chen YW, Lin CJ (2006) Combining SVMs with various feature selection strategies. In: Guyon I, Nikravesh M, Gunn S, Zadeh LA (eds) Feature extraction. Studies in fuzziness and soft computing, vol 207. Springer, Berling, Heidelberg

67. Breiman L, Friedman J (1985) Estimating optimal transformations for multiple regression and correlation. J Am Stat Assoc 391:1580–1598
68. Yan K, Park T, Yan G, Liu Z, Yang B, Chen C, Nemani R, Knyazikhin Y, Myneni R (2016) Evaluation of MODIS LAI/FPAR product collection 6. part 2: validation and intercomparison. Remote Sens 8(460)
69. Campos-Taberner M, j García-Haro F, Busetto L, Ranghetti L, Martínez B, Gilabert MA, Camps-Valls G, Camacho F, Boschetti M (2018) A critical comparison of remote sensing leaf area index estimates over rice-cultivated areas: from Sentinel-2 and Landsat-7/8 to MODIS, GEOV1 and EUMETSAT polar system. Remote Sens 10:763

Chapter 3
Deep Learning for Hyperspectral Image Analysis, Part I: Theory and Algorithms

Sebastian Berisha, Farideh Foroozandeh Shahraki, David Mayerich and Saurabh Prasad

Abstract Deep neural networks have emerged as a set of robust machine learning tools for computer vision. The suitability of convolutional and recurrent neural networks, along with their variants, is well documented for color image analysis. However, remote sensing and biomedical imaging often rely on hyperspectral images containing more than three channels for pixel-level characterization. Deep learning can facilitate image analysis in multi-channel images; however, network architecture and design choices must be tailored to the unique characteristics of this data. In this two-part series, we review convolution and recurrent neural networks as applied to hyperspectral imagery. Part I focuses on the algorithms and techniques, while Part II focuses on application-specific design choices and real-world remote sensing and biomedical test cases. These chapters also survey recent advances and future directions for deep learning with hyperspectral images.

3.1 Introduction

Hyperspectral imaging (HSI) combines spectroscopic instrumentation with imaging systems to provide spatially resolved spectroscopic data. HSI instrumentation can acquire hundreds or thousands of spectra in a $X \times Y \times Z$ data cube, where X and Y are spatial dimensions and Z describes spectral content (Fig. 3.1). Information encoded along the spectral dimension depends on modality, with the most common approaches being ultraviolet [1], visible [2], near-infrared [3], and vibrational [4]

S. Berisha · F. F. Shahraki · D. Mayerich · S. Prasad (✉)
University of Houston, Houston, TX, USA
e-mail: saurabh.prasad@ieee.org

S. Berisha
e-mail: sberisha@central.uh.edu

F. F. Shahraki
e-mail: fforoozandehshahraki@uh.edu

D. Mayerich
e-mail: mayerich@uh.edu

© Springer Nature Switzerland AG 2020   37
S. Prasad and J. Chanussot (eds.), *Hyperspectral Image Analysis*,
Advances in Computer Vision and Pattern Recognition,
https://doi.org/10.1007/978-3-030-38617-7_3

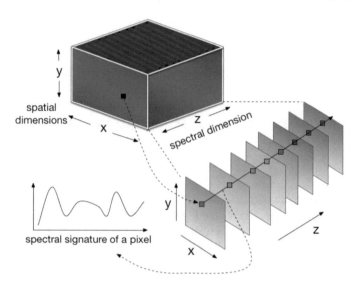

Fig. 3.1 Structure of an HSI data cube. The measured data in HSI can be visualized as a data cube. Each slice of the data cube contains an image of the scene at a particular wavelength. Each pixel is associated with a vector of spectral responses otherwise known as a spectral signature

spectroscopy. Non-optical methods include mass [5] and nuclear magnetic resonance (NMR) spectroscopy [6]. The encoded spectral signature provides insight into the material composition at each $[x, y]^T$ spatial location, where $x \in X$ and $y \in Y$. This spectral signature provides a fingerprint for material identification and quantifiable properties such as density, absorbance, and emission. HSI approaches have seen broad use in remote sensing [3], biomedicine [4], astronomy [7], agriculture and food quality [8, 9], and pharmaceuticals [10].

Interpreting spectra requires some form of analysis that can, at its most basic, include searches within a dictionary of known materials. Classification, regression, and object detection are becoming more common practices. Popular methods include unsupervised clustering using K-means [11–13] or hierarchical cluster analysis (HCA) [14, 15]. Supervised techniques are becoming more common for complex spectra composed of many molecular components. These include Bayesian classification [16, 17], random forests [18, 19], kernel classifiers such as support vector machines (SVMs) [20, 21], and linear discriminant analysis [22, 23]. SVMs are a particularly popular choice due to their simplicity, accuracy, and ability to classify high-dimensional data.

The emergence of deep learning has lead to more advanced feature extraction by combining both spatial and spectral information. Before we discuss the technical details of some popular deep learning architectures, we briefly summarize the history of deep learning in the context of computer vision, remote sensing, and biomedicine. We then describe the basic building blocks used to apply deep learning to HSI. Section 3.2 presents feed-forward neural network models. Popular deep neural network architectures are introduced in Sect. 3.3. Section 3.4 introduces convolutional neural networks (CNNs) and describes common CNN flavors used in HSI. In Sect. 3.5, we discuss some of the existing open-source software tools for deep learning. Section 3.6 concludes this chapter.

Mathematical notation—Throughout this chapter, we denote matrices and vectors as boldface uppercase and boldface lowercase letters (i.e., $\mathbf{Ax} = \mathbf{b}$) and scalars are denoted using a normal typeface (i.e., $ax = b$). All vectors are assumed to be column vectors, and the transpose operator is denoted by a superscript $\top$: $\mathbf{x} = [x_1, x_2, x_3, \ldots, x_n]^\top$.

3.1.1 History of Deep Learning in Computer Vision

Artificial neural networks (ANNs) were inspired by human brain architecture. A 1943 paper by the neuroscientist McCulloch and logician Pitts [24] studied the brain's ability to produce complex patterns using basic connected elements, called neurons. The authors presented a highly simplified model of a neuron, now known as a McCulloch–Pitts (MCP) neuron. A network of MCP neurons is the ancestor of the ANN. The neuropsychologist Donald Hebb introduced the Hebbian Learning Rule [25] in 1949, which postulates how biological neurons learn. According to Hebb, the synaptic connection between two neurons will strengthen if the linked cells are activated simultaneously. This work has immortalized Hebb a the father of modern ANNs. Based on Hebb's findings, Frank Rosenblatt modified the MCP neuron to create the first perceptron in 1958 [26]. Rosenblatt's perceptron was able to learn by modifying input weights and was instrumental to the later development of more complex networks. In 1974, Paul Werbos introduced the process of training neural networks through back-propagation of errors [27], which provided a deterministic approach for optimized learning.

In 1980, Kunihiko Fukushima introduced the neocognitron [28], which later inspired convolutional neural networks (CNNs). Fukushima's neocognitron is an ANN that consists of a feature extraction layer, S-layer, and a C-layer, which represent structured connections that organize extracted features. In 1985, Ackley et al. [29] invented the Boltzmann machine, which is a stochastic version of the Hopfield network, consisting of hidden and visible nodes. Boltzmann machines were the first networks capable of learning internal representations to solve difficult combinatorial problems. Smolensky invented the restricted Boltzmann machine (RBM) [30], originally known as the Harmonium, in 1986. The RBM is a version of the Boltzmann machine that eliminates connections between visible or hidden units to simplify training and implementation. Jordan introduced the modern definition of a recurrent neural network [31] (RNN) in 1985, referred to as a Jordan neural network, which contains one or more cycles (or loops).

That same year Rumelhart et al. [32] introduced autoencoders as a form of unsupervised learning. LeCun, inspired by the neocognitron, introduced LeNet [33] in

1990, which is thought of as the first CNN model and demonstrated the practical potential for deep neural networks. Hochreiter and Schmidhuber [34] introduced long short-term memory (LSTM) in 1997, which allowed RNNs to learn long-term dependencies. Specifically, LSTMs were designed to solve the problem of vanishing gradients. The current deep learning era started to flourish with the introduction of deep belief networks by Hinton et al. [35] in 2006, which consisted of multi-layered RBMs combined with a layer-wise pretraining algorithm. This pretraining strategy inspired the introduction of deep Boltzmann machines in 2009 [36]. The era of CNNs began with the AlexNet model [37], which demonstrated the effectiveness of CNNs on the challenging ImageNet dataset. AlexNet also spurred the development of numerous deep learning models to achieving even better performances.

3.1.2   History of Deep Learning for HSI Tasks

Various remote sensing applications are carried out by deep learning, including land use classification, target detection, change detection, and semantic segmentation. HSI classification is one of the most active research areas in remote sensing. Deep learning algorithms have demonstrated strong performance in these tasks because deep learning architectures with high-level features are highly robust to nonlinearities in HSI data.

The first deep learning architectures used for remote sensing HSI classification were multi-layer neural networks. In particular, feed-forward neural networks were proposed data by Subramanian et al. [38] in 1998 and Jimenez et al. [39] in 1999. Neural network models were used in 2009 for HSI spectral mixture analysis [40]. In 2010, Ratle et al. [41] proposed a semisupervised method for HSI classification based on neural networks. Licciardi et al. [42], in 2011, applied neural networks for the task of unmixing.

The first *modern* deep learning approaches were introduced by Lin et al. [43] in 2013 and extended by Chen et al. [44] in 2014. In both cases, the authors used autoencoders and stacked autoencoders (SAE) [45, 46] to extract deep features from hyperspectral data. In 2015, Tao et al. [47] used stacked sparse autoencoders for spectral–spatial feature learning. In the same year, a 1D CNN architecture was proposed for pixel-level classification of HSI data [48]. 2D CNNs were then exploited for the task of HSI classification by Makantasis et al. [49] and Yue et al. [50]. *Contextual deep learning* was also proposed [51] in 2015, which uses the SAE family to extract spectral–spatial features.

Another class of deep neural networks, *deep belief networks* (DBNs), was used for HSI classification [52] to extract spectral–spatial features. In 2016, Chen et al. [53] proposed a 3D CNN-based feature extraction framework combined with regularization to mitigate overfitting. DBNs were later used also by Zhou et al. [54] for feature extraction.

Li et al. [55] proposed a pixel-pair method to deal with a limited number of training samples, which is a common challenge in HSI. The authors used a deep CNN to learn pixel-pair features. The center and surrounding pixels are used to construct pixel pairs, which are then classified by the CNN. In 2018, Shu et al. [56] proposed another framework which the spectral cuboid is first preprocessed by PCA whitening, and then all the spectral patches are stacked to form a spectral quilt which is input to the two shallow CNNs for classification. Recurrent neural networks (RNNs) have also been recently used for HSI remote sensing data classification [57–59]. In [60], the authors demonstrate a framework that can use graph-based convolutional neural networks (GCNs) to effectively represent data residing on smooth manifolds, such as reflectance spectra of hyperspectral image pixels. In GCNs, a convolution operator is defined based on the graph Fourier transform to perform convolution/aggregation operations on feature vectors of its neighbors on the graph. In [61], the authors proposed a class of convolutional neural networks where convolutional filters are expressed as linear combinations from a predefined discrete directional filters inspired by the theory of shearlets and only the coefficients of the linear combination are learned during training.

In the context of biomedical applications, Goodacre et al. [62], in 1998, utilized pyrolysis mass spectrometry (PyMS), FTIR, and dispersive Raman microscopy data in combination with an artificial neural network (ANN) to discriminate clinically significant intact bacterial species. *Modern* deep learning methods have only appeared in biomedical HSI applications. In 2017, Halicek et al. [63] proposed a CNN classifier for detecting head and neck cancer from HSI data. Berisha et al. [64] utilized CNNs for cell identification in breast tissue biopsies. It was shown that CNNs outperform traditional spectral-based classifiers for FTIR image classification. Lotfollahi et al. [65] used deep learning to map infrared spectra to chemical stains to duplicate traditional histological images from label-free HSI data.

3.1.3 Challenges

Hyperspectral images contain more per-pixel data than traditional color imagery. This results in larger input vectors, producing networks with significantly more trainable parameters. A larger parameter space makes optimization much more memory intensive and increases the need for training examples to mitigate overfitting. Dimension reduction (DF) is a popular approach for reducing feature vector size (DR) [66–69]. DR methods reduce redundancy across spectral bands by mapping the input spectra to a lower dimensional subspace. Principal component analysis (PCA) is a common unsupervised DR method [44, 52, 53, 70, 71] and identifies input components by optimizing spectral variance. Most of the referenced research in this chapter leverages a combination of PCA, hierarchical feature learning, and linear regression.

3.2 Feed-Forward Neural Networks

Feed-forward neural networks form the basis for modern artificial neural networks [24, 72–74]. A feed-forward network approximates an *activation function* $f^*(\cdot)$ by optimizing based on a set of training samples. For example, a feed-forward network can capture the function $y = f^*(x)$, which maps an input x to an output value (or vector). Specifically, the feed-forward network defines a mapping $y = f(x; \Theta)$ where Θ is a vector of learned parameters that provide the best approximation to $f^*(x)$.

Feed-forward networks are so named because information flows in one direction through the network, from x to the output y, forming a directed acyclic graph. In its general form, a feed-forward network consists of an input layer, one or more hidden layers, and an output layer. Each node in a layer is an artificial neuron, where input is modified by a weight and summed across all other inputs. The resulting value is passed through a transfer function to one or more neurons in the next layer.

The feed-forward multi-layer perceptron (MLPs) [72, 74] is the most popular architecture for HSI applications and is commonly used as the final step in new deep neural networks. Applications leveraging MLPs include distinguishing between normal and injured fruit [75], classification of land and clouds in remote sensing [76], assessment of meat quality [77], microbial characterization [62], and identification of bacteria using Fourier transform infrared (FTIR) spectroscopy [78]. The input layer generally consists of a vector representing a spectrum, often after some data pre-processing, such as noise removal or dimension reduction.

3.2.1 Perceptron

A perceptron [26] or artificial neuron (Fig. 3.2) is the most basic processing unit of feed-forward neural networks. A perceptron can be modeled as a single-layer neural network with an input vector $x \in \mathbb{R}^n$, a bias b, a vector of trainable weights $w \in \mathbb{R}^n$, and an output unit y. Given the input x, the output y is computed by an activation function $f(\cdot)$ as follows:

$$y(x; \Theta) = f\left(\sum_{i=1}^{n} x_i w_i + b\right) = f(w^\top x + b), \tag{3.1}$$

where $\Theta = \{w, b\}$ represents the trainable parameter set. A perceptron in this form is a binary linear classifier. A logistic *sigmoid function* is commonly used for binary classification tasks.

The single output perceptron model can have multiple outputs (shown in Fig. 3.3) given an input $x \in \mathbb{R}^n$ by using an activation function $f(\cdot)$:

Fig. 3.2 A schematic view of a single output perceptron. Each input value x_i is multiplied by a weight factor w_i. The weighted sum added to the bias is then passed through an activation function to obtain the output y

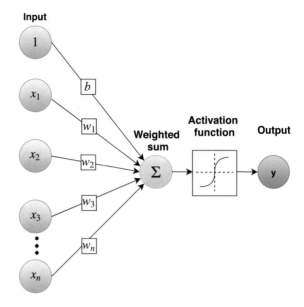

Fig. 3.3 A schematic view of a multiple output feed-forward neural network. Each input value x_i is multiplied by a weight factor W_{ij}, where W_{ij} denotes a connection weight between the input node x_i and the output node y_j. The weighted sum is added to the bias and then passed through an activation function to obtain the output y_j

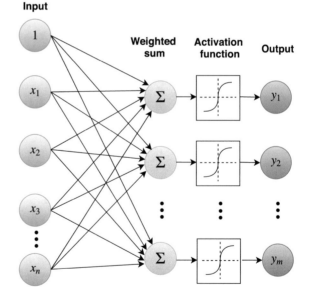

$$y_j(\boldsymbol{x}; \Theta) = f\left(\sum_{i=1}^{n} x_i W_{ij} + b_j\right) = f(\boldsymbol{w}_j^\top \boldsymbol{x} + b_j), \qquad (3.2)$$

where the parameter set here is $\Theta = \{\boldsymbol{W} \in \mathbb{R}^{n \times m}, \boldsymbol{b} \in \mathbb{R}^m\}$ and $\boldsymbol{w}_j$ denotes the jth column of $\boldsymbol{W}$.

A common activation function is used for multi-class applications *softmax*, which takes an input vector and produces an output vector of the same size containing scalar values between 0 and 1 that sum to 1. The output can therefore be interpreted as probability distribution.

3.2.2 Multi-layer Neural Networks

A single-layer perceptron network still represents a linear classifier, despite using nonlinear transfer functions. This limitation can be overcome by multi-layer neural networks (Fig. 3.4), which introduce one or more "hidden" layers between the input and output layers. Multi-layer neural networks are composed of several simple artificial neurons such that the output of one acts as the input of another. A multi-layer neural network can be represented by a composition function. For a two-layer network, the composition function can be written as

$$
y_j(\boldsymbol{x}; \Theta) = f^{(2)} \left(\sum_{k=1}^{h} W_{jk}^{(2)} * f^{(1)} \left(\sum_{i=1}^{n} W_{ki}^{(1)} * x_i + b_k^{(1)} \right) + b_j^{(2)} \right),
\tag{3.3}
$$

where h is the number of units in the hidden layer and the set of unknown parameters is $\Theta = \{W^{(1)} \in R^{h \times n}, W^{(2)} \in R^{1 \times h}\}$. In general, for $L - 1$ hidden layers the composition function, omitting the bias terms, can be written as

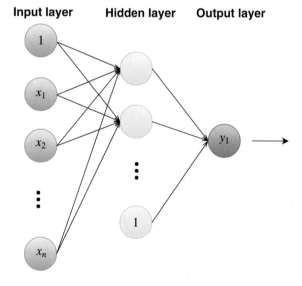

Fig. 3.4 Illustration of a feed-forward neural network composed of an input layer, a hidden layer, and an output layer. In this illustration, the multi-layer neural network has one input layer and one output unit. In most models, the number of hidden layers and output units is more than one

$$y_j(x; \Theta) = f^{(L)} \left(\sum_k W^L_{jk} * f^{L-1} \left(\sum_l W^{L-1}_{kl} * f^{L-2} \left(\cdots f^1 \left(\sum_i W^1_{zi} * x_i \right) \right) \right) \right).$$

(3.4)

Note that linear activation functions would result in a composite linear system. Nonlinear activation functions are generally chosen, with the same function used for each layer.

3.2.2.1 Activation Functions

The output of the perceptron, before applying an activation function, is an unbounded value summation. The activation function determines the (usually bounded) output interval. While both linear and nonlinear activation functions are used, nonlinear functions are the most practical for neural network applications. Several activation functions have been selected based on distinct advantages and disadvantages. In this section, we will discuss the most common activation functions for hyperspectral imaging tasks.

Binary step function—A simple threshold function can be used for binary activation of the neuron. The binary step function and its derivative are defined as

$$f(x) = \begin{cases} 1, & \text{if } x \geq 0, \\ 0, & \text{otherwise,} \end{cases} \quad f'(x) = 0, \ \forall \, x.$$

(3.5)

The constant gradient makes this function difficult to optimize during back-propagation. Furthermore, the binary step function cannot be used for multi-class output tasks.

Linear function—A linear function is defined as

$$f(x) = ax, \quad f'(x) = a,$$

(3.6)

which can be applied to multiple neurons and thus used for multi-class applications. A linear transfer function reduces the network to a linear system, simplifying training. However, the result will always produce a linear transformation and therefore is not suitable for nonlinear classification or regression tasks.

Sigmoid—The `sigmoid` function

$$\text{sigmoid}(x) = \frac{1}{(1 + e^{-x})}, \quad \text{sigmoid}'(x) = f(x) * (1 - f(x))$$

(3.7)

is smooth, continuously differentiable, and nonlinear. It ranges from 0 to 1 with an S shape. The gradient is higher between x values of -3 and 3 but flattens out for higher magnitude values. This implies that small changes in the input of range $[-3, \ 3]$

introduce large changes in output. This property is desirable for classification by providing a differential alternative to the binary step function. The dependence of the gradient on x allows back-propagation of errors so that weights can be updated during training. One disadvantage of the *sigmoid* function is that it is not symmetric around the origin and maps all input to positive values. This is undesirable since all inputs to downstream layers in multi-layer networks will have the same sign.

Hyperbolic tangent—The asymmetry of the *sigmoid* function is addressed with tanh, which is a scaled version of the *sigmoid*:

$$\tanh(x) = 2 \cdot \text{sigmoid}(2x) - 1, \quad \tanh'(x) = 1 - \tanh^2(x). \tag{3.8}$$

The *hyperbolic tangent (tanh)* has the same properties as the *sigmoid* function and is symmetric over the origin with ranges from -1 to 1. The tanh function has a steeper gradient than a sigmoid; however, the tanh are still very low for high-magnitude inputs resulting in a vanishing gradient.

Rectified linear unit (ReLU)—The ReLU activation function is widely used due to its computational simplicity and ability to facilitate fast training. The ReLU function and its gradient are given by

$$f(x) = \max(0, x), \quad f'(x) = \begin{cases} 1, & \text{if } x \geq 0, \\ 0, & \text{otherwise.} \end{cases} \tag{3.9}$$

ReLU enforces nonnegativity on the input, and thus only a few artificial neurons are activated during training, making the neural network sparse and more computationally efficient. However, since the gradient is zero for negative input, the corresponding weights are not updated during back-propagation. The neural network will therefore have dead neurons that are never activated after some training iteration.

Leaky ReLU—Leaky ReLU [79] addresses the issue of 0 gradients for negative inputs by mapping the negative input to a small linear component:

$$f(x) = \begin{cases} x, & \text{if } x \geq 0, \\ ax, & \text{otherwise,} \end{cases} \quad f'(x) = \begin{cases} 1, & \text{if } x \geq 0, \\ a, & \text{otherwise,} \end{cases} \tag{3.10}$$

thereby alleviating the issue of zero gradients leading to dead artificial neurons.

Parametric ReLU—The parametric ReLU [80] is identical to Eq. 3.10 with the exception that the a parameter is learned during training. This function is used when leaky ReLU fails to overcome the problem of dead neurons.

Softmax—The softmax function is extremely useful for classification functions. Unlike the sigmoid, which can only handle two-class problems, the softmax *function* handles multi-class classification by mapping the input to a range between 0 and 1 and dividing by the total sum, outputting a probability vector representing posteriors for each class:

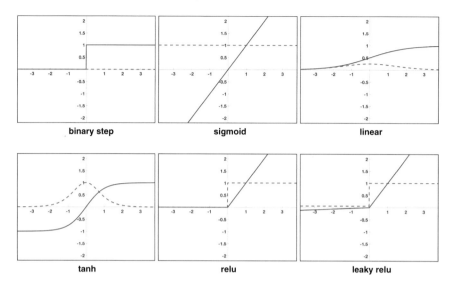

Fig. 3.5 Illustration of various types of activation functions (in −) and their corresponding gradients (in −−). There is a lack of precise rules/heuristics on how to choose an activation function. However, each activation function has certain properties that can help when deciding to make better choices for quicker convergence of a neural network

$$\text{softmax}(x_j) = \frac{e^{x_j}}{\sum_{i=1}^{C} e^{x_i}}, \quad j = 1, \dots, C, \tag{3.11}$$

where for classification problems C is the number of classes. The softmax function is usually used in the output layer of the classifier.

Several common activation functions are shown in Fig. 3.5. Many other variations have been explored to improve accuracy and provide faster convergence rates. We do not discuss them here, but the interested reader should explore the exponential linear unit (elu) [81], scaled exponential linear unit (SELU) [82], thresholded ReLU [83], concatenated ReLU [84], and rectified linear 6 [85].

3.2.3 Learning and Gradient Computation

The goal of training is to identify weights and biases that minimize some cost function based on a set of provided training data. The choice of cost function depends on the task at hand. Classification tasks commonly minimize classification error of the predicted labels using cross entropy (Eqs. 3.15 and 3.16). In regression problems, MSE (Eq. 3.12) is commonly used to minimize the difference between the ground truth and predicted value. In any case, the training goal is optimization of a cost function of many variables. A common technique used to solve such problems is the gradient descent algorithm.

Weights and the biases adjusted iteratively by applying some flavor of gradient descent, commonly known as *back-propagation* [86]. Partial derivatives of the cost function are calculated at each iteration with respect to weights and biases, and the parameters are correspondingly updated. Back-propagation provides insights into the overall behavior of the network and is crucial to learning in neural networks.

3.2.3.1   Loss Functions

Selection of a loss function is a crucial aspect of designing a neural network. The loss function, otherwise known as a cost or error function, measures the deviation of the neural network output $\hat{y}$ from expected values y. The loss function provides a nonnegative scalar value that decreases as prediction quality increases. Here, we discuss some of the most popular loss functions in deep learning applications.

Mean Squared Error (MSE), otherwise known as quadratic error, is a common performance measure in linear regression problems. This cost function is defined as a sum of squared differences across all N training samples:

$$\mathscr{L} = \frac{1}{N} \sum_{i=1}^{N} \left(y^{(i)} - \hat{y}^{(i)} \right)^2 , \tag{3.12}$$

where $\left(y^{(i)} - \hat{y}^{(i)} \right)$ is the residual. Minimizing the MSE amounts to minimizing the sum of the squared residuals, which is intuitive and easy to optimize. However, MSE applies more weight to larger differences, which may be undesirable in many applications.

Mean Absolute Error (MAE) measures the average absolute difference between predicted labels and observations:

$$\mathscr{L} = \frac{1}{N} \sum_{i=1}^{N} \left| y^{(i)} - \hat{y}^{(i)} \right| . \tag{3.13}$$

MAE residuals are given equal weight, making it more robust to outliers. Unfortunately, the MAE gradient is significantly more difficult to compute.

Mean Squared Logarithmic Error (MSLE) is a variation of MSE defined as the mean of the squared residuals of log-transformed true and predicted labels:

$$\mathscr{L} = \frac{1}{N} \sum_{i=1}^{N} \left(\log(y^{(i)} + 1) - \log(\hat{y}^{(i)} + 1) \right)^2 , \tag{3.14}$$

where 1 is added as a regularization term to avoid log(0). MSLE can be used when it is not desirable to penalize large errors but penalizes underestimates more than overestimates. Given these properties, it has been used in regression problems for predicting future house pricing.

Binary Cross Entropy is a common loss function in classification tasks that measures the performance of a classification model whose output is a probability between 0 and 1. In binary classification, where the number of classes is 2, the binary cross entropy is defined as

$$\mathcal{L} = -\frac{1}{N} \sum_{i=1}^{N} \left[y^{(i)} \log(\hat{y}^{(i)}) + (1 - y^{(i)}) \log(1 - \hat{y}^{(i)}) \right]. \qquad (3.15)$$

The value of cross entropy loss increases as the predicted probability diverges from the actual label. Thus, cross entropy is a measurement of the divergence between two probability distributions.

Categorical Cross Entropy provides a multi-class variant computed by calculating a separate cross entropy loss for each class:

$$\mathcal{L} = -\frac{1}{N} \sum_{i=1}^{N} \sum_{c=1}^{C} y^{i,c} \log(\hat{y}^{i,c}), \qquad (3.16)$$

where C is the number of classes, $y_{i,c}$ is 1 if class label c is the correct classification for the ith sample and 0 otherwise, and $\hat{y}^{i,c}$ is the predicted probability that the ith sample is of class c.

Kullback Leibler (KL) Divergence, otherwise known as relative entropy, measures the difference—or divergence—between two probability distributions:

$$\begin{aligned}
\mathcal{L} &= \frac{1}{N} \sum_{i=1}^{N} \mathcal{D}_{KL}\left(y^{(i)} || \hat{y}^{(i)}\right) \\
&= \frac{1}{N} \sum_{i=1}^{N} \left[y^{(i)} \log\left(\frac{y^{(i)}}{\hat{y}^{(i)}}\right) \right] \\
&= \underbrace{\frac{1}{N} \sum_{i=1}^{N} \left(y^{(i)} \log(y^{(i)})\right)}_{entropy} - \underbrace{\frac{1}{N} \sum_{i=1}^{N} \left(y^{(i)} \log(\hat{y}^{(i)})\right)}_{crossentropy},
\end{aligned} \qquad (3.17)$$

where $\mathcal{D}_{KL}$ denotes the KL divergence from $\hat{y}^{i}$ to $y^{(i)}$. The KL divergence is not commutative and thus it cannot be used as a distance metric.

We have provided a variety of Refs. [87–89] that the interested reader should be encouraged to explore for more exhaustive work on loss functions in neural networks.

3.2.3.2 Back-Propagation

Multi-layer networks are commonly trained by minimizing a loss function through back-propagation. Back-propagation is a powerful but simple method for training models which have a large number of trainable parameters.

The basic approach to learn an untrained network is to present a training pattern to the input layer, pass the signal through the network, and determine the predicted output at the output layer. The predicted outputs are compared to the actual output values, and any difference corresponds to an error which is a function of the network weights/parameters. When the back-propagation is based on gradient descent (c.f. Fig. 3.6), the weights (which are often initialized with random values) are changed in a direction that reduces the error. In multi-layer networks, back-propagating errors through the network allow computations of gradients of the loss function with respect to network weights to be computed in an efficient way. Given a training dataset, the standard gradient descent estimates the expectation value of the gradient of the loss function with respect to the network parameters by evaluating the loss and gradient over all available training samples. Stochastic gradient descent (SGD) eliminates the computation of an expectation of the gradient and utilizes a few samples that are randomly drawn from the training pool, thereby resulting in reduced variability during learning, and hence a more stable convergence.

Learning Rate—This is a hyperparameter in back-propagation which indicates the relative size of the change in weights, when they are being updated using gradient descent. Using too large or too small, a learning rate can cause the model to diverge or converge too slowly, respectively. Also, note that the loss function may not be convex with respect to the network weights, and we can end up in a "local minimum" especially when the learning rate is small. The learning rate hence must be appropriately chosen for a given learning task carefully.

Fig. 3.6 Forward-propagation and back-propagation flows in a four-layer network; Red arrows indicate the forward-propagation and activation value computation, and green arrows show back-propagation and error computation

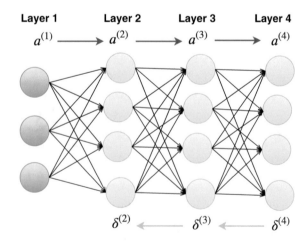

Vanishing gradient problem—During back-propagation, as the gradients propagate backward from the later layers in a chain, and as they approach the earlier layers, there is a possibility that the gradients eventually shrink exponentially until they vanish. This is called "vanishing gradient problem" and causes the nodes of earlier layer to learn very slowly as compared to the nodes in the later layers in the network. The earlier layers in the deep network are responsible for learning low-level features and can be thought of as building blocks of the deep networks. Thus, it is important that the early layers in the deep network be trained properly to lead to more accurate higher level features in later layers. This problem is particularly relevant when dealing with deep networks. Recent works with residual networks (ResNet) seek to address this problem [90, 91].

Exploding gradient problem—On the flip side, it may so happen that when the gradients reach the earlier layers, they get larger and larger to a point where they "explode"—a point at which the network would become unstable. This can also be avoided by carefully determining the network architecture and associate hyperparameters. It can also be kept in check by either clipping the gradient (not allowing the magnitude of the gradient to exceed a pre-determined threshold) or scaling it (rescaling the gradient such that it maintains a fixed norm).

3.3 Deep Neural Networks

Multi-layer feed-forward neural networks can approximate any multivariate continuous function with arbitrary accuracy [92–95]. It is also possible to approximate complex functions using *deep* architectures. In fact, it has been shown that deep models are required to learn highly varying functions representing high-level abstractions in applications vision, language, and other artificial intelligence tasks [96]. Shallow architectures can potentially require exponentially more hidden units with respect to the size of training data. Thus, the insufficient depth of a model can be detrimental for learning [96]. Deep architectures offer the possibility of having fewer hidden units per layer, which in turn can reduce the total number of model parameters. Most importantly, deep computational models learn representations of data automatically in a hierarchical manner with multiple levels of abstraction, from low-level/fine to high-level/abstract [97]. In this section, we introduce the basics of some deep neural network architectures that have been commonly investigated for HSI analysis tasks.

3.3.1 Autoencoders

An autoencoder (AE) [46, 86, 98–100] is an unsupervised artificial neural network that applies back-propagation to approximate the identity function, thereby learning an approximation $\hat{x}$ of the original input x. A general AE architecture (Fig. 3.7) con-

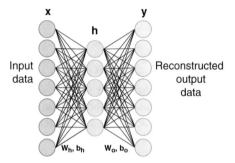

Fig. 3.7 Schematic diagram of an AE. The input x and the output y are of the same size. The network learns a representation of the input in the hidden layer. The network then tries to reconstruct the original input from the vector of hidden unit activations. If the number of hidden units is less than the size of the original input, then the network learns a *compressed* representation of the input. Even if the number of hidden units is larger than the input size, it is still possible to extract interesting features by imposing constraints on the network (such as *sparsity* on the hidden units)

sists of the input (x) layer, one or more hidden (h) layers, and an output (y) layer. The units between layers are connected via weighted connections. Figure 3.7 shows a sample AE architecture with only one hidden layer. This network consists of two parts: (1) an encoder that transforms the input into lower dimensional representation, and (2) a decoder that reconstructs the original vector. The architecture of the hidden layers applies constraints on the network that encourage learning useful features. In traditional applications, AEs have been used for dimensionality reduction, feature extraction, transfer learning [46], as well as numerous computer vision, image processing, and natural language processing tasks [101]. In HSI, autoencoders have been used for both classification [43, 102] and unmixing [103].

During the training of an AE, the input data are encoded using the weights between the input and the hidden layer, a bias, and an activation function:

$$h_i = f(W_h x_i + b_h), \quad (3.18)$$

where x_i is the ith input vector, W_h is the matrix of weights between the input and hidden layer, b_h is the bias vector of the hidden layer, f is a nonlinear activation function, and h_i is the feature vector extracted from the input vector x_i. In the case of an AE with one hidden layer, the input data are reconstructed in the output layer using the extracted features in the hidden layer:

$$y_i = f(W_o h_i + b_o), \quad (3.19)$$

where y_i is the ith reconstructed vector, W_o denotes the matrix of weights between the hidden and output layers, and b_o is the bias vector of the output layer. The training of an AE then becomes an optimization problem, which aims to find the optimal weights and biases that minimize the reconstruction error:

$$\boldsymbol{\Theta} = \underset{W,b}{\arg\min} \left(\frac{1}{N} \sum_{i=1}^{N} \|\boldsymbol{x}_i - \boldsymbol{y}_i\|_2^2 \right), \tag{3.20}$$

where $\boldsymbol{\Theta}$ denotes the optimum parameters that minimize the reconstruction error of N input vectors.

3.3.2 Stacked Autoencoders

A stacked autoencoder (SAE) [99, 104] is an ANN consisting of multiple autoencoder layers. The features extracted after each encoding and decoding phase are then sent to a softmax classifier. SAEs have been extensively applied to hyperspectral imaging in remote sensing [105] for dimension reduction and feature extraction.

SAE training is of unsupervised pretraining and supervised fine-tuning. During pretraining each AE is trained individually as described in the previous section. This is an unsupervised procedure since in this step the input labels are not used. After the training of each AE, the weights between the input and hidden layers are stored to be further used as initial values for the fine-tuning phase. In this step, multiple AEs are connected together in such a way that the hidden layer of AE in layer ℓ becomes the input of the AE in layer $\ell + 1$. In other words, the extracted features are passed to the next AE as input. This further implies that the output layer with its weights and biases is discarded. The fine-tuning step is supervised by attaching a classifier to the network. The whole network is trained using the input labels, and the weights obtained during the fine-tuning step are used as initial values in the optimization process. Figure 3.8 shows a diagram of a sample SAE.

3.3.3 Recurrent Neural Networks

Previously described models consider hyperspectral data as a collection of independent spectra, where each pixel is a point in an orderless three-dimensional feature space. These vector-based approaches lead to a loss of information in the spatial domain, since the spatial relationship between pixels is not considered. This has motivated the development and use of methods that leverage the spatial component of HSI data.

A recurrent neural network (RNN) [106] extends conventional feed-forward networks by introducing loops. RNNs process sequential inputs by utilizing recurrent hidden states, or memory cells, whose activations depend on previous steps. The retained states can represent information from arbitrarily long context windows, allowing the network to exhibit dynamic temporal behavior. This can lead to learning the dynamics of sequential input vectors over time.

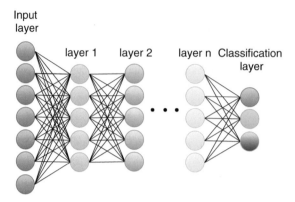

Input
layer

Fig. 3.8 Schematic diagram of a general SAE architecture for classification. The SAE network consists of multiple layers of AEs. The first layer is an AE, which is trained on raw inputs to learn primary features. The learned primary features are then fed to another AE layer that learns the secondary features. At the last AE layer, the extracted features are then given as input to a softmax classifier, which maps the features to class labels

By viewing the variability of a spectrum in the same way as a temporal signal, RNNs can characterize spectral correlation and band-to-band variability for a variety of applications, including multi-class classification [57–59].

To define an RNN, let $\{x_1, x_2, \ldots, x_T\}$ be a sequence of data, where x_i is the vector data at the ith time step. An RNN updates the recurrent hidden state, h_t, by

$$h_t = \begin{cases} 0, & \text{if } t = 0, \\ f(h_{t-1}, x_t), & \text{otherwise}, \end{cases} \tag{3.21}$$

where $f(\cdot)$ is a nonlinear activation function. In the standard RNN model, the update rule for the hidden state is implemented as

$$h_t = f(Wx_t + Uh_{t-1} + b), \tag{3.22}$$

where W is the weight matrix for the input at the current step and U is the weight matrix for hidden units at the previous step, and b is the bias. The output can be a sequence of vectors $\{y_1, y_2, \ldots, y_T\}$ computed as

$$\begin{aligned} o_i &= Vh_i + c, \\ y_i &= \text{softmax}(o_i), \end{aligned} \tag{3.23}$$

where V is a weight matrix shared across all steps, c is the bias, and softmax is an activation function computed as

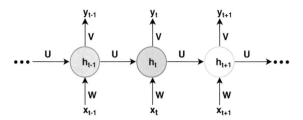

Fig. 3.9 The basic structure of a recurrent layer of an RNN. The recurrent layer takes as input one vector at the tth time step, x_t, and the previous hidden state h_{t-1}, and it returns a new hidden state h_t, which is computed by applying a nonlinear function to the linear operation of matrix-vector multiplication of x_t and h_{t-1} with their respective weight matrices. The output is then computed by applying the softmax function to the result of multiplication of h_t with a weight matrix

$$\text{softmax}(o_i) = \frac{1}{Z}\left[e^{o_i(1)}\ e^{o_i(2)}\ \cdots\ e^{o_i(n)}\right],$$
$$Z = \sum_{j=1}^{n} e^{o_i(j)},$$

$$(3.24)$$

where $o_i j$ is the jth component of o_i.

RNNs can model a probability distribution over the next element of the sequence data given its present state h_t. The sequence probability is

$$p(x_1, x_2, \ldots, x_T) = p(x_1) \cdots p(x_T | x_1, \ldots, x_{T-1}).$$
$$(3.25)$$

Each conditional probability distribution is modeled as

$$p(x_T | x_1, \ldots, x_{T-1}) = f(h_t).$$
$$(3.26)$$

Recurrent networks are typically trained using the *back-propagation through time (BPTT)* [107] algorithm—an approach that adapts the back-propagation idea to a sequential model. BPTT works by unfolding the recurrent network in time—it creates replicas of the network such that each temporal sample gets a copy of the network wherein all network copies have shared parameters. This then becomes a traditional network where back-propagation can be applied. Figure 3.9 shows an RNN with one recurrent layer, while a sample RNN architecture based on this basic building block that can be used for HSI data classification is shown in Fig. 3.10.

3.3.4 Long Short-Term Memory

RNNs have been successful in many machine learning and computer vision tasks; however, they introduce several training challenges. In particular, long-term

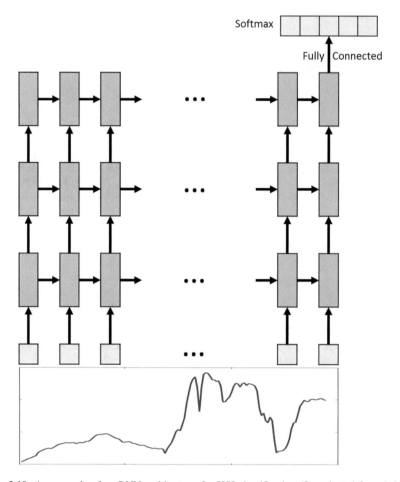

Fig. 3.10 An example of an RNN architecture for HSI classification (fig. adapted from [108]). Each layer has the architecture described in Fig. 3.9. The network when trained models the spectral envelope that contains potentially discriminative information about the classes of interest

sequences pose training difficulty due to vanishing or exploding gradients [109]. More sophisticated recurrent units have been developed to overcome this issue, including a recurrent hidden unit called long short-term memory (LSTM). LSTM is capable of learning long-term dependencies [34], replacing a recurrent hidden node with a memory cell containing a self-connected recurrent edge with fixed weight that ensures gradients can pass across several time steps without vanishing or exploding. LSTM networks have been applied for hyperspectral image classification using only pixel spectra vectors [57, 110].

An LSTM recurrent layer creates a memory cell c_t at step t consisting of an input gate, output gate, forget gate, and a new memory cell. LSTM activation is computed by

$$h_t = o_t \odot \tanh(c_t), \tag{3.27}$$

where o_t is the output gate that determines the exposed memory content, $\odot$ is an element-wise multiplication, and $\tanh(\cdot)$ is the hyperbolic tangent function. The output gate is updated by

$$o_t = \sigma\left(W^{(o)}x_t + U^{(o)}h_{t-1} + b_o\right), \tag{3.28}$$

where $\sigma(\cdot)$ is a logistic sigmoid function, $W^{(o)}$ and $U^{(o)}$ represent weight matrices of the output gate, and b_o is a bias. The memory cell c_t is updated by adding new content $\tilde{c}_t$ and discarding part of the present memory content:

$$c_t = i_t \odot \tilde{c}_t + f_t \odot c_{t-1}, \tag{3.29}$$

where i_t modulates the extent to which new information is added, and f_t determines the degree to which the current contents are forgotten. The memory cell $\tilde{c}_t$ contents are updated using

$$\tilde{c}_t = \tanh\left(W^{(c)}x_t + U^{(c)}h_{t-1} + b_c\right), \tag{3.30}$$

where $W^{(c)}$ and $U^{(c)}$ are weight matrices and b_c is a bias term. The input and the forget gates are computed as

$$\begin{aligned} i_t &= \sigma\left(W^{(i)}x_t + U^{(i)}h_{t-1} + b_i\right) \\ f_t &= \sigma\left(W^{(f)}x_t + U^{(f)}h_{t-1} + b_f\right), \end{aligned} \tag{3.31}$$

where $W^{(i)}, U^{(i)}, W^{(f)}, U^{(f)}$ are gate weight matrices and b_i, b_f are respective gate terms. Figure 3.11 shows a graphical illustration of an LSTM memory cell.

3.4   Convolutional Neural Networks

Convolutional neural networks (CNNs) are specialized feed-forward neural networks for processing data sampled on a uniform grid, such as an image. In the case of HSI, this can include 1D spectral sampling, 2D spatial sampling, or 3D sampling of the entire image tensor. Each CNN layer generates a higher level abstraction of the input data, generally called a "feature map", that preserves essential and unique information. CNNs are able to achieve superior performance by employing a deep hierarchy of layers and have recently become a popular deep learning method achieving significant success in hyperspectral pixel classification [48, 50, 70], scene understanding [111], target detection [112, 113], and anomaly detection [114].

Fig. 3.11 Illustration of an LSTM memory cell. i, f, and o are the input, forget, and output gates, respectively. $\tilde{c}$ denotes the new memory cell content

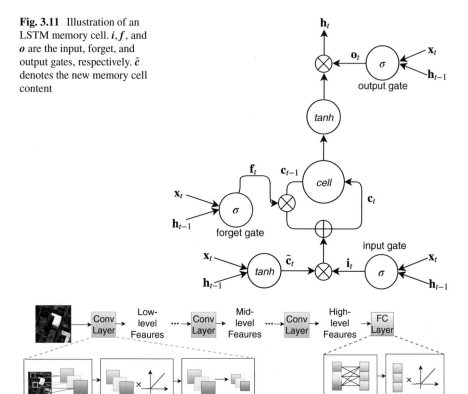

Fig. 3.12 An architecture of a convolutional neural network(CNN). Each convolution layer consists of convolution, nonlinearity, and pooling operations, which generates higher level feature of the input. The first few layers generate low-level features, middle layers generate mid-level features, and the last few layers generate high-level features which are fed into fully connected layer

3.4.1 Building Blocks of CNNs

CNNs are trainable multi-layer architectures composed of multiple feature extraction stages. Each stage consists of three layers: (1) a convolutional, (2) nonlinearity, and (3) pooling. A typical CNN is composed of some feature extraction stages followed by one or more fully connected layers and final classifier layer as shown in Fig. 3.12. Each part of a typical CNN is described in the following sections.

3.4.1.1 Convolutions

A CNN layer performs several convolutions in parallel to produce a set of linear activations. Convolutional layers are responsible for extracting local features at dif-

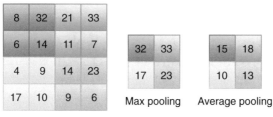

2×2 pooling, stride 2

Fig. 3.13 Various types of pooling operations. The max pooling function computes the maximum in the neighborhood of the window patches of size 2×2 with stride of 2. The average pooling takes average of the input elements in the window patch of size 2×2 with stride of 2

ferent positions using trainable kernels $W_{i,j}^{(l)}$ that act as connection weights between feature map i of layer $l - 1$ and feature map j of layer l. The units at convolution layer l compute activations X_j^l based on spatially contiguous units in the feature map X_i^{l-1} of layer $l - 1$ by convolving the kernels $W_{i,j}^l$:

$$X_j^{(l)} = f \left(\sum_{i=1}^{M^{(l-1)}} X_i^{(l-1)} * W_{i,j}^{(l)} + b_j^{(l)} \right), \tag{3.32}$$

where $M^{(l-1)}$ denotes the number of feature maps in the layer of $l - 1$, $*$ is convolution operator, $b_j^{(l)}$ is a bias parameter, and $f(\cdot)$ represents a nonlinear activation function.

3.4.1.2   Pooling

A *pooling function* reduces the dimensionality of a feature map and is applied to each data channel (or band) to reduce sensitivity to rotation, translation, and scaling (Fig. 3.13). Pooling functions also aggregate responses within and across feature maps. The pooling function combines a set of values within a receptive field (that defines a spatially local neighborhood, as set by the filter size) into fewer values and can be configured based on the size of the receptive field (e.g., 2×2) and selected pooling operation (e.g., max or average). The max pooling function applies a window function to the input patch and computes the maximum within the neighborhood to preserve texture information. The average pooling function calculates the mean of the input elements within a patch to preserve background information. Pooling is typically performed on non-overlapping blocks, however, some methods; however, this is not required [115]. In general, non-overlapping pooling is used for dimension reduction of the resulting feature map.

3.4.1.3 Fully Connected Layers

Fully connected ANN layers are typically used in the final stages of a CNN for classification or regression based on feature maps obtained through the convolutional filters. The output vector is then passed to a softmax function to obtain classification scores.

3.4.2 CNN Flavors for HSI

Hyperspectral image can be visually described as a three-dimensional data cube with spectral sampling along the z-axis. CNN can then be categorized into three groups: (a) 1D CNNs extracting spectral features, (b) 2D CNNs extracting spatial features, and (c) 3D CNNs extracting combined spectral–spatial features.

3.4.2.1 1D CNNs

A 1D CNN is responsible for pixel-level extraction of spectral features. As it is shown in Fig. 3.14, a hyperspectral vector is sent to the input layer and propagates through successive convolutional and pooling layers for feature extraction. Each convolutional layer has multiple convolutional filters, with sizes set using a hyperparameter. The output feature map of each convolutional layer is a 1D vector. 1D CNNs have been applied for multi-class pixel-level classification of HSI data [48]. We note that although such 1-D CNNs can be applied to pixel-level spectral reflectance data, they are not expected to be nearly as powerful in extracting abstract deep features as CNNs that operate on spatial information. This is because the one of the key drivers in the successful application of CNNs to imagery data stems from the multi-layer spatial convolutions that result in abstract deep spatial features representing the object morphology. Along a single dimension (a spectral reflectance profile of a pixel, for example), there are no such features of interest to learn (e.g., features representing edges or texture). At best, such 1-D CNNs can then better condition (e.g., through the series of filtering layers) the spectral reflectance data to make it more robust to variations before it is classified.

3.4.2.2 2D CNNs

Initial work on 2D CNNs (Fig. 3.15) made use of 2D CNN networks for classification of HSI data by taking a neighborhood window of size $w \times w$ around each labeled pixel and treat the whole window as a training sample [50, 70] to extract spatial features. Applying 2D CNN naively to HSI produces a feature map for each band. Since HSI data is often composed of several bands, this produces a large number of parameters that increase overfitting and computational requirements. Subspace learning methods

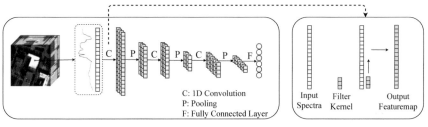

1D CNN Architecture **1D CNN Filtering Operation**

Fig. 3.14 An architecture of 1D convolutional neural network for hyperspectral images; A typical 1D CNN consist of 1D convolutional layers, pooling layers, and fully connected layers. In 1D CNNs, filter kernel is 1D which convolve the hyperspectral cube in spectral dimension

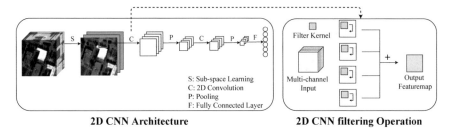

2D CNN Architecture **2D CNN filtering Operation**

Fig. 3.15 An architecture of 2D convolutional neural network for hyperspectral images; A typical 2D CNN consist of 2D convolutional layers, pooling layers, and fully connected layers. In 2D CNNs, filter kernel is 2D which convolve the hyperspectral cube in spatial dimension

are employed to reduce the spectral dimensionality prior to 2D feature extraction. Unsupervised methods such as principal component analysis (PCA) [49, 50, 70] have been exploited to reduce spectral dimensionality a practicable scale before 2D CNN training. However, the separate extraction of spectral and spatial features does not completely utilize spectral–spatial correlations within the data.

3.4.2.3   3D CNNs

After 1D CNN and 2D CNN which extract spectral features and local spatial features of each pixel, respectively, 3D CNN (Fig. 3.16) was introduced to learn the local signal changes in both the spatial and the spectral dimensions of the HSI data, and exploit important discrimination information. 3D CNN model takes advantage of the structural characteristics of the 3D HSI data and can exploit the joint spectral–spatial correlations information because the 3D convolution operation convolves the input data in both the spatial dimension and the spectral dimension simultaneously, while the 2D convolution operation convolves the input data in the spatial dimension. For the 2D convolution operation, regardless of whether it is applied to 2D data or 3D data, its output is 2D, while for 3D convolution operation, its output is also a cube.

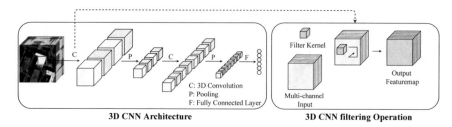

Fig. 3.16 An architecture of 3D convolutional neural network for hyperspectral images; A typical 3D CNN consist of 3D convolutional layers, pooling layers, and fully connected layers. In 3D CNNs, filter kernel is 3D which convolve the hyperspectral cube in both spatial and spectral dimensions

3D CNN network has been investigated to learn rich spectral–spatial information for hyperspectral data classification [116].

3.4.2.4   CNNs with RNNs (CRNNs)

A hybrid of convolutional and recurrent neural networks so-called CRNN (Convolutional recurrent neural network) [117, 118] is composed of several 1D convolutional and pooling layers followed by a few recurrent layers, as it is shown in Fig. 3.17. CRNN has the advantages of both convolutional and recurrent networks. First, the 1D convolutional layers are exploit to extract middle-level locally invariant features from the spectral sequence of the input. Second, the recurrent layers are used to obtain contextual information from the feature sequence obtained by the previous 1D CNN. Contextual information captures the dependencies between different bands in the hyperspectral sequence, which is useful for classification task. For the recurrent layers, the regular recurrent function or LSTM, which can capture very long dependencies, can be used. For cases with long length hyperspectral sequence which have long-term dependency, LSTM can be applied. At the end of this model, as in RNN, the last hidden state of the last recurrent layer will be fully connected to the classification layer. For training, as in CNN and RNN, the loss function is chosen as cross entropy, and mini-batch gradient descent is used to find the best parameters of the network. The gradients in the CNN part are calculated by the back-propagation algorithm, and gradients in the RNN part are calculated by the back-propagation through time (BPTT) algorithm [107]. CRNN have been used to learn discriminative features for hyperspectral data classification [108].

We note that an interesting variation of this idea would be a hybrid model where the first part of the network extracts spatial features from the data through per-channel (spatial) convolutional layers, and the latter part of the network models the evolution of the spectral envelope through a recurrent network. Although we study pixel-level use of recurrent networks to hyperspectral data (modeling spectral reflectance/absorbance evolution) in these chapters, modeling both spatial and spec-

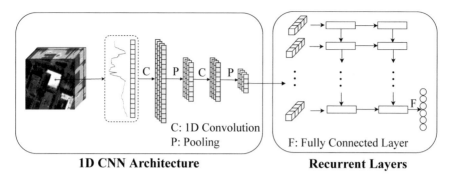

1D CNN Architecture   **Recurrent Layers**

Fig. 3.17 An architecture of convolutional recurrent neural network for hyperspectral images; First part: 1D CNN is exploit to extract middle-level locally invariant features from the spectral sequence of the input. Second part: The recurrent layers are used to obtain contextual information from the feature sequence obtained by the previous 1D CNN

tral information will enhance this idea significantly by jointly leveraging spatial and spectral information.

3.5 Software Tools for Deep Learning

The ubiquitous applications of deep learning in a wide variety of domains have resulted in the development of many deep learning software frameworks. Most of the existing frameworks are open source, and this fact has facilitated the implementation and sharing of models among the research community. Here, we describe some of the existing open-source deep learning frameworks categorized on languages supported, CNN and RNN modeling capability, ease of use in terms of architecture, and support for multiple GPUs. Note that this list is not meant to be exhaustive and only the most popular frameworks at the time of writing of this chapter are discussed.

TensorFlow [119] is a Python library which uses data flow graphs for numerical computation. TensorFlow provides utilities for efficient data pipelining and has built-in modules for the inspection, visualization, and serialization of models. It provides support for CNNs, RNNs, restricted Boltzmann machines, deep autoencoders, long short-term memory models, has multiple GPU support, and is considered to be the best documented open-source framework currently available. TensorFlow also provides visualization tools such as TensorBoard to facilitate understanding and debugging of TensorFlow implementations. However, TensorFlow implementations involve more low-level coding.

Theano [120] is a Python library for defining and managing mathematical expressions, which enables developers to perform numerical operations involving multi-dimensional arrays for large computationally intensive calculations. Theano provides support for CNNs and RNNs, and has multiple GPU support.

Caffe [121] is a C++ library with a Python interface that provides GPU support. Facebook has recently introduced the successor of Caffe, named Caffe2, which is designed for mobile and large-scale deployments in production environments. Caffe2 makes it easier to build demo apps by offering many pretrained models. It was built to be fast, scalable, and lightweight.

Keras [122] is a Python library that serves as a higher level interface to Tensorflow, Theano, Microsoft Cognitive Toolkit, and PlaidML. It has a large user base. Prototyping using Keras is simple and fast. It supports training on multiple GPUs, and it can be used for both CNN and RNN model implementations. Keras is extensible, i.e., it provides the option of adding user-defined functions such as layers, loss functions, or regularizers.

MXNet [123] supports various programming languages including R, Python, Julia, C++, JavaScript, and Scala. It also has advanced GPU support compared to the other frameworks. MXNet is characterized as relatively fast in terms of training and testing computational time for deep learning algorithms. MXNet supports CNN and RNN modeling and is also the framework of choice for the Amazon Web Services.

3.6 Conclusion

In this chapter, we reviewed the foundations of deep learning architectures that can benefit hyperspectral image analysis tasks, including variants of recurrent and convolutional neural networks. We note that many advanced variations of these flavors are emerging in the community, and hence the purpose of this chapter was to describe how basic elements of deep learning architectures can be deployed for multi-channel optical data—these are often the stepping stones to more advanced variants. Within the context of hyperspectral images, accounting for spectral and spatial information simultaneously and effectively is a key factor that differentiates the way in which such networks should be applied for this task, compared to how they are applied for color imagery. Sensor- and data-specific constraints must be kept in mind when designing the deep learning "recipes" around these networks. In Chap. 4, we will apply these networks to real-world hyperspectral data representing remote sensing and biomedical image analysis tasks. With the use of representative data in these applications, we show deep learning configurations that learn the underlying spatial and spectral properties effectively and discuss the nuances and challenges when applying such models to hyperspectral images.

References

1. Ozaki Y, Kawata S (2015) Far-and deep-ultraviolet spectroscopy. Springer
2. Gao L, Smith RT (2015) J Biophotonics 8(6):441

3. Van der Meer FD, Van der Werff HM, Van Ruitenbeek FJ, Hecker CA, Bakker WH, Noomen MF, Van Der Meijde M, Carranza EJM, De Smeth JB, Woldai T (2012) Int J Appl Earth Obs Geoinformation 14(1):112
4. Pahlow S, Weber K, Popp J, Bayden RW, Kochan K, Rüther A, Perez-Guaita D, Heraud P, Stone N, Dudgeon A et al (2018) Appl Spectrosc 72(101):52
5. van Hove ERA, Smith DF, Heeren RM (2010) J Chromatogr A 1217(25):3946
6. Le DB (1991) Magn Reson Q 7(1):1
7. Hearnshaw JB (1990) The analysis of starlight: one hundred and fifty years of astronomical spectroscopy. CUP Archive
8. Huang H, Yu H, Xu H, Ying Y (2008) J Food Eng 87(3):303
9. Yang D, Ying Y (2011) Appl Spectrosc Rev 46(7):539
10. Roggo Y, Chalus P, Maurer L, Lema-Martinez C, Edmond A, Jent N (2007) J Pharm Biomed Anal 44(3):683
11. Theiler JP, Gisler G (1997) In: Algorithms, devices, and systems for optical information processing, vol 3159. International Society for Optics and Photonics, pp 108–119
12. Lavenier D (2000) In: Los Alamos National Laboratory LAUR. Citeseer
13. Ly E, Piot O, Wolthuis R, Durlach A, Bernard P, Manfait M (2008) Analyst 133(2):197
14. Lee S, Crawford MM (2005) IEEE Trans Image Process 14(3):312
15. Yu P (2005) J Agric Food Chem 53(18):7115
16. Bhargava R, Fernandez DC, Hewitt SM, Levin IW (2006) Biochim Biophys Acta (BBA)-Biomembr 1758(7):830
17. Villa A, Benediktsson JA, Chanussot J, Jutten C (2011) IEEE Trans Geosci Remote Sens 49(12):4865
18. Großeruaschkamp F, Kallenbach-Thieltges A, Behrens T, Brüning T, Altmayer M, Stamatis G, Theegarten D, Gerwert K (2015) Analyst 140(7):2114
19. Mayerich DM, Walsh M, Kadjacsy-Balla A, Mittal S, Bhargava R (2014) In: Proceedings of SPIE—the international society for optical engineering, vol 9041, p 904107
20. Melgani F, Bruzzone L (2004) IEEE Trans Geosci Remote Sens 42(8):1778
21. Mercier G, Lennon M (2003) In: Proceedings of the 2003 IEEE international on geoscience and remote sensing symposium, 2003. IGARSS'03, vol 1. IEEE, pp 288–290
22. Bandos TV, Bruzzone L, Camps-Valls G (2009) IEEE Trans Geosci Remote Sens 47(3):862
23. Fu Z, Robles-Kelly A (2007) In: 2007 IEEE conference on computer vision and pattern recognition. IEEE, pp 1–7
24. McCulloch WS, Pitts W (1943) Bull Math Biophys 5(4):115
25. Hebb DO (1949) The organization of behaviour. Wiley, New York
26. Rosenblatt F (1958) Psychol Rev 65(6):386
27. Werbos P (1974) PhD dissertation, Harvard University
28. Fukushima K (1980) Biol Cybern 36(4):193
29. Ackley DH, Hinton GE, Sejnowski TJ (1985) Cogn Sci 9(1):147
30. Smolensky P (1986) Information processing in dynamical systems: foundations of harmony theory. Tech. rep., Colorado Univ. at Boulder Dept. of Computer Science
31. Jordan MI (1997) In: Advances in psychology, vol 121. Elsevier, pp 471–495
32. Rumelhart DE, Hinton GE, Williams RJ (1985) Learning internal representations by error propagation. Tech. rep., California Univ. San Diego La Jolla Inst. for Cognitive Science
33. LeCun Y, Boser BE, Denker JS, Henderson D, Howard RE, Hubbard WE, Jackel LD (1990) In: Advances in neural information processing systems, pp 396–404
34. Hochreiter S, Schmidhuber J (1997) Neural Comput 9(8):1735
35. Hinton GE, Osindero S, Teh YW (2006) Neural Comput 18(7):1527
36. Salakhutdinov R, Hinton G (2009) In: Artificial intelligence and statistics, pp 448–455
37. Krizhevsky A, Sutskever I, Hinton GE (2012) In: Advances in neural information processing systems, pp 1097–1105
38. Subramanian S, Gat N, Sheffield M, Barhen J, Toomarian N (1997) In: Algorithms for multispectral and hyperspectral imagery III, vol 3071. International Society for Optics and Photonics, pp 128–138

39. Jimenez LO, Morales-Morell A, Creus A (1999) IEEE Trans Geosci Remote Sens 37(3):1360
40. Plaza J, Plaza A, Perez R, Martinez P (2009) Pattern Recognit 42(11):3032
41. Ratle F, Camps-Valls G, Weston J (2010) IEEE Trans Geosci Remote Sens 48(5):2271
42. Licciardi GA, Del Frate F (2011) IEEE Trans Geosci Remote Sens 49(11):4163
43. Lin Z, Chen Y, Zhao X, Wang G (2013) In: 2013 9th international conference on information, communications & signal processing. IEEE, pp 1–5
44. Chen Y, Lin Z, Zhao X, Wang G, Gu Y (2014) IEEE J Sel Top Appl Earth Obs Remote Sens 7(6):2094
45. Deng L, Yu D et al (2014) Found Trends® Signal Process 7(3–4):197
46. Goodfellow I, Bengio Y, Courville A, Bengio Y (2016) Deep learning, vol 1. MIT Press, Cambridge
47. Tao C, Pan H, Li Y, Zou Z (2015) IEEE Geosci Remote Sens Lett 12(12):2438
48. Hu W, Huang Y, Wei L, Zhang F, Li H (2015) J Sens 2015
49. Makantasis K, Karantzalos K, Doulamis A, Doulamis N (2015) In: 2015 IEEE international geoscience and remote sensing symposium (IGARSS). IEEE, pp 4959–4962
50. Yue J, Zhao W, Mao S, Liu H (2015) Remote Sens Lett 6(6):468
51. Ma X, Geng J, Wang H (2015) EURASIP J Image Video Process 2015(1):20
52. Chen Y, Zhao X, Jia X (2015) IEEE J Sel Top Appl Earth Obs Remote Sens 8(6):2381
53. Chen Y, Jiang H, Li C, Jia X, Ghamisi P (2016) IEEE Trans Geosci Remote Sens 54(10):6232
54. Zhou X, Li S, Tang F, Qin K, Hu S, Liu S (2017) IEEE Trans Geosci Remote Sens 14(1):97
55. Li W, Wu G, Zhang F, Du Q (2017) IEEE Trans Geosci Remote Sens 55(2):844
56. Shu L, McIsaac K, Osinski GR (2018) IEEE Trans Geosci Remote Sens (99):1
57. Mou L, Ghamisi P, Zhu XX, Trans IEEE (2017) Geosci Remote Sens 55(7):3639
58. Liu B, Yu X, Yu A, Zhang P, Wan G (2018) Remote Sens Lett 9(12):1118
59. Guo Y, Han S, Cao H, Zhang Y, Wang Q (2018) Procedia Comput Sci 129:219
60. Shahraki FF, Prasad S (2018) In: 2018 IEEE global conference on signal and information processing (GlobalSIP). IEEE, pp 968–972
61. Labate D, Safari K, Karantzas N, Prasad S, Foroozandeh Shahraki F (2019) In: SPIE optical engineering + applications, San Diego, California, United States
62. Goodacre R, Burton R, Kaderbhai N, Woodward AM, Kell DB, Rooney PJ et al (1998) Microbiology 144(5):1157
63. Halicek M, Lu G, Little JV, Wang X, Patel M, Griffith CC, El-Deiry MW, Chen AY, Fei B (2017) J Biomed Opt 22(6):060503
64. Berisha S, Lotfollahi M, Jahanipour J, Gurcan I, Walsh M, Bhargava R, Van Nguyen H, Mayerich D (2019) Analyst
65. Lotfollahi M, Berisha S, Daeinejad D, Mayerich D (2019) Appl Spectrosc 0003702818819857
66. Lee C, Landgrebe DA (1993) IEEE Trans Geosci Remote Sens 31(4):792
67. Chang CI, Du Q, Sun TL, Althouse ML (1999) IEEE Trans Geosci Remote Sens 37(6):2631
68. Jimenez LO, Landgrebe DA (1999) IEEE Trans Geosci Remote Sens 37(6):2653
69. Bruce LM, Koger CH, Li J (2002) IEEE Trans Geosci Remote Sens 40(10):2331
70. Zhao W, Du S (2016) IEEE Trans Geosci Remote Sens 54(8):4544
71. Pan B, Shi Z, Xu X (2017) IEEE J Sel Top Appl Earth Obs Remote Sens 10(5):1975
72. Haykin S (1994) New York
73. Bishop CM et al (1995) Neural networks for pattern recognition. Oxford University Press
74. Svozil D, Kvasnicka V, Pospichal J (1997) Chemom Intell Lab Syst 39(1):43
75. ElMasry G, Wang N, Vigneault C (2009) Postharvest Biol Technol 52(1):1
76. Atkinson PM, Tatnall A (1997) Int J Remote Sens 18(4):699
77. Qiao J, Ngadi MO, Wang N, Gariépy C, Prasher SO (2007) J Food Eng 83(1):10
78. Udelhoven T, Naumann D, Schmitt J (2000) Appl Spectrosc 54(10):1471
79. Maas AL, Hannun AY, Ng AY (2013) In: Proceedings of the ICML, vol 30, p 3
80. He K, Zhang X, Ren S, Sun J (2015) In: Proceedings of the IEEE international conference on computer vision, pp 1026–1034
81. Clevert DA, Unterthiner T, Hochreiter S. arXiv preprint arXiv:1511.07289 (2015)

82. Klambauer G, Unterthiner T, Mayr A, Hochreiter S (2017) In: Advances in neural information processing systems, pp 971–980
83. Konda K, Memisevic R, Krueger D (2014) arXiv preprint arXiv:1402.3337
84. Shang W, Sohn K, Almeida D, Lee H (2016) In: International conference on machine learning, pp 2217–2225
85. Krizhevsky A, Hinton G (2010) Unpublished manuscript. 40(7)
86. Rumelhart DE, Hinton GE, Williams RJ (1986) Nature 323(6088):533
87. Rosasco L, Vito ED, Caponnetto A, Piana M, Verri A (2004) Neural Comput 16(5):1063
88. Janocha K, Czarnecki WM (2017) arXiv preprint arXiv:1702.05659
89. LeCun Y, Chopra S, Hadsell R, Ranzato M, Huang F (2006) Predicting structured data. 1(0)
90. Zhong Z, Li J, Ma L, Jiang H, Zhao H (2017) In: 2017 IEEE international geoscience and remote sensing symposium (IGARSS). IEEE, pp 1824–1827
91. Paoletti ME, Haut JM, Fernandez-Beltran R, Plaza J, Plaza AJ, Pla F (2018) IEEE Trans Geosci Remote Sens 57(2):740
92. Cybenko G (1989) Math Control Signals Syst 2(4):303
93. Hornik K (1991) Neural Netw 4(2):251
94. Hecht-Nielsen R (1992) In: Neural networks for perception. Elsevier, 1pp 65–93
95. Csáji BC (2001) Faculty of Sciences, Etvs Lornd University, Hungary, vol 24, p 48
96. Bengio Y et al (2009) Found Trends® Mach Learn 2(1):1
97. LeCun Y, Bengio Y, Hinton G (2015) Nature 521(7553):436
98. Baldi P, Hornik K (1989) Neural Netw 2(1):53
99. Bengio Y, Lamblin P, Popovici D, Larochelle H (2007) In: Advances in neural information processing systems, pp 153–160
100. Poultney C, Chopra S, Cun YL et al (2007) In: Advances in neural information processing systems, pp 1137–1144
101. Khan A, Baharudin B, Lee LH, Khan K (2010) J Adv Inf Technol 1(1):4
102. Ma X, Wang H, Geng J (2016) IEEE J Sel Top Appl Earth Obs Remote Sens 9(9):4073
103. Guo R, Wang W, Qi H (2015) In: 2015 7th workshop on hyperspectral image and signal processing: evolution in remote sensing (WHISPERS). IEEE, pp 1–4
104. Vincent P, Larochelle H, Lajoie I, Bengio Y, Manzagol PA (2010) J Mach Learn Res 11:3371
105. Zabalza J, Ren J, Zheng J, Zhao H, Qing C, Yang Z, Du P, Marshall S (2016) Neurocomputing 185:1
106. Lipton ZC, Berkowitz J, Elkan C (2015) arXiv preprint arXiv:1506.00019
107. Werbos PJ (1990) Proc IEEE 78(10):1550
108. Wu H, Prasad S (2017) Remote Sens 9(3):298
109. Hochreiter S, Bengio Y, Frasconi P, Schmidhuber J et al (2001) Gradient flow in recurrent nets: the difficulty of learning long-term dependencies
110. Zhou F, Hang R, Liu Q, Yuan X (2018) Neurocomputing
111. Zhang F, Du B, Zhang L (2016) IEEE Trans Geosci Remote Sens 54(3):1793. https://doi.org/10.1109/TGRS.2015.2488681
112. Vakalopoulou M, Karantzalos K, Komodakis N, Paragios N (2015) In: 2015 IEEE international geoscience and remote sensing symposium (IGARSS), pp 1873–1876. https://doi.org/10.1109/IGARSS.2015.7326158
113. Zhang L, Shi Z, Wu J (2015) IEEE J Sel Top Appl Earth Obs Remote Sens 8(10):4895. https://doi.org/10.1109/JSTARS.2015.2467377
114. Li W, Wu G, Du Q (2017) IEEE Geosci Remote Sens Lett 14(5):597. https://doi.org/10.1109/LGRS.2017.2657818
115. Li C, Yang SX, Yang Y, Gao H, Zhao J, Qu X, Wang Y, Yao D, Gao J (2018) Sensors 18:10. https://doi.org/10.3390/s18103587. http://www.mdpi.com/1424-8220/18/10/3587
116. Li Y, Zhang H, Shen Q (2017) Remote Sens 9(1):67
117. Zuo Z, Shuai B, Wang G, Liu X, Wang X, Wang B, Chen Y (2015) In: Proceedings of the IEEE conference on computer vision and pattern recognition workshops, pp 18–26
118. Xiao Y, Cho K (2016) arXiv preprint arXiv:1602.00367

119. Abadi M, Agarwal A, Barham P, Brevdo E, Chen Z, Citro C, Corrado GS, Davis A, Dean J, Devin M, Ghemawat S, Goodfellow I, Harp A, Irving G, Isard M, Jia Y, Jozefowicz R, Kaiser L, Kudlur M, Levenberg J, Mané D, Monga R, Moore S, Murray D, Olah C, Schuster M, Shlens J, Steiner B, Sutskever I, Talwar K, Tucker P, Vanhoucke V, Vasudevan V, Viégas F, Vinyals O, Warden P, Wattenberg M, Wicke M, Yu Y, Zheng X (2015) TensorFlow: large-scale machine learning on heterogeneous systems. https://www.tensorflow.org/. Software available from tensorflow.org

120. Al-Rfou R, Alain G, Almahairi A, Angermueller C, Bahdanau D, Ballas N, Bastien F, Bayer J, Belikov A, Belopolsky A, Bengio Y, Bergeron A, Bergstra J, Bisson V, Bleecher Snyder J, Bouchard N, Boulanger-Lewandowski N, Bouthillier X, de Brébisson A, Breuleux O, Carrier PL, Cho K, Chorowski J, Christiano P, Cooijmans T, Côté MA, Côté M, Courville A, Dauphin YN, Delalleau O, Demouth J, Desjardins G, Dieleman S, Dinh L, Ducoffe M, Dumoulin V, Ebrahimi Kahou S, Erhan D, Fan Z, Firat O, Germain M, Glorot X, Goodfellow I, Graham M, Gulcehre C, Hamel P, Harlouchet I, Heng JP, Hidasi B, Honari S, Jain A, Jean S, Jia K, Korobov M, Kulkarni V, Lamb A, Lamblin P, Larsen E, Laurent C, Lee S, Lefrancois S, Lemieux S, Léonard N, Lin Z, Livezey JA, Lorenz C, Lowin J, Ma Q, Manzagol PA, Mastropietro O, McGibbon RT, Memisevic R, van Merriënboer B, Michalski V, Mirza M, Orlandi A, Pal C, Pascanu R, Pezeshki M, Raffel C, Renshaw D, Rocklin M, Romero A, Roth M, Sadowski P, Salvatier J, Savard F, Schlüter J, Schulman J, Schwartz G, Serban IV, Serdyuk D, Shabanian S, Simon E, Spieckermann S, Subramanyam SR, Sygnowski J, Tanguay J, van Tulder G, Turian J, Urban S, Vincent P, Visin F, de Vries H, Warde-Farley D, Webb DJ, Willson M, Xu K, Xue L, Yao L, Zhang S, Zhang Y (2016) arXiv e-prints abs/1605.02688. http://arxiv.org/abs/1605.02688

121. Jia Y, Shelhamer E, Donahue J, Karayev S, Long J, Girshick R, Guadarrama S, Darrell T (2014) In: Proceedings of the 22Nd ACM international conference on multimedia, MM '14. ACM, New York, NY, USA, 2014, pp 675–678. https://doi.org/10.1145/2647868.2654889. http://doi.acm.org/10.1145/2647868.2654889

122. Chollet F et al (2015) Keras. https://keras.io

123. Chen T, Li M, Li Y, Lin M, Wang N, Wang M, Xiao T, Xu B, Zhang C, Zhang Z. CoRR abs/1512.01274 (2015). http://arxiv.org/abs/1512.01274

Chapter 4
Deep Learning for Hyperspectral Image Analysis, Part II: Applications to Remote Sensing and Biomedicine

Farideh Foroozandeh Shahraki, Leila Saadatifard, Sebastian Berisha, Mahsa Lotfollahi, David Mayerich and Saurabh Prasad

Abstract Deep neural networks are emerging as a popular choice for hyperspectral image analysis—compared with other machine learning approaches, they are more effective for a variety of applications in hyperspectral imaging. Part I (Chap. 3) introduces the fundamentals of deep learning algorithms and techniques deployed with hyperspectral images. In this chapter (Part II), we focus on application-specific nuances and design choices with respect to deploying such networks for robust analysis of hyperspectral images. We provide quantitative and qualitative results with a variety of deep learning architectures, and compare their performance to baseline state-of-the-art methods for both remote sensing and biomedical image analysis tasks. In addition to surveying recent developments in these areas, our goal in these two chapters is to provide guidance on how to utilize such algorithms for multichannel optical imagery. With that goal, we also provide code and example datasets used in this chapter.

F. F. Shahraki · L. Saadatifard · S. Berisha · M. Lotfollahi · D. Mayerich · S. Prasad (✉)
University of Houston, Houston, TX, USA
e-mail: saurabh.prasad@ieee.org

F. F. Shahraki
e-mail: fforoozandehshahraki@uh.edu

L. Saadatifard
e-mail: lsaadatifard@central.uh.edu

S. Berisha
e-mail: sberisha@central.uh.edu

M. Lotfollahi
e-mail: mlotfollahisohi@uh.edu

D. Mayerich
e-mail: mayerich@uh.edu

© Springer Nature Switzerland AG 2020
S. Prasad and J. Chanussot (eds.), *Hyperspectral Image Analysis*,
Advances in Computer Vision and Pattern Recognition,
https://doi.org/10.1007/978-3-030-38617-7_4

4.1 Introduction

In part I, we reviewed Hyperspectral Imaging (HSI) and the emergence of deep learning models increasingly used in remote sensing and biomedical HSI applications. We
also reviewed the foundations for deep learning as applied to multichannel optical
data, including autoencoders, recurrent neural networks (RNNs), convolutional neural networks (CNNs), and convolutional recurrent neural networks (CRNNs). This
chapter focuses on practical applications of deep learning in both remote sensing
and biomedical HSI. We will cover data collection and preprocessing, parameter
tuning, and practical considerations for selecting appropriate deep learning architectures. This chapter aims to guide future research in deep learning models for remote
sensing and biomedical hyperspectral image analysis.

Significant improvement in HSI sensors (miniaturization, lower costs, etc.) have
increased their effectiveness in a variety of applications including food quality and
safety assessment [1], crime scene analysis [2], archaeology and art conservation [3],
medical applications such as medical diagnosis [4] and image guided surgery [5],
remote sensing applications such as water resource management [6, 7], space surveillance [8], material identification, land cover/use classification [9–11], target detection [12, 13], and change detection [6, 14, 15].

Before the emergence of deep learning, computer vision relied heavily on hand-
crafted features to capture image texture, morphology, spatial, and spectral properties.
These features were then paired with a standard classifier, such as a support vector machine (SVM), random forest (RF), decision tree (DT), clustering, AdaBoost,
logistic regression or other traditional approaches. These algorithms, which can provide multivariate, nonlinear, nonparametric regression or classification have been
extensively studied for remote sensing and biomedical HSI data analysis [6, 16–23].
However, such machine learning approaches that depend on complex hand-crafted
features need a high level of domain knowledge to be extracted properly, and for complicated and irregular domains, extracting such robust features if often a challenging
task. Deep learning methods on the other hand have been shown to be very effective
for such tasks—one reason for this is such models often automatically extract high-
level abstract yet discriminative features from the data. They automatically learn
and construct a unique set of hierarchical high-level features optimized for a given
task. For example, the use of local connectivity patterns between neurons of adjacent
layers and weight sharing schemes make CNNs very effective. Deep learning algorithms have now emerged as a popular choice in HSI data analysis and compared
to conventional machine learning methods that are based on hand-crafted features
extracted from hypercubes, they have been shown to be very effective.

Deep learning is now being deployed for a variety of remotely sensed image
analysis tasks. In [24], the authors introduced deep learning based unsupervised
feature extraction for hyperspectral data classification using autoencoders. They show
that autoencoder extracted features increase accuracy of SVM and logistic regression
backend classifiers and obtain better accuracy than conventional feature extraction
such as PCA. 1D CNNs have been used to leverage spectral features in remote
sensing data [25]. In this method, due to a limited number of training samples,

a network with only one convolutional layer and one fully connected layer was used. 2D CNNs were applied to the first 10 to 30 principal components (applied on the per-pixel spectral reflectance features) of the hypercube to learn spectral and spatial properties of the HSI data for classification [26]. Recent studies have extracted spatial–spectral features using 3D CNNs, such as [27] to analyze HSI data, where the authors proposed a dense convolutional network that uses dilated convolutions [28] instead of scaling operations to learn features at different scales. Recurrent neural networks (RNNs) which are designed to handle sequential data have also been investigated as a tool for pixel-level analysis of spectral reflectance features [29–31]. Hyperspectral data are treated as spectral sequences, and an RNN is used to model the dependencies between different spectral bands. Sometimes, RNNs and CNNS are used together to make a robust pixel-level classification [31]. First, convolutional layers can extract middle-level, locally invariant features from the input sequence, and the following recurrent layers extract spectral-context. There is an alternate convolutional processing for data that resides on manifolds and graphs, and the resulting convolutional neural networks are referred to as graph-based convolutional neural networks (GCNs). In [9], the authors demonstrate a framework that can use GCNs to effectively represent data residing on smooth manifolds, such as reflectance spectra of hyperspectral image pixels. In GCNs, a convolution operator is defined based on the graph Fourier transform to perform convolution/aggregation operations on feature vectors of its neighbors on the graph. A key element to successfully deploy graph-based networks is construction of an effective affinity matrix. In this work, the authors proposed a semi-supervised affinity matrix construction that was able to leverage a few labeled samples along with a large quantity of unlabeled pixels.

For biomedical applications, many traditional artificial neural network (ANN) architectures have been used for classification and regression problems. However, they exhibit poor performance on independent testing data [32] due to overfitting from a large number of available parameters in a hyperspectral image. CNNs have become an effective deep learning technique for image analysis tasks [33, 34]. They are the current methods of choice for image classification [35, 36] because they are effective in exploiting spatial features by enforcing local patterns within the HSI image. Also, CNNs can extract correlations across the entire spectrum for a given pixel [37, 38].

Chapter outline—To successfully deploy deep learning architectures for hyperspectral image classification tasks, various domain-specific nuances must be considered—in this chapter, we seek to provide a comprehensive discussion of practical applications of deep learning to hyperspectral imaging. Sections 4.2 and 4.3 introduce the nature of remote sensing and biomedical HSI data and describe common data preprocessing strategies. Section 4.4 summarizes practical considerations relative to preparing data and deploying deep learning models such as CNNs, RNNs, and CRNN for HSI analysis. It also reviews other related works in the remote sensing and biomedical communities, where such models have been successfully deployed for HSI analysis. In Sect. 4.5, the general architecture and network parameters for deep learning models are discussed in the context of remote sensing and biomedical HSI classification tasks. Section 4.6 provides concluding remarks for the chapter.

4.2   Applications of Hyperspectral Imaging

4.2.1   Remote Sensing Case Study: Urban Land Cover Classification

Hyperspectral imagery is a popular imaging modality for remote sensing applications, such as for environmental and ecological monitoring, urban land cover classification and scene understanding, and defense applications. Over the years, several annotated datasets have been released within the remote sensing community to benchmark image analysis algorithms. Table 4.1 summarizes the properties of these datasets, and underscores a key problem one may encounter when applying deep learning approaches to HSI data—a limited number of annotated samples compared to that in optical imagery datasets, in addition to other issues such as variations in spectral content due to atmospheric conditions, sun-sensor-object geometry, variations in topography, etc. In this chapter, as a use-case, we will use an urban land cover classification task involving aerial hyperspectral imagery—the University of Houston 2013 dataset.

University of Houston 2013 dataset (UH 2013)—The 2013 University of Houston hyperspectral image was acquired over the University of Houston campus and the neighboring urban area—it was released as part of the 2013 IEEE GRSS Data Fusion Contest, and is now a well-known dataset in the community (details about this dataset are available here: http://hyperspectral.ee.uh.edu). The image contains 15 urban land-cover classes and has 144 spectral channels (bands) representing reflected at-sensor radiance over the 380–1050 nm wavelength range. It has a spatial dimension of 1905 × 349 pixels with a spatial resolution of 2.5 m. The data were acquired over the University of Houston campus and the surrounding urban area on June 23, 2012 from an aerial sensor (at an average height of 5500 ft above ground), and represents a land-cover classification task. Figure 4.1 shows the true color image of the UH dataset with the ground truth of the 15 urban land cover classes, and Fig. 4.2 represents the corresponding mean spectral reflectance signature of each class. As

Table 4.1 The main available HSI datasets used in remote sensing research community

Dataset	#Classes	#Pixels	#Labeled samples	#Bands	Spectral range (μm)	Spatial resolution (m)
UH 2013 (Aerial) [39]	15	664845	15029	144	0.36 − 1.05	2.5
UH 2018 (Aerial) [40]	20	5014744	547807	48	0.36 − 1.05	1
Indian Pines (Aerial) [41]	16	21025	10249	224	0.4 − 2.5	20
Salinas (Aerial) [42]	16	111104	54129	227	0.4 − 2.5	3.7
Pavia (Aerial) [43]	9	991040	50232	103	0.43 − 0.85	1.3
Kennedy SC (Aerial) [44]	13	314368	5211	176	0.4 − 2.5	18
Botswana (Satellite) [45]	14	377856	3248	145	0.4 − 2.5	30

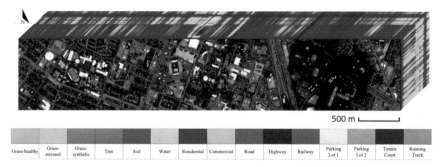

| Grass-healthy | Grass-stressed | Grass-synthetic | Tree | Soil | Water | Residential | Commercial | Road | Highway | Railway | Parking Lot 1 | Parking Lot 2 | Tennis Court | Running Track |

Fig. 4.1 The color image of the University of Houston (UH) 2013 dataset with the ground truth for 15 classes in different colors

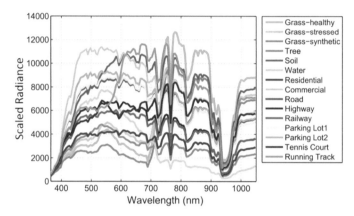

Fig. 4.2 Mean spectral signatures of 15 classes from the University of Houston (UH) 2013 dataset

can be seen from these spectral reflectance signatures, classes that represent different material properties often exhibit a distinct spectral reflectance response. Deep learning approaches that leverage spectral information in addition to spatial information would hence result in superior classification performance.

4.2.2 Biomedical Application: Tissue Histology

The current standard for cancer diagnosis is histopathology [46]. A diagnostic decision is made after several standardized steps which include (1) biopsy collection, (2) preprocessing, including fixation, embedding, and sectioning, (3) chemical staining, and finally (4) expert histological examination under a microscope. Common stains include hematoxylin and eosin (H&E), Masson's trichrome, and immunohistochemical stains that label specific proteins such as cytokeratin (Fig. 4.3). Pathologists

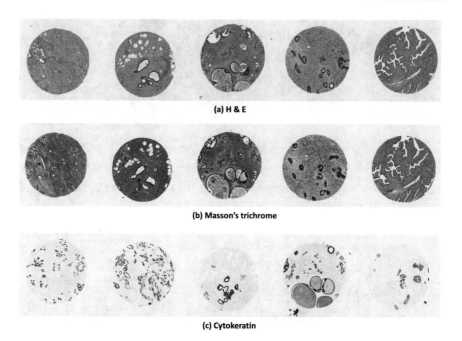

Fig. 4.3 Chemically stained images of patient breast biopsy cores from the BRC961 tissue array (US Biomax, Inc.). Tissue cores are shown stained with H & E (**a**), Masson's trichrome (**b**) and using immunohistochemistry to identify cytokeratin expression (**c**) [47]

leverage morphological details highlighted by these labels to make diagnostic and prognostic decisions. Since these datasets are relatively new to the community, we will describe them further and will also provide the typical data processing steps that are undertaken before analysis.

Histological labels are destructive and non-quantitative, and manual examination is time consuming, expensive, and prone to error. These challenges limit accuracy and scalability, making histological testing expensive and prolonging diagnosis. While computer vision in histopathology is an active area of research [48–50], chemical labeling is non-quantitative and inconsistent, placing a fundamental limit on cross-clinical accuracy.

Mid-infrared (IR) spectroscopic imaging is a quantitative alternative to histopathology, with the ability to extract molecular and morphological information without chemical stains [51]. The measured spectra represent molecular fingerprints tied to distinct biomolecules, such as proteins, lipids, DNA, collagen, glycogen, and carbohydrates. Images of individual bands, given in units of wavenumber (cm^{-1}) provide the spatial distribution of these molecules (Fig. 4.4).

Machine learning is frequently applied to spectroscopy and spectroscopic imaging to differentiate distinct tissue types [52–69], while recent methods attempt to map molecular spectra to conventional stains [70, 71] to produce images that can be interpreted by pathologists without additional training.

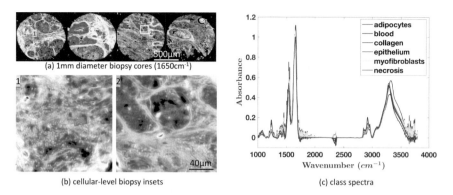

Fig. 4.4 **a** Mid-infrared images at band at $1650 \mathrm{cm}^{-1}$ of four breast biopsy cores from different patients with **b** high-resolution insets. The false color indicates the magnitude of the absorbance spectrum in arbitrary units. **c** Mean spectral signatures of 6 cancer-relevant cell types for **a** SD data and **b** HD data

4.2.2.1   Data Description

Tissue samples in this study were collected using Fourier-transform infrared (FTIR) spectroscopic imaging. Tissue samples are formalin fixed, paraffin embedded, and cut into 5 μm sections following normal histological guidelines. Tissue sections were then placed on IR-transparent barium fluoride (BaF2) substrates for imaging. This differs from traditional glass slides used in standard histopathology. Adjacent sections were identically cut and placed on standard glass slides for traditional histological staining. These adjacent sections facilitate annotation of the infrared images by expert histologists. Chemical stains used in this study include hematoxylin and eosin (H&E), Masson's trichrome, and a variety of immunohistochemical stains for cancer-relevant proteins such as cytokeratin and vimentin.

Two datasets were used in this study. The first dataset is the BRC961 high-definition breast cancer tissue microarray (TMA) purchased from Biomax US [72]. Images and annotations were provided by the Chemical Imaging and Structures Laboratory [47] and data are available online [73]. The second dataset consists of mouse kidney (MUSKIDNEY) collected and annotated by the authors using identical imaging methods [71].

4.2.2.2   Data Preprocessing

Both datasets were processed using standardized algorithms [74] for baseline correction and normalization to the Amide I ($1650 \mathrm{cm}^{-1}$) band. The following preprocessing steps were applied using the open-source SIproc software package [75]:

- Baseline correction: Scattering effects during FTIR imaging distort the acquired spectra [74]. Common techniques to resolve spectral distortions include

(1) baseline correction, (2) numerical differentiation to calculate first- or second-order derivatives, and (3) physically based modeling [76]. We use piecewise-linear "rubber band" baseline correction, which is among the simplest and most popular methods [77].

- Normalization: Infrared absorbance spectra scale linearly with concentration and path length. Therefore, molecular signatures must be based on the *shape* of the spectrum and independent of scale. Spectra are, therefore, normalized to minimize the effect of linear scaling between pixels. Common methods include vector normalization, where the spectrum is divided by the square root of its inner product, and band normalization, where the spectrum is divided by a common band. Normalization to a global protein bad, such as Amide I ($\sim 1650\,cm^{-1}$) or Amide II ($\sim 1550\,cm^{-1}$), is common [74, 78] when classifying baseline-corrected spectra. Vector normalization is more common when classifying spectral derivatives.

- Dimensionality reduction: Dimensionality reduction is applied to reduce the total number of parameters optimized during training, mitigating limitations in memory, processing time, and available training data. Common techniques include principal component analysis (PCA), linear discriminant analysis (LDA), manifold learning and their variants [79]. We use PCA for dimensionality reduction, which relies on the underlying assumption that important features exhibit high variance in the training data.

4.3 Practical Considerations and Related Work

4.3.1 Practical Considerations

Hyperspectral data are often large images (both due to their spatial dimensions and the number of spectral channels per pixel). For supervised learning, it is essential to acquire labeled training data (pixels or frames) to learn the classification model, and labeled test data (pixels or frames) to validate the classification models before they are deployed. When constructing such libraries of labeled samples, attention must be paid to the correlation within and between training and testing samples, the size of the labeled data pool and the mechanism through which labeled frames are extracted from large images.

- **Creation of training and testing datasets**—Ideal training and test sets are completely disjoint to ensure that there is no bias in reported accuracy. In many remote sensing tasks, a common approaches extract patches from a single large image—if training and test patches overlap, results may not be representative. Consequently, one may get an incomplete picture of the classifier's ability to generalize to new data. This may be unavoidable for some applications, but efforts should be made to minimize overlap between frames. In many works in the remote sensing community, it is customary to draw training and test samples (e.g., frames) randomly from the imagery/scene—in this setup, care must be taken to minimize or eliminate overlap between training and test frames that result. Remote sensing HSI data are also affected by clouds and other factors, such as sun-sensor-object geometry, variations in illumination, atmospheric conditions, and viewpoints. When constructing labeled libraries, efforts should be made to ensure this variability is represented in the training pool.

- **Extracting window patches/frames from HSI data for image-based classification**—When selecting spatial patch size, it is critical to account for the image resolution with respect to object sizes. If the spatial resolution is coarse, individual frames may contain multiple classes. If a large part of the resulting frame contains class information other than the class of the center pixel, it may not result in features that represent that class, but instead represent the background. When utilizing frames with 3D filters (such as in 3D CNNs), care should be chosen when setting the frame size—too large a frame size relative to the size of objects will result in the spectral direction of the spatial–spectral filters learn from heavily mixed spectra in the frame. Once a frame size is fixed, one way to further mitigate this issue is to apply a threshold on the window patch based on occupation (i.e., based on the dominant class in that frame). For example, at a threshold of 50%, we require that at least 50% of pixels in the frame are from the class belonging to the central pixel, otherwise, we do not include that frame in the dataset. The size of window patches is related to the resolution of the HSI image, and the size of convolutional filter should be chosen based on the size of the window patches.
- **Size of labeled samples**—Sample size plays a key role in deep learning applications. Deep learning networks such as CNNs often require a large number of labeled samples to ensure effective learning (and convergence) of the network parameters. Remote sensing data processing has been challenging because ground truth is often limited, often difficult and expensive to acquire. However, there exists abundant unlabeled data that can be leveraged as part of the training. Addressing this problem is an area of active research exploration within the areas of data augmentation [80], semi-supervised data analysis [81], and domain adaption [82]. These are discussed in Chap. 5.
- **Disjoint correlations across bands in absorbance spectra**—Chemical compounds are composed of molecular bonds represented at widely varying absorbance bands. Correlations between peaks in the spectral signature are, therefore, minimally dependent on distance in the spectrum. As a result, convolutional filters trained in the spectral dimension are of limited use in infrared spectroscopy. However, dimensionality reduction can be useful for capturing these correlations.

4.3.2 Related Developments in the Community

The low spatial resolution provided by commercial FTIR imaging systems has generally limited the application of deep learning approaches, which often rely on spatial filters for learning. Consequently, most IR classification approaches have leveraged spectral-level algorithms such as Naive Bayes classification [53] and Random Forests

[63, 67]. However 1D CNNs have been tested for spectral classification [83], and more recently 2D CNNs have demonstrated significant performance improvements for data collected using new high-definition imaging systems [69].

Deep learning has been more routinely applied in the remote sensing community using a variety of newer architectures:

- **CNNs**: Various CNN models have been designed for remote sensing HSI data classification and regression. In [84, 85], the authors investigate classification from pixel-level fusion [86] of HSI and Light Detection and Ranging (LiDAR) data using CNNs. They extracted material information from HSI data and information about the topography of objects in the scene from LiDAR data, and improved the accuracy of the CNN model by 10% compared to classification using only HSI data. In [87], the authors fuse both HSI and LiDAR data features at the level of image patches by introducing a two-stream densely connected convolutional neural network (DenseNet) architecture, which connects all preceding layers to a subsequent layer in a feed-forward manner. The first CNN stream exploits spatial and spectral features of HSI data using a 3D CNN model, and the second stream is responsible for extracting spatial information from LiDAR data using a 2D CNN. Following this, a fully connected network with two layers is employed to fuse high-level output feature maps of the two streams and achieve the complementary feature which is used for scene classification. In [88], the authors proposed a two-branch network to fuse Lidar and HSI information for classification. In this research work, the first branch which is responsible for extracting features from HSI data is a dual-tunnel CNN comprised of (1) 2D CNN tunnel which is responsible for extracting spatial features of HSI, and a (2) 1D CNN tunnel which extracts spectral features from the HSI pixels. The second branch is a CNN architecture which uses the LiDAR data to extract spatial features. The output featuremaps of each branch are combined to generate fused features which are then fed into a fully connected layer for classification. In [89], the authors propose a cascade network inspired by 'densenet' which is designed to combine features from different layers with a *shortcut path*. Furthermore, an asynchronous training strategy and a fine-tuning technique are adapted in the training phase. A deformable CNN approach—deformable HSI classification networks (DHCNet) is proposed in [90]. In regular CNNs, the sampling locations of the convolutions are fixed grids. DHC-Net introduces a deformable convolutional sampling location inspired by [91] to adaptively adjust the convolution kernel and pooling operators of CNNs based on object properties in the data. Compared to regular 2D CNNs, DHCNet improved the overall accuracy of UH 2013 and Pavia datasets by approximately 1%. In [92], a convolutional neural network is proposed where convolutional filters are expressed as linear combinations from a predefined library of sparse "basis" filters that are inspired by shearlets [93]. The motivation behind this design strategy is to impose a geometric sparsity constraint on the convolutional filters, enabling efficient and effective feature learning.
- **RNNs**: RNNs process HSI data at a pixel level to exploit the intrinsic sequential properties encoded in spectral reflectance signatures of hyperspectral pixels,

including spectral correlation and band to band variability [29]. In [29], the authors classify HSI datasets by introducing a novel RNN with a specially designed activation function and a modified gated recurrent unit (GRU) [94]. The newly introduced activation function in this paper is called "parametric rectified tanh (PRetanh)" which generalizes the rectified unit for the deep RNN and for this activation function higher learning rates (without the risk of divergence) were observed. GRU is similar to a long short-term memory (LSTM) with forget gates but has fewer parameters than LSTM, as it lacks an output gate [94]. Their methods exhibited a 10% improvement in classification accuracy compared to regular RNN and LSTM-based networks which use regular activation functions. A cascade RNN using GRUs was proposed in [30] to learn features of HSI images. In their model, they divide each pixel into different spectral sections, then consider each section as a sequence and feed it into an RNN layer for feature learning. Then, the learned features from all the sections are concatenated and the entire sequence is fed into another RNN layer for complementary feature learning. At the end of the network, the output of the second stage RNN is connected to a softmax layer for classification. The initial RNN layers aim to remove redundant information between adjacent spectral bands, and the subsequent RNN layer is applied to learn complementary information from nonadjacent spectral bands.

- **CRNNs**: In [31], the authors take advantage of both convolutional and recurrent networks together to classify remote sensing HSI data. This method is composed of a few convolutional layers followed by a few recurrent layers. The convolutional layers extract middle-level, locally invariant features from the input sequence, and pooling layers make the sequence shorter by downsampling. Depending on the data properties, e.g., number of spectral channels, RNNs can be replaced by other variants such as LSTM or GRU which result in "CLSTM" and "CGRU" models, respectively. In [30], the authors designed a Cascade RNN for HSI classification, they also proposed a CRNN-based model based on 2D CNN and RNN named "spectral–spatial cascaded RNN" model. They first split the small cube of the HSI data across the spectral domains, and then feed spatial properties of each band into several convolutional layers to learn spatial features which result in a feature sequence. Then, this sequence is divided into some subsequences, which are subsequently fed into the first-layer RNNs, respectively, to reduce redundancy of subsequences. Then, the outputs from the first-layer RNNs are combined to generate the sequence fed into the second-layer RNN to learn complementary information. Because it is deep and difficult to train, they used a transfer learning method to train convolutional layers separately and after fixing the weights of CNN part, they train the two-layer RNNs. At the end, the whole network is fine-tuned based on the learned parameters.

4.4 Experimental Setup

In this section, we present the deep learning architectures used for HSI classification in this chapter. Our objective of the following discussion is to present configurations that lead to discriminative models for HSI classification. The design choices we make for both the remote sensing and tissue histology datasets are empirically determined and will need to be adjusted when such models are applied to a different dataset, if the data properties change, but our goal here is to demonstrate the application of these models to two different use-cases for HSI analysis.

Notations used to describe deep learning models:

- "conv (receptive field size)—(number of filters)" denotes convolutional layers.
- "recu (feature dimension)" denotes recurrent layers.
- "max (pool size)" denotes pooling operation.

4.4.1 CNNs

CNNs for HSI Analysis in Remote Sensing—In this chapter, we study 1D, 2D, and 3D CNNs for multi-class hyperspectral image classification. In our experimental setup, 1D CNN has four convolutional layers (with pooling layers), 2D CNN has three convolutional layers (with pooling layers), and 3D CNN contains two convolutional layers (with pooling layers)—these parameters were experimentally determined. Table 4.2 represents the summary of network configurations of 1D, 2D, and 3D CNNs, respectively. We employ principal component analysis (PCA) applied on the spectral reflectance features to reduce the spectral dimensionality as a preprocessing for 2D CNNs. For pixel-based classification (1D CNNs) and 3D CNN models, the entire spectral reflectance signature is needed and hence PCA is not deployed with these. We implemented all CNN models using the Keras framework [95]. Experiments are carried out on a workstation with an NVIDIA(R) GeForce TitanX GPU and 3.0 GHz Intel(R) Core-i7-5960X CPU. For all CNN architectures, a Xavier uniform weight initializer [96] is used. For mini-batch gradient-based learning, the batch size is 128 for all the models, and the learning rate is set to 10^{-4} for 1D CNN and 2D CNNs, and for 3D CNNs, it is set to 10^{-5}. Based on the spatial resolution of the UH 2013 dataset, a window patch size of 5×5 is used for 2D and 3D CNN.

CNNs for HSI Analysis in Tissue Histology—Three different structures of the CNN network (1D, 2D, and 3D CNN) are used to classify the hyperspectral biomedical dataset. We used the first 372 bands of the dataset to feed to 1D and 3D CNN models. In 2D implementation, we first applied PCA to extract spectral information to 16 bands and then fed them to the 2D CNN model. All CNNs are implemented using stochastic gradient descent (SGD) on Keras. We ran these experiments on a workstation equipped with NVIDIA Tesla P100 GPU and Intel Xeon E5-2680v4 CPU. The input sample size rises significantly in 3D structure

Table 4.2 Summary of network configurations for 1D, 2D, and 3D CNN architectures used for classification of the UH 2013 dataset. The output of all the networks are fed into a fully connected layer followed by a softmax function. conv(x)-X: convolutional layer; x: receptive field size; X: number of filters; max(p): maxpooling layer; p:pool size

CNN 1D	CNN 2D	CNN 3D
conv(3) − 32	conv(3 × 3) − 32	conv(3 × 3 × 3) − 32
max(2)	conv(3 × 3) − 32	conv(3 × 3 × 3) − 64
conv(3) − 32	max(2,2)	max(2,2,2)
max(2)	conv(3 × 3) − 64	
conv(3) − 64	max(2,2)	
max(2)		
conv(3) − 64		
max(2)		

Table 4.3 1, 2, and 3D CNN structures that are implemented to classify the hyperspectral biomedical dataset. The output of all networks are flattened to feed to a fully connected layer and then another fully connected layer followed by a softmax layer predicts the classification results. conv(x)-X: convolutional layer; x: receptive field size; X: number of filters; max(p): maxpooling layer; p:pool size

CNN 1D	CNN 2D	CNN 3D
conv(13) − 16	conv(3 × 3) − 32	conv(3 × 3 × 7) − 16
max(2)	max(2,2)	conv(3 × 3 × 7) − 32
conv(11) − 32	conv(3 × 3) − 64	max(1,1,2)
max(2)	max(1,1)	conv(3 × 3 × 5) − 64
conv(9) − 64	conv(3 × 3) − 64	max(2,2,2)
max(2)	max(2,2)	conv(3 × 3 × 5) − 64
conv(7) − 128		max(2,2,2)
max(2)		
conv(5) − 256		
max(2)		
conv(3) − 512		
max(2)		

$(1D : (1 \times 1 \times 372) - 2D : (33 \times 33 \times 16) - 3D : (33 \times 33 \times 372))$ and together with the 3D network structure causes memory limitation. The batch size for 1D and 2D models set to 512 while the 3D batch size is 8. Table 4.3 defines CNN networks for implementing 1D, 2D and 3D CNNs.

4.4.2 RNNs

RNNs for HSI Analysis in Remote Sensing—Table 4.4 depicts the RNN architecture studied here for the UH 2013 dataset. The model is implemented in Keras and Xavier uniform weight initializer is used. The batch size is 128, and the learning rate is 10^{-4}. Experiments are carried out on a workstation with an NVIDIA(R) GeForce TitanX GPU and 3.0 GHz Intel(R) Core-i7-5960X CPU.

RNNs for HSI Analysis in Tissue Histology We implemented a 1D recurrent neural network to study the sequential correlations in hyperspectral pixels of the biomedical dataset. The RNN is implemented using stochastic gradient descent (SGD) on Keras. We ran this experiment on a workstation equipped with NVIDIA Tesla P100 GPU and Intel Xeon E5-2680v4 CPU. The input sample size is $(1 \times 1 \times 372)$ and the batch size is 512. Table 4.5 illustrates the architecture of the RNN network implemented for classification.

RNNs often have a smaller number of parameters than CNNs, but they are designed for very different types of analysis tasks. Where they can be successfully applied, in the case of limited number of samples, they will be less prone to convergence issues during training compared to CNNs [30]. Modern recurrent networks, such LSTM and GRU are designed to capture very long-term dependencies embedded in sequence data. In long-term sequences, the training procedure may face difficulty since the gradients tend to either vanish or explode [97]. By using gates in GRU and LSTM, the errors which are backpropagated through the sequence and layers can be preserved, enabling the recurrent network to learn the sequential information in spectral data without the risk of vanishing gradient.

Table 4.4 Summary of network configuration for RNN architecture used for UH 2013 dataset classification. The output of the network is followed by a softmax function. recu-D:recurrent layer; D:feature dimension

RNN		
recu-128	recu-256	recu-512

Table 4.5 The RNN network architecture for the tissue histology application. The output of the network is followed by a fully connected layer with the softmax activation function. recu-D:recurrent layer; D:feature dimension

RNN		
recu-1024	recu-1024	recu-1024

4.4.3 CRNNs

CRNNs for HSI Analysis in Remote Sensing—Convolutional and recurrent layers can be combined in a "Convolutional Recurrent Neural Network (CRNN)" [31]. The motivation behind deploying 1D CNN filtering layers prior to RNN layers is for the CNN layers to better condition the spectral reflectance signature, alleviating spectral variability and noise before the backend RNN learns the sequence information. Table 4.6 represents the summary of network configurations of CRNN used for the classification of the UH 2013 dataset. The model is implemented in Keras. Experiments are carried out on a workstation with an NVIDIA(R) GeForce TitanX GPU, and 3.0 GHz Intel(R) Core-i7-5960X CPU. The weights are initialized Xaviar unifrom, the batch size is 128, and the learning rate is 10^{-4}.

CRNNs for HSI Analysis in Tissue Histology—We implemented a 1D CRNN model (1D CNN followed by recurrent layers) to study the spectral dependencies in the tissue histology dataset. CRNN is implemented using stochastic gradient descent (SGD) on Keras. We ran this experiment on a workstation equipped with NVIDIA Tesla P100 GPU and Intel Xeon E5-2680v4 CPU. The input sample size is ($1 \times 1 \times 372$) and the batch size is 512. Table 4.7 illustrates the CRNN architecture used for this task.

Denoising and subsampling the HSI sequence via a 1D CNN before feeding it into recurrent can potentially accelerate the backpropagation through the recurrent layer and reduce the problem of vanishing gradients in gradient-based learning methods. Although not studied in this chapter, one can also combine 2D CNNs with RNNs where the 2D CNNs apply spatial filters on the images per channel, following which an RNN layer learns the underlying spectral information. In [30], the authors pro-

Table 4.6 Summary of network configurations for CRNN architecture used for UH 2013 dataset classification. The output of the network is followed by a softmax function. conv(x)-X: convolutional layer; max(p): maxpooling layer; recu-D: recurrent layer; x: receptive field; X: number of filters; D: feature dimension; p:pool size

CRNN					
conv(6)-32	max(2)	conv(6)-32	max(2)	recu-256	recu-512

Table 4.7 The CRNN architecture used for the tissue histology task. The output layer is a fully connected layer with the softmax activation function. Right arrows indicate the flow of the network. conv(x)-X: convolutional layer; max(p): maxpooling layer; recu-D: recurrent layer; x: receptive field; X: number of filters; D: feature dimension; p: pool size

CRNN					
conv(13)-16 →	max(2) →	conv(11)-32 →	max(2) →	conv(9)-64 →	max(2) →
conv(7)-128 →	max(2) →	conv(5)-256 →	max(2) →	recu-256 →	recu-512

posed a spectral–spatial cascaded RNN model that significantly reduced the noise and outliers, and retained more boundary details of objects compared to a 2D CNN. Since a 2D CNN does not exploit spectral information, its performance is not as it could be were spectral information leveraged as well—the spectral–spatial cascaded RNN addressed this by combining spatial CNNs with spectral RNNs.

4.5 Quantitative and Qualitative Results

4.5.1 Remote Sensing Results

Data Setup—Several existing works that deal with hyperspectral image classification validate classification performance by randomly drawing training and testing samples from the available pool of labeled samples. We contend that this may not provide rigorous insights on the discriminative potential of a machine learning model, because often with hyperspectral imagery, when training and test pixels are drawn at random, one has highly correlated frames in both the training and testing pools (e.g., frames that are very close to each other in space). This can lead to an unintentional bias in the classification performance that is reported which does not highlight the underlying generalization ability of the machine. To depict the sensitivity of models to the manner in which training and testing data are created, in this chapter, we systematically present remote sensing results with two variants of the UH 2013 dataset. In one variant, training and test data are drawn at random from the available pool of labeled samples, as is commonly done in the community, and in another variant, we prepare a disjoint dataset. Although this dataset is also drawn at random, we impose constraints ensuring that the training and testing frames do not impart a bias in the analysis.

Random Dataset—For image-level data analysis, randomly selecting samples from a single scene (e.g., a remotely sensed hyperspectral image) can result in high overlap and correlations between the extracted window patches. In this work, the samples are selected from the labeled pixels shown in Fig. 4.5b, and Table 4.8 represents the number of labeled samples extracted from the whole hyperspectral image per class. For preparing this "random dataset", first all the labeled pixels and their neighborhood pixels are extracted as window patches with a specific size (5×5 in this book chapter). Following this, patches are randomly split into training and test sets.

Disjoint Dataset—To prepare our "disjoint dataset", we seek to collect samples from different labeled regions of the hyperspectral image to minimize or in the best scenario avoid overlaps between test and train patches. We also consider window patch occupation of each sample by defining a threshold value depicting the percentage of a certain class that is occupied in the patch. If a large part of the window patch contains class information other than the class of central pixel (the pixel to be classified), it may not carry valuable information from a training and testing perspective,

(a) labeled pixels selected for disjoint datasets with test data (blue) and train data (red)

(b) all labeled pixels

Fig. 4.5 Labeled pixels extracted from UH 2013 dataset

especially for 3D filters like 3D CNN which analyze spatial and spectral information simultaneously. In this work, the window patch size for the UH dataset is set to 5×5 and the threshold is 50% which means 50% of the window patch pixels should have the same class label as that of the center pixel (and hence the label of the patch). After applying this constraint on window patch occupation and considering the minimum overlap between train and test samples, the pixels which are selected for the disjoint dataset are shown in Fig. 4.5a. To prepare our disjoint dataset, we seek to eliminate the overlap, but in some classes where we have a very small number of labeled samples, particularly classes that are sparse and only exist in one or a few regions in the scene (e.g., the grass-synthetic class), there may be a small overlap between window patches at the margins. All the pixels which are used for disjoint dataset are shown in Fig. 4.5a. Compared to the size of the original labeled samples, the sample size for the disjoint dataset decreases significantly because of the constraints that are imposed. Table 4.8 shows the number of samples per each class in the disjoint dataset.

We compare the performance of the deep learning models discussed above and compare it to that of several alternate machine learning techniques that have been commonly deployed for such tasks.

Experiment 1—Choosing the minimum training sample size which is required to train a model properly is difficult because it depends on various aspects of the experiment such as the complexity of the classification task, complexity of the learning algorithm, number of classes, number of input features, number of model parameters, use of batch normalization, pretrained weights in transfer learning, and so on. Often, a reasonable strategy is hence to determine this via empirical studies. In this experiment, we study the effect of training size on the classification performance. To show how training size can affect the accuracy, we run the experiments on disjoint dataset for training size ranging from 1 sample per class (an extreme case) to 100

Table 4.8 Disjoint and random datasets extracted from UH 2013 dataset. (left) Disjoint—5 × 5 window patch size—50% occupation. (right) Random—all labeled pixels for each class

Class	Size Train	Test		Class	Size
1-Grass-healthy	417	609		1-Grass-healthy	1251
2-Grass-stressed	171	575		2-Grass-stressed	1254
3-Grass-synthetic	176	489		3-Grass-synthetic	697
4-Tree	155	311		4-Tree	1244
5-Soil	565	298		5-Soil	1242
6-Water	111	90		6-Water	325
7-Residential	209	266		7-Residential	1268
8-Commercial	239	379		8-Commercial	1244
9-Road	234	292		9-Road	1252
10-Highway	355	517		10-Highway	1227
11-Railway	393	472		11-Railway	1235
12-Parking	317	548		12-Parking	1233
13-Parking	113	142		13-Parking	469
14-Tennis	164	222		14-Tennis	428
15-Running	154	411		15-Running	660
overall	3773	5621		overall	15029

samples per class, and fixed number of 80 samples per class for testing and compare the performance. We have 15 classes, hence the total number of training samples varies from 15 to 1500 samples, and the test dataset contains 1200 samples. For the random dataset, due to availability of more samples, the number of training samples are varied from 1 to 200 per class.

As can be seen in Fig. 4.6a, in general, increasing the sample size increases the performance of all methods, and most methods converge at or over 100 training samples per class. Among the CNNs, 2D CNN performs the best. Further, note the significant drop in performance of all methods when we switch from the random to disjoint dataset, highlighting bias embedded in the classification performance that is reported with the randomly sampled dataset without any constraints to minimize overlap. The randomly sampled results, although high, do not represent the true generalization ability of the learned models.

Experiment 2—Next, we study the effect of the depth of the network (number of convolutional and/or recurrent layers) on the classification accuracy. There is no deterministic way to ascertain the network depth/complexity—this must be determined empirically for the datasets at hand. The depth of the network will be a function of the available quality and quantity of the labeled training data, and the complexity of the underlying deep features that need to be extracted to result in a discriminative feature space.

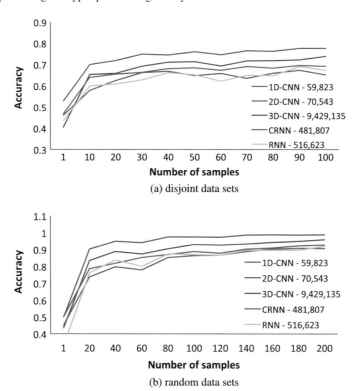

Fig. 4.6 Effect of training size on the deep learning networks accuracy for both random and disjoint datasets extracted from UH 2013

Table 4.9(top) and Table 4.9(bottom) show the effect of depth of the network on the classification accuracy for deep learning models for both the disjoint and random datasets, respectively. To have a fair comparison, we fixed the training and testing sizes for all the models. The training sizes are 100 samples per class and 200 samples per class for disjoint and random datasets, respectively. The testing size is also fixed to 80 samples per class. Based on these tables, we can conclude that increasing the number of layers does not improve the classification performance all the time. For example, in the 2D CNN model for disjoint dataset, by increasing the number of convolutional layers, the classification performance drops. It indicates that due to the small number of training samples (which is a common occurrence in remote sensing tasks) and classes that may not be very spatially complex, we do not need a deep network and the number of learnable parameters in the model (which is determined by the number of layers and the number of units per layer) should be decreased to prevent overfitting. Table 4.10 represents the number of trainable parameters for models as a function of depth. We notice a correlation between the number of parameters to the performance of the network. For example in Table 4.9(top), 2D CNN model with 1 convolutional layer works the best for disjoint dataset because it has the least number

Table 4.9 Effect of the depth (number of convolutional layers) of the deep learning networks on classification accuracy % (std) for both disjoint and random datasets extracted from UH 2013.(top) Disjoint datasets—100 samples per class for train and 80 samples per class for validation. (bottom) Random datasets—200 samples per class for train and 80 samples per class for validation

	Layer			
Method	1-conv	2-conv	3-conv	4-conv
1D CNN	70.34(1)	71.39(1)	69.55(1)	65.23(0.5)
2D CNN	78.86(0.8)	78.22(0.4)	77.61(0.4)	74.97(0.3)
3D CNN	72.11(0.7)	73.86(1)	72.77(0.2)	72.25(0.5)
	1-recu	2-recu	3-recu	
RNN	51.96(4)	66(2)	67.36(1)	
	1-conv 1-recu	2-conv 1-recu	1-conv 2-recu	2-conv 2-recu
CRNN	65.99(0.2)	67.33(1)	69(2)	69(0.3)
	Layer			
Method	1-conv	2-conv	3-conv	4-conv
1D CNN	91.01(0.2)	92.76(0.6)	91.74(0.9)	90.53(1.5)
2D CNN	98.33(0.3)	99.41(0.4)	99(0.3)	97.75(0.7)
3D CNN	91.8(0.1)	94.91(0.5)	95.19(0.1)	95(0.5)
	1-recu	2-recu	3-recu	
RNN	69.49(5)	85.61(3)	91.80(0.6)	
	1-conv 1-recu	2-conv 1-recu	1-conv 2-recu	2-conv 2-recu
CRNN	83.42(1.2)	87.58(0.5)	91.96(0.4)	92.66(0.4)

of trainable parameters among all the network depths. Thus, 100 training samples per class which are not strongly correlated to the test data can show the best performance in a model with the smallest number of parameters. Also, we can see in Table 4.10 and Table 4.9(bottom) that the model with depth of 2 convolutional layers has more number of parameters than that with 1 convolutional layer, but the accuracy improves in depth 2 for the random dataset. This indicates that the more number of training samples which are highly correlated a deeper network may appear to work better. We note that in the extreme case, accuracies from a randomly drawn training and test dataset would be similar to training accuracies due to the high overlap between the training and test patches, and hence caution must be exercised when drawing conclusions from such a dataset— although it may seem based on the results with the random dataset that a deeper network is preferred, we can see that does not hold true when the training and test data are more carefully crafted to minimize biases.

Figure 4.7 represents the validation and training loss for 2D CNN models as a function of different network depths for the disjoint dataset. We can see that by increasing the depth of the network, the validation loss is increasing which shows the models tend to overfit, and the fluctuation in the deeper network with 4 convolutional layer shows the model is not suitable for this amount and complexity of training samples.

Table 4.10 Number of trainable parameters in the deep learning models with various depths

Method	Layer			
	1-conv	2-conv	3-conv	4-conv
1D CNN	148,687	78,127	84,335	59,823
2D CNN	38,511	47,823	70,543	59,375
3D CNN	7,374,799	9,429,135	5,402,959	2,589,199
	1-recu	2-recu	3-recu	
RNN	18,575	119,055	516,623	
	1-conv 1-recu	2-conv 1-recu	1-conv 2-recu	2-conv 2-recu
CRNN	78,063	84,239	475,631	481,807

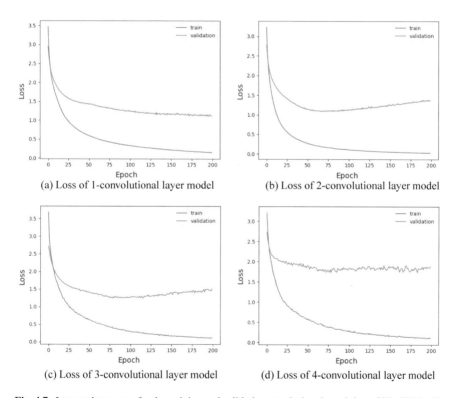

(a) Loss of 1-convolutional layer model

(b) Loss of 2-convolutional layer model

(c) Loss of 3-convolutional layer model

(d) Loss of 4-convolutional layer model

Fig. 4.7 Loss and accuracy for the training and validation sets during the training of 2D-CNN with different depths (number of convolutional layers) on the disjoint dataset

Experiment 3—Since the dataset is highly unbalanced (as our several datasets in remote sensing applications), to compare the classification performance of deep learning architectures for each class of the dataset, we run our experiments with the same number of training samples for each class. The results are shown in Table 4.11(top) and Table 4.11(bottom) for disjoint and random datasets, respectively. As can be seen Table 4.11(top), spatial properties of the commercial class helps in classification and improves the accuracy significantly. From Table 4.11(bottom), in some classes such as residential, railway, and parking lot 2, spatial proprieties provide discriminative information, hence favoring 2D and 3D CNNs. Compared to 1D CNN and 2D CNN which exploit spectral and spatial information of HSI data respectively, applying 3D CNN results in improved accuracy for grass-synthetic and commercial. This shows that exploiting both spatial and spectral information simultaneously can be helpful for classes that have distinct spectral and spatial properties. For most of the classes, CRNN achieved the better performance compared to 1D CNN which indicates that by combining convolutional and recurrent layers, CRNN model is able to extract more discriminative feature representations by exploiting dependencies between different spectral bands. From the CRNN performance, it can also be seen that the recurrent layers can extract the spectral dependencies more effectively from the middle-level features provided by convolutional layers.

Experiment 4—Window patch size for patch-based classification For patch (frame) -based classification methods like 2D and 3D CNNs, the size of window patches extracted from the hyperspectral image depends on the spatial resolution of the image and the size of the classes being analyzed. For high-resolution images, one can extract large patches to extract spatial contextual information as long as there are enough pixels of the class of interest (the center pixel, if patches are extracted around a center pixel) in each patch. In this experiment, we show the effect of varying window patch size on the classification performance for the 2D CNN model. Number of training samples for random and disjoint datasets are 200 samples per class and 100 samples per class, respectively. The number of test samples is 80 samples per class. As can be seen in Fig. 4.8, increasing only the window patch size while keeping the rest of configuration unchanged results in an improvement of classification accuracy for the random dataset, and 9×9 window patches result in the best classification result for this model. This can be misleading, because as the patch size is increased, there is, in fact, more overlap between test and train samples, particularly when the data are randomly drawn from the scene, which results in a biased estimate. As is observed from Fig. 4.8, for the disjoint dataset, there is a significant drop in performance compared to the random dataset. This underscores the need for a careful selection of training and test data when determining hyperparameters such as patch size.

Experiment 5—classification map—The classification maps when using the different classification approaches discussed in this chapter are shown in Fig. 4.9 and Fig. 4.10 for disjoint and random datasets respectively. The models used to generate these maps were trained using 200 samples per class and 100 samples per class for the random and disjoint datasets, respectively. It can be seen in Fig. 4.9 that when disjoint training and test data are used, the models struggle to generalize, particu-

Table 4.11 Per class classification accuracy % on deep learning models for disjoint and random dataset extracted from UH 2013 dataset. (top) Disjoint dataset—100 sample per class for train and 80 samples per class for test. (bottom) Random dataset—200 sample per class for train and 80 samples per class for test

Class	Method				
	1D CNN	2D CNN	3D CNN	CRNN	RNN
1-Grass-healthy	81.42	70.25	73.75	74.28	75.71
2-Grass-stressed	74.28	80	73.75	81.42	75.71
3-Grass-synthetic	100	99.25	97.5	100	100
4-Tree	57.14	90.5	90	88.57	90
5-Soil	82.85	99.5	73.75	87.14	58.57
6-Water	94.28	90.75	87.5	98.57	87.14
7-Residential	45.71	70.25	62.5	38.57	44.28
8-Commercial	2.85	59.75	58.75	4.28	10
9-Road	61.42	32	57.5	52.87	50
10-Highway	44.28	52.75	61.25	50	51.42
11-Railway	57.14	70.5	51.25	72.85	64.28
12-Parking Lot 1	1.42	50.1	15	18.57	21.42
13-Parking Lot 2	71.42	90	93.75	77.14	72.85
14-Tennis court	98.57	100	98.75	100	100
15-Running track	94.28	90.75	98.75	95.71	90
Average accuracy of all classes	64.47	76.42	72.91	69.33	66.1
Class	Method				
	1D CNN	2D CNN	3D CNN	CRNN	RNN
1-Grass-healthy	91.42	100	100	98.57	97.14
2-Grass-stressed	98.57	100	98.75	97.14	92.85
3-Grass-synthetic	100	99.5	100	98.57	100
4-Tree	95.71	99.75	98.75	97.14	100
5-Soil	95.71	100	100	98.57	97.14
6-Water	97.14	100	93.75	95.71	94.28
7-Residential	78.57	98.75	95	90	78.57
8-Commercial	92.85	90.5	92.5	95.71	92.85
9-Road	85.71	96.75	96.25	88.57	84.28
10-Highway	77.14	99.75	86.25	90	92.85
11-Railway	77.14	100	90	87.14	87.14
12-Parking Lot 1	84.28	99.5	87.5	88.57	97.14
13-Parking Lot 2	55.71	100	92.5	70	77.14
14-Tennis court	100	100	100	98.57	98.57
15-Running track	95.71	100	100	98.57	97.14
Average accuracy of all classes	88.37	98.96	95.41	92.85	92.47

Fig. 4.8 Effect of the size of window patches extracted from UH 2013 hyperspectral image on classification accuracy for both disjoint and random datasets

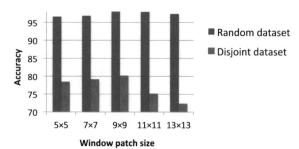

Table 4.12 Experimental classification results for various traditional machine learning models on random dataset extracted from UH 2013 dataset

Class	Method				
	SVM	Decision tree	Random forest	Naive bayes	kNN
1-Grass-healthy	94.28	94.28	95.71	84.28	100
2-Grass-stressed	88.57	91.42	95.71	74.28	98.57
3-Grass-synthetic	95.71	94.28	97.14	95.71	98.57
4-Tree	91.42	88.57	90	84.28	91.42
5-Soil	91.42	95.71	97.14	84.28	97.14
6-Water	81.42	85.71	90	80	88.57
7-Residual	50	70	75.71	32.85	68.57
8-Commercial	67.14	74.28	88.57	30	82.85
9-Road	35.71	58.57	70	78.57	67.14
10-Highway	47.14	75.71	88.57	1	87.14
11-Railway	37.14	60	78.57	40	77.14
12-Parking Lot 1	47.14	74.28	77.14	20	77.14
13-Parking Lot 2	57.14	57.14	41.42	4.28	40
14-Tennis court	97.14	98.57	97.14	92.85	100
15-Running track	98.57	97.14	94.28	91.42	95.71
Average	71.99	81.04	85.14	59.52	84.66

larly because there are no representative training samples in the shadow region (c.f. Fig. 4.9). Black rectangles in these maps highlight the classification performance in specific areas for different models where we want to highlight specific trends with respect to improved classification with CNN/CRNN and their variants.

Experiment 6—Baselines Table 4.12 represents experimental results for random dataset computed based on some traditional machine learning methods. As it is clear, compared to traditional machine learning methods listed in the Table 4.12, deep learning approaches show better performance.

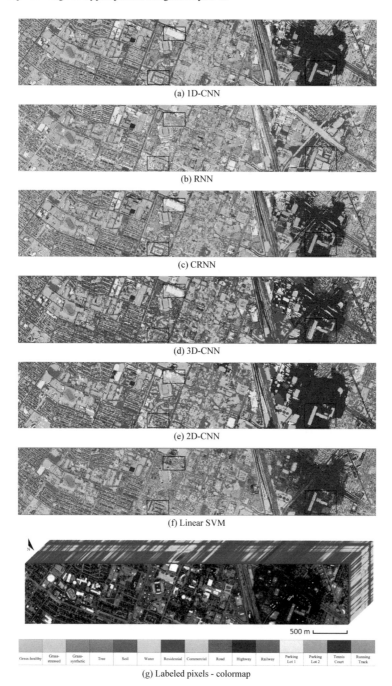

(a) 1D-CNN

(b) RNN

(c) CRNN

(d) 3D-CNN

(e) 2D-CNN

(f) Linear SVM

(g) Labeled pixels - colormap

Fig. 4.9 Classification maps of UH 2013 hyperspectral image computed from deep learning models trained by disjoint dataset extracted from UH 2013; All the models are trained using 100 samples per class

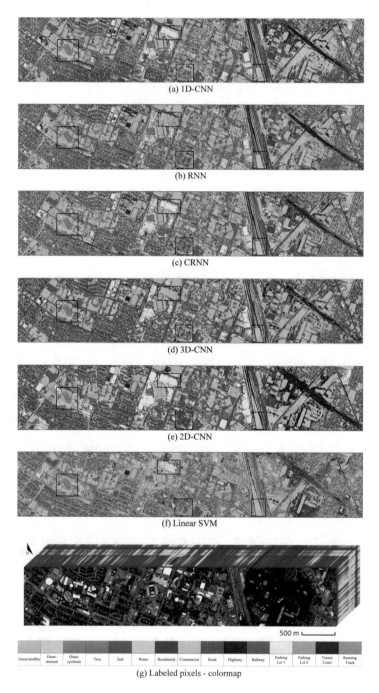

(a) 1D-CNN

(b) RNN

(c) CRNN

(d) 3D-CNN

(e) 2D-CNN

(f) Linear SVM

500 m

| Grass-healthy | Grass-stressed | Grass-synthetic | Tree | Soil | Water | Residential | Commercial | Road | Highway | Railway | Parking Lot 1 | Parking Lot 2 | Tennis Court | Running Track |

(g) Labeled pixels - colormap

Fig. 4.10 Classification maps of UH 2013 hyperspectral image computed from deep learning models trained by random dataset extracted from UH 2013; All the models are trained using 200 samples per class

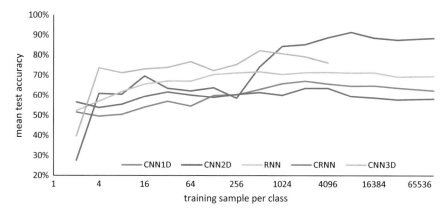

Fig. 4.11 The effect of the number of training samples on the test accuracy is indicated. The mean accuracy of three trained networks for each experiment is reported

4.5.2   Biomedical Results

4.5.2.1   Tissue-Type Classification

Most methods of FTIR classification leverage only individual pixels (spectra). Many traditional unsupervised and supervised approaches have been applied to FTIR data, including as k-means clustering [98], hierarchical cluster analysis (HCA) [99], Bayesian classifiers [55, 66, 100], random forest classifiers [63, 66, 67], support vector machines (SVMs) [66] and linear discriminant analysis [101], and ANNs [32, 66, 102]. Deep learning has not been fully explored due to the lack of spatial detail introduced by most FTIR imaging instruments, however recent advances in instrument resolution allows the application of CNNs [69, 71].

Experiment 1—The effect of the number of training samples on accuracy is studied in this experiment. CNN, RNN, and CRNN models are trained to identify five different cell types (adipocytes, collagen, epithelium, myofibroblasts, necrosis) from the BRC961 dataset. The training set consists of 100000 samples per class and the disjoint test set includes 30000 samples per class. We trained 1D and 2D CNNs, RNN, and CRNN on the range of 2–100000 samples per class, while the memory limited 3D CNN training to the range of 2–4096 samples per class. Test accuracy significantly increases with growing training size in 2D and 3D CNNs (Fig. 4.11). We ran each experiment three times to compute the average accuracy and standard deviation (Fig. 4.12).

Experiment 2—The effect of network depth is studied in this experiment. We increased the number of layers to extract deep features and investigate their performance for classifying BRC961. The number of trainable parameters increases for deeper models and affects convergence time and memory (Table 4.13). High numbers of trainable parameters in the 2D CNN, 3D CNN, and RNN makes them prone

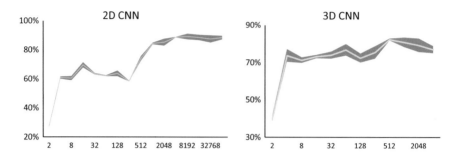

Fig. 4.12 The test accuracy variance from average for three runs of 2D and 3D convolutional neural networks is shown

Table 4.13 Classification networks were built with different number of layers for comparing their performance on predicting different cell types. The number of trainable parameters indicated in this table represents the depth and complexity of the network

Method	Layer				
	1-conv	2-conv	3-conv	4-conv	6-conv
1D CNN		387 k		460 k	970 k
2D CNN	3.7 M	5.9 M	5.2 M		
3D CNN		15.4 M	14 M	15.9 M	
	1-recu		2-recu		3-recu
RNN	1 M		3 M		5.2 M
	3-conv & 1-recu			5-conv & 2-recu	
CRNN	108 k			775 k	

k: kilo, M: million

to overfitting, so we introduced dropout to mitigate this issue. The average accuracy on test data is computed for all trained classification networks (Table 4.14). Deeper models improve the network performance, except in the case of the RNN, which confirms limited correlation between spectra in BRC961.

2D and 3D CNNs demonstrate more accurate classification results on this dataset. The high number of trainable parameters allows the 2D implementation to be well trained and generalizes to the test dataset. We randomly selected 10% of the test dataset as the valid set. The accuracy and cost on the train set and valid set during the training procedure are used to visualize the effect of deeper networks on extracting feature for 2D CNN (Fig. 4.13).

Experiment 3—The classification performance of each network for different classes is studied in this experiment. The same number of training samples for each class is used to train different architectures. Three models per network are trained to compute the average accuracy for each class (Table 4.15).

Experiment 4—The hyperspectral biomedical dataset is classified using traditional models to compare with the machine learning results (Table 4.16).

Table 4.14 Classification networks with different numbers of layers are trained to study the effectiveness of the network depth on the test accuracy. The average accuracy and standard deviation are reported. x-conv/recu: x number of convolutional/recurrent layers

Method	Layer				
	1-conv	2-conv	3-conv	4-conv	6-conv
1D CNN		57.25 ± 0.7		65.09 ± 3.9	67.13 ± 1.1
2D CNN	84.19 ± 3.3		84.73 ± 1	89.17 ± 2	
3D CNN		74.59 ± 1.4	80.51 ± 3.7	82.27 ± 0.6	
	1-recu		2-recu		3-recu
RNN	71.75 ± 0.8		68.94 ± 0.7	65.66 ± 0.6	
	3-conv & 1-recu			5-conv & 2-recu	
CRNN	61.29 ± 1.4			63.57 ± 1.5	

Table 4.15 The average and standard deviation for per class accuracy for classification networks on the hyperspectral biomedical dataset are presented. The same number of training samples for all classes are used to train machine learning models and test their performances

Class	Method				
	1D CNN	2D CNN	3D CNN	CRNN	RNN
Adipocytes	42.6	89.9	85.2	7.88	15.5
Collagen	96.6	97.7	96.4	95.5	98.7
Epithelium	88.5	89	85.6	90	91.2
Myofibroblasts	64	90.9	74.5	64.2	62.4
Necrosis	44	84.1	72.8	78.8	94.1
Average accuracy	67.1	90.3	82.9	66.4	72.4

Table 4.16 The traditional classification methods are used to classify hyperspectral biomedical dataset. Per class accuracy and average accuracy are reported for different methods

Class	Method				
	SVM	Decision tree	Random forest	Naive Bayes	kNN
Adipocytes	12.64	8.93	27	5.55	20.13
Collagen	94.07	89.99	96.56	97.88	96.98
Epithelium	84.44	80.36	89.62	87.81	84.62
Myofibroblasts	96.14	69.4	60.1	55.84	60.78
Necrosis	82.99	66.47	38.8	26.02	49.39
Average accuracy	74.05	63.03	62.41	54.62	62.38

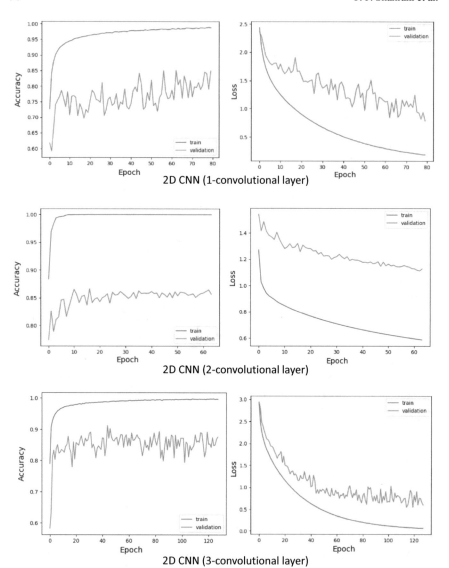

Fig. 4.13 Deeper models of the 2D CNN have more trainable parameters and it helps the network to minimize loss and increase the accuracy on valid and test dataset

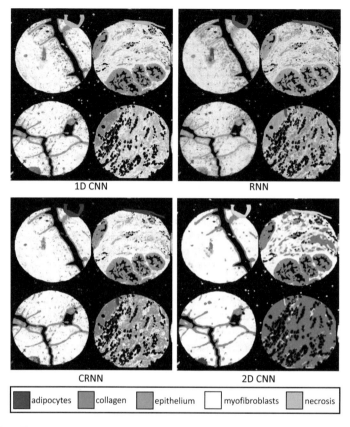

1D CNN

RNN

CRNN

2D CNN

| adipocytes | collagen | epithelium | myofibroblasts | necrosis |

Fig. 4.14 Different classes of the biomedical dataset are identified from the hyperspectral data by neural networks

Experiment 5—The classification maps for biomedical dataset are generated by trained models. The required memory limited us to compute the classification map for 3D CNN model. 5 classes are distinguished in 4 cores of the hyperspectral data using neural network (Fig. 4.14).

4.5.2.2   Digital Staining

In this section, we demonstrate a different direction for applying CNNs to map infrared spectra to traditional histological stains. The proposed CNN framework is described in Table 4.17. The input (IR spectra image) and target values (brightfield image) are first aligned to allow mapping between spectra and color value in the stained image. This requires the use of the same tissue sample for both IR spectroscopy and staining. The pixel-level alignments were performed manually using the GIMP open-source editing software [103] to apply affine transformations to sub-

Table 4.17 Summary of CNN configurations for digital staining. conv(x)-X: convolutional layer; max(p): maxpooling layer; FC(n): fully connected layer; x: receptive field; X: number of filters; p:pool size; n: number of nodes. The three neurons at the output layer provide the color channels for the output color image

CNN			
conv(3×3) $- 32 \rightarrow$	conv(3×3) $- 32 \rightarrow$	max(2×2) $\rightarrow$	conv(3×3) $- 32 \rightarrow$
conv(3×3) $- 32 \rightarrow$	max(2×2) $\rightarrow$	FC(128) $\rightarrow$	FC(3)

regions. Random samples from aligned subregions are used for training to avoid overfitting.

Weights were initialized with random values from a normal distribution with a 0 mean and standard deviation of 0.02. Training is carried out using batch optimization with a batch size of 128 and Adam optimizer with a learning rate of 0.01 to minimize the mean squared error (MSE) performance metric. L_2 regularization is also applied to reduce generalization error. We used softplus as a nonlinear activation function for each layer except the output layer which passes through a linear function to return the incoming tensor without changes.

The network was trained and tested on independent images from the same imaging system (Fig. 4.15). We used 16000 data points and extract a cropping window around them with the tensor size of $9 \times 9 \times 60$ pixels. Training and testing were performed using a Tesla k40M GPU. Training the network requires 33 s in average for 7 epochs and digital staining of a $512 \times 512 \times 60$ pixels hyperspectral image takes 20 s.

The proposed CNN reproduced high-resolution digital stains (Fig. 4.15c) and provide a great cellular-level detail when compared to the ground truth (Fig. 4.15d).

We compared our CNN with previously published pixel-level ANN (1 hidden layer with 7 nodes) [70]. The results (Fig. 4.15b) illustrate that applying the ANN to the same high-resolution data doesn't provide high-resolution results. Once the ANN framework maps only a single spectrum to corresponding digital color and the most chemical information of biological samples lies in the diffraction-limited region of spectrum (900–1500 cm^{-1}), the low resolution of this region limits the resolution of output image. We overcome this problem by leveraging the spatial features available at higher wavenumbers using CNN. CNN integrates non-diffraction-limited features from high-frequency region and chemical information from fingerprint region to produce more accurate staining patterns.

To evaluate the potential capability of the proposed CNN on different staining and spectroscopic imaging systems, some other experiments have been done. The proposed CNN was trained and tested on FTIR data from pig kidney to replicate DAPI staining (Fig. 4.16) and on prostate tissue imaged by Spero-QT (range 900–1800 cm^{-1}, pixel size 1.4 μm and FPA 480 ∗ 480) to resemble immunohistochemically staining (IHC) using Cytokeratin as a primary antibody (Fig. 4.17).

Experiment 1: Quantitative comparison—The results of digital staining are presented quantitatively. Figures 4.15, 4.16 and 4.17 illustrate that the generated images simply resemble the ground truth images visually. The synthesized images cannot be fully overlapped with the ground truths because first, the tissue sections undergo

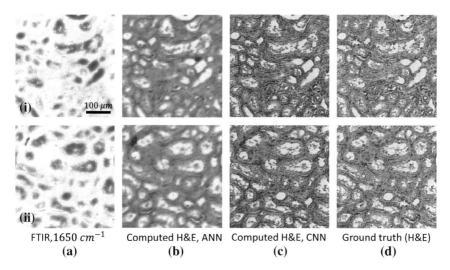

FTIR,1650 cm^{-1}	Computed H&E, ANN	Computed H&E, CNN	Ground truth (H&E)
(a)	**(b)**	**(c)**	**(d)**

Fig. 4.15 Molecular imaging reproduced by chemical imaging from FTIR on high magnification mode. **a** A single band of FTIR images. **b** Computed H&E from the CNN framework. **c** The physically stained H&E images as the ground truth

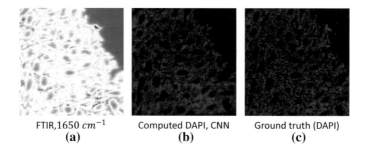

FTIR,1650 cm^{-1}	Computed DAPI, CNN	Ground truth (DAPI)
(a)	**(b)**	**(c)**

Fig. 4.16 Digital DAPI staining of pig kidney. **a** A single band of FTIR image. **b** Computed DAPI from CNN framework. **c** The physically- stained DAPI image under fluorescence microscope

staining process for generating the ground truth which causes local deformations on the tissues and also manual staining procedure can produce color variations which make the quantitative evaluations quite challenging.

We quantify the visibility of the digital H&E staining images generated by proposed CNN, ANN, and the ground truths using structural similarity (SSIM) index (Table 4.18). The SSIM index measures the similarity in case of luminance, contrast, and structure [104, 105].

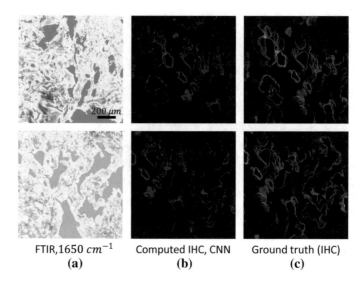

FTIR,1650 cm^{-1}	Computed IHC, CNN	Ground truth (IHC)
(a)	**(b)**	**(c)**

Fig. 4.17 IHC imaging reproduced by chemical imaging from SPERO. **a** A single band of SPERO images. **b** Computed Cytokeratin staining from the CNN framework. **c** The images of Cytokeratin-stained tissues

Table 4.18 SSIM index for the ANN and CNN digital staining outputs shown in Fig. 4.15

SSMI	High mag. ANN	High mag. CNN
Test image (i)	0.61	0.72
Test image (ii)	0.61	0.70

4.5.3 Source Code and Data

All the datasets and codes in both biomedical and remote sensing hyperspectral area provided for this book chapter can be found in detail as a Github project [106].

4.6 Design Choices and Hyperparameters

A critical part of designing deep learning models is the optimization over the hyperparameter space. The choice of hyperparameters significantly affects accuracy, speed of convergence, underfitting, and overfitting. Hyperparameters for deep learning models are often chosen by hand after iterative experimentations or are selected by a search algorithm. The choice of hyperparameters influences the structure of the model and heavily depends on the application and available data. Here, we discuss some of the most important hyperparameters that need to be fine-tuned when designing a deep learning architecture, and we provide a summary of design choices we have made for the remote sensing and tissue histology applications in the chapter.

4.6.1 Convolutional Layer Hyperparameters

Convolutional layers apply a filter kernel with a small receptive field [107–109] to extract spatial features. Required hyperparameters for optimization include filter size, padding, and stride.

The filter size is equivalent to the spatial region considered to contain a viable image feature. The size of the output of a convolutional layer can change compared to the original input image depending on the choice of boundary conditions or padding. Common choices for padding in deep learning models include valid and zero padding. In valid padding, the input is not padded and thus this results in a smaller output image size compared to the size of the input image. In zero padding, the input image is padded with zeros such that the input and output image will have the same size after convolution. The default choice for padding in most of deep learning software packages is valid padding. In general, there are many options for boundary conditions in the convolution operation and the choice usually depends on the application [110, 111].

The stride hyperparameter controls the shifting step size for the convolution filters. A shifting step size of one unit is generally the default. A stride of one means that the filter shifts by one each time as it convolves around the input volume. A higher stride number results in higher shifting of the receptive field and also in the shrinking of the output volume. The increase in the stride number can be used to reduce the overlapping of the receptive fields and also to decrease the output dimensions. This can be very useful in 3D CNN architectures to reduce the size of the third dimension throughout the convolutional layers.

4.6.2 Pooling Layer Hyperparameters

The normalized output of each convolution layer is passed through a pooling layer to summarize local features [112]. By using a stride size greater than 1, the pooling layer sub-samples the feature maps to merge semantically similar features to provide invariant representation [113, 114]. This further improves the computational time and mitigates overfitting by reducing the number of unknown parameters.

The pooling layer operates independently on every depth slice of the input and resizes it spatially. The depth dimension remains unchanged. If the feature map volume size is $m \times n \times d$, the pooling layer with filter dimension or pooling size of q and stride s produces an output with a volume size of $m' \times n' \times d$, where

$$m' = \frac{1}{s}(m - q) + 1$$

$$n' = \frac{1}{s}(n - q) + 1$$

Common pooling operations are max and average pooling. In max and average pooling, a local window of pixels is replaced by the maximum pixel in the window or an average of the pixels in that window, respectively. In most software packages the default hyperparameters for pooling are a pooling size or filters of size 2×2 and a stride of 2, which halves the size of the input in both width and height.

4.6.3 Training Hyperparameters

The training of deep learning models is essentially a large-scale complex optimization problem. Neural networks and their flavors are complicated functions that consist of millions of unknown parameters, which are optimized and learned during training. Common hyperparameters of any optimization method include the choice of a loss function, the learning rate at step t, the mini-batch size, and the number of iterations. Here, we describe some of these hyperparameters that have to be chosen when training neural networks.

Optimizers—Optimization algorithms for training neural networks aim to find the best network parameters, which minimize a loss or a cost function by computing its gradient [115]. There are many choices available for optimizers and all of them are essentially flavors of gradient descent. Popular optimizers that have been used in many deep learning models include stochastic gradient descent (SGD), Adam [116], Adagrad [117], and Adadelta [118].

SGD—Traditional gradient descent calculates the gradient of the entire dataset to update the parameters. However, in the case of large data sizes, it can be very slow and memory inefficient. SGD updates the network parameters using a single or a few (mini-batch) training samples. SGD training with the proper mini-batch size is faster and more memory efficient. In addition, the batches are periodically shuffled to avoid gradient bias.

Adam—The Adam optimizer computes the adaptive individual learning rates for each parameter using the estimations of first and second moments of the gradient.

Adagrad—The Adagrad optimizer also provides an adaptive learning rate for each feature. Adagrad introduces a decay factor based on the inverse square root of the cached value at each time step, which is well-suited for dealing with sparse data while not desirable on highly non-convex loss functions [119].

Adadelta—The Adadelta optimizer improves Adagrad by restricting the sum of gradients within a certain window. In addition, it reduces the aggressive decrease in the learning rate, which is a characteristic of Adagrad. Adadelta adapts the learning rate over time, removing the need for manual tuning.

Learning rate—The learning rate determines the size of the step toward to gradient direction. Very small learning rates lead to slow convergence of the network while a very large choice can lead to network divergence. In most of the optimizers, an initial learning rate needs to be chosen.

The initial learning rate is typically set in the range $[10^{-6}, 1]$. The default values are different depending on the software package used. However, for most optimizers, a good start value to try is 0.01 or 0.001. Nevertheless, the learning rate has to be tuned as it is critical to the convergence of the optimization method. In most optimizers, the learning rate is updated (most often decreased) throughout the iterations.

Loss functions—Loss functions are mathematical functions which evaluate the network performance with the current set of parameters by measuring the difference between network predictions and target values through a training set. An optimization method seeks to minimize the loss function by taking its derivative with respect to the unknown parameters. There exist a variety of loss functions (see Chap. 3). However, the most popular loss function used for classification tasks is categorical cross entropy while mean squared error is very commonly used for regression tasks.

Batch size—Using a mini-batch training strategy helps to reduce loss fluctuation, which occurs if a single sample is used during each training iteration. The choice of the batch size is mostly computational and it depends on the available system memory. Hence, systems with higher memory can potentially benefit from larger batch sizes. Parallelism can also be exploited for more efficient computation of the updates in the case of large batch sizes. Even though in general the choice of the batch size should not affect generalization capabilities of the network, a popular batch size for training is 128.

Training iterations—The number of training iterations or training epochs is generally chosen as a form of early stopping in order to avoid the semi-convergence behavior of iterative methods. The training is usually stopped when the performance of the network stops increasing on the validation set. One way to check the performance of the network on the validation set is by monitoring the validation loss throughout the iterations. Early stopping is a powerful way to prevent overfitting.

Data shuffling—It is very important to shuffle the training data. The purpose of data shuffling is to avoid feeding the network with mini-batches of highly correlated examples. The gradient of a mini-batch with highly correlated samples can lead to the estimation of the biased gradient. Shuffling of the data helps to get a more accurate estimation of the gradient and hence better updates of unknown parameters. In addition, it has been observed that the convergence speed of the network is improved when the data is shuffled.

4.6.4 General Model Hyperparameters

The architecture of deep learning models involves additional hyperparameters that need to be chosen for each layer, such as the number of hidden units per layer, the activation function, and the initialization of the weights and biases. The choice of each of these parameters can strongly affect model performance.

Weight initialization—The initialization of the weights is very important in the training of deep learning models. Proper initialization of the weights can help mitigate the local minimum trap problem. Popular initialization schemes that are provided in most deep learning software packages include initialization with a normal distribution, uniform distribution, truncated normal distribution, random orthogonal matrix, identity matrix, LeCun uniform or normal initializer, Xavier normal or uniform initializer, zeros initializer, ones initializer, constant initializer, and He normal or uniform initializer. Each one of these initializers have additional arguments that need to be chosen or default values provided by the software packages can be used. In general, biases are initialized to 0.

Activation functions—Activation functions apply a nonlinear transformation on the input and add more complexity to the network to improve the capability of the network for solving complex problems. Commonly used activation functions include the sigmoid, tanh, ReLU, leaky ReLU, and softplus. The sigmoid function limits the output to be between 0 and 1. However, it is not zero-centered and vanishes very low or very high gradient values. Tanh is very similar to sigmoid but is symmetric over the origin with range $[-1, 1]$. It still exhibits the problem of vanishing the gradient. ReLU has shown remarkable performance on deep networks [120]. ReLU is computationally efficient and has faster convergence. However, ReLU "kills" the neurons with negative values and thus the network can result in having units that are not activated anymore after some training iterations. Leaky ReLU assigns small linear values to the negative part to avoid zero gradients and enables updating parameters for negative input values. Softplus is the smooth version of ReLU with more stable estimations from both positive and negative inputs [121]. A more detailed discussion of activation functions is given in Chap. 1, Sect. 4.2.2.1.

Number of hidden units—The number of hidden units is a common hyperparameter that has to be chosen for each layer. This hyperparameter varies with the model and it mostly depends on the application, size of available training data, and system memory limitations. There are some general heuristics on choosing the number of hidden layers, such as usually the first hidden layer should be larger than the input layer. Some deep learning architectures tend to use increasing sizes for the number of hidden units in each layer and some use decreasing sizes (a 3D CNN could, for example, decrease the number of hidden units for data dimensionality reduction).

4.6.5 Regularization Hyperparameters

One of the major causes that delayed the application of deep learning algorithms is the problem of overfitting or the inability of the network to generalize well to unseen data. Over the years many regularization techniques have been developed to mitigate the problem of overfitting. Here, we discuss some common regularization methods, which can dramatically improve the performance of a deep learning model.

Dropout—Dropout is one of the most frequently used regularization techniques in deep learning. Throughout each training iteration, dropout randomly removes nodes from the network together with their incoming and outgoing connections. The probability threshold for choosing which nodes to keep drop is the dropout hyperparameter. A threshold of 0.5 is very common. Dropout increases the sparsity of the network and also makes the model have a different set of nodes throughout each training iteration. Dropout can also be applied to the input layer but it is usually mostly applied to the hidden layers.

Batch normalization—During training, as the weights are updated, the distribution of the inputs to the hidden layers change or shift around. The changes in the distribution of hidden unit values (known as internal covariate shift) make training challenging. This problem can be addressed by normalizing layer inputs using batch normalization [122]. Batch normalization reduces the covariate shift by normalizing the mean and variance of features across the examples in each mini-batch. It is similar to standard normalization applied to the input data. Batch normalization improves the stability of the network, reduces generalization error, avoids exploding or vanishing gradients, allows higher learning rates, and enables accelerated training.

Regularization of the weights—Another technique to reduce overfitting is to apply regularization on the network weights. The most popular types of weight regularization are $\ell 1$ and $\ell 2$ regularization. $\ell 2$ regularization, otherwise known as *weight decay*, is used to penalized large weight values while $\ell 1$ acts as a form of feature selection. $\ell 2$ regularization forces the weights to decay toward zero but not exactly zero while $\ell 1$ regularization can reduce the weights to become completely zero and thus result in a sparser network. The combination of $\ell 1$ and $\ell 2$ can be used for regularization as well. In both $\ell 1$ and $\ell 2$ regularization, a parameter has to be chosen. The regularization parameter determines the tradeoff between the original fit term of the loss function and the amount of regularization to be imposed. A regularization parameter close to zero yields the original loss function while a very large regularization parameter pushes more weights close to zero and can potentially lead to underfitting [123, 124].

Local response normalization—A layer that has been used in many deep learning architectures is local response normalization (LRN). This layer was introduced in order to provide an inhibition scheme aimed to be similar to the concept of *later inhibition* in neurobiology, which represents the capacity of an excited neuron to subdue its neighbors. The goal is to detect high-frequency features with a large response, which helps to create more contrast in a local area and subsequently increase the

Table 4.19 Summary of parameter and hyperparameter values used to train various deep learning models for UH 2013 dataset classification task

Parameters	1D CNN	RNN	CRNN	2D CNN	3D CNN
Optimizer	adam	adam	adam	adam	adam
Learning rate	1e-4	1e-4	1e-4	1e-4	1e-5
Loss function	Cross entropy	Cross entropy	Cross entropy	Cross entropy	Cross entropy
Weight initializer	Xavier uniform	Xavier uniform	Xavier uniform	Xavier uniform	Xavier uniform
Activation function	ReLU	tanh	ReLU-tanh	ReLU	ReLU
Batch size	128	128	128	128	128
Number of epochs	512	128	256	200	128
Drop out	20%	–	–	50%	–
Batch normalization	Yes	No	No	Yes	Yes

sensory perception. LRN normalizes the unbounded activations that can result after the application of some activation functions, such as ReLU [33]. The normalization around a local neighborhood of an excited artificial neuron makes it even more sensitive as compared to its neighbors. LRN also dampens the responses that are uniformly large in a local neighborhood. In general, it boosts artificial neurons with relatively larger activations. The normalization response, $b^i_{x,y}$, of an artificial neuron activity at location (x, y) after applying kernel i and activation function $g(\cdot)$, denoted as $a^i_{x,y}$ is computed as [33]

$$b^i_{x,y} = a^i_{x,y} \Bigg/ \left(k + \alpha \sum_{j=\max(0,i-n/2)}^{\min(N-1,i+n/2)} (a^j_{x,y})^2 \right)^{\beta},$$

where n is the number of adjacent feature maps around that kernel i. $k, n, \alpha,$ and β are constant hyperparameters. They are usually chosen to be $k = 1, n = 5, \alpha = 1$ and $\beta = 0.5$.

Data augmentation—A very simple way to attempt to make the model generalize better is to increase the size of the training data. This is especially useful in the case of image data. Common transformations that are applied for increasing the size of the training data include rotation, flipping, scaling, and shifting. Most of the existing deep learning software packages provide built-in implementations for data augmentation.

Remote Sensing dataset—We summarize the hyperparameters and parameters selected for remote sensing experiments of UH 2013 dataset in Table 4.19.

Biomedical dataset—Deep learning networks are trained and tested using different combinations of hyperparameters. The list of hyperparameters is summarized in Table 4.20.

Table 4.20 Different neural networks require different set of hyperparameters to be trained well. This table summarized best combinations that are used in this work

Parameters	1D CNN	RNN	CRNN	2D CNN	3D CNN
Optimizer	adam	adam	adam	adadelta	adam
Learning rate	1e-5	1e-4	1e-4	0.1	1e-5
Decay rate rho/epsilon	5e-3	0.1	0.5	0.95/1e-7	5e-3
Loss function	Cross entropy	Cross entropy	Cross entropy	Cross entropy	Cross entropy
Weight initializer	Xavier	Xavier	Xavier	Normal	Normal
Activation function	ReLU	tanh	ReLU	SoftPlus	ReLU
Drop out	–	35%	50%	50%	25%

4.7 Concluding Remarks

In this chapter, we applied foundational deep neural networks to two real-world hyperspectral image analysis tasks, representing remote sensing and tissue histology applications. We compare the performance of 1D, 2D, and 3D Convolutional Neural Networks, Recurrent Neural Networks and their variants for these applications, and discuss practical considerations and design choices for effective classification. For anyone looking to apply these to other datasets (e.g,. derived from different sensors), the hyperparameter choice may differ, but the approach presented in this chapter to study the performance of deep learning for multichannel optical imagery can be applied to other applications, sensors, and modalities. We note that our objective here was to review and study the fundamentals (building blocks) of deep learning models—in recent years, many advanced variants of these approaches are emerging, but they are all often building upon these building blocks. To conclude, we would like to highlight the following observations. The code and a sampling of data used in this chapter is available online.

- One must carefully approach the process of acquiring training and testing libraries for training and validating algorithms, to get an unbiased estimate with regards to generalization capacity and discriminative nature of features. This is particularly important with hyperspectral images, where training data are often manually labeled individually for every classification task. In particular, caution must be exercised to ensure the training and test frames do not overlap (when drawing these patches from a large scene over a wide geographical stretch, for example), and that the training and testing frames are representative of the sources of spectral variability commonly encountered in the application.
- Although recurrent networks have promise in modeling the spectral information, when comparing 1D (per-pixel spectral classification), recurrent networks performed better with the remote sensing tasks (where spectral reflectance properties encoded in the spectral envelope are being modeled) than with the tissue histol-

ogy task (where the spectral features of interests are localized absorption patterns instead of the shape of the spectral absorbance envelope).

- A crucial part of successfully deploying these networks for hyperspectral classification is the choice of the hyperparameters associated with the models—this choice will depend on the data characteristics and must be determined empirically for different applications/sensors/sensing platforms.

References

1. Gowen A, O'Donnell C, Cullen P, Downey G, Frias J (2007) Hyperspectral imaging an emerging process analytical tool for food quality and safety control. Trends Food Sci Technol 18(12):590–598
2. Schuler RL, Kish PE, Plese CA (2012) Preliminary observations on the ability of hyperspectral imaging to provide detection and visualization of bloodstain patterns on black fabrics. J Forensic Sci 57(6):1562–1569
3. Fischer C, Kakoulli I (2006) Multispectral and hyperspectral imaging technologies in conservation: current research and potential applications. Stud Conserv 51(sup1):3–16
4. Zhang Y, Chen Y, Yu Y, Xue X, Tuchin VV, Zhu D (2013) Visible and near-infrared spectroscopy for distinguishing malignant tumor tissue from benign tumor and normal breast tissues in vitro. J Biomed Opt 18(7):077003
5. Salzer R, Steiner G, Mantsch H, Mansfield J, Lewis E (2000) Infrared and raman imaging of biological and biomimetic samples. Fresenius' J Anal Chem 366(6–7):712–726
6. Liu K, Li X, Shi X, Wang S (2008) Monitoring mangrove forest changes using remote sensing and gis data with decision-tree learning. Wetlands 28(2):336
7. Everitt J, Yang C, Sriharan S, Judd F (2008) Using high resolution satellite imagery to map black mangrove on the texas gulf coast. J Coast Res 1582–1586
8. Yuen PW, Richardson M (2010) An introduction to hyperspectral imaging and its application for security, surveillance and target acquisition. Imaging Sci J 58(5):241–253
9. Shahraki FF, Prasad S (2018) Graph convolutional neural networks for hyperspectral data classification. In: 2018 IEEE global conference on signal and information processing (GlobalSIP), pp 968–972
10. Melgani F, Bruzzone L (2004) Classification of hyperspectral remote sensing images with support vector machines. IEEE Trans Geosci Remote Sens 42(8):1778–1790
11. Wu H, Prasad S (2016) Dirichlet process based active learning and discovery of unknown classes for hyperspectral image classification. IEEE Trans Geosci Remote Sens 54(8):4882–4895
12. Dong Y, Du B, Zhang L (2015) Target detection based on random forest metric learning. IEEE J Sel Top Appl Earth Obs Remote Sens 8:1830–1838 April
13. Zhang L, Zhang L, Tao D, Huang X, Du B (2014) Hyperspectral remote sensing image subpixel target detection based on supervised metric learning. IEEE Trans Geosci Remote Sens 52:4955–4965 Aug
14. Zhou X, Armitage AR, Prasad S (2016) Mapping mangrove communities in coastal wetlands using airborne hyperspectral data. In: 2016 8th workshop on hyperspectral image and signal processing: evolution in remote sensing (WHISPERS). IEEE, pp 1–5
15. Cui M, Prasad S (2016) Spectral-angle-based discriminant analysis of hyperspectral data for robustness to varying illumination. IEEE J Sel Top Appl Earth Obs Remote Sens 9(9):4203–4214
16. Xia J, Bombrun L, Berthoumieu Y, Germain C, Du P (2017) Spectral-spatial rotation forest for hyperspectral image classification. IEEE J Sel Top Appl Earth Obs Remote Sens 10(10):4605–4613

17. Tarabalka Y, Fauvel M, Chanussot J, Benediktsson JA (2010) Svm- and mrf-based method for accurate classification of hyperspectral images. IEEE Geosci Remote Sens Lett 7:736–740 Oct
18. Fauvel M, Benediktsson JA, Chanussot J, Sveinsson JR (2008) Spectral and spatial classification of hyperspectral data using svms and morphological profiles. IEEE Trans Geosci Remote Sens 46:3804–3814 Nov
19. Joelsson SR, Benediktsson JA, Sveinsson JR (2005) Random forest classifiers for hyperspectral data. In: Proceedings, 2005 IEEE international geoscience and remote sensing symposium 2005, IGARSS '05, vol 1, p 4
20. Zhang Y, Cao G, Li X, Wang B (2018) Cascaded random forest for hyperspectral image classification. IEEE J Sel Top Appl Earth Obs Remote Sens 11:1082–1094 April
21. Samaniego L, Bárdossy A, Schulz K (2008) Supervised classification of remotely sensed imagery using a modified k-nn technique. IEEE Trans Geosci Remote Sens 46(7):2112–2125
22. Chan JC-W, Paelinckx D (2008) Evaluation of random forest and adaboost tree-based ensemble classification and spectral band selection for ecotope mapping using airborne hyperspectral imagery. Remote Sens Environ 112(6):2999–3011
23. Li J, Bioucas-Dias JM, Plaza A (2010) Semisupervised hyperspectral image segmentation using multinomial logistic regression with active learning. IEEE Trans Geosci Remote Sens 48(11):4085–4098
24. Chen Y, Lin Z, Zhao X, Wang G, Gu Y (2014) Deep learning-based classification of hyperspectral data. IEEE J Sel Top Appl Earth Obs Remote Sens 7(6):2094–2107
25. Hu W, Huang Y, Wei L, Zhang F, Li H (2015) Deep convolutional neural networks for hyperspectral image classification. J Sens 2015
26. Makantasis K, Karantzalos K, Doulamis A, Doulamis N (2015) Deep supervised learning for hyperspectral data classification through convolutional neural networks. In: Geoscience and remote sensing symposium (IGARSS), 2015 IEEE International. IEEE, pp 4959–4962
27. Fang B, Li Y, Zhang H, Chan JC-W (2019) Hyperspectral images classification based on dense convolutional networks with spectral-wise attention mechanism. Remote Sens 11(2):159
28. Yu F, Koltun V (2015) Multi-scale context aggregation by dilated convolutions. arXiv:1511.07122
29. Mou L, Ghamisi P, Zhu XX (2017) Deep recurrent neural networks for hyperspectral image classification. IEEE Trans Geosci Remote Sens 55(7):3639–3655
30. Hang R, Liu Q, Hong D, Ghamisi P (2019) Cascaded recurrent neural networks for hyperspectral image classification. IEEE Trans Geosci Remote Sens
31. Wu H, Prasad S (2017) Convolutional recurrent neural networks forhyperspectral data classification. Remote Sens 9(3):298
32. Marini F, Bucci R, Magrì A, Magrì A (2008) Artificial neural networks in chemometrics: history, examples and perspectives. Microchem J 88(2):178–185
33. Krizhevsky A, Sutskever I, Hinton GE (2012) Imagenet classification with deep convolutional neural networks. In: Advances in neural information processing systems, pp 1097–1105
34. Sermanet P, Chintala S, LeCun Y (2012) Convolutional neural networks applied to house numbers digit classification. In: 2012 21st International conference on Pattern recognition (ICPR). IEEE, pp 3288–3291
35. LeCun Y, Bengio Y et al (1995) Convolutional networks for images, speech, and time series. Handb Brain Theory Neural Netw 3361(10):1995
36. LeCun Y, Bottou L, Bengio Y, Haffner P (1998) Gradient-based learning applied to document recognition. Proc IEEE 86(11):2278–2324
37. Chen Y, Jiang H, Li C, Jia X, Ghamisi P (2016) Deep feature extraction and classification of hyperspectral images based on convolutional neural networks. IEEE Trans Geosci Remote Sens 54(10):6232–6251
38. Li Y, Zhang H, Shen Q (2017) Spectral-spatial classification of hyperspectral imagery with 3D convolutional neural network. Remote Sens 9(1):67
39. 2013 ieee grss data fusion contest - fusion of hyperspectral and lidar data (2013). http://hyperspectral.ee.uh.edu/?page_id=459

40. 2018 ieee grss data fusion challenge - fusion of multispectral lidar and hyperspectral data (2018). http://hyperspectral.ee.uh.edu/?page_id=1075
41. Indian pines dataset. http://www.ehu.eus/ccwintco/index.php?title=Hyperspectral_Remote_Sensing_Scenes#Indian_Pines
42. Salinas dataset. http://www.ehu.eus/ccwintco/index.php?title=Hyperspectral_Remote_Sensing_Scenes#Salinas
43. Pavia dataset. http://www.ehu.eus/ccwintco/index.php?title=Hyperspectral_Remote_Sensing_Scenes#Pavia_Centre_and_University
44. Kennedy dataset. http://www.ehu.eus/ccwintco/index.php?title=Hyperspectral_Remote_Sensing_Scenes#Kennedy_Space_Center_.28KSC.29
45. Botswana dataset. http://www.ehu.eus/ccwintco/index.php?title=Hyperspectral_Remote_Sensing_Scenes#Botswana
46. Mittal S, Yeh K, Leslie LS, Kenkel S, Kajdacsy-Balla A, Bhargava R (2018) Simultaneous cancer and tumor microenvironment subtyping using confocal infrared microscopy for all-digital molecular histopathology. Proc Natl Acad Sci 115(25):E5651–E5660
47. Chemical imaging and structures laboratory. https://chemimage.illinois.edu/
48. Beck AH, Sangoi AR, Leung S, Marinelli RJ, Nielsen TO, Van De Vijver MJ, West RB, Van De Rijn M, Koller D (2011) Systematic analysis of breast cancer morphology uncovers stromal features associated with survival. Sci Transl Med 3(108):108ra113–108ra113
49. Cireşan DC, Giusti A, Gambardella LM, Schmidhuber J (2013) Mitosis detection in breast cancer histology images with deep neural networks. In: International conference on medical image computing and computer-assisted intervention. Springer, Berlin, pp 411–418
50. Araújo T, Aresta G, Castro E, Rouco J, Aguiar P, Eloy C, Polónia A, Campilho A (2017) Classification of breast cancer histology images using convolutional neural networks. PloS one 12(6):e0177544
51. Pahlow S, Weber K, Popp J, Bayden RW, Kochan K, Rüther A, Perez-Guaita D, Heraud P, Stone N, Dudgeon A et al (2018) Application of vibrational spectroscopy and imaging to point-of-care medicine: a review. Appl Spectrosc 72(101):52–84
52. Gazi E, Dwyer J, Gardner P, Ghanbari-Siahkali A, Wade A, Miyan J, Lockyer NP, Vickerman JC, Clarke NW, Shanks JH et al (2003) Applications of fourier transform infrared microspectroscopy in studies of benign prostate and prostate cancer. A pilot study. J Pathol 201(1):99–108
53. Fernandez DC, Bhargava R, Hewitt SM, Levin IW (2005) Infrared spectroscopic imaging for histopathologic recognition. Nat Biotechnol 23(4):469–474
54. Gazi E, Baker M, Dwyer J, Lockyer NP, Gardner P, Shanks JH, Reeve RS, Hart CA, Clarke NW, Brown MD (2006) A correlation of ftir spectra derived from prostate cancer biopsies with gleason grade and tumour stage. Eur Urol 50(4):750–761
55. Bhargava R, Fernandez DC, Hewitt SM, Levin IW (2006) High throughput assessment of cells and tissues: bayesian classification of spectral metrics from infrared vibrational spectroscopic imaging data. Biochimica et Biophysica Acta (BBA)-Biomembranes 1758(7):830–845
56. Srinivasan G, Bhargava R (2007) Fourier transform-infrared spectroscopic imaging: the emerging evolution from a microscopy tool to a cancer imaging modality. Spectroscopy (Santa Monica) 22(7):30–43
57. Bird B, Bedrossian K, Laver N, Miljković M, Romeo MJ, Diem M (2009) Detection of breast micro-metastases in axillary lymph nodes by infrared micro-spectral imaging. Analyst 134(6):1067–1076
58. Baker MJ, Gazi E, Brown MD, Shanks JH, Clarke NW, Gardner P (2009) Investigating ftir based histopathology for the diagnosis of prostate cancer. J Biophotonics 2(1–2):104–113
59. Šablinskas V, Urbonienė V, Ceponkus J, Laurinavicius A, Dasevicius D, Jankevičius F, Hendrixson V, Koch E, Steiner G (2011) Infrared spectroscopic imaging of renal tumor tissue. J Biomed Opt 16(9):096006
60. Walsh MJ, Holton SE, Kajdacsy-Balla A, Bhargava R (2012) Attenuated total reflectance fourier-transform infrared spectroscopic imaging for breast histopathology. Vib Spectrosc 60:23–28

61. Bergner N, Romeike BF, Reichart R, Kalff R, Krafft C, Popp J (2013) Tumor margin iden-
    tification and prediction of the primary tumor from brain metastases using ftir imaging and
    support vector machines. Analyst 138(14):3983–3990
62. Kallenbach-Thieltges A, Großerüschkamp F, Mosig A, Diem M, Tannapfel A, Gerwert K
    (2013) Immunohistochemistry, histopathology and infrared spectral histopathology of colon
    cancer tissue sections. J Biophotonics 6(1):88–100
63. Mayerich DM, Walsh M, Kadjacsy-Balla A, Mittal S, Bhargava R (2014) Breast histopathol-
    ogy using random decision forests-based classification of infrared spectroscopic imaging data.
    In: Proceedings of SPIE - The international society for optical engineering, vol 9041, p 904107
64. Nallala J, Diebold M-D, Gobinet C, Bouché O, Sockalingum GD, Piot O, Manfait M (2014)
    Infrared spectral histopathology for cancer diagnosis: a novel approach for automated pattern
    recognition of colon adenocarcinoma. Analyst 139(16):4005–4015
65. Ahmadzai AA, Patel II, Veronesi G, Martin-Hirsch PL, Llabjani V, Cotte M, Stringfellow
    HF, Martin FL (2014) Determination using synchrotron radiation-based fourier transform
    infrared microspectroscopy of putative stem cells in human adenocarcinoma of the intestine:
    corresponding benign tissue as a template. Appl Spectrosc 68(8):812–822
66. Mu X, Kon M, Ergin A, Remiszewski S, Akalin A, Thompson CM, Diem M (2015) Statistical
    analysis of a lung cancer spectral histopathology (SHP) data set. Analyst 140(7):2449–2464
67. Großerueschkamp F, Kallenbach-Thieltges A, Behrens T, Brüning T, Altmayer M, Stamatis
    G, Theegarten D, Gerwert K (2015) Marker-free automated histopathological annotation of
    lung tumour subtypes by FTIR imaging. Analyst 140(7):2114–2120
68. Kuepper C, Großerueschkamp F, Kallenbach-Thieltges A, Mosig A, Tannapfel A, Gerwert K
    (2016) Label-free classification of colon cancer grading using infrared spectral histopathology.
    Faraday Discuss 187:105–118
69. Berisha S, Lotfollahi M, Jahanipour J, Gurcan I, Walsh M, Bhargava R, Van Nguyen H,
    Mayerich D (2019) Deep learning for FTIR histology: leveraging spatial and spectral features
    with convolutional neural networks. Analyst 144(5):1642–1653
70. Mayerich D, Walsh MJ, Kadjacsy-Balla A, Ray PS, Hewitt SM, Bhargava R (2015) Stain-less
    staining for computed histopathology. Technology 3(01):27–31
71. Lotfollahi M, Berisha S, Daeinejad D, Mayerich D (2019) Digital staining of high-
    definition fourier transform infrared (FT-IR) images using deep learning. Appl Spectrosc,
    0003702818819857
72. US Biomax. https://www.biomax.us/tissue-arrays/Breast/BRC961. Accessed 30 Aug 2019
73. Scalable tissue imaging and modeling laboratory. https://stim.ee.uh.edu/
74. Baker MJ, Trevisan J, Bassan P, Bhargava R, Butler HJ, Dorling KM, Fielden PR, Fogarty
    SW, Fullwood NJ, Heys KA et al (2014) Using fourier transform IR spectroscopy to analyze
    biological materials. Nat Protoc 9(8):1771–1791
75. Berisha S, Chang S, Saki S, Daeinejad D, He Z, Mankar R, Mayerich D (2017) Siproc: an
    open-source biomedical data processing platform for large hyperspectral images. Analyst
    142(8):1350–1357
76. Bassan P, Sachdeva A, Kohler A, Hughes C, Henderson A, Boyle J, Shanks JH, Brown M,
    Clarke NW, Gardner P (2012) FTIR microscopy of biological cells and tissue: data analysis
    using resonant Mie scattering (RMieS) EMSC algorithm. Analyst 137(6):1370–1377
77. Derrick MR, Stulik D, Landry JM (2000) Infrared spectroscopy in conservation science. Getty
    Publications
78. Trevisan J, Angelov PP, Carmichael PL, Scott AD, Martin FL (2012) Extracting biological
    information with computational analysis of fourier-transform infrared (FTIR) biospectroscopy
    datasets: current practices to future perspectives. Analyst 137(14):3202–3215
79. Lunga D, Prasad S, Crawford MM, Ersoy O (2013) Manifold-learning-based feature extraction
    for classification of hyperspectral data: a review of advances in manifold learning. IEEE Signal
    Process Mag 31(1):55–66
80. Yu X, Wu X, Luo C, Ren P (2017) Deep learning in remote sensing scene classification: a data
    augmentation enhanced convolutional neural network framework. GIScience Remote Sens
    54(5):741–758

81. Wu H, Prasad S (2018) Semi-supervised deep learning using pseudo labels for hyperspectral image classification. IEEE Trans Image Process 27:1259–1270 March
82. Zhou X, Prasad S (2017) Domain adaptation for robust classification of disparate hyperspectral images. IEEE Trans Comput Imaging 3:822–836 Dec
83. Acquarelli J, van Laarhoven T, Gerretzen J, Tran TN, Buydens LM, Marchiori E (2017) Convolutional neural networks for vibrational spectroscopic data analysis. Anal Chim Acta 954:22–31
84. Morchhale S, Pauca VP, Plemmons RJ, Torgersen TC (2016) Classification of pixel-level fused hyperspectral and lidar data using deep convolutional neural networks. In: 2016 8th workshop on hyperspectral image and signal processing: evolution in remote sensing (WHISPERS), pp 1–5
85. Li H, Ghamisi P, Soergel U, Zhu X (2018) Hyperspectral and lidar fusion using deep three-stream convolutional neural networks. Remote Sens 10(10):1649
86. Pohl C, Van Genderen JL (1998) Review article multisensor image fusion in remote sensing: concepts, methods and applications. Int J Remote Sens 19(5):823–854
87. Wan Z, Yang R, You Y, Cao Z, Fang X (2018) Scene classification of multisource remote sensing data with two-stream densely connected convolutional neural network. In: Image and signal processing for remote sensing XXIV. International society for optics and photonics, vol. 10789, p 107890S
88. Xu X, Li W, Ran Q, Du Q, Gao L, Zhang B (2018) Multisource remote sensing data classification based on convolutional neural network. IEEE Trans Geosci Remote Sens 56(2):937–949
89. Huang G, Liu Z, Van Der Maaten L, Weinberger KQ (2017) Densely connected convolutional networks. In: Proceedings of the IEEE conference on computer vision and pattern recognition, pp 4700–4708
90. Zhu J, Fang L, Ghamisi P (2018) Deformable convolutional neural networks for hyperspectral image classification. IEEE Geosci Remote Sens Lett 15(8):1254–1258
91. Dai J, Qi H, Xiong Y, Li Y, Zhang G, Hu H, Wei Y (20147) Deformable convolutional networks. In: Proceedings of the IEEE international conference on computer vision, pp 764–773
92. Labate D, Safari K, Karantzas N, Prasad S, Foroozandeh Shahraki F (2019) Structured receptive field networks and applications to hyperspectral image classification. In: SPIE Optical Engineering + Applications. San Diego, California, United States
93. Labate D, Lim WQ, Kutyniok G, Weiss G (2005) Sparse multidimensional representation using shearlets. In: Wavelets XI, vol 5914. International Society for Optics and Photonics, p 59140U
94. Cho K, Van Merriënboer B, Gulcehre C, Bahdanau D, Bougares F, Schwenk H, Bengio Y (2014) Learning phrase representations using RNN encoder-decoder for statistical machine translation. arXiv:1406.1078
95. Chollet F, et al (2015) Keras. https://keras.io
96. Glorot X, Bengio Y (2010) Understanding the difficulty of training deep feedforward neural networks. In: Proceedings of the thirteenth international conference on artificial intelligence and statistics, pp 249–256
97. Chung J, Gulcehre C, Cho K, Bengio Y (2014) Empirical evaluation of gated recurrent neural networks on sequence modeling. arXiv:1412.3555
98. Ly E, Piot O, Wolthuis R, Durlach A, Bernard P, Manfait M (2008) Combination of FTIR spectral imaging and chemometrics for tumour detection from paraffin-embedded biopsies. Analyst 133(2):197–205
99. Yu P (2005) Applications of hierarchical cluster analysis (CLA) and principal component analysis (PCA) in feed structure and feed molecular chemistry research, using synchrotron-based fourier transform infrared (ftir) microspectroscopy. J Agric Food Chem 53(18):7115–7127
100. Tiwari S, Bhargava R (2015) Extracting knowledge from chemical imaging data using computational algorithms for digital cancer diagnosis. Yale J Biol Med 88(2):131–143

101. Yang H, Irudayaraj J, Paradkar MM (2005) Discriminant analysis of edible oils and fats by FTIR, FT-NIR and FT-raman spectroscopy. Food Chem 93(1):25–32
102. Fabian H, Thi NAN, Eiden M, Lasch P, Schmitt J, Naumann D (2006) Diagnosing benign and malignant lesions in breast tissue sections by using IR-microspectroscopy. Biochimica et Biophysica Acta (BBA)-Biomembranes 1758(7):874–882
103. Solomon RW (2009) Free and open source software for the manipulation of digital images. Am J Roentgenol 192(6):W330–W334
104. Wang Z, Bovik AC, Sheikh HR, Simoncelli EP (2004) Image quality assessment: from error visibility to structural similarity. IEEE Trans Image Process 13(4):600–612
105. Bhargava R (2007) Towards a practical Fourier transform infrared chemical imaging protocol for cancer histopathology. Anal Bioanal Chem 389:1155–1169 Sept
106. Github link of the book chapter. https://github.com/PrasadLab/DLOverviewHSI
107. Dong C, Loy CC, He K, Tang X (2016) Image super-resolution using deep convolutional networks. IEEE Trans Pattern Anal Mach Intell 38(2):295–307
108. Mobiny A, Moulik S, Van Nguyen H (2017) Lung cancer screening using adaptive memory-augmented recurrent networks. arXiv:1710.05719
109. Mobiny A, Lu H, Nguyen HV, Roysam B, Varadarajan N (2019) Automated classification of apoptosis in phase contrast microscopy using capsule network. IEEE Trans Med Imaging
110. Berisha S, Nagy JG (2014) Iterative methods for image restoration. In: Academic press library in signal processing, vol 4. Elsevier, pp 193–247
111. Mobiny A, Van Nguyen H (2018) Fast capsnet for lung cancer screening. In: International conference on medical image computing and computer-assisted intervention. Springer, Berlin, pp 741–749
112. Hinton GE, Srivastava N, Krizhevsky A, Sutskever I, Salakhutdinov RR (2012) Improving neural networks by preventing co-adaptation of feature detectors. arXiv:1207.0580
113. LeCun Y, Bengio Y, Hinton G (2015) Deep learning. Nature 521:436–444 May
114. Scherer D, Müller A, Behnke S (2010) Evaluation of pooling operations in convolutional architectures for object recognition. In: International conference on artificial neural networks. Springer, Berlin, pp 92–101
115. Goodfellow I, Bengio Y, Courville A (2016) Deep learning. MIT Press
116. Kingma DP, Ba J (2014) Adam: a method for stochastic optimization. arXiv:1412.6980
117. Duchi J, Hazan E, Singer Y (2011) Adaptive subgradient methods for online learning and stochastic optimization. J Mach Learn Res 12, 2121–2159
118. Zeiler MD (2012) Adadelta: an adaptive learning rate method. arXiv:1212.5701
119. Lipton ZC, Berkowitz J, Elkan C (2015) A critical review of recurrent neural networks for sequence learning. arXiv:1506.00019
120. Jin X, Xu C, Feng J, Wei Y, Xiong J, Yan S (2016) Deep learning with s-shaped rectified linear activation units. In: Thirtieth AAAI conference on artificial intelligence
121. Zheng H, Yang Z, Liu W, Liang J, Li Y (2015) Improving deep neural networks using softplus units. In: 2015 International joint conference on neural networks (IJCNN). IEEE, pp 1–4
122. Ioffe S, Szegedy C (2015) Batch normalization: accelerating deep network training by reducing internal covariate shift. arXiv:1502.03167
123. Liu J, Ye J (2010) Efficient L1/Lq norm regularization. arXiv:1009.4766
124. Xu Z, Zhang H, Wang Y, Chang X, Liang Y (2010) L 1/2 regularization. Sci China Inf Sci 53(6):1159–1169

Chapter 5
Advances in Deep Learning for Hyperspectral Image Analysis—Addressing Challenges Arising in Practical Imaging Scenarios

Xiong Zhou and Saurabh Prasad

Abstract Deep neural networks have proven to be very effective for computer vision tasks, such as image classification, object detection, and semantic segmentation—these are primarily applied to color imagery and video. In recent years, there has been an emergence of deep learning algorithms being applied to hyperspectral and multispectral imagery for remote sensing and biomedicine tasks. These multi-channel images come with their own unique set of challenges that must be addressed for effective image analysis. Challenges include limited ground truth (annotation is expensive and extensive labeling is often not feasible), and high dimensional nature of the data (each pixel is represented by hundreds of spectral bands), despite being presented by a large amount of unlabeled data and the potential to leverage multiple sensors/sources that observe the same scene. In this chapter, we will review recent advances in the community that leverage deep learning for robust hyperspectral image analysis despite these unique challenges—specifically, we will review unsupervised, semi-supervised, and active learning approaches to image analysis, as well as transfer learning approaches for multi-source (e.g., multi-sensor or multi-temporal) image analysis.

5.1 Deep Learning—Challenges presented by Hyperspectral Imagery

Since AlexNet [1] won the ImageNet challenge in 2012, deep learning approaches have gradually replaced traditional methods becoming a predominant tool in a variety of computer vision applications. Researchers have reported remarkable results with

X. Zhou
Amazon Web Service, Seattle, AI, USA
e-mail: xiongzho@amazon.com

S. Prasad (✉)
Hyperspectral Image Analysis Group, Department of Electrical
and Computer Engineering, University of Houston, Houston, TX, USA
e-mail: saurabh.prasad@ieee.org

© Springer Nature Switzerland AG 2020
S. Prasad and J. Chanussot (eds.), *Hyperspectral Image Analysis*,
Advances in Computer Vision and Pattern Recognition,
https://doi.org/10.1007/978-3-030-38617-7_5

117

deep neural networks in visual analysis tasks such as image classification, object detection, and semantic segmentation. A major differentiating factor that separates deep learning from conventional neural network based learning is the amount of parameters in a model. With hundreds of thousands even millions or billions of parameters, deep neural networks use techniques such as error backpropagation [2], weight decay [3], pretraining [4], dropout [5], and batch normalization [6] to prevent the model from overfitting or simply memorizing the data. Combined with the increased computing power and specially designed hardware such as Graphics Processing Units (GPU), deep neural networks are able to learn from and process unprecedented large-scale data to generate abstract yet discriminative features and classify them.

Although there is a significant potential to leverage from deep learning advances for hyperspectral image analysis, such data come with unique challenges which must be addressed in the context of deep neural networks for effective analysis. It is well understood that deep neural networks are notoriously data hungry insofar as training the models is concerned. This is attributed to the manner in which neural networks are trained. A typical training of a network comprises two steps: (1) pass data through the network and compute a task-dependent loss and (2) minimize the loss by adjusting the network weights by backpropagating the error [2]. During such a process, a model could easily end up overfitting [7], particularly if we do not provide sufficient training data. Data annotation has always been a major obstacle in machine learning research—and this requirement is amplified with deep neural networks. Acquiring extensive libraries such as ImageNet [8] for various applications may be very costly and time-consuming. This problem becomes even more acute when working with hyperspectral imagery for applications to remote sensing and biomedicine. Not only does one need specific domain expertise to label the imagery, annotation itself is challenging due to the resolution, scale, and interpretability of the imagery even by domain experts. For example, it can be argued that it is much more difficult to tell the different types of soil tillage apart by looking at a hyperspectral image than discerning everyday objects in color imagery. Further, the "gold-standard" in annotating remotely sensed imagery would be through field campaigns where domain experts verify the objects at exact geolocations corresponding to the pixels in the image. This can be very time-consuming and for many applications unfeasible. It is hence common in hyperspectral image analysis tasks to have a very small set of labeled ground truth data to train models from.

In addition to the label scarcity, the large inter-class variance of hyperspectral data also increases the complexity of the underlying classification task. Given the same material or object, the spectral reflectance (or absorbance) profiles from two hyperspectral sensors could be dramatically different because of the differences in wavelength range and spectral resolution. Even when the same sensor is used to collect images, one can get significant spectral variability due to the variation of view angle, atmospheric conditions, sensor altitude, geometric distortions, etc. [9]. Another reason for high spectral variability is mixed pixels arising from imaging platforms that result in low spatial resolution—as a result, the spectra of one pixel corresponds to more than one object on the ground [10].

For robust machine learning and image analysis, there are two essential components—deploying an appropriate machine learning model and leveraging a library of training data that is representative of the underlying inter-class and intra-class variability. For image analysis, specifically for classification tasks, deep learning models are variations of Convolution Neural Networks (CNNs) [11], which conduct a series of 2D convolutions between input images and (spatial) filters in a hierarchical fashion. It has been shown that such hierarchical representations are very efficient in recognizing objects in natural images [12]. When working with hyperspectral images, however, CNN-based features [13] such as color blobs, edges, shapes, etc. may not be the only features of interest for the underlying analysis. There is important information encoded in the spectral profile which can be very helpful for analysis. Unfortunately, in traditional applications of CNNs to hyperspectral imagery, modeling of spectral content in conjunction with spatial content is ignored. Although one can argue that spectral information could still be picked up when 2D convolutions are applied channel by channel or features from different channels are stacked together, such approaches would not constitute optimal modeling of spectral reflectance/absorbance characteristics. It is well understood that when the spectral correlations are explicitly exploited, spectral–spatial features are more discriminative—from traditional-wavelet-based feature extraction [14, 15] to modern CNNs [16–19]. In Chaps. 3 and 4, we have reviewed variations of convolutional and recurrent neural networks that model the spatial and spectral properties of hyperspectral data. In this chapter, we review recent works that specific address issues arising from deploying deep learning neural networks in challenging scenarios. In particular, our emphasis is on challenges presented by (1) limited labeled data, wherein one must leverage the vast amount of available unlabeled data in conjunction with limited data for robust learning and (2) multi-source optical data, wherein it is important to transfer models learned from one source (e.g., a specific sensor/platform/viewpoint/timepoint), and transfer the learned model to a different source (a different sensor/platform/viewpoint/timepoint), with the assumption that one source is rich in the quality and/or quantity of labeled training data while the other source is not.

5.2 Robust Learning with Limited Labeled Data

To address the labeled data scarcity, one strategy is to recruit resources (time and money, for example) with the goal of expanding the training library by annotating more data. However, for many applications, human annotation is neither scalable nor sustainable. An alternate (and more practical) strategy to address this problem is to design algorithms that do not require a large library of training data, but can instead learn from the extensive unlabeled data in conjunction with the limited amount of labeled data. Within this broad theme, we will review unsupervised feature learning, semi-supervised and active learning strategies. We will present results of several methods discussed in this chapter with three hyperspectral datasets—two of these are benchmark hyperspectral datasets, University of Pavia [20] and University of Houston [21], and represent urban land cover classification tasks. The University of Pavia dataset is a hyperspectral scene representing 9 urban land cover classes, with 103 spectral channels spanning the visible through near-infrared region. The 2013 University of Houston dataset is a hyperspectral scene acquired over the University of Houston campus and is representing 15 urban land cover classes. It has 144 spectral channels in the visible through near-infrared region. The third dataset is a challenging multi-source (multi-sensor/multi-viewpoint) hyperspectral dataset [22] that is particularly relevant in a transfer learning context—details of this dataset are presented later in Sect. 5.3.1.

5.2.1 Unsupervised Feature Learning

In contrast to the labeled data, unlabeled data are often easy and cheap to acquire for many applications, including remotely sensed hyperspectral imagery. Unsupervised learning techniques do not rely on labels and that makes this class of methods very appealing. Compared to supervised learning where labeled data are used as a "teacher" for guidance, models trained with unsupervised learning tend to learn relationships between data samples and estimate the data properties class-specific labelings of samples. In the sense that most deep networks can be comprised of two components—a feature extraction frontend and an analysis backend (e.g., undertaking tasks such as classification, regression, etc.), an approach can be completely unsupervised relative to the training labels (e.g., a neural network tasked with fusing sensors for super-resolution), or completely supervised (e.g., a neural network wherein both the features and the backend classifiers are learned with the end goal of maximizing inter-class discrimination). There are also scenarios wherein the feature extraction part of the network is unsupervised (where the labeled data are not used to train model parameters), but the backend (e.g., classification) component of the network is supervised. In this chapter, whenever the feature extraction component of a network is unsupervised (whether the backend model is supervised or unsupervised), we refer to this class of methods as carrying out "unsupervised feature learning".

The benefit of unsupervised feature learning is that we can learn useful features (in an unsupervised fashion) from a large amount of unlabeled data (e.g., spatial features representing the natural characteristics of a scene) despite not having sufficient labeled data to learn object-specific features, with the assumption that the features learned in an unsupervised manner can still positively impact a downstream supervised learning task.

In traditional feature learning (e.g., dimensionality reduction, subspace learning, or spatial feature extraction), the processing operators are often based on assumptions or prior knowledge about data characteristics. Optimization of feature learning to a task at hand is hence non-trivial. Deep learning-based methods address this problem in a data-adaptive manner, where the feature learning is undertaken in the context of the overall analysis task in the same network.

Deep learning-based strategies, such as autoencoders [23] and their variants, restricted Boltzmann machines (RBM) [24, 25], and deep belief networks (DBN) [26] have exhibited a potential to effectively characterize hyperspectral data. For classification tasks, the most common way to use unsupervised feature learning is to extract (*learn*) features from the raw data that can then be used to train classifiers downstream. Section 3.1 in Chap. 3 describes such a use of autoencoders for extracting features for tasks such as classification.

In Chen et al. [27], the effectiveness of autoencoder-derived features was demonstrated for hyperspectral image analysis. Although they attempted to incorporate spatial information by feeding the autoencoder with image patches, a significant amount of information is potentially lost due to the flattening process. To capture the multi-scale nature of objects in remotely sensed images, image patches with different sizes were used as inputs for a stacked sparse autoencoder in [28]. To extract similar multi-scale spatial–spectral information, Zhao et al. [29] applied a scale transformation by upsampling the input images before sending them to the stacked sparse autoencoder. Instead of manipulating the spatial size of inputs, Ma et al. [30] proposed to enforce the local constraint as a regularization term in the energy function of the autoencoder. By using a stacked denoising autoencoder, Xing et al. [31] sought to improve the feature stability and robustness with partially corrupted inputs. Although these approaches have been effective, they still require input signals (frames/patches) to be reshaped as one-dimensional vectors, which inevitably results in a loss of spatial information. To better leverage the spatial correlations between adjacent pixels, several works have been proposed to use the convolutional autoencoder to extract features from hyperspectral data [32–34].

Stacking layers have been shown to be an effective way to increase the representation power of an autoencoder model. The same principle applies to deep belief networks [35], where each layer is represented by a restricted Boltzmann machine. With the ability to extract a hierarchical representation from the training data, promising results have been shown for DBN/RBM for hyperspectral image analysis [36–42]. In recent works, some alternate strategies to unsupervised feature learning for hyperspectral image analysis have also emerged. In [43], a convolutional neural network was trained in a greedy layer-wise unsupervised fashion. A special learning criteria called Enforcing Population and Lifetime Sparsity (EPLS) [44] was utilized to ensure that the generated features are unique, sparse, and robust at the same time. In [45], the hourglass network [46], which shares a similar network architecture as an autoencoder, was trained for super-resolution using unlabeled samples in conjunction with noise. The reconstructed image was downsampled and compared with the real low-resolution image. The offset between these two was used as the loss function that was minimized to train the entire network. A minimized loss (offset) indicates the reconstruction from the network would be a good super-resolved estimate of the original image.

5.2.2   Semi-supervised learning

Although the feature learning strategy allows us to extract informative features from
unlabeled data, the classification part of the network still requires labeled training
samples. Methods that rely completely on unsupervised learning may not provide discriminative features from unlabeled data entirely for challenging classification tasks.
Semi-supervised deep learning is an alternate approach where unlabeled data are
used in conjunction with a small amount of labeled data to train deep networks (both
the feature extraction and classification components). It falls between supervised
learning and unsupervised learning and leverages benefits of both approaches. In the
context of classification, semi-supervised learning often provides better performance
compared to unsupervised feature learning, but without the annotation/labeling cost
needed for fully supervised learning [47].

Semi-supervised learning has been shown to be beneficial for hyperspectral image
classification in various scenarios [48–53]. Recent research [50] has shown that the
classification performance of a Multilayer Perceptron (MLP) can be improved by
adding an unsupervised loss. In addition to the categorical cross-entropy loss, a
symmetric decoder branch was added to the MLP and multiple reconstruction losses,
measured by the mean squared error of the encoder and decoder, were enforced to
help the network generate effective features. The reconstruction loss, in fact, served
as a regularizer to prevent the model from overfitting. A similar strategy has been
used with convolutional neural networks in [48].

A variant of semi-supervised deep learning, proposed by Wu and Prasad in [52],
entails learning a deep network that extracts features that are discriminative from the
perspective of the intrinsic clustering structure of data (i.e., these deep features can
discriminate between cluster labels—also referred to as pseudo-labels in this work)—
in short, the cluster labels generated from clustering of unlabeled data can be used
to boost the classification performance. To this end, a constrained Dirichlet Process
Mixture Model (DPMM) was used, and a variational inference scheme was proposed
to learn the underlying clustering from data. The clustering labels of the data were
used as *pseudo-labels* for training a convolutional recurrent neural network, where
a CNN was followed by a few recurrent layers (akin to a pretraining with pseudo-
labels). Figure 5.1 depicts the architecture of network. The network configuration is
specified in Table 5.1, where convolutional layers are denoted as "conv <filter size>-
<number of filters> and recurrent layers are denoted as "recur-<feature dimension>."

After pretraining with unlabeled data and associated pseudo-labels, the network
was fine-tuned with labeled data. This entails adding a few more layers to the previously trained network and learning only these layers from the labeled data. Compared
to traditional semi-supervised methods, the pseudo-label-based network, PL-SSDL,
achieved higher accuracy on the wetland data (a detailed description of this dataset
is provided in Sect. 3) as shown in Table 5.2. The effect of varying depth of the
pretrained network on the classification performance is shown in Fig. 5.2. Accuracy
increases as the model goes deeper, i.e., more layers. In addition to the environmental monitoring application represented by the wetland dataset, the efficacy of

Fig. 5.1 Architecture of the
convolutional recurrent
neural network. Cluster
labels are used for
pretraining the network.
(Source adapted from [52])

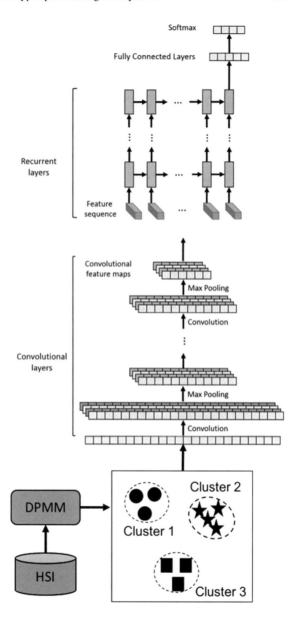

Table 5.1 Network configuration summary for the Aerial view wetland hyperspectral dataset. Every convolutional layer is followed by a max pooling layer, which is omitted for the sake of simplicity. (Source adapted from [52])

input-103 → conv3-32 → conv3-32 → conv3-64 → conv3-64
→ recur-256 → recur-512 → fc-64 → fc-64 → softmax-9

Table 5.2 Overall classification accuracies of different methods on the aerial view wetland dataset. (Source adapted from [52])

Methods	Label propagation	TSVM	SS-LapSVM	Ladder Networks	PL-SSDL
Accuracy	89.28 ± 1.04	92.24 ± 0.81	95.17 ± 0.85	93.17 ± 1.49	97.33 ± 0.48

PL-SSDL was also verified for urban land cover classification tasks using the University of Pavia [20] and the University of Houston [21] datasets, having 9 and 15 land cover classes, and representing spectra 103 and 144 spectral channels spanning the visible through near-infrared regions, respectively. As we can see from Fig. 5.3, features extracted with pseudo-label (middle column) are separated better than the raw hyperspectral data (left column), which implies pretraining with unlabeled data makes the features more discriminative. Compared to a network that is trained solely using labeled data, the semi-supervised method requires much less labeled samples due to the pretrained model. With only a few labeled samples per class, features are further improved by fine-tuning (right column) the network. Similar to this idea, Kang [53] later trained a CNN with pseudo-labels to extract spatial deep features through pretraining.

5.2.3 Active learning

Leveraging unlabeled data is the underlying principle of unsupervised and semi-supervised learning. Active learning, on the other hand, aims to improve the efficiency of acquiring labeling data as much as possible. Figure 5.4 shows a typical active learning flow, which contains four components: a labeled training set, a machine learning model, an unlabeled pool of data, and an oracle (a human annotator/domain expert). The labeled set is initially used for training the model. Based on the model's prediction, queries are then selected from the unlabeled pool and sent to the oracle for labeling. The loop is iterated until a pre-determined convergence criterion is met. The criteria used for selecting samples to query determines the efficiency of model training—efficiency here refers to the machine learning model reaching its full discriminative potential using as few queried labeled samples as possible. If every queried sample provides significant information to the model when labeled and incorporated into training, the annotation requirement will be small. A large part

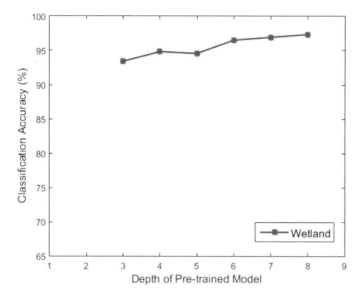

Fig. 5.2 Classification accuracy as a function of the depth of the pretrained model. (Source adapted from [52])

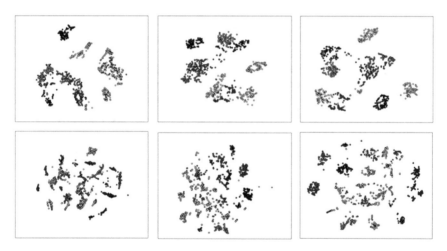

Fig. 5.3 t-SNE visualization of features at different training stages on the University of Pavia [20] (top row) and University of Houston [21] (bottom row) datasets. Left column represents raw image features, middle column represents features after unlabeled data pretraining, and right column represents feature after labeled data fine-tuning. (Source adapted from [52])

Fig. 5.4 Illustration of an
active learning system

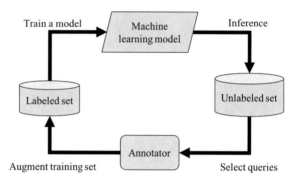

of active learning research is focused on designing suitable metrics to quantify the
information contained in an unlabeled sample that can be used for querying samples
from the data pool. A common thread in these works is the notion that choosing
samples that confuse the machine the most would result in a better (efficient) active
learning performance.

Active learning with deep neural networks has obtained increasing attention within
the remote sensing community in recent years [54–58]. Liu et al. [55] used features
produced by a DBN to estimate the representativeness and uncertainty of samples.
Both [56] and [57] explored using an active learning strategy to facilitate transferring
knowledge from one dataset to another. In [56], a stacked sparse autoencoder was
initially trained in the source domain and then fine-tuned in the target domain. To
overcome the labeled data bottleneck, an uncertainty-based metric was used to select
the most informative samples from the source domain for active learning. Similarly,
Lin et al. [57] trained two separate autoencoders from the source and target domains.
Representative samples were selected based on the density in the neighborhood of
the samples in the feature space. This allowed autoencoders to be effectively trained
using limited data. In order to transfer the supervision from source to target domain,
features in both domains were aligned by maximizing their correlation in a latent
space.

Unlike autoencoders and DBN, Convolution Neural Networks (CNNs) provide
an effective framework to exploit the spatial correlation between pixels in a hyperspectral image. However, when it comes to training with small data, CNNs tends
to overfit due to the large number of trainable network parameters. To solve this
problem, Haut [58] present an active learning algorithm that uses a special network
called Bayesian CNN [59]. Gal and Ghahramani [59] have shown that dropout in
neural network can be considered as an approximation to the Gaussian process, which
offers nice properties such as uncertainty estimation and robustness to overfitting. By
performing dropout after each convolution layer, the training of Bayesian CNN can
be cast as approximate Bernoulli variational inference. During evaluation, outputs
of a Bayesian CNN are averaged over several passes, which allows us to estimate
the model prediction uncertainty and the model suffers less from overfitting. Multiple uncertainty-based query criteria were then deployed to select samples for active
learning.

5.3 Knowledge Transfer Between Sources

Another common image analysis scenario entails learning with multiple sources, in particular, where one source is label "rich" (in the quantity and/or quality of labeled data) and the other source is label "starved". Sources in this scenario could imply different sensors, different sensing platforms (e.g., ground-based imagers, drones, or satellites), different time points, and different imaging viewpoints. In this situation, when it is desired to undertake analysis in the label-starved domain (often referred to as the target domain), a common strategy is to transfer knowledge from the label-rich domain (often referred to as the source domain).

5.3.1 Transfer Learning and Domain Adaptation

Effective training has always been a challenge with deep learning models. Besides requiring large amounts of data, the training itself is time-consuming and often comes with convergence and generalization problems. One major breakthrough of effective training of deep networks is the pretraining technique introduced by Hinton et al. [4], where a DBN was pretrained with unlabeled labeled data in a greedy layer-wise fashion, followed by a supervised fine-tuning. In particular, the DBN was trained one layer at a time by reconstructing outputs from the previous layer for the unsupervised pretraining. At the last training stage, all parameters were fine-tuned together by optimizing a supervised training criterion. In Erhan et al. in [60], the authors suggested that unsupervised pretraining works as a form of regularization. It not only provides a good initialization but also helps the generalization performance of the network. Similar to unsupervised pretraining, networks pretrained with supervision have also achieved huge success. In fact, using pretrained models as a starting point for new training has become a common practice for many analysis tasks [61, 62].

The main idea behind transfer learning is that knowledge gained from related tasks or a related data source can be transferred to a new task by fine-tuning on the new data. This is particularly useful when there is a data shortage in the new domain. In the computer vision community, a common approach to transfer learning is to initialize the network with weights that are pretrained for image classification on the ImageNet dataset [8]. The rationale for this is that ImageNet contains millions of natural images that are manually annotated and models trained with it tend to provide a "baseline performance" with generic and basic features commonly seen in natural images. Researchers have shown that features from lower layers of deep networks are color blobs, edges, shapes [13]. These basic features are usually readily transferable across datasets (e.g., data from different sources) [63].

In [64], Penatti et al. discussed the feature generalization in the remote sensing domain. Empirical results suggested that transferred features are not always better than hand-crafted features, especially when dealing with unique scenes in remote sensing images. Windrim et al. [65] unveiled valuable insights on transfer learning in the context of hyperspectral image classification. In order to test the effect of filter size and wavelength interval, multiple hyperspectral datasets were acquired with different sensors. The performance of transfer learning was examined through a comparison with training the network from scratch, i.e., randomly initializing network weights. Extensive experiments were carried out to investigate the impact of data size, network architecture, and so on. The authors also discussed the training convergence time and feature transferability under various conditions.

Despite the open questions which require more investigations, extensive studies have empirically shown the effectiveness of transfer learning on hyperspectral image analysis [39, 51, 66–76]. Marmanis et al. [66] introduced the pretrained model idea [1] for hyperspectral image classification. A pretrained AlexNet [1] was used as a fixed feature extractor and a two-layer CNN was attached for the final classification. Yang et al. [68] proposed a two-branch CNN for extracting spectral–spatial features. To solve the data scarcity problem, weights of lower layers were pretrained from another dataset and the entire network was then fine-tuned on the source dataset. Similar strategies have also been followed in [69, 74, 77].

Along with pretraining and fine-tuning, domain adaptation is another mechanism to transfer knowledge from one domain to another. Domain adaptation algorithms aim at learning a model from source data that can perform well on the target data. It can be considered as a sub-category of transfer learning, where the input distribution $p(X)$ changes while the label distribution $p(Y|X)$ remains the same across the two domains. Unlike the pretraining and fine-tuning method, which can be used when both distributions change, domain adaptation usually assumes the class-specific properties of the features within the two domains are correlated. This allows us to enforce stronger connections while transferring knowledge.

Othman et al. [70] proposed a domain adaptation network that can handle cross-scene classification when there is no labeled data in the target domain. Specifically, the network used three loss components for training: a classification loss (cross-entropy) in the source domain, a domain matching loss based on Maximum Mean Discrepancy (MMD) [78], and a graph regularization loss that aims to retain the geometrical structure of the unlabeled data in the target domain. The cross-entropy loss ensures that features produced by the network are discriminative. Having discriminative features in the original domain has also been found to be beneficial for the domain matching process [22]. In order to undertake domain adaptation, features from the two domains were aligned by minimizing the distribution difference. Zhou and Prasad [76] proposed to align domains (more specifically, features in these domains) based on Domain Adaptation Transformation Learning (DATL) [22]—DATL aligns class-specific features in the two domains by projecting the two domains onto a common latent subspace such that the ratio of within-class distance to between-class distance is minimized in that latent space.

Next, we briefly review how a projection such as DATL can be used to align deep networks for domain adaptation and present some results with multi-source hyperspectral data. Consider the distance between a source sample x_i^s and a target sample x_j^t in the latent space,

$$d(x_i^s, x_j^t) = \| f_s(x_i^s) - f_t(x_j^t) \|^2, \tag{5.1}$$

where f_s and f_t are feature extractors, e.g., CNNs that transform samples from both domains to a common feature space. To make the feature space robust to small perturbations in original source and target domains, the stochastic neighborhood embedding is used to measure classification performance [79]. In particular, the probability p_{ij} of the target sample x_j^t being the neighbor of the source sample x_i^s is given as

$$p_{ij} = \frac{\exp(-\| f_s(x_i^s) - f_t(x_j^t) \|^2)}{\sum_{x_k^s \in \mathcal{D}^S} \exp(-\| f_s(x_k^s) - f_t(x_j^t) \|^2)}, \tag{5.2}$$

where $\mathcal{D}^S$ is the source domain. Given a target sample with its label $(x_j^t, y_j^t = c)$, the source domain $\mathcal{D}^s$ can be split into a *same-class* set $\mathcal{D}_c^s = \{x_k^s | y_k = c\}$ and a *different-class* set $\mathcal{D}_{c'}^s = \{x_k^s | y_k \neq c\}$. In the classification setting, one wants to maximize the probability of making the correct prediction for x_j.

$$p_j = \frac{\sum_{x_i^s \in \mathcal{D}_c^s} \exp(-\| f_s(x_i^s) - f_t(x_j^t) \|^2)}{\sum_{x_k^s \in \mathcal{D}_{c'}^s} \exp(-\| f_s(x_k^s) - f_t(x_j^t) \|^2)}. \tag{5.3}$$

Maximizing the probability p_j is equivalent to minimizing the ratio of intra-class distances to inter-class distances in the latent space. This ensures that classes from the target domain and the source domain are aligned in the latent space. Note that the labeled data from the target domain (albeit limited) can be further used to make the feature more discriminative. The final objective function of DATL can then be written as

$$
\mathcal{L} = \beta \frac{\sum_{x_i^s \in \mathcal{D}_c^s} \exp(-\| f_s(x_i^s) - f_t(x_j^t) \|^2)}{\sum_{x_k^s \in \mathcal{D}_{c'}^s} \exp(-\| f_s(x_k^s) - f_t(x_j^t) \|^2)} \tag{5.4}
$$
$$
+ (1 - \beta) \frac{\sum_{x_i^t \in \mathcal{D}_c^t} \exp(-\| f_t(x_i^s) - f_t(x_j^t) \|^2)}{\sum_{x_k^t \in \mathcal{D}_{c'}^t} \exp(-\| f_t(x_k^s) - f_t(x_j^t) \|^2)}.
$$

The first term can be domain alignment term and the second term can be seen as a class separation term. β is the trade-off parameter that is data dependent. The greater the difference between source and target data, the larger value of β should be used to put more emphasis on domain alignment.

Depending on the feature extractors, Eq. 5.3 can be either solved using conjugate gradient-based optimization [22] or treated as a loss and solved using stochastic gradient descent [80]. DATL has been shown to be effective for addressing large domain shifts such as between street view and satellite hyperspectral images [22] acquired with different sensors and imaged with different viewpoints.

Figure 5.5 shows the architecture of Feature Alignment Neural Network (FANN) that leverages DATL. Two Convolutional Recurrent Neural Networks (CRNN) were

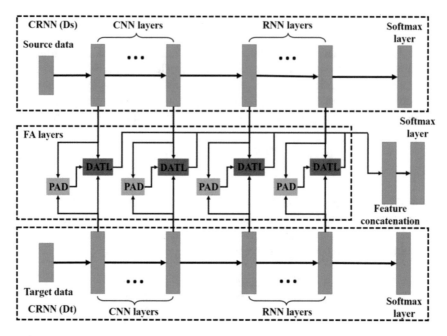

Fig. 5.5 The architecture of feature alignment neural network. (Source adapted from [76])

trained separately for the source and target domains. Features from corresponding layers were connected through an adaptation module, which is composed of a DATL term and a trade-off parameter that balances the domain alignment and the class separation. Specifically, the trade-off parameter β is automatically estimated by a proxy A-distance (PAD) [81].

$$\beta = \text{PAD}/2 = 1 - 2\epsilon, \tag{5.5}$$

where PAD is defined as $\text{PAD} = 2(1 - 2\epsilon)$ and $\epsilon \in [0, 2]$ is the generalization error of a linear SVM trained to discriminate between two domains. Aligned features were then concatenated and fed to a final softmax layer for classification (Table 5.3).

The performance of FANN was evaluated on a challenging domain adaptation dataset introduced in [22]. See Fig. 5.6 for the true color images for the source and target domains. The dataset consists of hyperspectral images of ecologically sensitive wetland vegetation that were collected by different sensors from two viewpoints— "Aerial" and "Street view" (and using sensors with different spectral characteristics) in Galveston, TX. Specifically, the aerial data were acquired using the ProSpecTIR VS sensor aboard an aircraft and has 360 spectral bands ranging from 400 nm to 2450 mm with a 5 nm spectral resolution. The aerial view data were radiometrically calibrated and corrected. The resulting reflectance data has a spatial coverage of 3462×5037 pixels at a 1 m spatial resolution. On the other hand, the Street view data were acquired through the Headwall Nano-Hyperspec sensor on a different date

Table 5.3 Network configuration summary for the Aerial and Street view wetland hyperspectral dataset (A-S view wetland). (Source adapted from [76])

FANN (A-S view wetland)
CRNN (Street) → DATL ← CRNN (Aerial)
(conv4-128 + maxpooling) → DATL ← (conv5-512 + maxpooling)
(conv4-128 + maxpooling) → DATL ← (conv5-512 + maxpooling)
(conv4-128 + maxpooling) → DATL ← (conv5-512 + maxpooling)
(conv4-128 + maxpooling) → DATL ← (conv5-512 + maxpooling)
(conv4-128 + maxpooling) → DATL ← (conv5-512 + maxpooling)
recur-64 → DATL ← recur-128
Fully connected-12

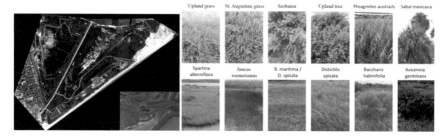

Fig. 5.6 Aerial and Street view wetland hyperspectral dataset. Left: Aerial view of the wetland data (target domain). Right: Street view of wetland data (source domain). (Source adapted from [76])

and represents images acquired by operating the sensor on a tripod and imaging the vegetation in the field during ground-reference campaigns. Unlike the Aerial view data, the street view data represent at-sensor radiance data with 274 bands spanning 400 nm and 1000 nm at 3 nm spectral resolution. As can be seen from Fig. 5.7, spectral signatures for the same class are very different between the source and target domains. With very limited labeled data in the aerial view, FANN achieved significant classification improvement compared to traditional domain adaptation methods (see Table 5.4).

As can be seen from Fig. 5.8, raw hyperspectral features from source and target domains are not aligned with each other. Due to the limited labeled data in Aerial view data, the mixture of classes happens in a certain level. The cluster structures are improved slightly by CRNN, see Fig. 5.8c, d for comparison. On the contrary, the source data, i.e., Street view data, have well-separated cluster structure. However, the classes are not aligned between the two domains; therefore, labels from the source domain cannot be used to directly train a classifier for the target domain. After passing all samples through the FANN, the two domains are aligned class by class in the latent space, as shown in Fig. 5.8e.

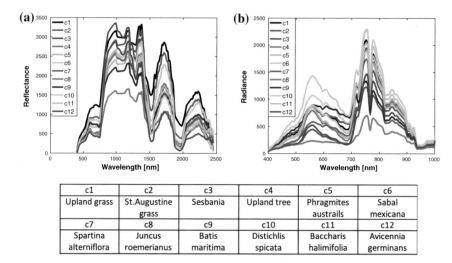

c1	c2	c3	c4	c5	c6
Upland grass	St.Augustine grass	Sesbania	Upland tree	Phragmites austrails	Sabal mexicana
c7	c8	c9	c10	c11	c12
Spartina alterniflora	Juncus roemerianus	Batis maritima	Distichlis spicata	Baccharis halimifolia	Avicennia germinans

Fig. 5.7 Mean spectral signature of the Aerial view (target domain) wetland data (**a**) and the Street view (source domain) wetland data (**b**). Different wetland vegetation species (classes) are indicated by colors. (Source adapted from [76])

Table 5.4 Overall classification accuracies of different domain adaptation methods on the Aerial and Street view wetland dataset. (Source adapted from [76])

Methods	SSTCA	KEMA	D-CORAL	FANN
Accuracy	85.3 ± 5.6	87.3 ± 1.7	92.5 ± 1.9	95.8 ± 1.1

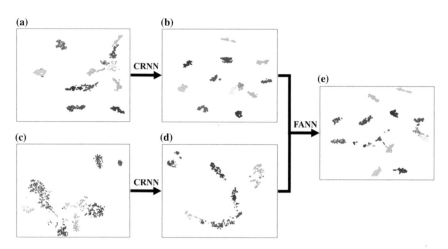

Fig. 5.8 t-SNE feature visualization of the Aerial and Street view wetland hyperspectral data at different stages of FANN. **a** Raw spectral features of Street view data in source domain. **b** CRNN features of Street view data in source domain. **c** Raw spectral features of Aerial view data in source domain. **d** CRNN features of Aerial view data in source domain. **e** FANN features for both domains in the latent space. (Source adapted from [76])

Table 5.5 Overall accuracy of the features of alignment layers and concatenated features for the Aerial and Street view wetland dataset. (Source adapted from [76])

Layer	FA-1	FA-2	FA-3	FA-4	FA-5	FA-6	FANN
OA	88.1	86.2	83.9	75.7	72.0	86.4	95.8

To better understand the feature adaptation process, features from all layers were investigated individually and compared to the concatenated features. Performance of each alignment layer is shown in Table 5.5. Consistent with observations in [63], accuracies drop from the first layer to the fifth layer as features become more and more specialized toward the training data. Therefore, the larger domain gap makes domain adaptation challenging. Although the last layer (FA-6) was able to mitigate this problem, this is because the recurrent layer has the ability to capture contextual information along the spectral direction of the hyperspectral data. Features from the last layer are the most discriminative ones, which allow the aligning module (DATL) to put more weight on the domain alignment (c.f. β in Eqs. 5.4 and 5.5). The concatenated features obtained the highest accuracy compared to individual layers. As mentioned in [76], an improvement of this idea would be to learn a combination weights for different layers instead of a simple concatenation.

5.3.2 Transferring Knowledge—Beyond Classification

In addition to image classification/semantic segmentation tasks, the notion of transferring knowledge between sources and datasets has also been used for many other tasks, such as object detection [67], image super-resolution [71], and image captioning [72].

Compared to image-level labels, training an object detection model requires object-level labels and corresponding annotations (e.g., through bounding boxes). This increases the labeling requirement/costs for efficient model training. Effective feature representation is hence crucial to the success of these methods. As an example, in order to detect aircraft from remote sensing images, Zhang et al. [67] proposed to use the UC Merced land use dataset [82] as a background class to pretrain Faster RCNN [83]. By doing this, the model gained an understanding of remote sensing scenes which facilitated robust object detection. The underlying assumption in such an approach is that even though the foreground objects may not be the same, the background information remains largely unchanged across the sources (e.g., datasets) and can, hence, be transferred to a new domain.

Another important application of remote sensed images is pansharpening, where a panchromatic image (which has a coarse/broad spectral resolution, but very high spatial resolution) is used to improve the spatial resolution of multi/hyperspectral image. However, a high-resolution panchromatic image is not always available for the same area that is covered by the hyperspectral images. To solve this problem,

Yuan et al. [71] pretrained a super-resolution network with natural images and applied the model to the target hyperspectral image band by band. The underlying assumption in this work is that the spatial features in both the high- and low-resolution images are the same in both domains irrespective of the spectral content.

Traditional visual tasks like classification, object detection, and segmentation interpret an image at either pixel or object level. Image captioning takes this notion a step further and aims to summarize a scene in a language that can be interpreted easily. Although many image captioning methods have been proposed for natural images, this topic has not been rigorously developed in the remote sensing domain. Shi et al. [72] proposed satellite image captioning by using a pretrained Fully Convolutional Network (FCN) [84]. The base network was pretrained for image classification on ImageNet. To understand the images, three losses were defined at the object, environment, and landscape scale, respectively. Predicted labels at different levels were then sent to a language generation model for captioning. In this work, the task in target domain is very different from the one in the source domain. Despite that, pretrained model still provides features that are generic enough to help understanding the target domain images.

5.4 Data Augmentation

Flipping and rotating images usually do not affect the class labels of objects within the image. A machine learning model can benefit if the training library is augmented with samples with these simple manipulations. By changing the input training images in a way that does not affect the class, it allows algorithms to train from more examples of the object, and the models hence generalize better to test data. Data generation and augmentation share the same philosophy—to generate synthetic or transformed data that is representative of real-world data and can be used to boost the training.

Data augmentation such as flipping, rotation, cropping, and color jittering have been shown to be very helpful for training deep neural networks [1, 85, 86]. These operations, in fact, have become common practice when training models for natural image analysis tasks. Despite the differences between hyperspectral and natural images, standard augmentation methods like rotation, translation, and flipping have been proven to be useful in boosting the classification accuracy of hyperspectral image analysis tasks [87] and [88]. To simulate the variance in the at-sensor radiance and mixed pixels during the imaging process, Chen et al. [16] created *virtual samples* by multiplying random factors to existing samples and linearly combining samples with random weights, respectively. Li et al. [89] showed the performance can be further improved by integrating spatial similarity through pixel block pairs, in which a 3×3 window around the labeled pixel was used as a block and different blocks were paired together based on their labels to augment the training set. A similar spatial constraint was also used by Feng et al. [90], where unlabeled pixels were assigned labels for training if their k-nearest neighbors (in both spatial and spectral domains) belong to the same class. Haut et al. [91] used a random occlusion idea to augment data in the spatial domain. It randomly erases regions from the hyperspectral images during the training. As a consequence, the variance in the spatial domain increased and led to a model that generalized better.

Some flavors of data fusion algorithms can be thought of as playing the role of data augmentation, wherein supplemental data sources are helping the training of the models. For instance, a building roof and a paved road both can be made from similar materials—in such a case, it may be difficult for a model to tell differential these classes from the reflectance spectra alone. However, this distinction can be easily made by comparing their topographic information (e.g., using LiDAR data). A straightforward approach to fuse hyperspectral and LiDAR data would be training separate networks—one for each source/sensor and combining their features either through concatenation [92, 93] or some other schemes such as a composite kernel [94]. Zhao et al. [95] presented data fusion of multispectral and panchromatic images. Instead of applying CNN to the entire image, features were extracted for superpixels that were generated from the multispectral image. Particularly, a fixed size window around each superpixel was split into multiple regions and the image patch in each region was feed into a CNN for extracting local features. These local features were sent to an autoencoder for fusion and a softmax layer was added at the end for prediction. Due to its relatively high spatial resolution, the panchromatic image can produce spatial segments at a finer scale than the multispectral image. This was leveraged to refine the predictions by further segmenting each superpixel based on panchromatic image.

Aside from augmenting the input data, generating synthetic data that resembles real-life data is another approach to increase training samples. Generative adversarial network (GAN) [96] introduced a trainable approach to generate new synthetic samples. GAN consists of two sub-networks, a generator and a discriminator. During the training, two components play a game with each other. The generator is trying to fool the discriminator by producing samples that are as realistic as possible, and the discriminator is trying to discern whether a sample is synthetically generated or belongs to the training data. After the training process converges, the generator will be able to produce samples that look similar to the training data. Since it does not require any labeled data, there has been an increasing interest in using GAN for data augmentation in many applications. This has been applied to hyperspectral image analysis in recent years [49, 73, 97, 98]. Both [49] and [98] used GAN for hyperspectral image classification, where a softmax layer was attached to the discriminator. Fake data were treated as an additional class in the training set. Since a large amount of unlabeled was used for training the GAN, the discriminator became good at classifying all samples. A transfer learning idea was proposed for super-resolution in [73], where a GAN is pretrained on a relatively large dataset and fine-tuned on the UC Merced land use dataset [82].

5.5 Future Directions

In this chapter, we reviewed recent advances in deep learning for hyperspectral image analysis. Although a lot of progress has been made in recent years, there is still a lot of open problems and related research opportunities. In addition to making advances in algorithms and network architectures (e.g., networks for multi-scale, multi-sensor data analysis, data fusion, image super-resolution, etc.), there is a need for addressing fundamental issues that arise from insufficient data and the fundamental nature of the data being acquired. Toward this end, the following directions are suggested:

- Hyperspectral ImageNet: We have witnessed the immense success brought about in part by the ImageNet dataset for traditional image analysis. The benefit of building a similar dataset for hyperspectral image is compelling. If such libraries can be created for various image analysis tasks (e.g., urban land cover classification, ecosystem monitoring, material characterization, etc.), they will enable learning truly deep networks that learn highly discriminative spatial–spectral features.
- Interdisciplinary collaboration: Developing an effective model for analyzing hyperspectral data requires a deep understanding of both the properties of the data itself and machine learning techniques. With this in mind, networks that reflect the optical characteristics of the sensing modalities (e.g., inter-channel correlations) and variability caused in acquisition (e.g., varying atmospheric conditions) should add more information for the underlying analysis tasks compared to "black-box" networks.

References

1. Krizhevsky A, Sutskever I, Hinton GE (2012) Imagenet classification with deep convolutional neural networks. In: Advances in neural information processing systems. pp 1097–1105
2. Rumelhart DE, Hinton GE, Williams RJ et al (1988) Learning representations by back-propagating errors. Cogn Model 5(3):1
3. Krogh A, Hertz JA (1992) A simple weight decay can improve generalization. In: Advances in neural information processing systems, pp 950–957
4. Hinton GE, Osindero S, Teh Y-W (2006) A fast learning algorithm for deep belief nets. Neural Comput 18(7):1527–1554
5. Srivastava N, Hinton G, Krizhevsky A, Sutskever I, Salakhutdinov R (2014) Dropout: a simple way to prevent neural networks from overfitting. J Mach Learn Res 15(1):1929–1958
6. Ioffe S, Szegedy C, Batch normalization: accelerating deep network training by reducing internal covariate shift. arXiv:1502.03167
7. Caruana R, Lawrence S, Giles CL (2001) Overfitting in neural nets: backpropagation, conjugate gradient, and early stopping. In: Advances in neural information processing systems, pp 402–408
8. Deng J, Dong W, Socher R, Li L-J, Li K, Fei-Fei L (2009) Imagenet: a large-scale hierarchical image database. In: IEEE conference on computer vision and pattern recognition. IEEE, pp 248–255
9. Camps-Valls G, Tuia D, Bruzzone L, Benediktsson JA (2014) Advances in hyperspectral image classification: earth monitoring with statistical learning methods. IEEE Signal Process Mag 31(1):45–54

10. Bioucas-Dias JM, Plaza A, Dobigeon N, Parente M, Du Q, Gader P, Chanussot J (2012) Hyperspectral unmixing overview: geometrical, statistical, and sparse regression-based approaches. IEEE J Select Topi Appl Earth Observ Remote Sens 5(2):354–379
11. LeCun Y, Kavukcuoglu K, Farabet C (2010) Convolutional networks and applications in vision. In: Proceedings of 2010 IEEE international symposium on circuits and systems. IEEE, pp 253–256
12. Boureau Y-L, Bach F, LeCun Y, Ponce J (2010) Learning mid-level features for recognition. In: IEEE computer society conference on computer vision and pattern recognition. Citeseer, pp 2559–2566
13. Yosinski J, Clune J, Fuchs T, Lipson H (2015) Understanding neural networks through deep visualization. In: In ICML workshop on deep learning. Citeseer
14. Shen L, Jia S (2011) Three-dimensional gabor wavelets for pixel-based hyperspectral imagery classification. IEEE Trans Geosci Remote Sens 49(12):5039–5046
15. Zhou X, Prasad S, Crawford MM (2016) Wavelet-domain multiview active learning for spatial-spectral hyperspectral image classification. IEEE J Select Top Appl Earth Observ Remote Sens 9(9):4047–4059
16. Chen Y, Jiang H, Li C, Jia X, Ghamisi P (2016) Deep feature extraction and classification of hyperspectral images based on convolutional neural networks. IEEE Trans Geosci Remote Sens 54(10):6232–6251
17. Li Y, Zhang H, Shen Q (2017) Spectral-spatial classification of hyperspectral imagery with 3D convolutional neural network. Remote Sens 9(1):67
18. Paoletti M, Haut J, Plaza J, Plaza A (2018) A new deep convolutional neural network for fast hyperspectral image classification. ISPRS J Photogrammetry Remote Sens 145:120–147
19. Zhong Z, Li J, Luo Z, Chapman M (2018) Spectral-spatial residual network for hyperspectral image classification: A 3-d deep learning framework. IEEE Trans Geosci Remote Sens 56(2):847–858
20. Pavia university hyperspectral data. http://www.ehu.eus/ccwintco/index.php?title=Hyperspectral_Remote_Sensing_Scenes
21. University of Houston hyperspectral data. http://hyperspectral.ee.uh.edu/?page_id=459
22. Zhou X, Prasad S (2017) Domain adaptation for robust classification of disparate hyperspectral images. IEEE Trans Comput Imaging 3(4):822–836
23. Rumelhart DE, Hinton GE, Williams RJ (1985) Learning internal representations by error propagation. Tech. rep., California Univ San Diego La Jolla Inst for Cognitive Science
24. Smolensky P (1986) Information processing in dynamical systems: foundations of harmony theory. Tech. rep, Colorado Univ at Boulder Dept of Computer Science
25. Hinton GE (2002) Training products of experts by minimizing contrastive divergence. Neural Comput 14(8):1771–1800
26. Hinton GE, Salakhutdinov RR (2006) Reducing the dimensionality of data with neural networks. Science 313(5786):504–507
27. Chen Y, Lin Z, Zhao X, Wang G, Gu Y (2014) Deep learning-based classification of hyperspectral data. IEEE J Select Top Appl Earth Observ Remote Sens 7(6):2094–2107
28. Tao C, Pan H, Li Y, Zou Z (2015) Unsupervised spectral-spatial feature learning with stacked sparse autoencoder for hyperspectral imagery classification. IEEE Geosci Remote Sens Lett 12(12):2438–2442
29. Zhao C, Wan X, Zhao G, Cui B, Liu W, Qi B (2017) Spectral-spatial classification of hyperspectral imagery based on stacked sparse autoencoder and random forest. Eur J Remote Sens 50(1):47–63
30. Ma X, Wang H, Geng J (2016) Spectral-spatial classification of hyperspectral image based on deep auto-encoder. IEEE J Select Top Appl Earth Observ Remote Sens 9(9):4073–4085
31. Xing C, Ma L, Yang X (2016) Stacked denoise autoencoder based feature extraction and classification for hyperspectral images. J Sens
32. Kemker R, Kanan C (2017) Self-taught feature learning for hyperspectral image classification. IEEE Trans Geosci Remote Sens 55(5):2693–2705

33. Ji J, Mei S, Hou J, Li X, Du Q (2017) Learning sensor-specific features for hyperspectral images via 3-dimensional convolutional autoencoder. In IEEE international geoscience and remote sensing symposium (IGARSS). IEEE, pp 1820–1823

34. Han X, Zhong Y, Zhang L (2017) Spatial-spectral unsupervised convolutional sparse auto-encoder classifier for hyperspectral imagery. Photogram Eng Remote Sens 83(3):195–206

35. Le Roux N, Bengio Y (2008) Representational power of restricted boltzmann machines and deep belief networks. Neural Comput 20(6):1631–1649

36. Li T, Zhang J, Zhang Y (2014) Classification of hyperspectral image based on deep belief networks. In: IEEE international conference on image processing (ICIP). IEEE, pp 5132–5136

37. Midhun M, Nair SR, Prabhakar V, Kumar SS (2014) Deep model for classification of hyper-spectral image using restricted boltzmann machine. In: Proceedings of the 2014 international conference on interdisciplinary advances in applied computing. ACM, p. 35

38. Chen Y, Zhao X, Jia X (2015) Spectral-spatial classification of hyperspectral data based on deep belief network. IEEE J Select Top Appl Earth Observ Remote Sens 8(6):2381–2392

39. Tao Y, Xu M, Zhang F, Du B, Zhang L (2017) Unsupervised-restricted deconvolutional neural network for very high resolution remote-sensing image classification. IEEE Trans Geosci Remote Sens 55(12):6805–6823

40. Zhou X, Li S, Tang F, Qin K, Hu S, Liu S (2017) Deep learning with grouped features for spatial spectral classification of hyperspectral images. IEEE Geosci Remote Sens Lett 14(1):97–101

41. Li C, Wang Y, Zhang X, Gao H, Yang Y, Wang J (2019) Deep belief network for spectral-spatial classification of hyperspectral remote sensor data. Sensors 19(1):204

42. Tan K, Wu F, Du Q, Du P, Chen Y (2019) A parallel gaussian-bernoulli restricted boltzmann machine for mining area classification with hyperspectral imagery. IEEE J Select Top Appl Earth Observ Remote Sens 12(2):627–636

43. Romero A, Gatta C, Camps-Valls G (2016) Unsupervised deep feature extraction for remote sensing image classification. IEEE Trans Geosci Remote Sens 54(3):1349–1362

44. Romero A, Radeva P, Gatta C (2015) Meta-parameter free unsupervised sparse feature learning. IEEE Trans Pattern Anal Mach Intell 37(8):1716–1722

45. Haut JM, Fernandez-Beltran R, Paoletti ME, Plaza J, Plaza A, Pla F (2018) A new deep generative network for unsupervised remote sensing single-image super-resolution. IEEE Trans Geosci Remote Sens 99: 1–19

46. Newell A, Yang K, Deng J (2016) Stacked hourglass networks for human pose estimation. In: European conference on computer vision. Springer, pp 483–499

47. Chapelle O, Scholkopf B, Zien A (2009) Semi-supervised learning (chapelle, O. etal., eds.; 2006)[book reviews]. IEEE Trans Neural Netw 20(3):542–542

48. Liu B, Yu X, Zhang P, Tan X, Yu A, Xue Z (2017) A semi-supervised convolutional neural network for hyperspectral image classification. Remote Sens Lett 8(9):839–848

49. He Z, Liu H, Wang Y, Hu J (2017) Generative adversarial networks-based semi-supervised learning for hyperspectral image classification. Remote Sens 9(10):1042

50. Kemker R, Luu R, Kanan C (2018) Low-shot learning for the semantic segmentation of remote sensing imagery. IEEE Trans Geosci Remote Sens 99:1–10

51. Niu C, Zhang J, Wang Q, Liang J (2018) Weakly supervised semantic segmentation for joint key local structure localization and classification of aurora image. IEEE Trans Geosci Remote Sens 99:1–14

52. Wu H, Prasad S (2018) Semi-supervised deep learning using pseudo labels for hyperspectral image classification. IEEE Trans Image Process 27(3):1259–1270

53. Kang X, Zhuo B, Duan P (2019) Semi-supervised deep learning for hyperspectral image clas-sification. Remote Sens Lett 10(4):353–362

54. Sun Y, Li J, Wang W, Plaza A, Chen Z, Active learning based autoencoder for hyperspec-tral imagery classification. In: IEEE international geoscience and remote sensing symposium (IGARSS). IEEE, pp. 469–472

55. Liu P, Zhang H, Eom KB (2017) Active deep learning for classification of hyperspectral images. IEEE J Select Top Appl Earth Observ Remote Sens 10(2):712–724

56. Deng C, Xue Y, Liu X, Li C, Tao D, Active transfer learning network: A unified deep joint spectral-spatial feature learning model for hyperspectral image classification. IEEE Trans Geosci Remote Sens

57. Lin J, Zhao L, Li S, Ward R, Wang ZJ (2018) Active-learning-incorporated deep transfer learning for hyperspectral image classification. IEEE J Select Top Appl Earth Observ Remote Sens 11(11):4048–4062

58. Haut JM, Paoletti ME, Plaza J, Li J, Plaza A (2018) Active learning with convolutional neural networks for hyperspectral image classification using a new Bayesian approach. IEEE Trans Geosci Remote Sens 99:1–22

59. Gal Y, Ghahramani Z, Bayesian convolutional neural networks with bernoulli approximate variational inference. arXiv:1506.02158

60. Erhan D, Bengio Y, Courville A, Manzagol P-A, Vincent P, Bengio S (2010) Why does unsupervised pre-training help deep learning? J Mach Learn Res 11:625–660

61. Sermanet P, Eigen D, Zhang X, Mathieu M, Fergus R, LeCun Y, Overfeat: integrated recognition, localization and detection using convolutional networks. arXiv:1312.6229

62. Donahue J, Jia Y, Vinyals O, Hoffman J, Zhang N, Tzeng E, Darrell T (2014) Decaf: a deep convolutional activation feature for generic visual recognition. In: International conference on machine learning, 2014, pp 647–655

63. Yosinski J, Clune J, Bengio Y, Lipson H (2014) How transferable are features in deep neural networks? In: Advances in neural information processing systems, pp 3320–3328

64. Penatti OA, Nogueira K, Dos JA (2015) Santos, do deep features generalize from everyday objects to remote sensing and aerial scenes domains? In: Proceedings of the IEEE conference on computer vision and pattern recognition workshops, pp 44–51

65. Windrim L, Melkumyan A, Murphy RJ, Chlingaryan A, Ramakrishnan R (2018) Pretraining for hyperspectral convolutional neural network classification. IEEE Trans Geosci Remote Sens 56(5):2798–2810

66. Marmanis D, Datcu M, Esch T, Stilla U (2016) Deep learning earth observation classification using imagenet pretrained networks. IEEE Geosci Remote Sens Lett 13(1):105–109

67. Zhang F, Du B, Zhang L, Xu M (2016) Weakly supervised learning based on coupled convolutional neural networks for aircraft detection. IEEE Trans Geosci Remote Sens 54(9):5553–5563

68. Yang J, Zhao Y-Q, Chan JC-W (2017) Learning and transferring deep joint spectral-spatial features for hyperspectral classification. IEEE Trans Geosci Remote Sens 55(8):4729–4742

69. Mei S, Ji J, Hou J, Li X, Du Q (2017) Learning sensor-specific spatial-spectral features of hyperspectral images via convolutional neural networks. IEEE Trans Geosci Remote Sens 55(8):4520–4533

70. Othman E, Bazi Y, Melgani F, Alhichri H, Alajlan N, Zuair M (2017) Domain adaptation network for cross-scene classification. IEEE Trans Geosci Remote Sens 55(8):4441–4456

71. Yuan Y, Zheng X, Lu X (2017) Hyperspectral image superresolution by transfer learning. IEEE J Select Top Appl Earth Observ Remote Sens 10(5):1963–1974

72. Shi Z, Zou Z (2017) Can a machine generate humanlike language descriptions for a remote sensing image? IEEE Trans Geosci Remote Sens 55(6):3623–3634

73. Ma W, Pan Z, Guo J, Lei B (2018) Super-resolution of remote sensing images based on transferred generative adversarial network. In: IGARSS 2018–2018 IEEE international geoscience and remote sensing symposium. IEEE, pp 1148–1151

74. Liu X, Chi M, Zhang Y, Qin Y (2018) Classifying high resolution remote sensing images by fine-tuned vgg deep networks. In: IGARSS 2018–2018 IEEE international geoscience and remote sensing symposium. IEEE, pp 7137–7140

75. Sumbul G, Cinbis RG, Aksoy S (2018) Fine-grained object recognition and zero-shot learning in remote sensing imagery. IEEE Trans Geosci Remote Sens 56(2):770–779

76. Zhou X, Prasad S (2018) Deep feature alignment neural networks for domain adaptation of hyperspectral data. IEEE Trans Geosci Remote Sens 99:1–10

77. Xie M, Jean N, Burke M, Lobell D, Ermon S (2016) Transfer learning from deep features for remote sensing and poverty mapping. In: Thirtieth AAAI conference on artificial intelligence

78. Fortet R, Mourier E (1953) Convergence de la r?eparation empirique vers la réparation théorique. In: Ann Scient École Norm Sup 70:266–285
79. Hinton GE, Roweis ST (2003) Stochastic neighbor embedding. In: Advances in neural information processing systems, pp 857–864
80. Xu X, Zhou X, Venkatesan R, Swaminathan G, Majumder O (2019) d-sne: Domain adaptation using stochastic neighborhood embedding. In: Proceedings of the IEEE conference on computer vision and pattern recognition, pp 2497–2506
81. Ben-David S, Blitzer J, Crammer K, Pereira F (2007) Analysis of representations for domain adaptation. In: Advances in neural information processing systems, pp 137–144
82. Yang Y, Newsam S (2010) Bag-of-visual-words and spatial extensions for land-use classification. In: Proceedings of the 18th SIGSPATIAL international conference on advances in geographic information systems. ACM, pp 270–279
83. Ren S, He K, Girshick R, Sun J (2015) Faster r-cnn: towards real-time object detection with region proposal networks. In: Advances in neural information processing systems, pp 91–99
84. Long J, Shelhamer E, Darrell T (2015) Fully convolutional networks for semantic segmentation. In: Proceedings of the IEEE conference on computer vision and pattern recognition, pp 3431–3440
85. Liu W, Anguelov D, Erhan D, Szegedy C, Reed S, Fu C-Y, Berg AC (2016) Ssd: Single shot multibox detector. In: European conference on computer vision. Springer, pp 21–37
86. Chen L-C, Papandreou G, Kokkinos I, Murphy K, Yuille AL (2017) Deeplab: semantic image segmentation with deep convolutional nets, atrous convolution, and fully connected crfs. IEEE Trans Pattern Anal Mach Intell 40(4):834–848
87. Lee H, Kwon H (2017) Going deeper with contextual cnn for hyperspectral image classification. IEEE Trans Image Process 26(10):4843–4855
88. Yu X, Wu X, Luo C, Ren P (2017) Deep learning in remote sensing scene classification: a data augmentation enhanced convolutional neural network framework. GISci Remote Sens 54(5):741–758
89. Li W, Chen C, Zhang M, Li H, Du Q (2019) Data augmentation for hyperspectral image classification with deep cnn. IEEE Geosci Remote Sens Lett 16(4):593–597
90. Feng J, Chen J, Liu L, Cao X, Zhang X, Jiao L, Yu T, Cnn-based multilayer spatial–spectral feature fusion and sample augmentation with local and nonlocal constraints for hyperspectral image classification. IEEE J Select Top Appl Earth Observ Remote Sens
91. Haut JM, Paoletti ME, Plaza J, Plaza A, Li J, Hyperspectral image classification using random occlusion data augmentation. IEEE Geosci Remote Sens Lett
92. Xu X, Li W, Ran Q, Du Q, Gao L, Zhang B (2017) Multisource remote sensing data classification based on convolutional neural network. IEEE Trans Geosci Remote Sens 56(2):937–949
93. Li H, Ghamisi P, Soergel U, Zhu X (2018) Hyperspectral and lidar fusion using deep three-stream convolutional neural networks. Remote Sens 10(10):1649
94. Feng Q, Zhu D, Yang J, Li B (2019) Multisource hyperspectral and lidar data fusion for urban land-use mapping based on a modified two-branch convolutional neural network. ISPRS Int J Geo-Inform 8(1):28
95. Zhao W, Jiao L, Ma W, Zhao J, Zhao J, Liu H, Cao X, Yang S (2017) Superpixel-based multiple local cnn for panchromatic and multispectral image classification. IEEE Trans Geosci Remote Sens 55(7):4141–4156
96. Goodfellow I, Pouget-Abadie J, Mirza M, Xu B, Warde-Farley D, Ozair S, Courville A, Bengio Y (2014) Generative adversarial nets. In: Advances in neural information processing systems, pp 2672–2680
97. Zhan Y, Hu D, Wang Y, Yu X (2018) Semisupervised hyperspectral image classification based on generative adversarial networks. IEEE Geosci Remote Sens Lett 15(2):212–216
98. Zhu L, Chen Y, Ghamisi P, Benediktsson JA (2018) Generative adversarial networks for hyperspectral image classification. IEEE Trans Geosci Remote Sens 56(9):5046–5063

Chapter 6
Addressing the Inevitable Imprecision: Multiple Instance Learning for Hyperspectral Image Analysis

Changzhe Jiao, Xiaoxiao Du and Alina Zare

Abstract In many remote sensing and hyperspectral image analysis applications, precise ground truth information is unavailable or impossible to obtain. Imprecision in ground truth often results from highly mixed or sub-pixel spectral responses over classes of interest, a mismatch between the precision of global positioning system (GPS) units and the spatial resolution of collected imagery, and misalignment between multiple sources of data. Given these sorts of imprecision, training of traditional supervised machine learning models which rely on the assumption of accurate and precise ground truth becomes intractable. Multiple instance learning (MIL) is a methodology that can be used to address these challenging problems. This chapter investigates the topic of hyperspectral image analysis given imprecisely labeled data and reviews MIL methods for hyperspectral target detection, classification, data fusion, and regression.

6.1 Motivating Examples for Multiple Instance Learning in Hyperspectral Analysis

In standard supervised machine learning, each training sample is assumed to be coupled with the desired classification label. However, acquiring accurately labeled training data can be time consuming, expensive, or at times infeasible. Challenges with obtaining precise training labels and location information are pervasive throughout many remote sensing and hyperspectral image analysis tasks. A learning methodol-

C. Jiao
Xidian University, Xi'an, China
e-mail: cjiao@xidian.edu.cn

X. Du
University of Michigan, Ann Arbor, USA
e-mail: xiaodu@umich.edu

A. Zare (✉)
University of Florida, Gainesville, USA
e-mail: azare@ufl.edu

© Springer Nature Switzerland AG 2020
S. Prasad and J. Chanussot (eds.), *Hyperspectral Image Analysis*,
Advances in Computer Vision and Pattern Recognition,
https://doi.org/10.1007/978-3-030-38617-7_6

Multiple Instance Learning:

Negative Bags
Label = 0

Positive Bags
Label = +1

Fig. 6.1 In multiple instance learning, data is labeled at the bag level. A bag is labeled as a *positive* bag if it contains at least one target instance. The number of target versus nontarget instances in each positive bag is unknown. A bag is labeled as a *negative* bag if it contains only nontarget instances. In this figure, blue points correspond to nontarget instances where red points correspond to target instances. Source: © [2019] IEEE. Reprinted, with permission, from [62]

ogy to address imprecisely labeled training data is multiple instance learning (MIL). In MIL, data is labeled at the *bag* level where a bag is a multi-set of data points as illustrated in Fig. 6.1. In standard MIL, bags are labeled as "positive" if they contain any instances representing a target class whereas bags are labeled as "negative" if they contain only nontarget instances. Generating labels for a bag of points is often much less time consuming and aligns with the realistic scenarios encountered in remote sensing applications as outlined in the following motivating examples.

- *Hyperspectral Classification*: Training a supervised classifier requires accurately labeled spectra for the classes of interest. In practice, this is often accomplished by creating a ground truth map of a hyperspectral scene (scenes which frequently contain hundreds of thousands of pixels or more). Generation of ground truth maps is challenging due to labeling ambiguity that naturally arises due to relatively coarse resolution and compound diversity of the remotely sensed hyperspectral scene. For example, an area that is labeled as vegetation may contain both plants and bare soil, making the training label inherently ambiguous. Furthermore, labeling each pixel of the hyperspectral scene is tedious and annotator performance is generally inconsistent from person to person or over time. Due to these challenges, "ground-rumor" may be a more appropriate term than "ground-truth" for the maps that are generated. These ambiguities naturally map to the MIL framework by allowing an annotator to label spatial regions if it contains a class of interest (corresponding to positive bags) and negative bags for spatial regions known to exclude those classes. For instance, an annotator can easily mark (e.g., circle on a

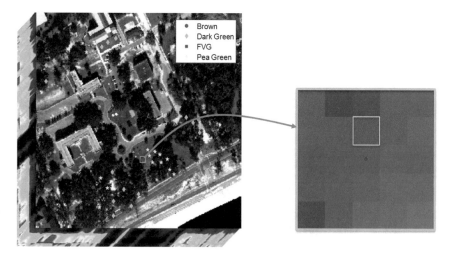

Fig. 6.2 Illustration of inaccurate coordinates from GPS: one target denoted as brown by GPS has one pixel drift. Source: © [2018] Elsevier Reprinted, with permission, from [45]

map) positive bag regions that contain vegetation and then mark regions of only bare soil and building/man-made materials for negative bags when vegetation is the class of interest.

- *Sub-pixel Target Detection*: Consider the hyperspectral target detection problem illustrated in Fig. 6.2. This hyperspectral scene was collected over the University of Southern Mississippi-Gulfpark Campus [1] and includes many emplaced targets. These targets are cloth panels of four colors (Brown, Dark Green, Faux Vineyard Green, and Pea Green) varying from 0.5 m × 0.5 m, 1 m × 1 m, and 3 m × 3 m in size. The ground sample distance of this hyperspectral data set is 1m. Thus, the 0.5 m × 0.5 m targets are, at best, a quarter of a pixel in size; the 1 m × 1 m targets are, at best, exactly one pixel in size; and the 3 m × 3 m targets cover multiple pixels. However, the targets are rarely aligned with the pixel grid, resulting in the 0.5 m × 0.5 m and 1 m × 1 m target responses often straddling multiple pixels and being sub-pixel. The scene also had heavy tree coverage and resulted in targets being heavily occluded by the tree canopy. The sub-pixel nature of the targets and occlusion by the tree canopy causes this to be a challenging target detection problem and one in which manual labeling of target location by visual inspection is impractical. Ground truth locations of the targets in this scene were collected by a GPS unit with 2–5 m accuracy. Thus, the ground truth is only accurate up to some spatial region (as opposed to the pixel level). For example, the region highlighted in Fig. 6.2 contains one brown target. From this highlighted region, one can clearly see that the GPS coordinate of this brown target (denoted by the red dot) is shifted one pixel from the actual brown target location (denoted by the yellow rectangle). This is a rare example where we can visually see the brown target. Most of the targets are difficult to distinguish visibly. Developing a classifier or extracting a

Fig. 6.3 An example of 3D scatterplot of LiDAR data over the University of Southern Mississippi-Gulfpark campus. The LiDAR points were colored by the RGB imagery provided by HSI sensors over the scene. Source: © [2020] IEEE. Reprinted, with permission, from [86]

pure prototype for the target class given incomplete knowledge of the training data is intractable using standard supervised learning methods. This also directly maps to the MIL framework since each positive bag can correspond to the spatial region associated with each ground truth point and its corresponding range of imprecision and negative bags can correspond to spatial regions that do not overlap with any ground truth point or its associated halo of uncertainty.

- *Multi-sensor Fusion*: When fusing information obtained by multiple sensors, each sensor may provide complementary information that can aid scene understanding and analysis. Figure 6.3 shows a three-dimensional scatter plot of the LiDAR (Light Detection And Ranging) point cloud data over the University of Southern Mississippi-Gulfpark Campus collected simultaneously with the hyperspectral imagery (HSI) described above. In this data set, the hyperspectral and LiDAR data can be leveraged jointly for scene segmentation, ground cover classification, and target detection. However, there are challenges that arise during fusion. The HSI and LiDAR data are of drastically different modalities and resolutions. HSI is collected natively on a pixel grid with a 1 m ground sample distance whereas the raw LiDAR data is a point cloud with a higher resolution of 0.60 m cross track and 0.78 m along track spot spacing. Commonly, before fusion, data is co-registered onto a shared pixel grid. However, image co-registration and rasterization may introduce inaccuracies [2, 3]. In this example, consider the edges of the buildings with gray roofs in Fig. 6.3. Some of the hyperspectral pixels of the buildings have been inaccurately mapped to LiDAR points corresponding to the neighboring grass pixels on the ground. Similarly, some hyperspectral points corresponding to sidewalk and dirt roads have been inaccurately mapped to high elevation values similar to nearby trees and buildings. Directly using such inaccurate measurements for fusion can cause further inaccuracy or error in classification, detection, or prediction. Therefore, it is beneficial to develop a fusion algorithm that is able to handle such inaccurate/imprecise measurements. Imprecise co-registration can also be mapped to the MIL framework by considering a *bag* of points from a local region in one sensor (e.g., LiDAR) to be candidates for fusion in each pixel in the other sensors (e.g., hyperspectral).

These examples illustrate that remote-sensing data and applications are often plagued with inherent spatial imprecision in ground truth information. Multiple instance learning is a framework that can alleviate the issues that arise due to this imprecision. Therefore, although imprecise ground truth plagues instance-level labels, bags (i.e., spatial regions) can be labeled readily and analyzed using MIL approaches.

6.2 Introduction to Multiple Instance Classification

MIL was first proposed by Dietterich et al. [4] for the prediction of drug activity. The effectiveness of a drug is determined by how tightly the drug molecule binds to a larger protein molecule. Although a molecule may be determined to be effective, it can have variants called "conformations" of which only one (or a few) actually binds to the desired target binding site. In this task, the learning objective is to infer the correct shape of the molecule that actually has tight binding capacity. In order to solve this problem, Dietterich et al. introduced the definition of "bags." Each molecule was treated as a bag and each possible conformation of the molecule was treated as an instance in that bag. This directly induces the definition of multiple instance learning. A positively labeled bag contains at least one positive instance (but, also, some number of negative instances) and negatively labeled bags are composed of entirely negative instances. The goal is to uncover the true positive instances in each positive bag and what characterizes positive instances.

Although initially proposed for this drug activity application, the multiple instance learning framework is extremely relevant and applicable to a number of remote-sensing problems arising from imprecision in ground truth information. By labeling data and operating at the bag level, ground truth imprecision inherent in remote sensing problems are addressed and accounted for within a multiple instance learning framework.

6.2.1 Multiple Instance Learning Formulation

The multiple instance learning framework can be formally described as follows. Let $\mathbf{X} = [\mathbf{x}_1, \ldots, \mathbf{x}_N] \in \mathbb{R}^{n \times N}$ be training data instances where n is the dimensionality of an instance and N is the total number of training instances. The data are grouped into K bags, $\mathbf{B} = \{\mathbf{B}_1, \ldots, \mathbf{B}_K\}$, with associated binary bag-level labels, $L = \{L_1, \ldots, L_K\}$ where $L_i \in \{0, 1\}$ for two-class classification. A bag, $\mathbf{B}_i$, is termed *positive* with $L_i=1$ if it contains at least one positive instance. The exact number or identification of positive and negative instances in each positive bag is unknown. A bag is termed *negative* with $L_i=0$ when it contains only negative instances. The instance $\mathbf{x}_{ij} \in \mathbf{B}_i$ denotes the jth instance in bag $\mathbf{B}_i$ with the (unknown) instance-level label $l_{ij} \in \{0, 1\}$.

In standard supervised machine learning methods, all instance level labels are known for the training data. However, in multiple instance learning, only the bag-level labels are known. Given this formulation, the fundamental goal of an MIL method is to determine what instance-level characteristics are common across all positive bags and cannot be found in any instance in any negative bag.

6.2.2 Axis-Parallel Rectangles, Diverse Density, and Other General MIL Approaches

Many general MIL approaches have been developed in the literature. Axis-parallel rectangles (APR) [4] algorithms were the first set of MIL algorithms proposed by Dietterich et al. for drug activity prediction in the 1990s. An axis-parallel rectangle can be viewed as a region of true positive instances in the feature space. In APR algorithms, a lower and upper bound encapsulating the positive class is estimated in each feature dimension. Three APR algorithms, greedy feature selection elimination count (GFS elim-count), greedy feature selection kernel density estimation (GFS kde), and iterated discrimination (iterated-discrim) algorithms were investigated and compared in [4]. As an illustration, GFS elim-count APR refers to finding an APR in a greedy manner starting from a region that exactly covers all of the positive instances. Figure 6.4 shows the "all-positive APR" as a solid line bounding box of the instances, where the unfilled markers represent positive instances and filled markers represent negative instances. As shown in the figure, the all-positive APR may contain several negative examples. The algorithm proceeds by greedily eliminating all negative instances within the APR while maintaining as many positive instances as possible. The dashed box in Fig. 6.4 indicates the final APR identified by the GFS elim-count algorithm by iteratively excluding the "cheapest" negative instance, determined by requiring the minimum number of positive instances that need to be removed from the APR to exclude that negative instance.

Diverse density (DD) [5, 6] was one of the first multiple instance learning algorithms that estimated a *positive concept*. The positive concept is a representative of

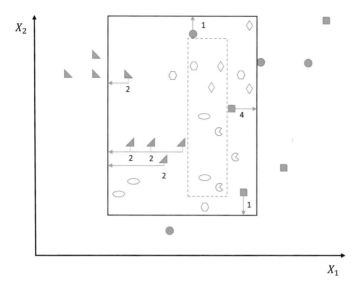

Fig. 6.4 Illustration of the GFS elim-count procedure for excluding negative instances. The "all-positive APR" is indicated by a solid box. The unfilled markers represent positive instances and filled markers represent negative instances. The final APR is indicated by the dashed box [4]

the positive class. This representative is estimated in DD by identifying a representative feature vector that is close to the intersection of all positive bags and far from every negative instance. In other words, the target concept represents an area that preserves both a high density of target points and a low density of nontarget points, called *diverse density*. This is accomplished in DD by maximizing the likelihood function in Eq. (6.1),

$$\arg\max_{d} \prod_{i=1}^{K^+} \Pr(\mathbf{d} = \mathbf{s}|\mathbf{B}_i^+) \prod_{i=K^++1}^{K^++K^-} \Pr(\mathbf{d} = \mathbf{s}|\mathbf{B}_i^-), \qquad (6.1)$$

where $\mathbf{s}$ is the assumed true positive concept, $\mathbf{d}$ is the concept representative to be estimated, K^+ is the number of positive bags and K^- is the number of negative bags. The first term in Eq. (6.1), which is used for all positive bags, is defined by the noisy-or model,

$$\mathbf{Pr}(\mathbf{d} = \mathbf{s}|\mathbf{B}_i^+) = \mathbf{Pr}(\mathbf{d} = \mathbf{s}|\mathbf{x}_{i1}, \mathbf{x}_{i2}, \dots, \mathbf{x}_{iN_i}) = 1 - \prod_{j=1}^{N_i} (1 - \mathbf{Pr}(\mathbf{d} = \mathbf{s}|\mathbf{x}_{ij} \in \mathbf{B}_i^+)),$$

$$(6.2)$$

where $\mathbf{Pr}(\mathbf{d} = \mathbf{s}|\mathbf{x}_{ij}) = exp(-\|\mathbf{x}_{ij} - \mathbf{d}\|^2)$. The term in (6.2) can be interpreted as requiring there be at least one instance in positive bag $\mathbf{B}_i^+$ that is close to the positive representative $\mathbf{d}$. This can be understood by noticing that (6.2) evaluates to 1 if there is at least one instance in the positive bag that is close to the representative (i.e., $exp(-\|\mathbf{x}_{ij} - \mathbf{d}\|^2) \to 1$ which implies $1 - \mathbf{Pr}(\mathbf{d} = s|\mathbf{x}_{ij} \in \mathbf{B}_i^+) \to 0$, resulting in $1 - \prod_{j=1}^{N_i}(1 - \mathbf{Pr}(\mathbf{d} = s|\mathbf{x}_{ij} \in \mathbf{B}_i^+)) \to 1$). In contrast, (6.2) evaluates to 0 if all points in a positive bag are far from the positive concept.

The second term is defined by

$$\mathbf{Pr}(\mathbf{d} = s|\mathbf{B}_i^-) = \prod_{j=1}^{N_i}(1 - \mathbf{Pr}(\mathbf{d} = s|\mathbf{x}_{ij} \in \mathbf{B}_i^-)). \tag{6.3}$$

which encourages positive concepts to be far from all negative points. The noisy-or model, however, is highly non-smooth and there are several local maxima in the solution space. This is alleviated in practice by performing gradient ascent repeatedly with starting points from every positive instance to maximize the proposed log-likelihood function. Alternatively, an expectation maximization version of diversity density (EM-DD) [7] was proposed by Zhang et al. in order to improve the computation time of DD [5, 6]. EM-DD assumes there exists only one instance per bag corresponding to the bag-level label and treats the knowledge of the key-point instance corresponding to the bag-level label as a hidden latent variable. EM-DD starts with an initial estimate of the positive concept $\mathbf{d}$ and iterates between an expectation step (E-step) that selects one point per bag as the representative point of that bag and then performs a quasi-newton optimization (M-step) [8] on the single-instance DD problem. In practice, EM-DD is much more computationally efficient than DD. However, the computational benefits are traded-off with potential inferior performance accuracy to DD [9].

Since the development of the APR and DD, many MIL approaches have been developed and published in the literature. These include prototype-based methods such as the dictionary-based multiple instance learning (DMIL) algorithm [10] and its generalization, generalized dictionaries for multiple instance Learning (GDMIL) [11] which propose to optimize the noisy-or model using dictionary learning approaches by learning a set of discriminative positive dictionary atoms to describe the positive class [12–14]. The Max-Margin Multiple-Instance Dictionary Learning (MMDL) methods [15] adopts the bag of words concept [16] and trains a set of linear SVMs as a codebook. The novel assumption of MMDL is that the positive instances could belong to many different categories. For example, the positive class "computer room" may have image patches containing a desk, a screen, and a keyboard. The MILIS algorithm [17] alternates between the selection of an instance per bag as a prototype that represents its bag and training a linear SVM on these prototypes.

Additional support vector machine-based methods include the MILES (Multiple-Instance Learning via Embedded Instance Selection) approach [18] which embeds each training and testing bag into a high-dimensional space and then performs classification in the mapping space using a one-norm support vector machine (SVM) [19]. Furthermore, the mi-SVM and MI-SVM methods model the MIL problem as a generalized mixed integer formulation of the support vector machine [20]. MissSVM algorithm [21] solves the MIL problem using a semi-supervised SVM with the constraint that at least one point from each positive bag must be classified as positive. Hoffman et al. [22] jointly exploit the image-level and bounding box labels and achieve state-of-the-art results in object detection. Li and Vasconcelos [23] further investigate MIL problem with labeling noise in negative bags and use "top instances" as the representatives of "soft bags", then proceed with bag-level classification via latent-SVM [24].

Meng et al. [25] integrate the self-paced learning (SPL) [26] into MIL and propose SP-MIL for co-saliency detection. The Citation-kNN [27] algorithm adapts the k nearest neighbor (kNN) method [28] to MIL problems by using the Hausdorff distance [29] to compute distance between two bags and assigns bag-level labels based on the nearest neighbor rules. Extensions of Citation-kNN include Bayesian Citation-kNN [30] and Fuzzy-Citation-kNN [31, 32]. Furthermore, a large number of MIL neural network methods such as [33] (often called "weak" learning methods) have also been developed. Among the vast literature of MIL research, very few methods focus on remote sensing and hyperspectral analysis. These methods are reviewed in the following sections.

6.3 Multiple Instance Learning Approaches for Hyperspectral Target Characterization and Sub-pixel Target Detection

Hyperspectral target detection refers to the task of locating all instances of a target given a known spectral signature within a hyperspectral scene [34–36]. Hyperspectral target detection is challenging for a number of reasons: (1) *Class Imbalance*: The number of training instances from the positive target class is small compared to that of the negative training data such that training a standard classifier is difficult; (2) *Sub-pixel Targets*: Due to the relatively low spatial resolution of hyperspectral imagery and the diversity of natural scenes, one single pixel may also contain different ground materials, resulting in sub-pixel targets of interest; and (3) *Imprecise Labels*: As outlined in Sect. 6.1, precise training labels are often difficult to obtain. For these reasons, signature-based hyperspectral target detection [34] is commonly used as opposed to a two-class classifier. However, the performance of a signature-based detector depends on the target signature and obtaining an effective target signature is challenging. In the past, this was commonly accomplished by measuring target signatures for materials of interest in the lab or using point-spectrometers in the field. However, this approach may introduce error due to changing environmental and atmospheric conditions that impact spectral responses.

In this section, algorithms for multiple instance target characterization (i.e., estimation of target concepts) from training data with label ambiguity are presented. The aim is to estimate the target concepts from highly mixed training data that are effective for target detection. Since these algorithms extract target concepts from training data assumed to have the same environmental context, influence from background materials, environmental and atmospheric conditions are addressed during target concept estimation.

6.3.1 Extended Function of Multiple Instances

The extended Function of Multiple Instances (eFUMI) approach [37, 38] is motivated by the linear mixing model in hyperspectral analysis. eFUMI assumes each data point is a convex combination of target and/or nontarget concepts (i.e., endmembers) and performs linear unmixing (i.e., decomposing spectra into endmembers and the proportion of each endmember found in the associated pixel spectra) to estimate positive and negative concepts. The approach also addresses label ambiguity by incorporating a latent variable which indicates whether each instance of a positively labeled bags is a true target.

More formally, the goal of eFUMI is to estimate a target concept, $\mathbf{d}_T$, nontarget concepts, $\mathbf{d}_k$, $\forall k = 1, \ldots M$, the number of needed nontarget concepts, M, and the abundances, $\mathbf{a}_j$, which define the convex combination of the concepts for each data point $\mathbf{x}_j$ from labeled bags of hyperspectral data. If a bag B_i is positive, there is at least one data point in B_i containing target,

$$\text{if } L_i = 1, \exists \mathbf{x}_j \in B_i \text{ s.t. } \mathbf{x}_j = \alpha_{jT}\mathbf{d}_T + \sum_{k=1}^{M} \alpha_{jk}\mathbf{d}_k + \boldsymbol{\varepsilon}_j, \alpha_{jT} > 0. \tag{6.4}$$

However, the exact number of data points in a positive bag with a target contribution (i.e., $\alpha_{jT} > 0$) and target proportions are unknown. Furthermore, if B_i is a negative bag, this indicates that none of the data in this bag contains target,

$$\text{if } L_i = 0, \forall \mathbf{x}_j \in B_i, \mathbf{x}_j = \sum_{k=1}^{M} \alpha_{jk}\mathbf{d}_k + \boldsymbol{\varepsilon}_j. \tag{6.5}$$

Given this framework, the eFUMI objective function is shown in (6.7). The three terms in this objective function were motivated by the sparsity promoting iterated constrained endmember (SPICE) algorithm [39]. The first term computes the squared error between the input data and its estimate found using the current target and nontarget signatures and proportions. The parameter u is a constant controlling the relative importance of various terms. The scaling value w, which aids in the data imbalance issue by weighting the influence of positive and negative data, is shown in (6.6),

$$w_{l(\mathbf{x}_j)} = \begin{cases} 1, & \text{if } l(\mathbf{x}_j) = 0; \\ \frac{\alpha N^-}{N^+}, & \text{if } l(\mathbf{x}_j) = 1. \end{cases} \tag{6.6}$$

where N^+ is the total number of points in positive bags and N^- is the total number of points in negative bags.

The second term of the objective encourages target and nontarget signatures to provide a tight fit around the data by minimizing the squared difference between each signature and the global data mean, μ_0. The third term is a sparsity promoting term used to determine M, the number of nontarget signatures needed to describe the input data where $\gamma_k = \frac{\Gamma}{\sum_{j=1}^{N} a_{jk}^{(t-1)}}$ and Γ is a constant parameter that controls the degree sparsity is promoted. Higher values of Γ generally result in a smaller estimate M value. The $a_{jk}^{(t-1)}$ values are the proportion values estimated in the previous iteration of the algorithm. Thus, as the proportions for a particular endmember decrease, the weight of its associated sparsity promoting term increases.

$$F = \frac{1}{2}(1-u)\sum_{j=1}^{N} w_j \left\| (\mathbf{x}_j - z_j a_{jT} \mathbf{d}_T - \sum_{k=1}^{M} a_{jk}\mathbf{d}_k) \right\|_2^2 + \frac{u}{2}\sum_{k=T,1}^{M} \left\| \mathbf{d}_k - \mu_0 \right\|_2^2 + \sum_{k=1}^{M} \gamma_k \sum_{j=1}^{N} a_{jk}$$
(6.7)

$$E[F] = \sum_{z_j \in \{0,1\}} \left[\frac{1}{2}(1-u)\sum_{j=1}^{N} w_j P(z_j|\mathbf{x}_j, \boldsymbol{\theta}^{(t-1)}) \left\| \mathbf{x}_j - z_j a_{jT} \mathbf{d}_T - \sum_{k=1}^{M} a_{jk}\mathbf{d}_k \right\|_2^2 \right]$$
$$+ \frac{u}{2}\sum_{k=T,1}^{M} \|\mathbf{d}_k - \mu_0\|_2^2 + \sum_{k=1}^{M} \gamma_k \sum_{j=1}^{N} a_{jk}$$
(6.8)

The difference between (6.7) and the SPICE objective is the inclusion of a set of hidden, latent variables, z_j, $j = 1, \ldots, N$, accounting for the unknown instance-level labels $l(\mathbf{x}_j)$. To address the fact that the z_j values are unknown, the expected values of the log likelihood with respect to z_j is taken as shown in (6.8). In (6.8), $\boldsymbol{\theta}^t$ is the set of parameters estimated at iteration t and $P(z_j|\mathbf{x}_j, \boldsymbol{\theta}^{(t-1)})$ is the probability of individual points containing any proportion of target or not. The value of the term $P(z_j|\mathbf{x}_j, \boldsymbol{\theta}^{(t-1)})$ is determined given the parameter set estimated in the previous iteration and the constraints of the bag-level labels, L_i, as shown in (6.9),

$$P(z_j|\mathbf{x}_j, \boldsymbol{\theta}^{(t-1)}) = \begin{cases} e^{-\beta r_j}, & \text{if } z_j = 0, L_i = 1; \\ 1 - e^{-\beta r_j}, & \text{if } z_j = 1, L_i = 1; \\ 0, & \text{if } z_j = 1, L_i = 0; \\ 1, & \text{if } z_j = 0, L_i = 0; \end{cases}$$
(6.9)

where β is a scaling parameter and $r_j = \left\| \mathbf{x}_j - \sum_{k=1}^{M} a_{jk}\mathbf{d}_k \right\|_2^2$ is the approximation residual between $\mathbf{x}_j$ and its representation using only background endmembers. The definition of $P(z_j|\mathbf{x}_j, \boldsymbol{\theta}^{(t-1)})$ in (6.9) indicates that if a point $\mathbf{x}_j$ is a nontarget point, it should be fully represented by the background endmembers with very small residual r_j; thus, $P(z_j = 0|\mathbf{x}_j, \boldsymbol{\theta}^{(t-1)}) = e^{-\beta r_j} \rightarrow 1$. Otherwise, if $\mathbf{x}_j$ is a target point, it may not be well represented by only the background endmembers, so the residual r_j must be large and $P(z_j = 1|\mathbf{x}_j, \boldsymbol{\theta}^{(t-1)}) = 1 - e^{-\beta r_j} \to 1$. Note, z_j is unknown only for the positive bags; in the negative bags, z_j is fixed to 0. This constitutes the *E-step* of the EM algorithm.

The *M-step* is performed by optimizing (6.8) for each of the desired parameters. The method is summarized in Algorithm 6.1.[1] Please refer to [37] for detailed discussion of the optimization approach and derivation.

Algorithm 6.1 *e*FUMI EM algorithm

1: Initialize $\boldsymbol{\theta}^0 = \{\mathbf{d}_T, \mathbf{D}, \mathbf{A}\}, t = 1$
2: **repeat**
3:        *E-step*: Compute $P(z_j|\mathbf{x}_j, \boldsymbol{\theta}^{(t-1)})$ given $\boldsymbol{\theta}^{t-1}$
4:        *M-step*:
5:        Update $\mathbf{d}_T$ and $\mathbf{D}$ by maximizing (6.8) wrt. $\mathbf{d}_T, \mathbf{D}$
6:        Update $\mathbf{A}$ by maximizing (6.8) wrt. $\mathbf{A}$ s.t. the sum-to-one and non-negative constraints
7:        Prune each $\mathbf{d}_k, k = 1, \ldots, M$ if $\max_j(a_{jk}) \leq \tau$ where τ is a fixed threshold (e.g. $\tau = 10^{-6}$)
8:        $t \leftarrow t + 1$
9: **until** Convergence
10: **return** $\mathbf{d}_T, \mathbf{D}, \mathbf{A}$

6.3.2 Multiple Instance Spectral Matched Filter and Multiple Instance Adaptive Coherence/Cosine Detector

The *e*FUMI algorithm described above can be viewed as a semi-supervised hyperspectral unmixing algorithm, where the endmembers of the target and nontarget materials are estimated. Since *e*FUMI minimizes the reconstruction error of the data, it is a *representative* algorithm that learns target concepts that are representatives for (and have similar shape to) the target class. Significant challenges in applying the *e*FUMI algorithm in practice are the large number of parameters that need to be set and the fact that all positive bags are combined in the algorithm, neglecting the MIL concept that each positive bag contains at least one target instance.

In contrast, the multiple instance spectral matched filter (MI-SMF) and multiple instance adaptive coherence/cosine detector (MI-ACE) [41] learn *discriminative* target concepts that maximize the SMF or ACE detection statistics, which preserves bag structure and does not require tuning parameter settings. These goals are accomplished by optimizing the following objective function,

$$\arg\max_{\mathbf{s}} \frac{1}{K^+} \sum_{i:L_i=1} \Lambda(\mathbf{x}_i^*, \mathbf{s}) - \frac{1}{K^-} \sum_{i:L_i=0} \frac{1}{N_i^-} \sum_{\mathbf{x}_{ij} \in B_i^-} \Lambda(\mathbf{x}_{ij}, \mathbf{s}), \qquad (6.10)$$

---

[1]The *e*FUMI implementation is available at: https://github.com/GatorSense/FUMI [40].

where $\mathbf{s}$ is the target signatures, $\Lambda(\mathbf{x}, \mathbf{s})$ is the detection statistics of data point $\mathbf{x}$ given target signature $\mathbf{s}$, and $\mathbf{x}_i^*$ is the selected representative instance from the positive bag B_i^+, K^+ is the number of positive bags and K^- is the number of negative bags.

$$\mathbf{x}_i^* = \arg\max_{\mathbf{x}_{ij} \in B_i^+} \Lambda(\mathbf{x}_{ij}, \mathbf{s}). \tag{6.11}$$

This general objective can be applied to any target detection statistics. However, consider the ACE detector, $\Lambda_{ACE}(\mathbf{x}, \mathbf{s}) = \dfrac{\mathbf{s}^T \boldsymbol{\Sigma}_b^{-1}(\mathbf{x}-\boldsymbol{\mu}_b)}{\sqrt{\mathbf{s}^T \boldsymbol{\Sigma}_b^{-1}\mathbf{s}}\sqrt{(\mathbf{x}-\boldsymbol{\mu}_b)^T \boldsymbol{\Sigma}_b^{-1}(\mathbf{x}-\boldsymbol{\mu}_b)}}$, where $\boldsymbol{\mu}_b$ is the mean of the background and $\boldsymbol{\Sigma}_b$ is the background covariance. This detection statistic can be viewed as an inner product in a whitened coordinate space

$$
\begin{aligned}
\Lambda_{ACE}(\mathbf{x}, \mathbf{s}) &= \frac{\mathbf{s}^T \boldsymbol{\Sigma}_b^{-1}(\mathbf{x}-\boldsymbol{\mu}_b)}{\sqrt{\mathbf{s}^T \boldsymbol{\Sigma}_b^{-1}\mathbf{s}}\sqrt{(\mathbf{x}-\boldsymbol{\mu}_b)^T \boldsymbol{\Sigma}_b^{-1}(\mathbf{x}-\boldsymbol{\mu}_b)}} \\
&= \frac{\mathbf{s}^T \mathbf{U}\mathbf{V}^{-\frac{1}{2}}\mathbf{V}^{-\frac{1}{2}}\mathbf{U}^T(\mathbf{x}-\boldsymbol{\mu}_b)}{\sqrt{\mathbf{s}^T \mathbf{U}\mathbf{V}^{-\frac{1}{2}}\mathbf{V}^{-\frac{1}{2}}\mathbf{U}^T\mathbf{s}}\sqrt{(\mathbf{x}-\boldsymbol{\mu}_b)^T \mathbf{U}\mathbf{V}^{-\frac{1}{2}}\mathbf{V}^{-\frac{1}{2}}\mathbf{U}^T(\mathbf{x}-\boldsymbol{\mu}_b)}} \\
&= \left(\frac{\hat{\mathbf{s}}}{\|\hat{\mathbf{s}}\|}\right)^T \left(\frac{\hat{\mathbf{x}}}{\|\hat{\mathbf{x}}\|}\right) \\
&= \hat{\mathbf{s}}^T \hat{\mathbf{x}}, \tag{6.12}
\end{aligned}
$$

where $\hat{\mathbf{x}} = \mathbf{V}^{-\frac{1}{2}}\mathbf{U}^T(\mathbf{x}-\boldsymbol{\mu}_b)$, $\hat{\mathbf{s}} = \mathbf{V}^{-\frac{1}{2}}\mathbf{U}^T\mathbf{s}$, $\mathbf{U}$ and $\mathbf{V}$ are the eigenvectors and eigenvalues of the background covariance matrix $\boldsymbol{\Sigma}_b$, respectively, $\overset{\wedge}{\hat{\mathbf{s}}} = \frac{\hat{\mathbf{s}}}{\|\hat{\mathbf{s}}\|}$, and $\overset{\wedge}{\hat{\mathbf{x}}} = \frac{\hat{\mathbf{x}}}{\|\hat{\mathbf{x}}\|}$. It is clear from Eq. (6.12) that the ACE detector response is the cosine value between a test data point, $\mathbf{x}$, and a target signature, $\mathbf{s}$, after whitening. Thus, the objective function (6.10) for MI-ACE can be rewritten as

$$\arg\max_{\overset{\wedge}{\hat{\mathbf{s}}}} \frac{1}{K^+}\sum_{i:L_i=1} \overset{\wedge}{\hat{\mathbf{s}}}^T \overset{\wedge}{\hat{\mathbf{x}}}_i^* - \frac{1}{K^-}\sum_{i:L_i=0} \frac{1}{N_i^-}\sum_{\mathbf{x}_{ij} \in B_i^-} \overset{\wedge}{\hat{\mathbf{s}}}^T \overset{\wedge}{\hat{\mathbf{x}}}_{ij}, \text{ such that } \overset{\wedge}{\hat{\mathbf{s}}}^T \overset{\wedge}{\hat{\mathbf{s}}} = 1. \tag{6.13}$$

The l_2 norm constraint, $\overset{\wedge}{\hat{\mathbf{s}}}^T \overset{\wedge}{\hat{\mathbf{s}}} = 1$, is resulted from the normalization term in Eq. (6.12). The optimum for (6.13) can be derived by solving the Lagrangian optimization problem for the target signature

$$\overset{\wedge}{\hat{\mathbf{s}}} = \frac{\mathbf{t}}{\|\mathbf{t}\|}, \text{ where } \mathbf{t} = \frac{1}{K^+}\sum_{i:L_i=1} \overset{\wedge}{\hat{\mathbf{x}}}_i^* - \frac{1}{K^-}\sum_{i:L_i=0} \frac{1}{N_i^-}\sum_{\mathbf{x}_{ij} \in B_i^-} \overset{\wedge}{\hat{\mathbf{x}}}_{ij}. \tag{6.14}$$

A similar approach can be applied for the spectral matched filter detector,

$$\Lambda_{SMF}(\mathbf{x}, \mathbf{s}) = \frac{\mathbf{s}^T \mathbf{\Sigma}_b^{-1}(\mathbf{x} - \boldsymbol{\mu}_b)}{\sqrt{\mathbf{s}^T \mathbf{\Sigma}_b^{-1} \mathbf{s}}}, \tag{6.15}$$

resulting in the following update equation for MI-SMF:

$$\hat{\mathbf{s}} = \frac{\mathbf{t}}{\|\mathbf{t}\|}, \text{ where } \mathbf{t} = \frac{1}{K^+} \sum_{i:L_i=1} \hat{\mathbf{x}}_i^* - \frac{1}{K^-} \sum_{i:L_i=0} \frac{1}{N_i^-} \sum_{\mathbf{x}_{ij} \in B_i^-} \hat{\mathbf{x}}_{ij}. \tag{6.16}$$

Algorithm 6.2 MI-SMF/MI-ACE

1: Compute $\boldsymbol{\mu}_b$ and $\mathbf{\Sigma}_b$ as the mean and covariance of all instances in the negative bags
2: Subtract the background mean and whiten all instances, $\hat{\mathbf{x}} = \mathbf{V}^{-\frac{1}{2}} \mathbf{U}^T (\mathbf{x} - \boldsymbol{\mu}_b)$
3: If MI-ACE, normalize: $\hat{\mathbf{x}} = \frac{\hat{\mathbf{x}}}{\|\hat{\mathbf{x}}\|}$
4: Initialize $\hat{\mathbf{s}}$ using the instance in a positive bag resulting in largest objective function value
5: **repeat**
6:     Update the selected instances, $\mathbf{x}_i^*$, for each positive bag, $\mathbf{B}_i^+$ using (6.11)
7:     Update $\hat{\mathbf{s}}$ using (6.14) for MI-ACE or (6.16) for MI-SMF
8: **until** Stopping Criterion Reached
9: **return** $\mathbf{s} = \frac{\mathbf{t}}{\|\mathbf{t}\|}$, where $\mathbf{t} = \mathbf{U}\mathbf{V}^{\frac{1}{2}}\hat{\mathbf{s}}$

The MI-SMF and MI-ACE algorithms alternate between the two steps: (1) selecting representative instances from each positive bag and (2) updating the target concept **s**. The MI-SMF and MI-ACE methods stop when there is no change in the selection of instances from positive bags across subsequent iterations. Similar to [7], since there exists a finite set of possible selection of positive instances given a finite training bags, the convergence of MI-SMF and MI-ACE is guaranteed. In the experiments shown in [41], MI-SMF and MI-ACE generally converged with less than seven iterations. The MI-SMF/MI-ACE algorithm is summarized in Algorithm 6.2.[2] Please refer to [41] for a detailed derivation of the algorithm.

6.3.3 Multiple Instance Hybrid Estimator

Both *e*FUMI and the MI-ACE/MI-SMF methods are limited in that they only estimate a single target concept. However, in many problems, the target class has significant spectral variability [43]. The Multiple Instance Hybrid Estimator (MI-HE) [44, 45] was developed to fill this gap and estimate multiple target concepts simultaneously.

---

[2]The MI-SMF and MI-ACE implementations are available at: https://github.com/GatorSense/MIACE [42].

The proposed MI-HE algorithm maximizes the responses of the hybrid sub-pixel detector [46] within the MIL framework. This is accomplished by maximizing the following objective function:

$$
J = \ln \prod_{i=1}^{K^+} \left(\frac{1}{N_i} \sum_{j=1}^{N_i} \Pr(l_{ij} = 1|\mathbf{B}_i)^b \right)^{\frac{1}{b}} \prod_{i=K^++1}^{K} \prod_{j=1}^{N_i} \Pr(l_{ij} = 0|\mathbf{B}_i)
$$

$$
= -\sum_{i=1}^{K^+} \frac{1}{b} \ln \left(\frac{1}{N_i} \sum_{j=1}^{N_i} \exp\left(-\beta \frac{\|\mathbf{x}_{ij} - \mathbf{Da}_{ij}\|^2}{\|\mathbf{x}_{ij} - \mathbf{D}^-\mathbf{p}_{ij}\|^2} \right)^b \right)
$$

$$
+ \rho \sum_{i=K^++1}^{K} \sum_{j=1}^{N_i} \|\mathbf{x}_{ij} - \mathbf{D}^-\mathbf{p}_{ij}\|^2
$$

$$
+ \frac{\alpha}{2} \sum_{i=K^++1}^{K} \sum_{j=1}^{N_i} \left((\mathbf{D}^+\mathbf{a}_{ij}^+)^T \mathbf{x}_{ij} \right)^2, \tag{6.17}
$$

where the first term corresponds to a generalized mean (GM) term [47], which can approximate the *max* operation as b approaches $+\infty$. This term can be interpreted as determining a representative positive instance in each positive bag by identifying the instance that maximizes the hybrid sub-pixel detector (HSD) [46] statistic, $\exp\left(-\beta \frac{\|\mathbf{x}_{ij} - \mathbf{Da}_{ij}\|^2}{\|\mathbf{x}_{ij} - \mathbf{D}^-\mathbf{p}_{ij}\|^2} \right)$. In the HSD, each instance is modeled as a sparse linear combination of target and/or background concepts $\mathbf{D}$, $\mathbf{x} \approx \mathbf{Da}$, where $\mathbf{D} = \left[\mathbf{D}^+ \ \mathbf{D}^- \right] \in \mathbb{R}^{d \times (T+M)}$, $\mathbf{D}^+ = [\mathbf{d}_1, \ldots, \mathbf{d}_T]$ is the set of T target concepts and $\mathbf{D}^- = \left[\mathbf{d}_{T+1}, \ldots, \mathbf{d}_{T+M} \right]$ is the set of M background concepts, β is a scaling parameter, and $\mathbf{a}_{ij}$ and $\mathbf{p}_{ij}$ are the sparse representation of $\mathbf{x}_{ij}$ given the entire concept set $\mathbf{D}$ and background concept set $\mathbf{D}^-$, respectively. The second term in the objective function is viewed as the background data fidelity term, which is based on the assumption that minimizing the least squares of all negative points provides a good description of the background. The scaling factor ρ is usually set to be smaller than one to control the influence of negative bags. The third term is the cross incoherence term (motivated by the Dictionary Learning with Structured Incoherence [48] and the Fisher discrimination dictionary learning (FDDL) algorithm [49, 50]) that encourages positive concepts to have distinct spectral signatures from negative points.

The initialization of target concepts in $\mathbf{D}$ is conducted by computing the mean of T random subsets drawn from the union of all positive training bags. The vertex component analysis (VCA) [53] method was applied to the union of all negative bags and the M cluster centers (or vertices) were set as the initial background concepts. The pseudocode of the MI-HE algorithm is presented in Algorithm 6.3.[3] Please refer to [44] for a detailed optimization derivation.

---

[3]The MI-HE implementation is available at: https://github.com/GatorSense/MIHE [54].

Algorithm 6.3 MI-HE algorithm

Input: MIL training bags $\mathbf{B} = \{\mathbf{B}_1, \ldots, \mathbf{B}_K\}$, MI-HE parameters
1: Initialize $\mathbf{D}^0$, $iter = 0$
2: **repeat**
3:        **for** $t = 1, \ldots, T$ **do**
4:              Solve $\mathbf{a}_{ij}$, $\mathbf{p}_{ij}$, $\forall i \in \{1, \ldots, K\}$, $j \in \{1, \ldots, N_i\}$ using the iterative shrinkage-
                 thresholding algorithm [51, 52]
5:              Update $\mathbf{d}_t$ using gradient descent
6:              $\mathbf{d}_t \leftarrow \frac{1}{\|\mathbf{d}_t\|_2} \mathbf{d}_t$
7:        **end for**
8:        **for** $k = T + 1, \ldots, T + M$ **do**
9:              Solve $\mathbf{a}_{ij}$, $\mathbf{p}_{ij}$, $\forall i \in \{1, \ldots, K\}$, $j \in \{1, \ldots, N_i\}$ using the iterative shrinkage-
                 thresholding algorithm [51, 52]
10:             Update $\mathbf{d}_k$ using gradient descent
11:             $\mathbf{d}_k \leftarrow \frac{1}{\|\mathbf{d}_k\|_2} \mathbf{d}_k$
12:       **end for**
13:       $iter \leftarrow iter + 1$
14: **until** Stopping criterion reached
15: **return** $\mathbf{D}$

6.3.4 Multiple Instance Learning for Multiple Diverse Hyperspectral Target Characterizations

The multiple instance learning of multiple diverse characterizations for SMF (MILMD-SMF) and ACE detector (MILMD-ACE) [55] is an extension of MI-ACE and MI-SMF that learns multiple target signatures for characterization of the variability in hyperspectral target concepts. Different from the MI-HE method explained above, the MILMD-SMF and MILMD-ACE methods do not model target and background signatures explicitly. Instead, the MILMD-SMF and MILMD-ACE methods focus on maximizing the detection statistics of the positive bags and capturing the characteristics of the training data using a set of diverse target signatures, as shown below:

$$\mathbf{S}^* = \arg \max_{\mathbf{S}} \prod_i P(\mathbf{S}|B_i, L_i = 1) \prod_i P(\mathbf{S}|B_i, L_i = 0), \qquad (6.18)$$

where $\mathbf{S} = \left\{ \mathbf{s}^{(1)}, \mathbf{s}^{(2)}, \ldots \mathbf{s}^{(K)} \right\}$ is the K assumed target signatures and $P(\mathbf{S}|B_i, L_i = 1)$ and $P(\mathbf{S}|B_i, L_i = 0)$ denote the probabilities given the positive and negative bags, respectively. The authors consider the following equivalent form of (6.18) for multiple target characterization can be shown as

$$\mathbf{S}^* = \arg \max_{\mathbf{S}} \{C_1(\mathbf{S}) + C_2(\mathbf{S})\}, \qquad (6.19)$$

$$C_1(\mathbf{S}) = \frac{1}{N^+} \sum_{i:L_i=1} \Omega(D, \, X_i^*, \mathbf{S}), \qquad (6.20)$$

$$C_1(\mathbf{S}) = -\frac{1}{N^-} \sum_{i:L_i=0} \Upsilon(D, X_i, \mathbf{S}), \tag{6.21}$$

where $\Omega(\cdot)$ and $\Upsilon(\cdot)$ are defined to capture the detection statistics of the positive and negative bags, $D(\cdot)$ is detection response of the given ACE or SMF detectors and $\mathbf{X}_i^* = \{\mathbf{x}_i^{(1)*}, \mathbf{x}_i^{(2)*}, \dots, \mathbf{x}_i^{(K)*}\}$ is the subset of the ith positive bag of selected instances with maximum detection responses corresponding to one of the target signatures $\mathbf{s}^k$ such that

$$x_i^{(k)*} = \arg\max_{\mathbf{x}_n \in \mathbf{B}_i, L_i=1} D(\mathbf{x}_n, \mathbf{s}^{(k)}). \tag{6.22}$$

The term $\Omega(D, X_i^*, \mathbf{S})$ is the global detection statistics term for the positive bags whose ACE form is shown in

$$\Omega_{ACE}(D, X_i^*, \mathbf{S}) = \frac{1}{K} \sum_k \hat{\mathbf{s}}^{(k)T} \hat{\mathbf{x}}_i^{(k)*}. \tag{6.23}$$

Similar to [41], $\hat{\mathbf{s}}^{(k)}$ and $\hat{\mathbf{x}}^{(k)}$ are the transformed kth target signature and correspond instance after whitening using the background information and normalization. The global detection term $\Omega_{ACE}(D, X_i^*, \mathbf{S})$ provides an average detection statistics over the positive bags given a set of learned target signatures. Of particular note for this method, in contrast with MI-HE, is the approach assumes that each positive bag contains a representative for each variation of the positive concept.

On the other hand, the global detection term $\Upsilon_{ACE}(D, X_i, \mathbf{S})$ for negative instances should be small and thus suppresses the background as shown in Eq. (6.24). This definition means if the maximum responses of target signature set $\mathbf{S}$ over the negative instances are minimized, the estimated target concepts can effectively discriminate nontarget training instances

$$\Upsilon_{ACE}(D, X_i, \mathbf{S}) = \frac{1}{N_{i,L_i=0}} \sum_{\mathbf{x}_n \in \mathbf{B}_i, L_i=0} \max_k \hat{\mathbf{s}}^{(k)T} \hat{\mathbf{x}}_n. \tag{6.24}$$

In order to explicitly apply the normalization constraint and encourage diversity in the estimated multiple target concepts, [55] also includes two terms, a normalization term by pushing the inner product of the estimated signatures to 1 and a diversity promoting term by maximizing the difference between estimated target concepts as shown in (6.25), and (6.26), respectively.

$$C^{div}(\mathbf{S}) = -\frac{2}{K(K-1)} \sum_{k,l,k \neq l} \hat{\mathbf{s}}^{(k)T} \hat{\mathbf{s}}^{(l)}, \tag{6.25}$$

$$C^{con}(\mathbf{S}) = -\frac{1}{K} \sum_k \left| \hat{\mathbf{s}}^{(k)T} \hat{\mathbf{s}}^{(k)} - 1 \right|. \tag{6.26}$$

Combining the global detection statistics, the diversity promoting and normalization constraint terms, the final cost function is shown as (6.27).

$$
C_{ACE} = \frac{1}{N^+} \sum_{i:L_i=1} \sum_k \frac{1}{K} \hat{\mathbf{s}}^{(k)T} \hat{\mathbf{x}}_i^{(k)*} - \frac{1}{N^-} \sum_{i:L_i=0} \frac{1}{N_{i,L_i=0}} \sum_{\mathbf{x}_n \in \mathbf{B}_i, L_i=0} \max_k \hat{\mathbf{s}}^{(k)T} \hat{\mathbf{x}}_n
$$
$$
- \frac{2\alpha}{K(K-1)} \sum_{k,l,k \neq l} \hat{\mathbf{s}}^{(k)T} \hat{\mathbf{s}}^{(l)} - \frac{\lambda}{K} \sum_k \left| \hat{\mathbf{s}}^{(k)T} \hat{\mathbf{s}}^{(k)} - 1 \right|. \tag{6.27}
$$

The objective for SMF can be similarly derived, where the only difference is the use of training data without normalization. For the optimization of Eq. (6.27), gradient descent is applied. Since the $max(\cdot)$ and $|\cdot|$ operators are not differentiable at zero, the noisy-or function is adopted as an approximation for $max(\cdot)$ and a subgradient method is performed to compute the gradient of $|\cdot|$. Please refer to [55] for a detailed optimization derivation.

6.3.5 Experimental Results for MIL in Hyperspectral Target Detection

In this section, several MIL learning methods on both simulated and real hyperspectral detection tasks are evaluated to illustrate the properties of these algorithms and provide insight into how and when these methods are effective.

For the experiments conducted in this paper, the parameter settings of the comparison algorithms were optimized using a grid search on the first task of each experiment and then applied to the remaining tasks. For example, for mi-SVM classifier on the Gulfport Brown target task, the γ value of the RBF kernel was firstly varied from 0.5 to 5 at a step size of 0.5, and then a finer search around the current best value (with the highest AUC) at a step of 0.1 was performed. For algorithms with stochastic result, e.g., EM-DD, eFUMI, each parameter setting was run five times and the median performance was selected. Finally the optimal parameters that achieve the highest AUC for the brown target were selected and used for the other three target types.

6.3.5.1 Simulated Data

As discussed in Sect. 6.3.1, the eFUMI algorithm combines all positive bags as one big positive bag and all negative bags as one big negative bag and learns target concept from the big positive bag that is different from the negative bag. Thus, if the negative bags contain incomplete knowledge of the background, e.g., some nontarget concept appears only in the subset of positive bags, eFUMI will perform poorly. However, the discriminative MIL algorithms, e.g., MI-HE, MI-ACE, and MI-SMF, maintain bag structure and can distinguish the target.

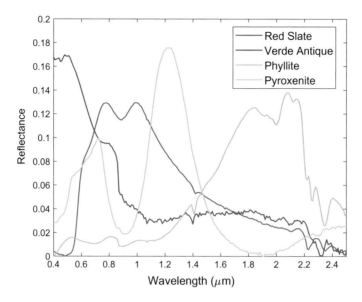

Fig. 6.5 Signatures from ASTER library used to generate simulated data

Given this hypothesis, simulated data was generated from four spectra selected from the ASTER spectral library [56]. Specifically, the Red Slate, Verde Antique, Phyllite, and Pyroxenite spectra from the rock class with 211 bands and wavelengths ranging from 0.4 to 2.5 μm (as shown in Fig. 6.5 in solid lines) were used as endmembers to generate hyperspectral data. Red Slate was labeled as the target endmember.

Four sets of highly mixed noisy data with varied mean target proportion value (α_{t_mean}) were generated, a detailed generation process can be found in [37]. Specifically, this synthetic data has 15 positive and 5 negative bags with each bag having 500 points. If it is a positively labeled bag, there are 200 highly mixed target points containing mean target (Red Slate) proportion from 0.1 to 0.7, respectively, to vary the level of target presence from weak to high. Gaussian white noise was added so that signal-to-noise ratio of the data was set to 20 dB. To highlight the ability of MI-HE, MI-ACE and MI-SMF to leverage individual bag-level labels, we use different subsets of background endmembers to build synthetic data as shown in Table 6.1.

Table 6.1 List of constituent endmembers for synthetic data with incomplete background Knowledge

Bag no.	Bag label	Target endmember	Background endmember
1–5	+	Red slate	Verde Antique, Phyllite, Pyroxenite
6–10	+	Red slate	Phyllite, Pyroxenite
11–15	+	Red slate	Pyroxenite
16–20	−	N/A	Phyllite, Pyroxenite

Table 6.1 shows that the negatively labeled bags only contain two negative endmembers and there exists one confusing background endmember in the first 5 positive bags which is Verde Antique. It is expected that the discriminative MIL algorithms, MI-HE, MI-ACE, and MI-SMF, should be able to perform well in this experiment configuration.

The aforementioned MI-HE [44, 45], eFUMI [37, 38], MI-SMF and MI-ACE [41], DMIL [10, 11] and mi-SVM [9] are multiple instance target concept learning methods. The mi-SVM algorithm performs a comparison of MIL approach that does not rely on estimating a target signature. Figure 6.6a shows the estimated target signature from data with 0.3 mean target proportion value. It clearly shows that eFUMI is always confused with another nontarget endmember, Verde Antique, that exists in some positive bags but is excluded from the background bags. It also shows the other comparison algorithms can estimate a target concept close to the ground truth Red Slate spectrum. One thing need to be explained here is since MI-ACE and MI-SMF are discriminative concept learning methods that try to minimize the detection response of negative bags, they are not expected to recover the true target signature.

For simulated detection analysis, estimated target concepts from the training data were then applied to the test data generated separately following the same generating procedure. The detection was performed using the HSD [46] or ACE [57] detection statistic. For MI-HE and eFUMI, both methods were applied since those two algorithms can come out as a set of background concept from training simultaneously; for MI-SMF, both SMF and ACE were applied since MI-SMF's objective is maximizing the multiple instance spectral matched filter; for the rest multiple instance target concept learning algorithms, MI-ACE, DMIL, only ACE was applied. For the testing procedure of mi-SVM, a regular SVM testing process was performed using LIBSVM [58], and the decision values (signed distances to hyperplane) of test data determined from trained SVM model were taken as the confidence values. For the signature-based detectors, the background data mean and covariance were estimated from the negative instances of the training data.

For quantitative evaluation, Fig. 6.6b shows the receiver operating characteristic (ROC) curves using estimated target signature, where it can be seen that the eFUMI is confused with the testing Verde Antique data at very low PFA (probability of false alarms) rate. Table 6.2 shows the area under the curve (AUC) of proposed MI-HE and comparison algorithms. The results reported are the median results over five runs of the algorithm on the same data. From Table 6.2, it can be seen that for MI-HE and MI-ACE, the best performance on detection was achieved using ACE detector, which is quite close to the performance of using the ground truth target signature (denoted as values with stars). The reason that MI-HE's detection using HSD detector is a little worse is that HSD relies on knowing the complete background concept to properly represent each nontarget testing data, the missing nontarget concept (Verde Antique) makes the nontarget testing data containing Verde Antique maintain a relatively large reconstruction error, and thus large detection statistic.

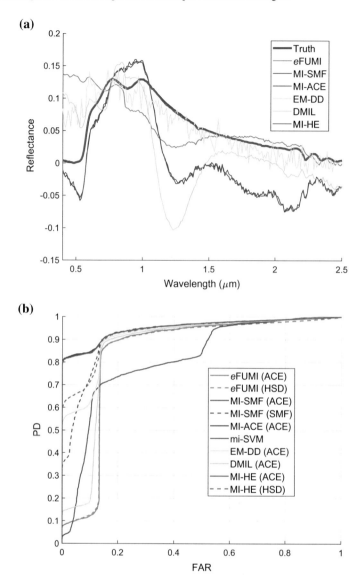

Fig. 6.6 MI-HE and comparisons on synthetic data with incomplete background knowledge, $\alpha_{t\_mean} = 0.3$. MI-SMF and MI-ACE are not expected to recover the true signature. **a** Estimated target signatures for Red Slate and comparison with ground. **b** ROC curves cross validated on test data

Table 6.2 Area under the ROC curves for MI-HE and comparison algorithms on simulated hyperspectral data with incomplete background knowledge. Best results shown in bold, second best results underlined, and ground truth shown with an asterisk

Algorithm	α_{t_mean}			
	0.1	0.3	0.5	0.7
MI-HE (HSD)	0.743	<u>0.931</u>	0.975	0.995
MI-HE (ACE)	<u>0.763</u>	**0.952**	**0.992**	**0.999**
eFUMI [37] (ACE)	0.675	0.845	0.978	<u>0.998</u>
eFUMI [37] (HSD)	0.671	0.564	0.978	<u>0.998</u>
MI-SMF [41] (SMF)	0.719	0.923	0.972	0.993
MI-SMF [41] (ACE)	0.735	**0.952**	**0.992**	**0.999**
MI-ACE [41] (ACE)	**0.764**	**0.952**	**0.992**	**0.999**
mi-SVM [9]	0.715	0.815	0.866	0.900
DMIL [10, 11] (ACE)	0.687	0.865	0.971	0.996
Ground Truth (ACE)	0.765*	0.953*	0.992*	0.999*

6.3.5.2 MUUFL Gulfport Hyperspectral Data

The MUUFL Gulfport hyperspectral data set collected over the University of Southern Mississippi-Gulfpark Campus was used to evaluate the target detection performance across various MIL classification methods. This data set contains 325×337 pixels with 72 spectral bands corresponding to wavelengths from 367.7 to 1043.4 nm at a 9.5−9.6 nm spectral sampling interval. The ground sample distance of this hyperspectral data set is 1 m [1]. The first four and last four bands were removed due to sensor noise. Two sets of this data (Gulfport Campus Flight 1 and Gulfport Campus Flight 3) were selected as cross-validated training and testing data for these two data sets have the same altitude and spatial resolution. Throughout the scene, there are 64 man-made targets in which 57 were considered in this experiment which are cloth panels of four different colors: Brown (15 examples), Dark Green (15 examples), Faux Vineyard Green (FVGr) (12 examples), and Pea Green (15 examples). The spatial location of the targets are shown as scattered points over an RGB image of the scene in Fig. 6.7. Some of the targets are in the open ground and some are occluded by the live oak trees. Moreover, the targets also vary in size, for each target type, there are targets that are $0.25\,m^2$, $1\,m^2$, and $9\,m^2$ in area, respectively, resulting a very challenging, highly mixed sub-pixel target detection problem.

MUUFL Gulfport Hyperspectral Data, Individual Target Type Detection
For this part of the experiments, each individual target type was treated as a target class, respectively. For example, when "Brown" is selected as target class, a 5×5 rectangular region corresponding to each of the 15 ground truth locations denoted by GPS was grouped into a positive bag to account for the drift coming from GPS. This size was chosen based on the accuracy of the GPS device used to record the ground truth locations. The remaining area that does not contain a brown target was

●	Brown
◆	Dark Green
■	FVG
	Pea Green

Fig. 6.7 MUUFL Gulfport data set RGB image and the 57 target locations

grouped into a big negative bag. This constructs the detection problem for "Brown" target. Similarly, there are 15, 12, and15 positive labeled bags for Dark Green, Faux Vineyard Green, and Pea Green, respectively.

The comparison algorithms were evaluated on this data using the Normalized Area Under the receiver operating characteristic curve (NAUC) in which the area was normalized out to a false alarm rate (FAR) of 1×10^{-3} false alarms/m^2 [59]. During detection on the test data, the background mean and covariance were estimated from the negative instances of the training data. The results reported are the median results over five runs of the algorithm on the same data.

Figure 6.8a shows the estimated target concept by all comparisons for Dark Green target type training on flight 3. We can see that the eFUMI and MI-HE are able to recover the target concept quite close to ground truth spectra manually selected from the scene. Figure 6.8b shows the detection ROCs given target spectra estimated on flight 3 and cross validated on flight 1. Table 6.3 shows the NAUCs for all comparison algorithms cross validated on all four types of target, where it can be seen that MI-HE generally outperforms the comparisons for most of the target types and achieves close to the performance of using ground truth target signatures. Since MI-HE is a discriminative target concept learning framework that aims to distinguish one target instance from each positively labeled bag, MI-HE had a lower performance for the pea green target because of the relatively large occlusion of those targets causing difficulty in distinguishing pea green signature from each of the positive bag.

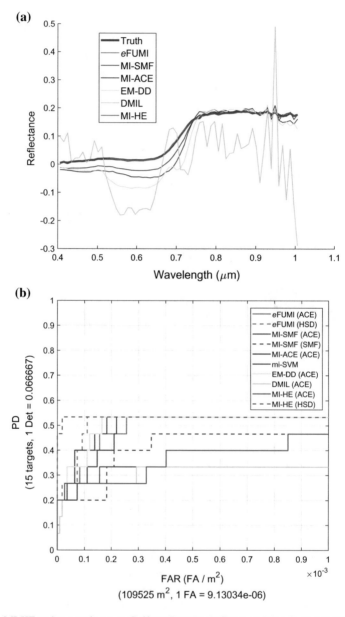

Fig. 6.8 MI-HE and comparisons on Gulfport Data Dark Green, training flight 3 testing flight 1. **a** Estimated target signatures from flight 3 for Brown and comparison with ground truth. **b** ROC curves cross validated on flight 1

Table 6.3 Area under the ROC curves for MI-HE and comparison algorithms on Gulfport data with individual target type. Best results shown in bold, second best results underlined, and ground truth shown with an asterisk

Alg.	Train on Flight 1; Test on Flight 3				Train on Flight 3; Test on Flight 1			
	Brown	Dark Gr.	Faux Vine Gr.	Pea Gr.	Brown	Dark Gr.	Faux Vine Gr.	Pea Gr.
MI-HE (HSD)	**0.499**	**0.453**	**0.655**	0.267	**0.781**	**0.532**	**0.655**	0.350
MI-HE (ACE)	0.433	0.379	0.104	0.267	0.710	0.360	0.111	0.266
eFUMI [37] (ACE)	0.423	0.377	0.654	0.267	0.754	0.491	0.605	0.393
eFUMI [37] (HSD)	0.444	0.436	0.653	0.267	0.727	0.509	0.500	0.333
MI-SMF [41] (SMF)	0.419	0.354	0.533	0.266	0.657	0.405	0.650	0.384
MI-SMF [41] (ACE)	0.448	0.382	0.579	0.316	0.760	0.501	0.613	0.388
MI-ACE [41] (ACE)	0.474	0.390	0.485	**0.333**	0.760	0.483	0.593	0.380
mi-svm [9]	0.206	0.195	0.412	0.265	0.333	0.319	0.245	0.274
EM-DD [7] (ACE)	0.411	0.381	0.486	0.279	0.760	0.503	0.541	**0.416**
DMIL [10, 11] (ACE)	0.419	0.383	0.191	0.009	0.743	0.310	0.081	0.083
Ground Truth (ACE)	0.528*	0.429*	0.656*	0.267*	0.778*	0.521*	0.663*	0.399*

MUUFL Gulfport Hyperspectral Data, All Four Target Types Detection

For training and detection for the four target types together, the positive bags were generated by grouping each of the 5×5 regions denoted by the ground truth that it contains any of the four types of target. Thus, for each flight there are 57 target points and 57 positive bags were generated. The remaining area that does not contain any target was grouped into a big negative bag. Table 6.4 summarizes the NAUCs as a quantitative comparison, which shows that the detection statistic by the proposed MI-HE using HSD is significantly better than the comparison algorithms.

Table 6.4 Area under the ROC curves for MI-HE and comparison algorithms on Gulfport data with all four target types. Best results shown in bold, second best results underlined, and ground truth shown with an asterisk

Alg.	Test Flight 3	Test Flight 1	Alg.	Test Flight 3	Test Flight 1
MI-HE (HSD)	**0.304**	**0.449**	MI-SMF [41] (ACE)	0.219	0.327
MI-HE (ACE)	<u>0.257</u>	0.254	MI-SMF [41] (SMF)	0.198	0.277
eFUMI [37] (ACE)	0.214	<u>0.325</u>	mi-SVM [9]	0.235	0.269
eFUMI [37] (HSD)	0.256	0.331	EM-DD [7] (ACE)	0.211	0.310
MI-ACE [41] (ACE)	0.226	0.340	DMIL [10, 11] (ACE)	0.198	0.225
Ground Truth (ACE)	0.330*	0.490*			

6.4 Multiple Instance Learning Approaches for Classifier Fusion and Regression

Although more extensively studied for the case of sub-pixel hyperspectral target detection, the Multiple Instance Learning approach can be used in other hyperspectral applications including fusion with other sensors and regression, in addition to two-class classification and detection problems discussed in previous sections. In this section, algorithms for multiple instance classifier fusion and regression are presented and their applications to hyperspectral and remote sensing data analysis are discussed.

6.4.1 Multiple Instance Choquet Integral Classifier Fusion

The multiple instance Choquet integral (MICI) algorithm[4] [61, 62] is a multiple instance classifier fusion method to integrate different classifier outputs with imprecise labels under the MIL framework. In MICI, the Choquet integral [63, 64] was used under the MIL framework to fuse outputs from multiple classifiers or sensors for improving the accuracy and accounting for imprecise labels for hyperspectral classification and target detection.

The Choquet integral (CI) is an effective nonlinear information aggregation method based on the fuzzy measure. Assume there exists m sources, $C = \{c_1, c_2, \ldots, c_m\}$, for fusion. These "sources" can be the decision outputs by different classifiers or data collected by different sensors. The power set of C is denoted as 2^C, which

---

[4]The MICI implementation is available at: https://github.com/GatorSense/MICI [60].

contains all possible (crisp) subsets of C. A monotonic and normalized fuzzy measure, $\mathbf{g}$, is a real valued function that maps $2^C \to [0, 1]$. It satisfies the following properties:

1. $g(\emptyset) = 0$; *empty set*
2. $g(C) = 1$; *normalization property*
3. $g(A) \leq g(B)$ if $A \subseteq B$ and $A, B \subseteq C$. *monotonicity property.*

Let $h(c_k; \mathbf{x}_n)$ denote the output of the kth classifier, c_k, on the nth instance, $\mathbf{x}_n$. The discrete Choquet integral of instance $\mathbf{x}_n$ given C (m sources) is computed using

$$C_{\mathbf{g}}(\mathbf{x}_n) = \sum_{k=1}^{m} \left[h(c_k; \mathbf{x}_n) - h(c_{k+1}; \mathbf{x}_n) \right] g(A_k), \qquad (6.28)$$

where the sources are sorted so that $h(c_1; \mathbf{x}_n) \geq h(c_2; \mathbf{x}_n) \geq \cdots \geq h(c_m; \mathbf{x}_n)$ and $h(c_{m+1}; \mathbf{x}_n)$ is defined to be zero. The fuzzy measure element value $g(A_k)$ corresponds to the subset $A_k = \{c_1, c_2, \ldots, c_k\}$.

In a classifier fusion problem, given training data and fusion sources, $h(c_m; \mathbf{x}_n)$ $\forall m, n$ are known. The desired bag-level labels for sets of $C_{\mathbf{g}}(\mathbf{x}_n)$ values are also known (positive label "+1", negative label "0"). Then, the goal of the MICI algorithm is to learn all the element values of the unknown fuzzy measure $\mathbf{g}$ from the training data and bag-level (imprecise) labels. The MICI method includes three variations to formulate the fusion problem under the MIL framework to address label imprecision. The variations include the noisy-or model, the min-max model, and the generalized-mean model.

The MICI noisy-or model follows the Diverse Density formulation (see Sect. 6.2.2) and uses a noisy-or objective function

$$J_N = \sum_{a=1}^{K^-} \sum_{i=1}^{N_b^-} \ln \left(1 - \mathcal{N} \left(C_{\mathbf{g}}(\mathbf{x}_{ai}^-) | \mu, \sigma^2 \right) \right)$$
$$+ \sum_{b=1}^{K^+} \ln \left(1 - \prod_{j=1}^{N_b^+} 1 - \mathcal{N} \left(C_{\mathbf{g}}(\mathbf{x}_{bj}^+) | \mu, \sigma^2 \right) \right), \qquad (6.29)$$

where K^+ denotes the total number of positive bags, K^- denotes the total number of negative bags, N_b^+ is the total number of instances in positive bag b, and N_a^- is the total number of instances in negative bag a. Each data point/instance is either positive or negative, as indicated by the following notation: $\mathbf{x}_{ai}^-$ is the ith instance in the ath negative bag and $\mathbf{x}_{bj}^+$ is the jth instance in the bth positive bag. The $C_{\mathbf{g}}$ is the Choquet integral output given measure $\mathbf{g}$ computed using (6.28). The μ and σ^2 are the mean and variance of the Gaussian function $\mathcal{N}(\cdot)$, respectively. In practice, the parameter μ can be set to 1 or a value close to 1 for two-class classifier fusion problems, in order to encourage the CI values of positive instances to be 1 and the CI values of negative instances to be far from 1. The variance of the Gaussian σ^2 controls how sharply the CI values are pushed to 0 and 1, and thus controls the weighting of the two terms in the objective function. By maximizing the objective function (6.29), the CI values of all the points in the negative bag are encouraged to be zero (first term) and the CI values of at least one instance in the positive bag are encouraged to be one (second term), which follows the MIL assumption.

The MICI min-max model applies the min and max operators to the negative and positive bags, respectively. The min-max model follows the MIL formulation without the need to manually set parameters such as the Gaussian variance in the noisy-or model. The objective function of the MICI min-max model is

$$
J_M = \sum_{a=1}^{K^-} \max_{\forall \mathbf{x}_{ai}^- \in \mathbf{B}_a^-} \left(C_{\mathbf{g}}(\mathbf{x}_{ai}^-) - 0 \right)^2 + \sum_{b=1}^{K^+} \min_{\forall \mathbf{x}_{bj}^+ \in \mathbf{B}_b^+} \left(C_{\mathbf{g}}(\mathbf{x}_{bj}^+) - 1 \right)^2, \tag{6.30}
$$

where $\mathbf{B}_a^-$ denotes the ath negative bag, and $\mathbf{B}_b^+$ denotes the bth positive bag. The remaining terms follow the same notation as in (6.29). The first term of the objective function encourages the CI values of all instances in the negative bag to be zero, and the second term encourages the CI values of at least one instance in the positive bag to be one. By minimizing the objective function in (6.30), the MIL assumption is satisfied.

Instead of selecting only one instance from each bag as a "prime instance" that determines the bag-level label as does the min-max model, the MICI generalized-mean model allows more instances to contribute toward the classification of bags. The MICI generalized-mean objective function is written as

$$
J_G = \sum_{a=1}^{K^-} \left[\frac{1}{N_a^-} \sum_{i=1}^{N_a^-} \left(C_{\mathbf{g}}(\mathbf{x}_{ai}^-) - 0 \right)^{2p_1} \right]^{\frac{1}{p_1}} + \sum_{b=1}^{K^+} \left[\frac{1}{N_b^+} \sum_{j=1}^{N_b^+} \left(C_{\mathbf{g}}(\mathbf{x}_{bj}^+) - 1 \right)^{2p_2} \right]^{\frac{1}{p_2}},
$$
$$\tag{6.31}$$

where p_1 and p_2 are the exponential factors controlling the generalized-mean operation. When $p_1 \to +\infty$ and $p_2 \to -\infty$, the generalized-mean terms becomes equivalent to the min and max operators, making the generalized-mean model equivalent to the min-max model. By adjusting the p value, the generalized-mean term can act as varying other aggregating operators, such as arithmetic mean ($p = 1$) or quadratic mean ($p = 2$). For another interpretation, when $p \geq 1$, the generalized-mean can be rewritten as the l_p norm [65].

The MICI models can be optimized by sampling-based evolutionary algorithms, where the element values of fuzzy measure $\mathbf{g}$ are sampled and selected through a truncated Gaussian distribution either based on valid interval (how much the element value can change without violating the monotonicity property of the fuzzy measure), or based on the counts of times a measure element is used in all training instances. A more detailed optimization process and psuedocode of the MICI models can be seen in [62, 66]. The MICI models have been used for hyperspectral sub-pixel target detection [61, 62] and were effective in fusing multiple detector inputs (e.g., the ACE detector) and can yield competitive classification results.

6.4.2 Multiple Instance Regression

Multiple instance regression (MIR) handles multiple instance problems where the prediction values are real-valued, instead of binary class labels. The MIR methods have been used in remote sensing literature for applications such as aerosol optical depth retrieval [67, 68] and crop yield prediction [62, 68–70].

Prime-MIR was one of the earliest MIR algorithms, proposed by Ray and Page in 2001 [71]. Prime-MIR is based on the "primary instance" assumption, which assumes there is only one primary instance per bag that contributes to the real-valued bag-level label. Prime-MIR assumes a linear regression hypothesis and the goal is to find a hyperplane $\mathbf{Y} = \mathbf{Xb}$ such that

$$\mathbf{b} = \arg\min_b \sum_{i=1}^{n} L\left(y_i, X_{ip}, \mathbf{b}\right),$$ (6.32)

where X_{ip} is the primary instance in bag i, and L is some error function, such as the squared error. An expectation–maximization (EM) algorithm was used to iteratively solve for the ideal hyperplane. First, a random hyperplane was initialized. For each instance j in each bag i, the error L of the instance X_{ij} to the hyperplane $\mathbf{Y} = \mathbf{Xb}$ was computed. In the E-step, the instance with the lowest error L was selected as the "primary instance." In the M-step, a new hyperplane was constructed by performing a multiple regression over all the primary instances selected in the E-step. The two steps were repeated until the algorithm converges and the best hyperplane solution was returned. In [71], Prime-MIR showed the benefits of using multiple instance regression over ordinary regression, especially when the non-primary instances in the bag were not correlated with the primary instances.

The MI k-NN approach and its variations [72] extends the Diverse Density, kNN, and Citation-kNN for real-valued multiple instance learning. The minimal Hausdorff distance from [27] was used to measure the distance between two bags. Given two sets of points $A = a_1, \ldots a_m$ and $B = b_1, \ldots, b_n$, the Hausdorff distance is defined as

$$H(A, B) = \max\{h(A, B), h(B, A)\},$$ (6.33)

where $h(A, B) = \max_{a \in A} \min_{b \in B} \|a - b\|$, $\|a - b\|$ is the Euclidean distance between points a and b. In the MI k-NN algorithm, the prediction made for a bag B is the average label of the k closest bags, measured in Hausdorff metric. In the MI citation-kNN algorithm, the prediction made for a bag B is the average label of the R closest bag neighbors of B measured in Hausdorff metric and C-nearest citers, where the "citers" include the bags where B is a one of their C-nearest neighbors. It is generally recommended that $C = R + 2$ [72]. The third variant, a diverse density approach for the real-valued setting, maximizes

$$\prod_{i=1}^{K} Pr(r|B_i) \tag{6.34}$$

where $Pr(t|B_i) = (1 - |l_i - Label(B_i|t)|)/Z$, K is the total number of bags, t is the target point, l_i is the label for the ith bag, and Z is a normalization constant. The results in [72] showed good prediction performance of all three variants on a benchmark Musk Molecules data set [4], but the performance of both the nearest neighbor and diverse density algorithms were sensitive to the number of relevant features, as expected based on the sensitivity of the Hausdorff distance to outliers.

A real-valued multiple instance on-line model proposed by Goldman and Scott [73] uses MIR for learning real-valued geometric patterns, motivated by landmark matching problem in robot navigation and vision applications. This algorithm associates a real-valued label with each point and uses the Hausdorff metric to help classify a bag as positive, if the points in the bag are within some Hausdorff distance from target concept points. This algorithm differs from the supervised MIR in that the standard supervised MIR learns from a given set of training bags and bag-level training labels, while [73] applies an online agnostic model [74–76] where the learners make predictions as the bag $\mathbf{B}_t$ is presented at iteration t. Wang et al. [77] also used the idea of online MIR, i.e., to use the latest arrived bag with its training label to update the current predictive model. This work was also extended in [78].

A regularization framework for MIR proposed by Cheung and Kwok [79] defines a loss function that takes into consideration both training bags and training instances. The first part of the loss function computes the error (loss) between training bags label and its predictions and the second part considers the loss between the bag label prediction and all the instances in the bag. This work still adopted the "primary instance" assumption but simplified to assume the primary instance was the instance with the highest prediction output value. This model provided comparable or better performance on the synthetic Musk Molecules data set [72] as citation-kNN [27] and Multiple Instance kernel-based SVM [79, 80].

Most MIR methods discussed above only provided theoretical discussions or results on synthetic regression data sets. More recently, MIR methods have been applied to real-world hyperspectral and remote sensing data analysis. Wagstaff et al. in [69, 70] investigated using MIR to predict crop yield from remotely sensed data collected over California and Kansas. In [69], a novel method for inferring the "salience" of each instance was proposed with regard to the real-valued bag label. The salience of each instance, i.e., its "relevance" with respect to all other instances in the bag to predict the bag label, is the weight associated with each instance. The salience values were defined to be nonnegative and sum to one for all instances in each bag. Like Ray and Page [71], Wagstaff et al. followed the "primary-instance" assumption but their primary instance, or "exemplar" of a bag, is the weighted average of all the points in the bag instead of one single instance from the bag. Given training bags and instances, a set of salience values are solved based on a fixed linear regression model and given the estimated salience, the regressor is updated and the algorithm reiterates until convergence. This work did not intend to provide predictions over new data, but instead focused on understanding the contents (the salience) of each training instance.

Wagstaff et al. then made use of the salience learned to provide predictions for new, unlabeled bags by proposing an MI-ClusterRegress algorithm (or sometimes referred to as the Cluster-MIR algorithm) [70] that mapped instances onto (hidden) cluster labels. The main assumption of MI-ClusterRegress is that the instances from a bag are drawn (with noise) from a set of underlying clusters and one of the clusters is "relevant" to the bag-level labels. After obtaining k clusters for each bag by EM-based Gaussian mixture models (or any other clustering method), a local regression model is constructed for each cluster. MI-ClusterRegress then selects the best-fit model and use it to predict labels for test bags. A support vector regression learner [81] is used for regression prediction. Results on simulated and predicting crop yield data sets show that modeling the bag structure when the structure (cluster) is present is effective for regression prediction, especially when the cluster number k is equal to or larger than what is actually present in the bags.

In Chap. 2, Moreno-Martínez et al. proposed a kernel distribution regression (KDR) model for MIR by embedding the bag distribution in a high-dimensional Hilbert space and performing standard least squares regression on the mean embedded data. This kernel method exploits the rich structure in bags by considering all higher order moments of the bag distributions and performing regression with the bag distributions directly. This kernel method also allows to combine bags with different number of instances per bag by summarizing the bag feature vectors with a set of mean map embeddings of instances in the bag. The KRD model was shown to outperform standard regression models such as the least squares regularized linear regression model (RLR) and the (nonlinear) kernel ridge regression (KRR) method for crop yield applications.

Wang et al. [67, 68] proposed a probabilistic and generalized mixture model for MIR based on the primary-instance assumption (sometimes referred to as the EM-MIR algorithm). It is assumed that the bag label is a noisy function of the primary instance, and the conditional probability $p(y_i|\mathbf{B}_i)$ for predicting label y_i for the ith bag is dependent entirely on the primary instance. A binary random variable z_{ij} is defined such that $z_{ij} = 1$ if the jth instance in the ith bag is the primary instance and $z_{ij} = 0$ if otherwise. The mixture model for each bag i is written as

$$p(y_i|\mathbf{B}_i) = \sum_{j=1}^{N_i} p(z_{ij} = 1|\mathbf{B}_i)p(y_i|\mathbf{x}_{ij}) \tag{6.35}$$

$$= \sum_{j=1}^{N_i} \pi_{ij} p(y_i|\mathbf{x}_{ij}), \tag{6.36}$$

where π_{ij} is the (pior) probability that the jth instance in the ith bag is the primary instance, $p(y_i|\mathbf{x}_{ij})$ is the label probability given the primary instance $\mathbf{x}_{ij}$ and N_i is the total number of instances in the ith bag $\mathbf{B}_i$. Therefore, the learning problem is transformed to learning the mixture weights π_{ij} and $p(y_i|\mathbf{x}_{ij})$ from training data and an EM algorithm is used to optimize the parameters. This work discussed several methods to set the prior π_{ij}, including using deterministic function, or as a Gaussian function of prediction deviation, or as a parametric function (in this case a feed-forward neural network). It was discussed in [68] that several algorithms discussed above, including Prime-MIR [71] and Pruning-MIR [67], are in fact the special case of the mixture model. The mixture model MIR shows better performance on simulated data as well as for predicting aerosol optical depth (AOD) from remote sensing data and predicting crop yield applications, compared with the Cluster-MIR [70] and Prime-MIR [71] algorithms described above.

Two baseline methods for MIR have also been described in [68], Aggregate-MIR, and Instance-MIR. In Aggregate-MIR, a "meta-instance" is obtained for each bag by averaging all the instances in that bag, and a regression model can be trained using the bag-level labels and the meta-instances. In Instance-MIR, all instances in a bag are assumed to have the same label as the bag-level label, and a regression model can be trained by combining all instances from all bags. Then, in testing, the label for a test bag is the average of all the instance-level labels in that test bag. The Aggregate-MIR and Instance-MIR methods belong to the "input summary" and "output expansion" approaches as described in Chap. 2, Sect. 2.3.1. These two methods are straightforward and easy to implement, and have been used as basic comparison methods for a variety of MIR applications.

The robust fuzzy clustering for MIR (RFC-MIR) algorithm was proposed by Trabelsi and Frigui [82] to incorporate data structure in MIR. The RFC-MIR algorithm uses fuzzy clustering methods such as the fuzzy c-means (FCM) and possibilistic c-means (PCM) [83] to cluster the instances and fit multiple local linear regression models to the clusters. Similar to Cluster-MIR, the RFC-MIR method combines all instances from all training bags for clustering. However, Cluster-MIR performs clustering in an unsupervised manner without considering bag-level labels, while RFC-MIR uses instance features as well as labels in clustering. Validation results of RFC-MIR show improved accuracy on crop yield prediction and drug activity prediction applications [84], and the possibilistic memberships obtained from the RFC-MIR algorithm can be used to identify the primary and irrelevant instances in each bag.

In parallel with the multiple instance classifier fusion models described in Sect. 6.4.1, a Multiple Instance Choquet Integral Regression (MICIR) model[5] has been proposed to accommodate real-valued predictions for remote sensing applications [62]. The objective function of the MICIR model is written as

$$\min \sum_{i=1}^{K} \left[\min_{\forall j, x_{ij} \in \mathbf{B}_i} (C_g(\mathbf{x}_{ij}) - o_i)^2 \right], \tag{6.37}$$

where o_i is the desired training labels for bag $\mathbf{B}_i$. Note that MICIR is able to fuse real-valued outputs from regression models as well as from classifiers. When o_i is

---

[5]The MICIR implementation is available at: https://github.com/GatorSense/MICI [60].

binary, MICIR reduces to the MICI min-max model for two-class classifier fusion. The MICIR algorithm also follows the primary instance assumption by minimizing the error between the CI value of one primary instance and the given bag-level labels, while allowing imprecision in other instances. Similar to MICI classifier fusion models, an evolutionary algorithm can be used to sample the fuzzy measure **g** from the training data.

Overall, Multiple Instance Regression methods have been studied in the literature for nearly two decades and most studies are based on the primary-instance assumption proposed by Ray and Page in 2001. Linear regression models were used in most MIR methods if a regressor was used and experiments have shown effective results of using MIR on crop yield prediction and aerosol optical depth retrieval applications given remote sensing data.

6.4.3 Multiple Instance Multi-resolution and Multi-modal Fusion

Previous MIL classifier fusion and regression methods, such as the MICI and the MICIR models, can only be applied if the fusion sources have the same number of data points and the same resolution across multiple sensors. As motivated in Sect. 6.1, in remote sensing applications, sensor outputs often have different resolutions and modalities, such as rasterized hyperspectral imagery versus LiDAR point cloud data. To address multi-resolution and multi-modal fusion under imprecision, the multiple instance multi-resolution fusion (MIMRF) algorithm[6] was developed to fuse multi-resolution and multi-modal sensor outputs while learning from automatically generated, imprecisely labeled data [66, 86].

In multi-resolution and multi-modal fusion, there can be a set of *candidate* points from a local region from one sensor that corresponds to one point from another sensor, due to sensor measurement inaccuracy and different data resolutions and modalities. Take hyperspectral imagery and LIDAR point cloud data fusion, for example, for each pixel H_i in the HSI imagery, there may exist a set of $\{L_{i1}, L_{i2}, \ldots, L_{il}\}$ points from the LiDAR point cloud that corresponds to the area covered by the pixel H_i. The MIMRF algorithm first constructs such correspondences by writing the *collection* of the sensor outputs for pixel i as

$$\mathbf{S}_i = \begin{bmatrix} H_i & L_{i1} \\ H_i & L_{i2} \\ \vdots & \vdots \\ H_i & L_{il} \end{bmatrix}. \tag{6.38}$$

---

[6]The MIMRF implementation is available at: https://github.com/GatorSense/MIMRF [85].

This notation can extend to any number of correspondences l by row, and multiple sensors by column. The MIMRF assumes that, at least one point in all candidate LiDAR points is accurate, but it is unknown which one. One of the goals of the MIMRF algorithm is to automatically select the correct points with accurate measurement and correspondence information. To achieve this goal, the CI fusion for the collection of the sensor outputs of the ith negative data point is written as

$$C_{\mathbf{g}}(\mathbf{S}_i^-) = \min_{\forall \mathbf{x}_k^- \in \mathbf{S}_i^-} C_{\mathbf{g}}(\mathbf{x}_k^-), \tag{6.39}$$

and the CI fusion for the collection of the sensor outputs values of the jth positive data point is written as

$$C_{\mathbf{g}}(\mathbf{S}_j^+) = \max_{\forall \mathbf{x}_l^+ \in \mathbf{S}_j^+} C_{\mathbf{g}}(\mathbf{x}_l^+), \tag{6.40}$$

where $\mathbf{S}_i^-$ is the collection of sensor outputs for the ith negative data point and $\mathbf{S}_j^+$ is the collection of sensor outputs for the jth positive data point; $C_{\mathbf{g}}(\mathbf{S}_i^-)$ is the Choquet integral output for $\mathbf{S}_i^-$ and $C_{\mathbf{g}}(\mathbf{S}_j^+)$ is the Choquet integral output for $\mathbf{S}_j^+$. In this way, the min and max operators automatically select one data point (which is assumed to be the data point with correct information) from each negative and positive bag to be used for fusion, respectively.

Moreover, the MIMRF is designed to handle bag-level imprecise labels. Recall that the MIL framework assumes a bag is labeled positive if at least one instance in the bag is positive and a bag is labeled negative if all the instances in the bag are negative. Thus, the objective function for MIMRF algorithm is proposed as

$$J = \sum_{a=1}^{K^-} \max_{\forall \mathbf{S}_{ai}^- \in \mathbf{B}_a^-} \left(C_{\mathbf{g}}(\mathbf{S}_{ai}^-) - 0 \right)^2 + \sum_{b=1}^{K^+} \min_{\forall \mathbf{S}_{bj}^+ \in \mathbf{B}_b^+} \left(C_{\mathbf{g}}(\mathbf{S}_{bj}^+) - 1 \right)^2$$

$$= \sum_{a=1}^{K^-} \boxed{\max_{\forall \mathbf{S}_{ai}^- \in \mathbf{B}_a^-}} \left(\min_{\forall \mathbf{x}_k^- \in \mathbf{S}_{ai}^-} C_{\mathbf{g}}(\mathbf{x}_k^-) - 0 \right)^2 + \sum_{b=1}^{K^+} \boxed{\min_{\forall \mathbf{S}_{bj}^+ \in \mathbf{B}_b^+}} \left(\max_{\forall \mathbf{x}_l^+ \in \mathbf{S}_{bj}^+} C_{\mathbf{g}}(\mathbf{x}_l^+) - 1 \right)^2, \tag{6.41}$$

where K^+ is the total number of positive bags, K^- is the total number of negative bags, $\mathbf{S}_{ai}^-$ is the collection of ith instance set in the ath negative bag and similar for $\mathbf{S}_{bj}^+$. $C_{\mathbf{g}}$ is the Choquet integral given fuzzy measure $\mathbf{g}$, $\mathbf{B}_a^-$ is the ath negative bag, and $\mathbf{B}_b^+$ is the bth positive bag. The term $\mathbf{S}_{ai}^-$ is the collection of input sources for the ith pixel in the ath negative bag and $\mathbf{S}_{bj}^+$ is the collection of input sources for the jth pixel in the bth positive bag.

In (6.41), the min and max operators outside the squared errors (the boxed terms) are comparable to the MICI min-max model. The max operator encourages the Choquet integral of all the points in the negative bag to be 0 and the min operator encourages the Choquet integral of at least one point in the positive bag to be 1 (second term), which satisfies the MIL assumption. The min and max operators inside the squared error terms come from (6.39) and (6.40), which selects one correspondence for each collection of candidates. By minimizing the objective function in (6.41), the first term encourages the fusion output of all the points in the negative bag to the desired negative label 0, and the second term encourages the fusion output of at least one of the points in the positive bag to the desired positive label $+1$. This satisfies the MIL assumption while addressing label imprecision for multi-resolution and multi-modal data. The MIMRF algorithm has been used to fuse rasterized hyperspectral imagery and un-rasterized LiDAR point cloud data over urban scenes and have shown effective fusion results for land cover classification [66, 86].

Here is a small example to illustrate the performance of the MIMRF algorithm using the MUUFL Gulfport hyperspectral and LiDAR data set collected over the University of Southern Mississippi-Gulfpark Campus [1]. An illustration of the rasterized hyperspectral imagery and the LiDAR data over the complete scene can be seen in Figs. 6.2 and 6.3 in Sect. 6.1. The task here is to fuse hyperspectral and LiDAR data to perform building detection and classification. The simple linear iterative clustering (SLIC) algorithm [87, 88] was used to segment the hyperspectral imagery. The SLIC algorithm is a widely used, unsupervised superpixel segmentation algorithm that can produce spatially coherent regions. Each superpixel from the segmentation is treated as a "bag" in our learning process and all pixels in each superpixel are all the instances in the bag. The bag-level labels in this data set were generated from OpenStreetMap (OSM), a third-party, crowd-sourced online map [89]. OSM provides map information for urban regions around the world. Figure 6.9c shows the map extracted from Open Street Map (OSM) over the study area based on the ground cover tags available, such as "highway", "footway", "building", etc. Information from Google Earth [90], Google Maps [91], and geo-tagged photographs from a digital camera taken at the scene were also be used as auxiliary data to assist the labeling process. This way, reliable bag-level labels can be automatically generated with minimal human intervention. These bag-level labels will then be used in the MIMRF objective function (6.41) to learn the unknown fuzzy measure **g** for HSI-LiDAR fusion. Figure 6.9 shows the RGB imagery, the SLIC segmentation, and the OSM map labels for the MUUFL Gulfport hyperspectral imagery.

Three multi-resolution and multi-modal sensor outputs were used as fusion sources, one generated from HSI imagery and two from raw LiDAR point cloud data. The first fusion source is the ACE detection map on buildings based on the mean spectral signature of randomly sampled building points from the scene. The ACE detection map for buildings is shown in Fig. 6.10a. As shown, the ACE confidence map highlights most buildings, but also highlights some roads which have similar spectral signature (similar construction material, such as asphalt). The ACE detector also failed to detect the top right building due to the darkness of the roof. Two other fusion sources were generated from LiDAR point cloud data according to the building height profile, with the rasterized confidence maps shown in Fig. 6.10b and Fig. 6.10c. Note that in MIMRF fusion, the LiDAR sources will be point clouds and Figs. 6.10b and c are provided for visualization and comparison purposes only.

As shown in Fig. 6.10, each HSI and LiDAR sensor output contains certain building information. The goal is to use MIMRF to fuse all three sensor outputs and perform accurate building classification. We randomly sampled 50% the bags (the

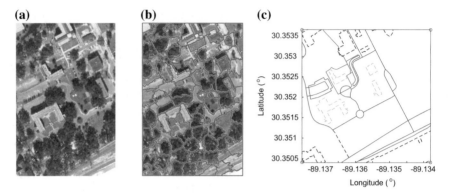

Fig. 6.9 The RGB image (**a**), SLIC segmentation (**b**), and the OSM map for the MUUFL Gulfport hyperspectral imagery (**c**). In the OSM map, the blue lines correspond to road and highway. The magenta lines correspond to sidewalk/footway. The green lines marks buildings. Here, the "building" tag is specific to the buildings with a grey (asphalt) roof. The black lines correspond to "other" tags. Source: © [2020] IEEE. Reprinted, with permission, from [86]

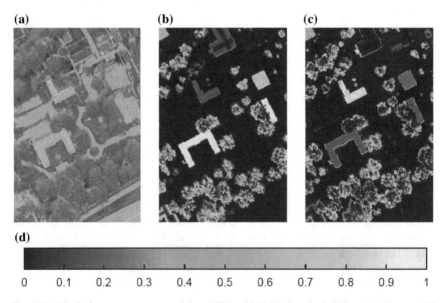

Fig. 6.10 The fusion sources generated from HSI and LiDAR data for building detection. **a** ACE detection map from HSI data. **b**, **c** LiDAR building detection map from two LiDAR flights. The colorbar can be seen in **d**. Source: © [2020] IEEE. Reprinted, with permission, from [86]

Fig. 6.11 An example of ROC curve results for building detection across all methods

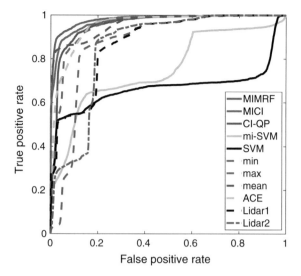

superpixels) and use these to learn a set of fuzzy measures for the MIMRF algorithm. We conducted such random sampling three times by using the MATLAB *randperm( )* function and call these the three random runs. The sampled bags are different at each random run. In each random run, the MIMRF algorithm is applied to learn a fuzzy measure from the randomly sampled 50% bags, and fusion results are evaluated on the remaining 50% data on a pixel level. Note that there will be two sets of results in each run—learn from the first sampled 50% bags (denoted "Half1") and perform fusion on the second half of data (denoted "Half2"), and vice versa. The fusion results of MIMRF were compared with previously discussed MIL algorithms such as MICI and mi-SVM and the CI-QP approach. The CI-QP (Choquet integral-quadratic programming) approach [64] is a CI fusion method that learns a fuzzy measure for the Choquet integral by optimizing a least squares error objective using quadratic programming. Note that these comparison methods only work with rasterized LiDAR imagery, while the MIMRF algorithm can directly handle raw LiDAR point cloud data. The fusion results of MIMRF were also compared with commonly used fusion methods, such as min, max, and mean operators and a support vector machine, as well as the ACE and LiDAR sensor sources before fusion.

Figure 6.11 shows an example of the receiver operating characteristic (ROC) curve results for building detection across all comparison methods. Table 6.5 shows the area under curve (AUC) results across all methods in all random runs. Table 6.6 shows the root mean square error (RMSE) results across all methods in all random runs. The AUC evaluates how well the method detects the buildings (the higher AUC the better) and the RMSE shows how the detection results on both the building and nonbuilding points differ from the ground truth (the lower the RMSE the better). We observed from the tables that the MIMRF method was able to achieve high AUC detection results and low RMSE compared to other methods, and the MIMRF is stable across different randomizations. The MICI classifier fusion method also did well in detection (high AUC), but has higher RMSE compared to MIMRF, possibly due to MICI's inability to handle multi-resolution data. The min operator did well in RMSE due to the fact that it places low confidence everywhere, but was unable to have high detection results. The ACE detector did well in detection, which shows that the hyperspectral signature is effective at distinguishing building roof materials. However, it also places high confidence on other asphalt materials such as road, and thus yields a high RMSE value.

Figures 6.12 and 6.13 shows a qualitative comparison of our fusion performance. Figure 6.12 shows an example of our randomly sampled bags. All the semi-transparent bags marked by the red lines in Fig. 6.12a were used to learn a fuzzy measure in our method, and we evaluate pixel-level fusion results against the "test" ground truth shown in Fig. 6.12b. Note that the MIMRF is a self-supervised method that learns a fuzzy measure from bag-level labels and produces pixel-level fusion results. Although standard training and testing scheme does not apply here, this experiment is set up using cross validation to show that the MIMRF algorithm is able to utilize the fuzzy measure learned from one part of the data and apply that fuzzy measure to perform fusion on new test data, even when the learned bags were excluded from testing.

Table 6.5 The AUC results of building detection using MUUFL Gulfport HSI and LiDAR data across three random runs. (The higher the AUC the better.) The best two results with the highest AUC were **bolded** and underlined, respectively. "Half1" refers to the results of learning a fuzzy measure from the first 50% of the bag-level labels from campus 1 data and perform pixel-level fusion on the second half. "Half2" refers to the results of learning a fuzzy measure from the second 50% of the bag-level labels from campus 1 data and perform pixel-level fusion on the first half. The ACE, Lidar1, and Lidar2 rows show results from the individual HSI and LiDAR sources before fusion; the methods below the dotted line show fusion results for all comparison methods. The standard deviations of MICI and MIMRF methods are computed across three runs (three random fuzzy measure initializations) and are shown in parentheses. Same notation is applied for the RMSE table below as well

	First Random run		Second Random run		Third Random run	
	Half1	Half2	Half1	Half2	Half1	Half2
ACE	0.954	0.961	0.938	0.967	0.963	0.947
Lidar1	0.874	0.914	0.879	0.904	0.920	0.874
Lidar2	0.855	0.813	0.879	0.796	0.830	0.848
SVM	0.670	0.854	0.791	0.918	0.928	0.823
min	0.872	0.863	0.890	0.849	0.870	0.872
max	0.946	0.945	0.953	0.939	0.948	0.945
mean	0.963	0.952	0.969	0.947	0.959	0.960
mi-SVM	0.752	0.886	0.795	0.942	0.930	0.923
CI-QP	0.955	0.959	0.959	0.939	0.962	0.964
MICI	0.972(0.001)	**0.963(0.000)**	**0.976(0.000)**	0.960(0.000)	0.968(0.000)	**0.971(0.000)**
MIMRF	**0.978(0.003)**	**0.963(0.002)**	0.972(0.000)	**0.971(0.001)**	**0.973(0.000)**	**0.971(0.002)**

Table 6.6 The RMSE results of building detection using MUUFL Gulfport HSI and LiDAR data across three random runs. (The lower the RMSE the better.) The best two results with the highest AUC were **bolded** and <u>underlined</u>, respectively. "Half1" refers to the results of learning a fuzzy measure from the first 50% of the bag-level labels from campus 1 data and perform pixel-level fusion on the second half. "Half2" refers to the results of learning a fuzzy measure from the second 50% of the bag-level labels from campus 1 data and perform pixel-level fusion on the first half. The ACE, Lidar1, and Lidar2 rows show results from the individual HSI and LiDAR sources before fusion; the methods below the dotted line show fusion results for all comparison methods. The standard deviations of MICI and MIMRF methods are computed across three runs (three random fuzzy measure initializations) and are shown in parentheses

	First Random run		Second Random run		Third Random run	
	Half1	Half2	Half1	Half2	Half1	Half2
ACE	0.345	0.339	0.348	0.307	0.334	0.350
Lidar1	0.291	0.255	0.278	0.268	0.266	0.280
Lidar2	0.294	0.270	0.267	0.297	0.269	0.295
SVM	0.348	0.332	0.437	<u>0.250</u>	0.409	0.284
min	<u>0.265</u>	<u>0.235</u>	<u>0.248</u>	0.255	**0.240**	0.263
max	0.417	0.417	0.419	0.413	0.423	0.420
mean	0.307	0.291	0.296	0.298	0.298	0.302
mi-SVM	0.425	0.459	0.432	0.253	0.406	<u>0.232</u>
CI-QP	0.403	0.377	0.405	0.413	0.388	0.397
MICI	0.356(0.002)	0.348(0.002)	0.374(0.001)	0.336(0.001)	0.356(0.000)	0.350(0.000)
MIMRF	**0.238(0.024)**	**0.192(0.025)**	**0.244(0.002)**	**0.208(0.011)**	<u>0.255(0.002)</u>	**0.177(0.001)**

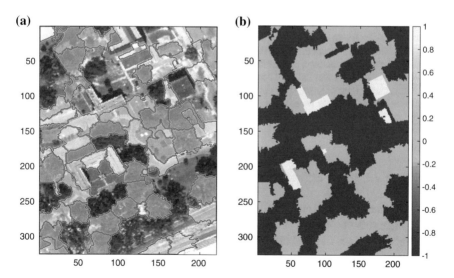

(a)   (b)

Fig. 6.12   a An illustration for the 50% randomly sampled bags from one of our random runs. The MIMRF algorithm learns a fuzzy measure from the red-labeled, transparent bags. **b** The ground truth for the the other 50% data [92]. The yellow and green regions are building and nonbuilding ground truth locations in the "test" data. The dark blue (labeled "−1") regions denote the 50% of the bags that were used in MIMRF learning and therefore not included in the testing process

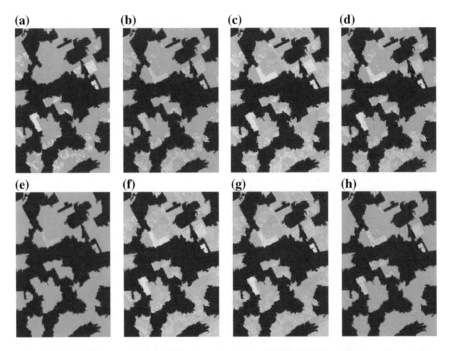

Fig. 6.13 The fusion results for building detection in the MUUFL Gulfport data set, learned from the randomly drawn bags shown in Fig. 6.12a and evaluated on the remaining regions against the ground truth shown in Fig. 6.12b. Note that the MIMRF method learns a set of fuzzy measures from bag-level data and produced per-pixel fusion results on the fusion regions. The subplots show fusion results by **a** SVM; **b** min operator; **c** max operator; **d** mean operator; **e** mi-SVM; **f** CI-QP; **g** MICI; **h** MIMRF. The yellow highlights where the fusion algorithm places high detection confidence and green indicates low confidence, and the dark blue indicates the regions not used in the evaluation. This plot uses the same color bar as in Fig. 6.10d. It is desirable that high confidence (yellow color) was placed on buildings for building detection. As shown, the MIMRF algorithm in **h** was able to detect all buildings (yellow color) in the regions that were evaluated and have low confidence (green color) on nonbuilding areas. The other comparison methods either missed some buildings, or have many more false positives in non-building regions, such as tree canopy

Figure 6.13 shows all fusion results on the test regions across all methods. As shown, the MIMRF algorithm in Fig. 6.13h was able to detect all buildings (yellow) in the evaluation regions well while having low confidence (green) on nonbuilding areas. The other comparison methods either missed some buildings, or have many more false positives in non-building regions. Other randomizations yielded similar effects.

To summarize, the above experimental results show that the MIMRF method was able to successfully perform detection and fusion with high detection accuracy and low root mean square error for multi-resolution and multi-modal data sets. This experiment further demonstrated the effectiveness of the self-supervised learning approach used by the MIMRF method at learning a fuzzy measure from one part of the data (using only bag-level labels) and perform pixel-level fusion on other regions. Guided by publicly available crowd-sourced data such as the OpenStreetMap, the MIMRF algorithm is able to automatically generate imprecise bag-level labels instead of the traditional manual labeling process. Moreover, [86] has shown effective results of MIMRF fusion on agricultural applications as well, in addition to hyperspectral and LiDAR analysis. We envision the MIMRF as an effective fusion method to perform pixel-level classification and produce fusion maps with minimal human intervention for a variety of multi-resolution and multi-modal fusion applications.

6.5 Summary

This chapter introduced the Multiple Instance Learning framework and reviewed MIL methods for hyperspectral classification, sub-pixel target detection, classifier fusion, regression, and multi-resolution multi-modal fusion. Given imprecise (bag-level) ground truth information in the training data, the MIL methods are effective in addressing the inevitable imprecision observed in remote-sensing data and applications.

- Imprecise training labels are omnipresent in hyperspectral image analysis, due to unreliable ground truth information, sub-pixel targets, and occlusion, and heterogeneous sensor outputs. MIL methods can handle bag-level labels instead of requiring pixel-perfect labels in training, which enables easier annotation and more accurate data analysis.
- Multiple instance target characterization algorithms were presented, including *e*FUMI, MI-ACE/MI-SMF, and MI-HE algorithms. These algorithms can estimate target concepts from the data given imprecise labels, without obtaining target signature a priori.
- Multiple instance classifier fusion and regression algorithms were presented. In particular, the MICI method is versatile in that it can perform classifier fusion and regression with minor adjustments in the objective function.
- The MIMRF algorithm extends upon MICI to multi-resolution and multi-modal sensor fusion on remote sensing data with label uncertainty. To our knowledge, this is the first algorithm that can handle HSI imagery and LiDAR point cloud fusion without co-registration or rasterization, considering imprecise labels.
- Various optimization strategies exist to optimize an MIL problem, such as expectation maximization, sampling-based evolutionary algorithm, and gradient descent.

The algorithms discussed in this chapter covers the state-of-the-art MIL approaches and provides an effective solution to address the imprecision challenges in hyperspectral image analysis and remote-sensing applications. There are several challenges in these current approaches that warrant future work. For example, current MI regression methods often rely on the "primary instance" assumption, which may not hold in all applications; or that MIL assumes no contamination (of positive points) in negative bags, but in practice this is often not the case. Future study in more flexible MIL frameworks (such as using kernel embedding as described in Chap. 2) can be conducted in relaxing these assumptions.

References

1. Gader P, Zare A, Close R et al (2013) MUUFL gulfport hyperspectral and lidar airborne data set. Technical report, University of Florida, Gainesville, FL, REP-2013-570. Data and code. https://github.com/GatorSense/MUUFLGulfport and Zenodo. https://doi.org/10.5281/zenodo.1186326
2. Brigot G, Colin-Koeniguer E, Plyer A, Janez F (2016) Adaptation and evaluation of an optical flow method applied to coregistration of forest remote sensing images. IEEE J Sel Topics Appl Earth Observ 9(7):2923–2939
3. Cao S, Zhu X, Pan Y, and Yu Q (2014) A stable land cover patches method for automatic registration of multitemporal remote sensing images. IEEE J Sel Topics Appl Earth Observ 7(8):3502–3512
4. Dietterich TG, Lathrop RH, Lozano-Pérez T et al (1997) Solving the multiple instance problem with axis-parallel rectangles. Artif Intell 89(1–2):31–71
5. Maron O, Lozano-Perez T (1998) A framework for multiple-instance learning. In: Advances in neural information processing systems (NIPS), pp 570–576
6. Maron O, Ratan AL (1998) Multiple-instance learning for natural scene classification. In: International conference on machine learning, vol 98, pp 341–349
7. Zhang Q, Goldman SA (2002) EM-DD: an improved multiple-instance learning technique. In: Advances in neural information processing systems (NIPS), vol 2, pp 1073–1080
8. Press WH, Flannery BP, Teukolsky SA (1992) Numerical recipes in C: the art of scientific programming. Cambridge University Press, Cambridge
9. Andrews S, Tsochantaridis I, Hofmann T (2002) Support vector machines for multiple-instance learning. In: Advances in neural information processing systems (NIPS) 561–568
10. Shrivastava A, Pillai JK, Patel VM, Chellappa R (2014) Dictionary-based multiple instance learning. In: IEEE international conference on image processing (ICIP), pp 160–164
11. Shrivastava A, Patel VM, Pillai JK, Chellappa R (2015) Generalized dictionaries for multiple instance learning. Int J Comput Vis 114(2–3):288–305
12. Mallat SG, Zhang Z (1993) Matching pursuits with time-frequency dictionaries. IEEE Trans Signal Process 41(12):3397–3415
13. Aharon M, Elad M, Bruckstein A (2006) K-SVD: an algorithm for designing overcomplete dictionaries for sparse representation. IEEE Trans Signal Process 54(11):4311–4322
14. Mairal J, Bach F, Ponce J (2012) Task-driven dictionary learning. IEEE Trans Pattern Anal Mach Intell 34(4):791–804
15. Wang X, Wang B, Bai X, Liu W, Tu Z (2013) Max-margin multiple-instance dictionary learning. In: International conference on machine learning, pp 846–854
16. Blei DM, Ng AY, Jordan MI (2003) Latent dirichlet allocation. J Mach Learn Res 3:993–1022
17. Fu Z et al (2011) MILIS: multiple instance learning with instance selection. IEEE Trans Pattern Anal Mach Intell 33(5):958–977
18. Chen Y, Bi J, Wang JZ (2006) MILES: multiple-instance learning via embedded instance selection. IEEE Trans Pattern Anal Mach Intell 28(12):1931–1947

19. Zhu J, Rosset S, Hastie T, Tibshirani R (2004) 1-norm support vector machines. In: Advances in neural information processing systems (NIPS), vol 16, pp 49–56
20. Schölkopf B, Smola AJ (2002) Learning with kernels: support vector machines, regularization, optimization, and beyond. MIT Press
21. Zhou Z, Xu J (2007) On the relation between multi-instance learning and semi-supervised learning. In: Proceedings of the 24th international conference on machine learning, pp 1167–1174
22. Hoffman J et al (2015) Detector discovery in the wild: joint multiple instance and representation learning. In: IEEE conference on computer vision and pattern recognition (CVPR), pp 2883–2891
23. Li W, Vasconcelos N (2015) Multiple instance learning for soft bags via top instances. In: IEEE conference on computer vision and pattern recognition (CVPR), pp 4277–4285
24. Felzenszwalb PF, Girshick RB, McAllester D, Ramanan D (2010) Object detection with discriminatively trained part-based models. IEEE Trans Pattern Anal Mach Intell 32(9):1627–1645
25. Zhang D, Meng D, Han J (2017) Co-saliency detection via a self-paced multiple-instance learning framework. IEEE Trans Pattern Anal Mach Intell 39(5):865–878
26. Kumar MP, Packer B, Koller D (2010) Self-paced learning for latent variable models. In: Advances in neural information processing systems (NIPS), pp 1189–1197
27. Wang J (2000) Solving the multiple-instance problem: a lazy learning approach. In: Proceedings of the 17th international conference on machine learning, pp 1119–1125
28. Duda RO, Hart PE (1973) Pattern classification and scene analysis. Wiley, New York
29. Huttenlocher DP et al (1993) Comparing images using the Hausdorff distance. IEEE Trans Pattern Anal Mach Intell 15(9):850–863
30. Jiang L, Cai Z, Wang D et al (2014) Bayesian citation-KNN with distance weighting. Int J Mach Learn Cybern 5(2):193–199
31. Ghosh D, Bandyopadhyay S (2015) A fuzzy citation-kNN algorithm for multiple instance learning. In: IEEE international conference on fuzzy systems, pp 1–8
32. Villar P, Montes R, Sánchez A et al (2016) Fuzzy-Citation-KNN: a fuzzy nearest neighbor approach for multi-instance classification. In: IEEE international conference on fuzzy systems, pp 946–952
33. Wang X, Yan Y, Tang P et al (2018) Revisiting multiple instance neural networks. Pattern Recognit 74:15–24
34. Nasrabadi NM (2014) Hyperspectral target detection: an overview of current and future challenges. IEEE Signal Process Mag 31(1):34–44
35. Manolakis D, Marden D, Shaw GA (2003) Hyperspectral image processing for automatic target detection applications. Linc Lab J 14(1):79–116
36. Manolakis D, Truslow E, Pieper M, Cooley T, Brueggeman M (2014) Detection algorithms in hyperspectral imaging systems: an overview of practical algorithms. IEEE Signal Process Mag 31(1):24–33
37. Jiao C, Zare A (2015) Functions of multiple instances for learning target signatures. IEEE Trans Geosci Remote Sens 53(8):4670–4686
38. Zare A, Jiao C (2014) Extended functions of multiple instances for target characterization. In: IEEE workshop hyperspectral image signal process: evolution in remote sensing (WHISPERS), pp 1–4
39. Zare A, Gader P (2007) Sparsity promoting iterated constrained endmember detection for hyperspectral imagery. IEEE Geosci Remote Sens Lett 4(3):446–450
40. Jiao C, Zare A (2019) GatorSense/FUMI: initial release (Version v1.0). Zenodo. https://doi.org/10.5281/zenodo.2638304
41. Zare Jiao C, Glenn T (2018) Discriminative multiple instance hyperspectral target characterization. IEEE Trans Pattern Anal Mach Intell 65(10):2634–2648
42. Zare A, Jiao C, Glenn T (2018). GatorSense/MIACE: version 1 (Version v1.0). Zenodo. https://doi.org/10.5281/zenodo.1467358
43. Zare A, Ho KC (2014) Endmember variability in hyperspectral analysis: addressing spectral variability during spectral unmixing. IEEE Signal Process Mag 31(1):95–104

44. Jiao C, Zare A (2017) Multiple instance hybrid estimator for learning target signatures. In: IEEE international geoscience and remote sensing symposium, pp 1–4
45. Jiao C et al (2018) Multiple instance hybrid estimator for hyperspectral target characterization and sub-pixel target detection. ISPRS J Photogramm Remote Sens 146:232–250
46. Broadwater J, Chellappa R (2007) Hybrid detectors for subpixel targets. IEEE Trans Pattern Anal Mach Intell 29(11):1891–1903
47. Babenko B, Dollár P, Tu Z, Belongie S (2008) Simultaneous learning and alignment: multi-instance and multi-pose learning. In: Workshop on faces in 'Real-Life' images: detection, alignment, and recognition
48. Ramirez I, Sprechmann P, Sapiro G (2010) Classification and clustering via dictionary learning with structured incoherence and shared features. In: IEEE conference on computer vision and pattern recognition, pp 3501–3508
49. Yang M, Zhang L, Feng X, Zhang D (2014) Sparse representation based fisher discrimination dictionary learning for image classification. Int J Comput Vis 109(3):209–232
50. Yang M, Zhang L, Feng X, Zhang D (2011) Fisher discrimination dictionary learning for sparse representation. In: International conference on computer vision, pp 543–550
51. Figueiredo MAT, Nowak RD (2003) An EM algorithm for wavelet-based image restoration. IEEE Trans Image Process 12(8):906–916
52. Daubechies I, Defrise M, De Mol C (2003) An iterative thresholding algorithm for linear inverse problems with a sparsity constraint. Commun Pure Appl Math 57(11):1413–1457
53. Nascimento JMP, Dias JMB (2005) Vertex component analysis: a fast algorithm to unmix hyperspectral data. IEEE Trans Geosci Remote Sens 43(4):898–910
54. Jiao C, Zare A (2018) GatorSense/MIHE: initial release (Version 0.1). Zenodo. https://doi.org/10.5281/zenodo.1320109
55. Zhong P, Gong Z, Shan J (2019) Multiple instance learning for multiple diverse hyperspectral target characterizations. IEEE Trans Neural Netw Learn Syst 31(1): 246–258
56. Baldridge AM, Hook SJ, Grove CI, Rivera G (2009) The ASTER spectral library version 2.0. Remote Sens Environ 113(4):711–715
57. Kraut S, Scharf LL (1999) The CFAR adaptive subspace detector is a scale-invariant GLRT. IEEE Trans Signal Process 47(9):2538–2541
58. Chang C, Lin C (2011) LIBSVM: a library for support vector machines. ACM Trans Intell Syst Technol 2(3):1–27
59. Glenn T, Zare A, Gader P, Dranishnikov D (2013) Bullwinkle: scoring code for sub-pixel targets (Version 1.0) [Software]. https://github.com/GatorSense/MUUFLGulfport/
60. Du X, Zare A (2019) GatorSense/MICI: initial release (Version v1.0). Zenodo. https://doi.org/10.5281/zenodo.2638378
61. Du X, Zare A, Keller JM, Anderson DT (2016) Multiple Instance Choquet integral for classifier fusion. IEEE Congr Evol Comput 1054–1061
62. Du X, Zare A (2019) Multiple instance Choquet integral classifier fusion and regression for remote sensing applications. IEEE Trans Geosci Remote Sens 57(5):2741–2753
63. Choquet G (1954) Theory of capacities. Ann L'Institut Fourier 5:131–295
64. Keller JM, Liu D, Fogel DB (2016) Fundamentals of computational intelligence: neural networks, fuzzy systems and evolutionary computation. IEEE press series on computational intelligence, Wiley
65. Rolewicz S (2013) Functional analysis and control theory: linear systems. Springer Science & Business Media, Dordrecht, The Netherlands
66. Du X (2017) Multiple instance choquet integral for multiresolution sensor fusion. Doctoral dissertation, University of Missouri, Columbia, MO, USA
67. Wang Z, Radosavljevic V, Han B et al (2008) Aerosol optical depth prediction from satellite observations by multiple instance regression. In: Proceedings of the SIAM international conference on data mining, pp 165–176
68. Wang Z, Lan L, Vucetic S (2012) Mixture model for multiple instance regression and applications in remote sensing. IEEE Trans Geosci Remote Sens 50(6):2226–2237

69. Wagstaff KL, Lane T (2007) Salience assignment for multiple-instance regression. In: International conference on machine learn, workshop on constrained optimization and structured output spaces
70. Wagstaff KL, Lane T, Roper A (2008) Multiple-instance regression with structured data. In: IEEE international conference on data mining workshops, pp 291–300
71. Ray S, Page D (2001) Multiple instance regression. In: Proceedings of the 18th international conference on machine learning, vol 1, pp 425–432
72. Dooly DR, Zhang Q, Goldman SA, Amar RA (2002) Multiple-instance learning of real-valued data. J Mach Learn Res 3:651–678
73. Goldman SA, Scott SD (2003) Multiple-instance learning of real-valued geometric patterns. Ann Math Artif Intell 39(3):259–290
74. Haussler D (1992) Decision theoretic generalizations of the PAC model for neural net and other learning applications. Inf Comput 100(1):78–150
75. Kearns MJ, Schapire RE, Sellie LM (1994) Toward efficient agnostic learning. Mach Learn 17(2–3):115–141
76. Kivinen J, Warmuth MK (1997) Exponentiated gradient versus gradient descent for linear predictors. Inf Comput 132(1):1–63
77. Wang ZG, Zhao ZS, Zhang CS (2013) Online multiple instance regression. Chin Phys B 22(9):098702
78. Dooly DR, Goldman SA, Kwek SS (2006) Real-valued multiple-instance learning with queries. J Comput Syst Sci 72(1):1–5
79. Cheung PM, Kwok JT (2006) A regularization framework for multiple-instance learning. In: Proceedings of the 23rd international conference on machine learning, pp 193–200
80. Gärtner T, Flach PA, Kowalczyk A, Smola AJ. Multi-instance kernels. In: Proceedings of the 19th international conference on machine learning, vol 2, no 3, pp 179–186
81. Gunn SR (1998) Support vector machines for classification and regression. ISIS Tech Rep 14(1):5–16
82. Trabelsi M, Frigui H (2019) Robust fuzzy clustering for multiple instance regression. Pattern Recognit
83. Krishnapuram R, Keller JM (1993) A possibilistic approach to clustering. IEEE Trans Fuzzy Syst 1(2):98–110
84. Davis J, Santos Costa V, Ray S, Page D (2007) Tightly integrating relational learning and multiple-instance regression for real-valued drug activity prediction. In: Proceedings on international conference on machine learning, vol 287
85. Du X, Zare A (2019) GatorSense/MIMRF: initial release (Version v1.0). Zenodo. https://doi.org/10.5281/zenodo.2638382
86. Du X, Zare A (2019) Multiresolution multimodal sensor fusion for remote sensing data with label uncertainty. IEEE Trans Geosci Remote Sens, In Press
87. Achanta R, Shaji A, Smith K, Lucchi A, Fua P, Süsstrunk S (2010) Slic superpixels. Ecole Polytechnique Fédéral de Lausssanne (EPFL). Tech Rep 149300:155–162
88. Achanta R, Shaji A, Smith K, Lucchi A, Fua P, Süsstrunk S (2012) SLIC superpixels compared to state-of-the-art superpixel methods. IEEE Trans Pattern Anal Mach Intell 34(11):2274–2282
89. OSM contributors (2018) Open street map. https://www.openstreetmap.org
90. Google (2018) Google earth. https://www.google.com/earth/
91. Google (2018) Google maps. https://www.google.com/maps/
92. Du X, Zare A (2017) Technical report: scene label ground truth map for MUUFL gulf-port data set. University of Florida, Gainesville, FL, Tech Rep 20170417. http://ufdc.ufl.edu/IR00009711/00001

Chapter 7
Supervised, Semi-supervised, and Unsupervised Learning for Hyperspectral Regression

Felix M. Riese and Sina Keller

Abstract In this chapter, we present an entire workflow for hyperspectral regression based on supervised, semi-supervised, and unsupervised learning. Hyperspectral regression is defined as the estimation of continuous parameters like chlorophyll a, soil moisture, or soil texture based on hyperspectral input data. The main challenges in hyperspectral regression are the high dimensionality and strong correlation of the input data combined with small ground truth datasets as well as dataset shift. The presented workflow is divided into three levels. (1) At the data level, the data is pre-processed, dataset shift is addressed, and the dataset is split reasonably. (2) The feature level considers unsupervised dimensionality reduction, unsupervised clustering as well as manual feature engineering and feature selection. These unsupervised approaches include autoencoder (AE), t-distributed stochastic neighbor embedding (t-SNE) as well as uniform manifold approximation and projection (UMAP). (3) At the model level, the most commonly used supervised and semi-supervised machine learning models are presented. These models include random forests (RF), convolutional neural networks (CNN), and supervised self-organizing maps (SOM). We address the process of model selection, hyperparameter optimization, and model evaluation. Finally, we give an overview of upcoming trends in hyperspectral regression. Additionally, we provide comprehensive code examples and accompanying materials in the form of a hyperspectral dataset and Python notebooks via GitHub [98, 100].

F. M. Riese · S. Keller (✉)
Karlsruhe Institute of Technology, Karlsruhe, Germany
e-mail: sina.keller@kit.edu

F. M. Riese
e-mail: felix.riese@kit.edu

© Springer Nature Switzerland AG 2020
S. Prasad and J. Chanussot (eds.), *Hyperspectral Image Analysis,*
Advances in Computer Vision and Pattern Recognition,
https://doi.org/10.1007/978-3-030-38617-7_7

187

7.1   Introduction to Hyperspectral Regression

Precise information about the spatial and temporal distribution of continuous physical parameters is of great importance in many scopes of environmental applications. A physical parameter, herein, describes the characteristics and conditions of a physical object, state, or process. One example of such a physical parameter is soil moisture.

When monitoring such continuous physical parameters directly on site, conventional point-wise measurement techniques are most widely used. These techniques measure continuous values of the respective parameter by analyzing in situ probes. Such in situ measurements are precise at a specific location but are often inefficient for covering a large area.

For monitoring physical parameters over large areas, hyperspectral remote sensing techniques are applied as complementary solutions (see, e.g., [29]). Hyperspectral sensors mounted on satellites, unmanned aerial vehicles (UAVs), or handhelds record this kind of data with different spectral, temporal, and spatial resolutions depending on the applied platform. In general, the recorded hyperspectral data contains, to a certain extent, spectral information related to the physical parameter to be considered. Hyperspectral remote sensing aims to retrieve this spectral information out of the recorded hyperspectral data. To estimate physical parameters with hyperspectral data, a model is required to link the hyperspectral data and the information about the physical parameter. Such a model can be obtained based on a training dataset containing hyperspectral data and reference data of the respective physical parameter. The reference data or ground truth is measured in situ. Hyperspectral and in situ measured data both characterize the physical parameter. However, they differ, for example, in terms of sampling time and spatial coverage which poses a challenge when combining such data in one dataset.

As an example, we consider the estimation of soil moisture over a large area with hyperspectral data (e.g., [62]). Soil moisture is a physical parameter which is relevant, for example, in hydrological modeling of river catchments. Hyperspectral data may be available from field campaigns. As a reference, in situ measurement data of soil moisture at specific points of the same area is needed. The underlying task, now, is to estimate soil moisture by linking soil moisture reference data to the hyperspectral data.

By definition, hyperspectral data is high-dimensional. The linkage of such high-dimensional input data with 1-dimensional (1D) soil moisture data represents a non-linear regression problem. In this chapter, we focus on data-driven machine learning (ML) models since they are capable of dealing with these kinds of regression tasks [62]. We introduce the term **hyperspectral regression** which refers to ML regression solely based on hyperspectral data. Note that some studies use the term *parameter estimation* as a synonym for hyperspectral regression. However, in the broad field of ML, the definition of the term parameter estimation varies depending on the applied context.

There is no single ML model which is equally suited on all regression tasks [108]. Instead, a framework of different approaches is used which can adapt to various regression tasks. In this chapter, we demonstrate all aspects of a typical hyperspectral regression workflow on a soil moisture dataset [99]. The dataset is introduced in [99] and can be downloaded from [98]. Note that this is a 1D dataset with single pixels instead of 2-dimensional (2D) images. It consists of 679 datapoints with one soil moisture value and 125 hyperspectral bands per datapoint. The presented models can also be combined with other types of datasets. First regression results [99] demonstrate very precise estimations and imply a very low Bayes error, which is the lowest possible estimation error of the given ML task. In summary, the dataset represents an optimal benchmark dataset for applying ML models in hyperspectral regression.

Figure 7.1 illustrates the **structure** of this chapter. The structure is based on a typical hyperspectral regression workflow and is divided into three levels, data level, feature level, and model level. At first, we give an overview of different learning techniques and definitions of technical terms which we rely on later (see Sect. 7.2). Subsequently, we introduce the concept of pre-processing, the challenge of dataset shift, and dataset splitting in Sect. 7.3. Section 7.4 deals with any kind of pre-processing aspects. Finally, we have a detailed look at the different ML models in Sect. 7.5. Additionally, we provide comprehensive code examples and accompanying materials in the form of data and Python notebooks online on GitHub ($\rightarrow$ Notebooks 1 to 7 (https://github.com/felixriese/hyperspectral-regression)) [100].

In this chapter, we present the essential steps of a hyperspectral regression workflow in detail and in an application-oriented way. Further exemplary applications of hyperspectral regression are also presented in Chap. 2. In summary, the **objectives** of this chapter are

- to understand the **possibilities** and **challenges** in hyperspectral regression,
- to gain an overview of a typical hyperspectral regression **workflow**,
- to analyze and **pre-process** the used dataset,
- to understand the challenge of **dataset shift**,
- to understand and to apply different **dataset splitting** approaches,
- to **generate new features** with unsupervised dimensionality reduction, unsupervised clustering, and manual feature engineering,
- to **select** the most important input **features** for a hyperspectral regression task,
- to understand the different strengths and weaknesses of the most relevant **ML models** for hyperspectral regression,
- to **select** the most appropriate supervised or semi-supervised **ML model** for a given hyperspectral regression task,
- to understand the possible applications of **active learning** models for hyperspectral regression,
- to **optimize** and finally to **evaluate** a selected ML model based on the given regression task,
- to apply **Python packages** in the context of hyperspectral regression with our **best practices** and **implementation** examples [100], and
- to gain an overview of the upcoming **trends** in hyperspectral regression.

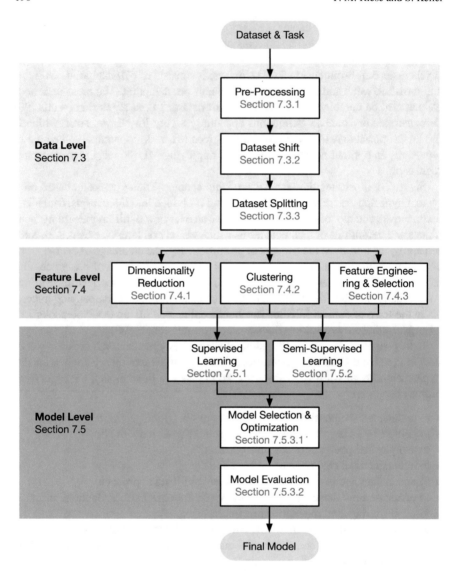

Fig. 7.1 Typical hyperspectral regression workflow on three levels: model level, feature level, and data level

7.2 Fundamentals of Hyperspectral Regression

In recent years, the hyperspectral remote sensing community mainly focused on classification tasks. Classification refers to the estimation of discrete classes, for example, to distinguish between land cover classes like *water*, *vegetation*, *road*, and *building*. Both classification and regression are about building predictive models. The difference is that, in classification, the target space is discrete (e.g., land cover classes), whereas in regression, the targets are continuous (e.g., soil moisture).

In the context of ML regression, different approaches can be applied depending on the objective and the availability of reference data. In Fig. 7.2, these different approaches are visualized schematically. We can distinguish four cases. In case (a), reference data, meaning labels containing the ground truth, is available for all (hyperspectral) input datapoints. In this context, **supervised learning** models are suitable. A supervised model is able to learn from all available input–output data pairs. In case (b), we have an incompletely labeled dataset. That is, some of the samples are missing the correct ground truth labels. In this context, we can rely on **semi-supervised learning** models. They learn from the complete input–output pairs as well as from the datapoints without labels. One extension of semi-supervised learning is **active learning**. In this case, case (c), the active learning model is able to suggest the user, for example, a human, which missing labels would increase the estimation performance the most. Active learning is of use when collecting reference data (labels) is expensive and time-consuming. Finally, in the case (d) when no labels are available, **unsupervised learning** can be applied. Unsupervised learning is useful, for example, for dimensionality reduction and clustering.

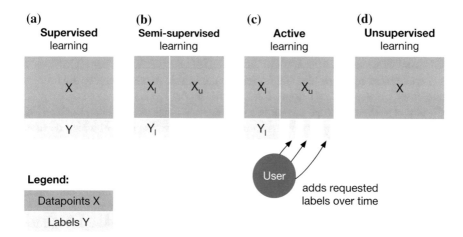

Fig. 7.2 Depending on the availability of labels for our training data, we can distinguish four types of learning algorithms: Overview of the availability of labels in four ML approaches: **a** supervised learning, **b** semi-supervised learning, **c** active learning, and **d** unsupervised learning

The **mathematical notation conventions** used in this chapter are consistent with [25]: $X = (x_1, \ldots x_n)$ is a set of n input datapoints $x_i \in X$ for all $i \in [n] := \{1, \ldots n\}$. Every datapoint x_i consists of m input features. In hyperspectral regression, the input features represent the m hyperspectral bands and it is $X \subset \mathbb{R}^m$. In supervised learning, case (a), $y_i \in Y$ with $Y = (y_1, \ldots, y_n)$ are the labels of the datapoints x_i and the training set is given as pairs (x_i, y_i). In semi-supervised and active learning, cases (b) and (c), the dataset X is divided into two parts. The first part consists of the datapoints $X_l := (x_1, \ldots, x_l)$ with the corresponding labels $Y_l := (y_1, \ldots, y_l)$ and the second part consists of the datapoints $X_u := (x_{l+1}, \ldots, x_{l+u})$ without any labels. It is $l + u = n$. Again, we have $y_i \in Y$ for $i = 1, \ldots, l$. For regression, the labels are continuous in the cases (a) to (c) which means $Y \subset \mathbb{R}$. Note that also more-dimensional labels can be used in regression. In the d-dimensional case, it is $Y \subset \mathbb{R}^d$. Within the scope of this chapter, we will stick to 1D labels, meaning $d = 1$. We refer to this combination of hyperspectral input data and desired output data as **datapoint**. In the unsupervised case (d), the dataset only consists of input datapoints X without any labels.

In the field of hyperspectral remote sensing and in the analysis of hyperspectral data, there are many **applications for ML**. Relevant examples of hyperspectral regression with ML are clustered according to their respective target variables in Table 7.1. A general overview of remote sensing image processing with a focus on traditional ML models and physical models is given in [23]. Most current studies address ML classification with hyperspectral data (e.g., overview in [45]), whereas only few studies focus on hyperspectral regression (e.g., [3, 119]). The ML models used for the respective regression tasks are described in Sect. 7.5.

Depending on the hyperspectral regression task, we need to select an appropriate ML model [108]. At best, the selected ML model is able to learn all relevant nuances of the training dataset (low bias) and is able to generalize well on unknown datasets (low variance). Accomplishing low bias and low variance at the same time is impossible. Thus, a **trade-off between bias and variance** [41, 43, 81] has to be addressed while selecting an appropriate model (see Sect. 7.5.3). When an ML model is characterized by a low bias (high variance), it is able to adapt well to the training dataset which also includes noise. Such an ML model tends toward **overfitting**. An ML model with low variance is more robust against noise and outliers. Such an ML model is not able to adapt well to the nuances of the training dataset which is called **underfitting**.

When focusing on hyperspectral regression, several **challenges** need to be handled (e.g., [14]). For instance, hyperspectral data is characterized by high dimensionality and narrow bandwidths. As a result, hyperspectral bands are highly correlated. The intrinsic, virtual dimensionality of the hyperspectral data, therefore, is much smaller. Determining the virtual dimensionality is a difficult challenge [24]. Most approaches suggest dimensionality reduction of the data which often is performed in an unsupervised fashion (see Sect. 7.4.1).

With the high dimensionality of the hyperspectral input data, ML models suffer from the **curse of dimensionality** [10]. The increasing number of input dimensions leads to an exponential growth of the feature-space volume. Therefore, more

Table 7.1 Examples for hyperspectral regression with ML

Target variable	References
Background estimation	[118]
Biomass	[3, 27, 84]
CDOM, diatoms, green algae, turbidity	[61, 75]
Chlorophyll *a* concentration	[22, 61, 75–77]
Nitrogen content	[2, 69, 70, 82, 133]
Soil moisture	[3, 62, 82, 97, 99, 113]
Soil organic/inorganic content	[19, 70, 82, 92]
Soil texture: sand, silt, clay content	[19, 70, 71]
Vegetation pigment content	[26]

training data is needed to cover this volume with the same density. Increasing the size of the training dataset is one possible and often applied solution to handle this challenge. Since new developments in optical remote sensing have emerged over the last decades, the technical possibility exists to record large hyperspectral datasets. For instance, satellites and UAVs are capable of recording hyperspectral data on a large scale. Additionally, the processing of large datasets is computationally expensive.

Another important challenge is the measurement of reference data over large areas. Continuous physical parameters (see Table 7.1) like soil moisture and soil texture need to be measured manually. This is time-consuming and expensive. It is, therefore, important to be able to work with **small datasets or with incompletely labeled datasets** (see Sect. 7.5.2). In addition, the dataset shift poses a further challenge in the context of hyperspectral regression. Dataset shift is caused by differences between the training dataset and new, for the ML model unknown, datasets. Accompanying issues and possible solutions are pointed out in Sect. 7.3.2.

7.3   Hyperspectral Regression at the Data Level

The data level is the first level of the presented hyperspectral regression workflow. At the data level, pre-processing as the first part focuses on collecting, validating, and preparing data (see Sect. 7.3.1). The second part addresses the challenges of dataset shift and provides possible approaches to cope with it (see Sect. 7.3.2). We conclude by introducing several approaches of dataset splitting for the evaluation of ML models in Sect. 7.3.3.

7.3.1 Pre-processing

The first part of the presented hyperspectral regression workflow is pre-processing [23, 124]. We divide the pre-processing into three steps: reading in data, preparing data, and validating data. First, we need to **read in the data**. In Python, datasets can be conveniently read in using existing and established software packages such as *Pandas* [80] or *TensorFlow* [1].

The second step of pre-processing is the **validation of the data**. We highly recommend to explore the dataset before further processing (see → Notebook 1.1 (https://github.com/felixriese/hyperspectral-regression)). The exploration procedure could include a check of the value range of the input features and the target variable. The data validation can be achieved by an analysis of the datasets statistics and by a visualization of the dataset. Thus, we obtain an overview of the used dataset. Additionally, we recognize possible challenges in the dataset such as outliers, missing values or labels as well as dataset shift at an early stage. The latter is addressed in detail in Sect. 7.3.2. A useful example to motivate the investigation of the dataset with statistical methods and visualizations is given in [78].

The last step of pre-processing is the **preparation of the data**. Depending on the results of the data validation and the applied ML models (see Sect. 7.5), the dataset might need to be normalized or transformed. The data normalization makes the training less dependent on the scale of the input data. Typical normalization techniques scale the numerical data, for example, linearly between 0 and 1, or around 0 with a standard deviation of 1. Additionally, it might be necessary to transform categorical data to numerical values since some ML models like artificial neural networks (ANN, see Sect. 7.5.1.5) only work with numerical data. A common way to achieve this is *one hot encoding*. Each categorical feature is represented by one entry in a binary vector.

Exemplary implementations of pre-processing and resulting plots can be found in → Notebook 1.1 (https://github.com/felixriese/hyperspectral-regression). In the following, we summarize three **best practices**:

- Read in data with **existing packages** like the Python package *Pandas* [80] or pre-existing functions in *TensorFlow* [1] which support many common file formats.
- **Visualize** the dataset *and* generate **statistics** about it. Use perceptually uniform colormaps, for example, *viridis* [72]. **Understand** your data!
- Use data **normalization** or transformation if the applied ML model requires it.

7.3.2 Dataset Shift

Most ML models rely on the *independent and identically distributed* (i.i.d.) assumption. The i.i.d. assumption refers to the independent collection of the training dataset and new, unknown datasets (see Sect. 7.3.3) which are identically distributed.

In this context, the term *training dataset* refers to the dataset that is available during the training of the ML model. For example, in hyperspectral regression of soil moisture, the hyperspectral data as well as the ground truth labels of soil moisture should cover all (in reality) possible values. Otherwise, **dataset shift** occurs and the estimation performance might suffer [83, 94]. In general, three main types of dataset shift exist:

- **Covariate shift** [109, 115] is defined as a change of the input feature distribution $P(X)$. It is the best studied type of dataset shift in the literature. For example, in hyperspectral regression of soil moisture, rainfall events between two measurement days affect the input feature distribution of this two-day dataset.
- **Prior probability shift** [36, 94] is defined as a change of the target variable distribution $P(Y)$ without a change in X. This change mostly occurs in the application of generative models. For example, in the hyperspectral regression of soil moisture, the distribution of soil moisture can vary due to the underlying soil structure while the soil surface remains unchanged.
- **Concept shift** [127] or concept drift is a change in the relationship between the input data and the target variable. The concept shift is the most challenging type of dataset shift to handle. For example, in hyperspectral regression of chlorophyll *a* concentration, the relationship between hyperspectral input data and chlorophyll *a* concentration as target variable can change due to undetectable hydrochemical processes.

In the following, we present an **example of covariate shift** in hyperspectral regression of soil moisture which is the most relevant type of dataset shift for this application. The distributions of hyperspectral and reference data are shifted between the training dataset and an unknown dataset. The distribution of the exemplary soil moisture reference data is presented in Fig. 7.3. After training on the training dataset, the ML model should be able to estimate soil moisture on a new, unknown dataset. However, as a result of the covariate shift, the ML model is not able to estimate soil moisture reasonably on the given, unknown dataset ($R^2 \approx 0$, see → Notebook 1.2 (https://github.com/felixriese/hyperspectral-regression) and Sect. 7.5.3.2).

Several **causes of dataset shift** exist. One cause is the **sample selection bias**. Sample selection bias can occur in the scope of different data measurements. In hyperspectral regression, it often occurs as a result of the parallel use of different hyperspectral sensors and changes of the measuring site. Another cause for dataset shift is **non-stationary environments**. Non-stationary environments appear when the training environment differs from the test environment. This distinction can be temporal or spatial. Since hyperspectral satellites record data at different locations and during different seasons, dataset shift commonly occurs.

Various ways exist to deal with the challenges of dataset shift. In most ML studies for hyperspectral regression, dataset shift is simply ignored. In this case, the applied model is static with regard to the dataset shift. Such models can be used further as a baseline model allowing the detection of dataset shift and enabling the evaluation of approaches aiming at the reduction of the effects of dataset shift.

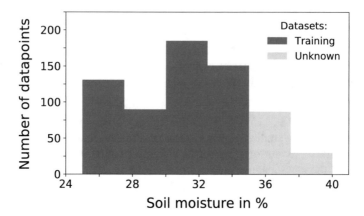

Fig. 7.3 Target variable distribution of the presented soil moisture dataset [99] with exemplary dataset shift from training subset to prediction subset

A first approach to reduce the effects of dataset shift is to **re-fit or update** the ML model to new data. In the case of time series, this means re-fitting or updating the ML model on more recent data. In the case of 2D areal data, this means re-fitting on more training areas. Another approach is to **re-weight** the training dataset based on temporal (time series) or spatial (2D data) features. For example, training data of time series can be re-weighted so that newer datapoints are more important in the training than preceding ones. Further, the ML model can be set up to inherently learn temporal changes to reduce the bias of seasonality and timing. In the following, we summarize our **best practices** on how to deal with dataset shift:

- **Visualize** your data and use simple baseline models to detect dataset shift.
- If possible, **update** (otherwise re-fit) your ML model regularly using new data.

7.3.3 Dataset Splitting

To evaluate the **generalization** abilities of an ML model, the full available dataset needs to be split into smaller datasets. In general, dataset splitting should meet the i.i.d. assumption (see Sect. 7.3.2). In Fig. 7.4, the two most commonly applied split types are illustrated. In the first type, the full dataset is split into two subsets: **training** and **test**. In the second type, the three subsets training, **validation**, and test are generated. In both split types, the training dataset is used repeatedly to train the ML model. The test dataset is used only once to evaluate the final ML model. The split types differ with respect to the way the ML models are optimized (see Sect. 7.5.3.1). In the 3-subset split, the validation dataset is repeatedly used for the evaluation of the generalization abilities of the ML model in the optimization process. In the 2-subset split, the training dataset is used for both training and evaluation in the optimization

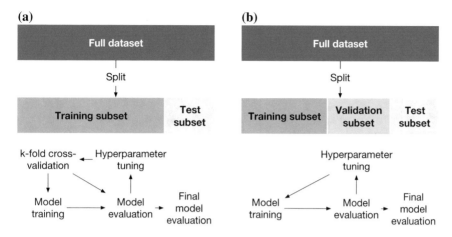

Fig. 7.4 **a** Dataset splitting into the two subsets: training and test. For the model optimization, a k-fold cross-validation is applied. **b** Dataset splitting into three subsets: training, validation, and test. The model optimization is performed on the training and validation dataset

process by applying a **k-fold cross-validation**. Within the k-fold cross-validation, the training dataset is randomly partitioned into k subsets of similar size. One of the k subsets is then used for the evaluation of the ML model, while the remaining $k - 1$ subsets are used for the training of the ML model. This selection is repeated so that every subset is used once as validation subset. Note that it is not trivial to apply k-fold cross-validation on time series due to possible casual relationships.

After deciding on the number of dataset subsets, the splitting approach needs to be defined. In the following, we present several dataset splitting approaches which are illustrated in Fig. 7.5. They are described in detail in [114] with their respective strengths and weaknesses. The most commonly applied splitting approach is a **random split** or random sampling (e.g., [39]). The subsets are randomly sampled which leads to subsets with relatively similar target variable distributions. However, for spatially or temporarily correlated data like 2D hyperspectral image data or time series, a pixel-wise random split can lead to biased subsets. Since a significant number of datapoints in one subset have direct spatial or temporal neighbors, the datapoints are highly correlated in between the subsets. Training on one datapoint and evaluating the model performance on a neighboring datapoint leads to highly biased results.

Another splitting approach is **systematic splitting** or systematic sampling. For this approach, every kth datapoint is used for the test subset, with k defined as

$$k = \frac{n}{n_{\text{test}}}. \tag{7.1}$$

The total number of datapoints is given as n and the number of datapoints for the test dataset is given as n_{test}. Systematic splitting does not rely on a random number generator which simplifies the implementation. Overall, systematic splitting is more

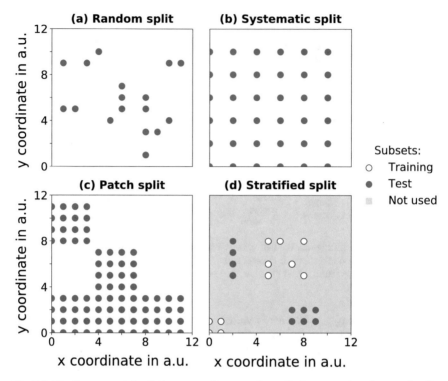

Fig. 7.5 The four presented splitting approaches: **a** random split, **b** systematic split, **c** patch split, and **d** stratified split

robust against spatially or temporally clustered regression inaccuracies. One of the shortcomings, though, is the assumption that the data is homogeneously distributed. If the dataset shows periodical patterns, systematic splitting generally performs badly.

A further approach to split datasets in ML is **patch splitting** or patch sampling. In patch splitting, the data is split into patches or blocks. In the case of hyperspectral 2D images, an image is split into a chessboard-like pattern. Time series data is split into time blocks. The split into the different subsets is randomly performed, similar to the random split but patch-wise instead of pixel-wise. This splitting approach reduces the spatial and temporal bias, while maintaining similar distributions of the different subsets is more difficult.

The last presented splitting approach is called **stratified splitting** or stratified sampling. It combines a pre-partitioning of the data into distinct areas with random splitting on each of these areas individually [114]. The assumption of stratified splitting is that the pre-partitioning generates representative but distinct areas to separate training and test subset as well as similar distributions (see *i.i.d. assumption* in Sect. 7.3.2). For example, 2D hyperspectral data can be partitioned according to geographical areas. In a hyperspectral classification, we can, for example, partition the data based on land use classes.

Table 7.2 Strengths and weaknesses of different splitting methods

Split	Strengths	Weaknesses
Random	• Similar distributions • Simple	• Spatial & temporal bias
Systematic	• Similar distributions • Simple	• Spatial & temporal bias • Homogeneous data only
Patch	• Less bias	• Different distributions
Stratified	• Similar distributions • Smaller dataset → faster model training	• Partition-able data only • Pre-partitioning needs time

The strengths and weaknesses of the presented dataset splitting approaches are listed in Table 7.2. Implementation examples can be found in → Notebook 1.3 (https://github.com/felixriese/hyperspectral-regression). In the following, we list the most important **best practices** for dataset splitting:

- **Split** your data. Without dataset splitting, meaningful model evaluation (see Sect. 7.5.3.2) is not possible.
- Use **existing random number generators** like the Python package *Numpy* [126] for randomization (see → Notebook 1.3 (https://github.com/felixriese/hyperspectral-regression)).
- Use **splitting ratios** of ≈ 70/30 in the 2-subset split and ≈ 60/20/20 in the 3-subset split for small datasets between 100 and 1000 datapoints. For larger datasets with 1 000 000 and more datapoints, we recommend a 3-subset split with a splitting ratio of 98/1/1.
- Try random splitting for **1D hyperspectral data**.
- Use stratified splitting for **2D hyperspectral data** instead of random splitting to avoid spatial bias.
- Try patch splitting for **time series** data.

7.4 Hyperspectral Regression at the Feature Level

The **feature level** follows the data level in our presented hyperspectral regression workflow. It consists of three parts: unsupervised dimensionality reduction (Sect. 7.4.1), unsupervised clustering (Sect. 7.4.2), and feature engineering as well as feature selection (Sect. 7.4.3). The definition of unsupervised learning is explained in Sect. 7.2.

7.4.1 Dimensionality Reduction

Since correlations and redundancies between input features can occur, the virtual dimensionality of a dataset is often smaller than the given dimensionality [24]. The term **dimensionality reduction** refers to the reduction of the dimension m of the input data to a smaller dimension $m_r \leq m$ toward the virtual dimensionality. In addition, the term **compression** focuses on the reduction of the dimension m of the data to the smallest possible $m_{\min} \leq m_r \leq m$. In most cases after applying dimensionality reduction, it is only possible to reconstruct *similar* data, not the original input data. The topic of dimensionality reduction and compression in general is reviewed in detail in [54, 121]. Note that the term *feature extraction* is often used instead of dimensionality reduction (see, e.g., [59]).

We discuss in the following the most relevant approaches of dimensionality reduction in hyperspectral regression (see Table 7.3). A commonly applied approach is the **Principal Component Analysis** (PCA) [90]. The PCA transforms the input data orthogonally based on the variance along newly found axes. These new axes are referred to as principal components. The principal components are sorted by decreasing variance. That is, the first principal component has the largest variance. Therefore, the set of the first few principal components contain most of a dataset's variance and at best, most of the information contained in the dataset.

A further approach of dimensionality reduction is called **Maximum Noise Fraction** (MNF) [50]. MNF applies PCA, but rather than maximizing the variance along the principal components, it maximizes the signal-to-noise ratio. Note that in several studies the MNF is called *minimum* noise fraction.

Autoencoder (AE) [55] is an artificial neural network (ANN) approach for dimensionality reduction. An AE consists of an input layer of input dimension m, followed by several hidden layers with smaller dimension $m_{\text{hidden}} < m$ and an output layer of size m. The dimension reduction of input to hidden layers is called *encoding*. In the encoding, the AE finds a lower dimensional representation of the input data. The dimension increase of the encoded data in the original dimension m is called *decod-*

Table 7.3 Overview of unsupervised learning approaches for dimensionality reduction

Approach	Implementation	Reference	Exemplary applications
PCA	→ 2.1.1 (https://github.com/felixriese/ hyperspectral-regression)	[90]	[61, 137]
MNF		[50]	[67]
AE	→ 2.1.2 (https://github.com/felixriese/ hyperspectral-regression)	[55]	[117, 128]
t-SNE	→ 2.1.3 (https://github.com/felixriese/ hyperspectral-regression)	[73]	[40, 136]
UMAP	→ 2.1.4 (https://github.com/felixriese/ hyperspectral-regression)	[79]	[111]

ing. The AE is trained in an unsupervised manner with the hyperspectral data for both input and (desired) output data. Then, the encoding part of the trained AE can be used for dimensionality reduction on the (hyperspectral) input data. In sum, the full AE with encoding and decoding can also be used for noise removal (denoising) of the hyperspectral input data. More details about ANN are presented in Sect. 7.5.1.5. Since an AE consists of many free parameters, large training datasets are necessary for the training. Note that only the number of hyperspectral input datapoints needs to be large for the AE. The dataset that includes ground truth labels can be small. Since the combination of large input data and small ground truth data is characteristic for multi- and hyperspectral satellite data, AE is well suited in this context. Additional details of AE and PCA applied in hyperspectral image analysis are discussed in Chaps. 3 and 13.

Finally, we list two additional dimensionality reduction approaches. The **t-distributed stochastic neighbor embedding** (t-SNE) [73] is a non-linear approach which reduces high-dimensional input data to a dataset with the dimension $m_r \in \{2, 3\}$. Therefore, this approach is well suited not only for dimensionality reduction but for the visualization of a dataset as well. A recently presented dimensionality reduction approach is called **uniform manifold approximation and projection** (UMAP) [79]. UMAP is comparable with the t-SNE algorithm incorporating several advantages in terms of speed and performance. Since UMAP is a relatively new approach, it has to be investigated further in context of hyperspectral regression.

In Table 7.4, the strengths and weaknesses of the five presented dimensionality reduction approaches are listed. Exemplary visualizations of the first two components of PCA, AE, t-SNE, and UMAP are shown in Fig. 7.6. All four approaches show different distributions of visible clusters of datapoints with similar soil moisture values. The implementations for the presented algorithms can be found in → Notebook 2.1 (https://github.com/felixriese/hyperspectral-regression). In the following, we list several best practices for dimensionality reduction in hyperspectral regression:

- **Normalize** the data before applying dimensionality reduction.
- Use PCA a **simple and fast baseline** for further approaches.
- Apply AE on datasets with **sufficient input data** like satellite images.
- **Try UMAP** as a relatively new approach generating promising results.

7.4.2 Clustering

Clustering a dataset means the grouping datapoints with respect to a pre-defined similarity metric. Datapoints are clustered, mostly in an unsupervised manner, based on the input features such as hyperspectral bands. When clustering is included in the hyperspectral regression workflow, the resulting cluster information is added as a new input feature used to train the ML model. In this section, we discuss the most commonly applied clustering algorithms in hyperspectral regression (see Table 7.5).

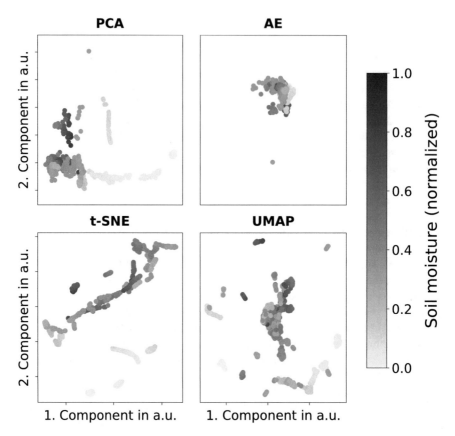

Fig. 7.6 Exemplary visualization of the introduced hyperspectral dataset [99] for the first two components of PCA, AE, t-SNE, and UMAP. The color of the datapoints corresponds to the normalized soil moisture

Table 7.4 Strengths and weaknesses of dimensionality reduction methods

	Strengths	Weaknesses
PCA	• Fast and simple • Many extensions	• Difficult to interpret • Variance $\neq$ information • Linearity assumption
MNF	• Focus on signal versus noise	• Noise estimation needed • No Python implementation
AE	• Strong performance • (Deep) architecture	• Big (input) data required • Slow
t-SNE	• Non-linear, powerful • Visualization	• Only 2D and 3D output • Slow
UMAP	• Fast and powerful	• Not (yet) established

Table 7.5 Overview of unsupervised learning approaches for clustering

Approach	Implementation	Reference	Exemplary applications
k-means	→ 2.2.1 (https://github.com/felixriese/ hyperspectral-regression)	[74]	[7, 116, 125]
DBSCAN	→ 2.2.2 (https://github.com/felixriese/ hyperspectral-regression)	[35]	[31]
SOM	→ 2.2.3 (https://github.com/felixriese/ hyperspectral-regression)	[65]	[99, 102]

In **k-means** clustering [74], the datapoints are grouped into a fixed number of k clusters. Each cluster is defined by its cluster center which is found by minimizing the sum of distances of the datapoints to their respective nearest cluster center. An example of a distance metric is the Euclidean distance d which is defined for the vectors $a, b \in \mathbb{R}^m$ as

$$d(a, b) = \sqrt{\sum_{i=1}^{m}(a_i - b_i)^2}. \tag{7.2}$$

Another clustering algorithm is called density-based spatial clustering of applications with noise (**DBSCAN**) [35]. In the DBSCAN algorithm, clusters are defined as areas of higher density in the feature space. Higher density in this context means that the density of the respective areas in the feature space is higher than the average density of the dataset. Thus, no pre-definition of the number of clusters as in k-means is necessary. Another difference to the k-means algorithm is that some datapoints are not assigned to a cluster. However, based on their localization in low-density areas, they are considered *outliers.*

Clustering can also be performed by **self-organizing maps** (SOM) [65] (see also Sect. 7.5.1.6). A SOM is, in general, an unsupervised learning approach for data visualization and clustering. It is a type of neural network with one input layer and a 2D grid as output layer. Output and input layers are fully connected. Neurons on the output layer are linked by a neighborhood relationship. The SOM adapts to a dataset by adapting the neuron weights through a distance measure (e.g., Euclidean distance, see Eq. (7.2)). Besides the neuron with the smallest distance value, the neurons in the neighborhood are adapted as well. Overfitting is reduced by this neighborhood relationship.

In Table 7.6, the strengths and weaknesses of clustering algorithms in hyperspectral regression are listed. This overview enables the reader to choose a suitable clustering algorithm for the task at hand. All clustering implementations can be found in → Notebook 2.2 (https://github.com/felixriese/hyperspectral-regression).

Table 7.6 Strengths and weaknesses of clustering algorithms

	Strengths	Weaknesses
k-means	• Fast • Every datapoint in one "hard" cluster	• Pre-defined k
DBSCAN	• Finds number of clusters • Most datapoints in one "hard" cluster • Outlier detector	• Difficult tuning • Slow
SOM	• Preserved topology on 2D grid • Supervised extension (Sect. 7.5.1.6)	• Difficult tuning • Slow • Datapoints in "soft" clusters

7.4.3 Feature Engineering and Feature Selection

In Sects. 7.4.1 and 7.4.2, we apply dimensionality reduction and clustering to generate new features. In contrast to these data-driven approaches, **feature engineering** is based on prior knowledge. The generated features can be categorized as spectral features or spatial features. The engineering of **spectral features** is inspired by physical processes. Spectral features are commonly characterized by a ratio or the normalized difference of hyperspectral bands. The most popular example in hyperspectral regression is the normalized difference vegetation index (NDVI) [103] which corresponds to photosynthesis processes. **Spatial features** are often generated based on contextual information of neighboring pixels (datapoints). Examples for spatial features are objects, edges, and contours. They are generally created by the application of filters. Note that spatial features can only be generated from hyperspectral images when their corresponding spatial resolution is adequate.

In contrast to feature engineering, **feature selection** describes the process of selecting a subset of all available input features which can be used as input data for supervised ML models. In context of hyperspectral regression, the term *band selection* is often used instead of the term feature selection. The main advantage of feature selection over feature engineering or dimensionality reduction is that the features (hyperspectral bands) are physically meaningful. For example, principal components cannot be interpreted physically (see Sect. 7.4.1). Therefore, feature selection applied on data of one sensor can be transferred to data of another sensor with slightly different hyperspectral bands.

Three main approaches exist in feature selection: filter methods, wrapper methods, and embedded methods [15, 64]. **Filter** methods select features based on quality measures like the correlations between features and target variable as well as correlations between individual features. The main disadvantage of filter methods is that they only consider relationships between *two* variables, either "feature-to-target" or "feature-to-feature". **Wrapper** methods select feature subsets based on the relationship of these feature subsets with the target variable. A third option is provided by some supervised learning models such as tree-based models which have their own built-in feature selection included in the estimation process (see *feature importance* in Sect. 7.5.1.2). These built-in feature selections are called **embedded** methods.

An overview of feature selection is presented in [51]. A review on several applications of feature engineering and feature selection in the context of remote sensing image processing is provided in [23]. An example application of feature selection is given in Chap. 11. Exemplary implementations of feature engineering and feature selection are shown in → Notebook 2.3 (https://github.com/felixriese/hyperspectral-regression). The use of feature selection and feature engineering depends on the dataset as well as on the applied supervised ML model. We recommend in the following several **best practices** in hyperspectral regression:

- **Consider** feature engineering or feature selection when working with small datasets, especially for supervised ML models like ANNs (see Sect. 7.5.1.5).
- **Do not use feature engineering or selection** for supervised ML models like random forest (see Sect. 7.5.1.2). Also, do not use feature selection with deep ANNs (Sect. 7.5.1.5) in the case of large datasets since they are able to learn new by themselves.

7.5 Hyperspectral Regression at the Model Level

Regression is defined as the estimation of continuous parameters with input data. This estimation is based on mapping input data with desired output data with a specific ML model. In *supervised learning*, there exists an output for every input in the training dataset. In Sect. 7.5.1, we introduce several supervised learning models. *Semi-supervised learning* refers to the case when only a few complete input–output pairs are available. The rest of the input data is missing a desired target output. In Sect. 7.5.2, several approaches for semi-supervised learning with hyperspectral data are presented. Finally, we give an overview about model selection strategies, model optimization, and model evaluation in Sect. 7.5.3.

7.5.1 Supervised Learning Models

In practice, solving a regression problem often requires more than just a single ML model. To evaluate the model performance, appropriate task-specific metrics must be used to compare and finally select the best ML model for the problem at hand. Hence, we focus on the most relevant ML models in hyperspectral regression: linear and partial least squares regression (Sect. 7.5.1.1), tree-based models (Sect. 7.5.1.2), support vector machines (Sect. 7.5.1.3), k-nearest neighbors (Sect. 7.5.1.4), artificial neural networks (Sect. 7.5.1.5), and self-organizing maps (Sect. 7.5.1.6). Table 7.7 briefly lists these models with their respective references as well as the references to relevant applications.

Table 7.7 Overview of different supervised learning models

Model	Section	Implementation	Reference	Exemplary applications
LIN	7.5.1.1	→ 3.1 (https://github.com/felixriese/hyperspectral-regression)		
PLS	7.5.1.1	→ 3.1 (https://github.com/felixriese/hyperspectral-regression)	[129]	[19, 27, 69, 70]
RF	7.5.1.2	→ 3.2 (https://github.com/felixriese/hyperspectral-regression)	[17]	[2, 75, 84]
ET	7.5.1.2	→ 3.2 (https://github.com/felixriese/hyperspectral-regression)	[44]	[101]
GTB	7.5.1.2	→ 3.2 (https://github.com/felixriese/hyperspectral-regression)	[16, 42]	[19, 70]
SVM	7.5.1.3	→ 3.3 (https://github.com/felixriese/hyperspectral-regression)	[122]	[22, 82, 92, 113]
k-NN	7.5.1.4	→ 3.4 (https://github.com/felixriese/hyperspectral-regression)	[4]	[62, 118]
ANN	7.5.1.5	→ 3.5 (https://github.com/felixriese/hyperspectral-regression)	e.g., [41]	[5, 26, 133]
CNN	7.5.1.5	→ 3.5 (https://github.com/felixriese/hyperspectral-regression)	[66]	[71, 134]
RNN, LSTM	7.5.1.5	→ 3.5 (https://github.com/felixriese/hyperspectral-regression)	[107], [56]	[134]
Supervised SOM	7.5.1.6	→ 3.6 (https://github.com/felixriese/hyperspectral-regression)	[65, 99, 102]	[61, 62]

7.5.1.1 Linear and Partial Least Squares Regression

One of the simplest machine learning models for estimating physical parameters is the **linear regression** (LIN). One formulation of the underlying mathematical model is assigning one coefficient β_j to every dimension of the input data, in this case per hyperspectral band. In addition, β_0 is often added as offset term. In combination with the error term ϵ_i which corresponds to the estimation error of each input–output pair (x_i, y_i), the LIN can be formulated as

$$y_i = \beta_0 \cdot 1 + \beta_1 x_{i1} + \beta_2 x_{i2} \ldots + \epsilon_i =: x_i^T \beta + \epsilon_i$$
$$\rightarrow \quad \epsilon_i = y_i - x_i^T \beta. \tag{7.3}$$

LIN aims to find values for all β_i which minimize the error term ϵ_i for all datapoints (x_i, y_i). One common minimization technique is least squares. The sum of squared residuals S is defined as

$$S = \sum_{i=1}^{n} \epsilon_i^2. \tag{7.4}$$

Then, the factors β_i are modified to minimize S. Note that in the presented formulation of LIN, there is exactly one analytical solution for a given regression task. The final model is LIN estimation for the applied dataset.

The explained LIN is rarely applied on hyperspectral data. Two studies have compared this model to nine other regression models [61, 62]. In conclusion, LIN is unable to solve non-linear regression problems such as the estimation of soil moisture or chlorophyll a concentration.

If the dimension of x_i is significantly larger than the number of datapoints n, as it is in the case of hyperspectral data, it is difficult to apply LIN. Another challenge for LIN is multicollinearity, which means the strong linear relationship between hyperspectral bands. **Partial least squares** (PLS) regression [129] as a bilinear factor model can handle this high dimensionality and the multicollinearity by projecting X and Y into new spaces. The aim is to find the direction in the X space that corresponds to the direction of the maximum variance in the Y space. PLS has similarities to PCA (see Table 7.3) but also includes the target variable space Y. The mathematical model is described in detail, for example, in [130].

In contrast to LIN, PLS is widely used in hyperspectral regression. In a study of the USDA National Soil Survey Center [19], PLS is applied, for example, to estimate clay content, soil organic content, and inorganic content based on VNIR spectroscopy. According to this study, PLS is easy to use, but it is outperformed by other approaches like tree-based models (see Sect. 7.5.1.2). Additionally, PLS is used to estimate biomass with HyMap airborne images [27]. In their study, the authors test different combinations of pre-processing such as band selection as well as the use of indices like the normalized difference vegetation index (NDVI). They find, the PLS performance cannot be increased by band selection. This finding is an example of the robustness of the regression with the PLS model and high-dimensional input data. Other studies focus on the estimation of canopy nitrogen content in winter wheat [69] and further soil spectroscopy [70]. The latter introduces the combination of PLS with tree-based methods (see Sect. 7.5.1.2) which turns out to be a successful concept. The strengths and weaknesses of LIN and PLS regression are summarized in Table 7.8.

The implementations of the LIN and the PLS regression are given in → Notebook 3.1 (https://github.com/felixriese/hyperspectral-regression). In the following, we share **best practices** when applying LIN and PLS in hyperspectral regression:

- Use LIN as a **baseline** model to benchmark more sophisticated models.
- For PLS, tune the **number of components** to keep. This number can be optimized with a method proposed in Sect. 7.5.3.1. Alternatively, it can be chosen based on visualizing the regression performance with the number of components.
- **Normalize** the input data before training the PLS model with this data.

Table 7.8 Strengths and weaknesses of LIN and PLS regression

	Strengths	Weaknesses
LIN	• Simple to understand • Many processes in nature linear	• Sensitive to outliers • Only linear relationships
PLS	• Good for high-dim. and strongly correlated data • Easy to tune • Useful for small datasets	• Often mediocre performance

7.5.1.2 Tree-Based Models

Tree-based regression is based on **decision trees** (DTs). DTs consist of a root node and leave nodes connected by branches. The basic idea is to split the training dataset at every branch into subsets based on the input features, for example, hyperspectral bands. In the best case, this split leads to leaves at the end of the branches containing similar values of the respective physical parameter to be estimated. The algorithm of DT regression is defined as follows [18]:

1. Start with the root node.
2. Start with the most significant input feature (hyperspectral band) of the training data, for example, according to the Gini impurity.
3. Divide the input data with a (binary) cut c_1 on that input feature x_i, for example, according to the Gini impurity.
4. Divide data along the next best feature on cut c_j for $j = 2, 3, \ldots$ which are calculated similarly to step 3.
5. Stop if a condition is met, for example, maximum number of nodes, maximum depth, or maximum purity.
6. Then, the ground truth labels of the datapoints are averaged for every individual leaf. Finally, every leaf contains one output value.

In the context of regression, the trained DT is applied for the estimation of the physical parameter. Every input datapoint is mapped onto a leaf containing the respective output value. In steps 2 and 4, the DT algorithm finds the most important feature at each branch in order to divide the dataset into more homogeneous subsets. For this reason, most software implementations of the DT algorithm return a trained estimator and an importance ranking of each input feature. This ranking is called **feature importance**. In the case of regression with hyperspectral data, the importance ranking refers to the hyperspectral bands. The implementation of the feature importance differs depending on the applied software. For example, the feature importance can be based on the permutation of the respective values of each input feature. The bigger the influence of an input feature on the regression performance, the more important it is.

In Table 7.9, the most important **strengths and weaknesses** of the DT algorithm are summarized. To address the issue of overfitting of a single DT, an ensemble of

Table 7.9 Strengths and weaknesses of decision tree (DT) regression

	Strengths	Weaknesses
DT	• Easy to interpret • No data preparation • Numerical *and* categorical data • Good on large datasets • Feature importance	• Weak estimation performance • Not very robust • Large trees tend to overfit

trees can be used. In the following, we focus on two ensembling techniques: bagging and boosting.

The main idea of bootstrap aggregation, or **bagging**, is to average over a number of estimators trained on slightly different training datasets. In case of tree-based regression, the average is calculated over multiple DTs with different setups or training datasets. The trees are trained in parallel. **Random forest** (RF) is one implementation of bagging with DTs [17]. Its algorithm is defined in the initialization (step 1) and three repeated steps (steps 2 to 4):

1. Initialization: Set the number of trees B, the number of features is m.
2. Bootstrap: Sample learning batch containing n datapoints with replacement from a dataset with n datapoints. There should be $n_1 \approx 2/3n$ different samples.
3. Feature bagging: At every node, a random subset of $m_{\text{bag}} = \sqrt{m}$ features is used for the splitting. This leads to a decreasing correlation between the different trees.
4. Regression: See DT algorithm above.

Every tree only uses between 60 and 70% of the datapoints for the training process. The remaining 30 to 40% of the datapoints can be used to evaluate the estimation performance of the respective trees. The regression error of these trees on their ignored datapoints is called **out-of-bag error** and is a good estimate for the generalization error of the ML model. This reduces the need for an extra validation dataset.

Extremely randomized trees (ET) are a modification of the RF algorithm [44]. Compared to the RF algorithm, the splitting process for each node is modified. Randomized thresholds are calculated for each feature of the random subset. Finally, the best threshold is used for the split in the respective node. This modification leads to less variance and increases the bias. In Table 7.10, strengths and weaknesses of DT bagging algorithms are presented.

Boosting is another technique to improve the regression based on DTs [34, 106]. It relies on learning multiple estimators which are incrementally generated and improved. **Gradient tree boosting** (GTB) as an example of a boosting algorithm applies a gradient descent optimization [16, 42]. Shallow trees are fitted iteratively on the negative gradient of the loss function.

With respect to **bias and variance**, the two tree ensembling techniques differ. In bagging, fully grown DTs are used. By decreasing the correlation between the trees, the variance of the estimator also decreases. The bias remains unchanged.

Table 7.10 Strengths and weaknesses of DT bagging and DT boosting for regression

	Strengths	Weaknesses
Bagging	• Good performance • Little overfitting & variance • Out-of-bag estimate *bullet* Highly parallel • Minor optimization needed	• Less intuitive than one DT • Time-costly prediction
Boosting	• Less bias	• Less intuitive than one DT • More overfitting than bag. • Time-costly training • Tuning difficult

In boosting, relatively shallow trees are used which implies a high bias and a low variance. The ensemble of these shallow trees, the boosted model, reduces mostly the bias. An overview of strengths and weaknesses of tree-based boosting algorithms is given in Table 7.10.

In **hyperspectral regression**, tree bagging techniques such as RF or extremely randomized trees are one of the most frequently used regression models. For example, RF models are applied to estimate biomass with a smaller dataset of WorldView-2 satellite images [84]. The feature importance was used to create new input features. A similar approach was pursued for the estimation of sugarcane leaf nitrogen concentration based on EO-1 Hyperion hyperspectral data [2]. Compared with RF models, ET perform consistently better several regression tasks such as estimating, for example, soil moisture and chlorophyll *a* concentration [61, 62, 75–77, 97]. According to these studies, the additional randomization seems to improve the regression performance.

It appears that boosting is less common in hyperspectral regression. Gradient tree boosting, for example, is applied in context of soil characterization with hyperspectral data in the range of 350 to 2500 nm [19, 70]. To optimize the application of gradient tree boosting, a combination of gradient boosting with PLS is introduced for high-dimensional data [70].

As a conclusion of the section on tree-based regression with hyperspectral data, we provide an overview of **best practices**. This list contains essential aspects excerpted from literature and own studies. It makes no claim of completeness, but it will be updated together with the implementation example in → Notebook 3.2 (https://github.com/felixriese/hyperspectral-regression).

- Choose tree-based **ensemble** techniques over single DT estimators.
- Select the **number of estimators** for bagging techniques as trade-off between time consumption and estimation performance. A good start is often a value between 100 and 1000.
- Use the **out-of-bag estimate** in the training of a bagging estimator to speed up the training.
- **Tune** bagging approaches by optimizing the most important hyperparameters: the tree size parameters like the maximum tree depth.

- Use shallow trees for boosting with **depth**, for example, of 1 or 2.

7.5.1.3 Support Vector Machines

The aim of **support vector machines** (SVMs) for regression is to find a function (model) that approximates the given training data with at most a deviation of ϵ from the given labels [122]. A detailed explanation of SVM regression is given in [112]. The linear function is one example of such a function:

$$f(x) = w \cdot x + b \text{ with } w \in X, b \in \mathbb{R}, \tag{7.5}$$

with the dot product $w \cdot x$ in X. At the same time, the function is set up as flat as possible. This means that the norm of w needs to be as small as possible which is achieved by

$$
\begin{aligned}
&\text{minimize} \quad \frac{1}{2}||w||^2, \\
&\text{subject to} \quad
\begin{cases}
y_i - w \cdot x_i - b \leq \epsilon \\
w \cdot x_i + b - y_i \leq \epsilon
\end{cases},
\end{aligned}
\tag{7.6}
$$

with the datapoints x_i and the respective labels y_i. Additionally, the SVM can be set up to adapt to non-linear functions with the **kernel trick**.

SVM regression is a widely applied tool in **hyperspectral regression**. For example, the ocean chlorophyll concentration has been estimated based on Medium Resolution Imaging Spectrometer (MERIS) data [22]. According to this study, SVM regression provides accurate and robust estimations with little bias compared to other regression models, especially in the case of small datasets. The combination of SVM regression with different feature selections through pre-processing is tested in [92]. Herein, soil organic carbon is estimated based on VIS/NIR spectroscopy with accurate results. In the estimation of soil moisture based on airborne hyperspectral input data, the SVM performs well and shows good generalization properties [113]. A result of this study is that appropriate atmospheric corrections are needed for the SVM model to perform properly. A special type of SVM regression, the least squares SVM, is applied to estimate soil properties [82].

Overall, SVMs are well suited for high-dimensional regression problems. Their strengths and weaknesses are summarized in Table 7.11. To eliminate existing disadvantages, several variations of the SVM algorithm can be used. Examples of implementations are given in → Notebook 3.3 (https://github.com/felixriese/hyperspectral-regression). In the following, we list **best practices** for the use of SVM in hyperspectral regression:

- Use automated **hyperparameter optimization** with cross-validation as described in Sect. 7.5.3.1.
- **Normalize** the input data to improve the SVM estimation performance.

Table 7.11 Strengths and weaknesses of SVM for regression

	Strengths	Weaknesses
SVM	• Good for high-dimensional data • Robust, little overfitting • Strong theoretical foundation	• Extensive tuning needed • Difficult to interpret • Slow training for large datasets

Table 7.12 Strengths and weaknesses of k-NN for regression

	Strengths	Weaknesses
k-NN	• No training needed • Low bias • Easy to understand	• Slow prediction • Sensitive to outliers • No data understanding • Difficult for high-dim. data

7.5.1.4 K-Nearest Neighbors

Regression with **k-nearest neighbors** (k-NN) [4] is applied for several decades. The k-NN algorithm relies on a distance measure, for example, the Euclidean distance d which is defined in Eq. (7.2). One implementation of k-NN for regression is described in the following. To estimate the target variable (e.g., soil moisture) from the input datapoint x_j (hyperspectral bands), the following steps are taken with a pre-set k:

1. Calculate the distances (e.g., Euclidean distance, see Eq. (7.2)) between x_j and every datapoint of the training dataset.
2. Order the training datapoints from smallest to largest distance.
3. Calculate average y over k closest datapoints weighted by inverse distance. This means that closer neighbors contribute more to the average than neighbors further away.

As a remark, the number of datapoints to be used in the averaging k needs to be tuned for every dataset, for example, using cross-validation. In Table 7.12, the strengths and weaknesses of the k-NN regression algorithm are summarized. As with the SVM regression, further implementations of the k-NN regression algorithm exist which can resolve weaknesses.

In hyperspectral regression, the **application** of k-NN is relatively rare for solving regression tasks. As one example, k-NN has been used for background estimation on a variety of images like eight-band WorldView-2 and the 126-band HyMap imagery [118]. The k-NN regression performs well without intensive tuning. In addition, the k-NN has been included in an ML framework of ten models and has been evaluated in the estimation of soil moisture as well as several water quality parameters [61, 62]. Compared to more common models such as RF (see Sect. 7.5.1.2) and SVM (see Sect. 7.5.1.3), the k-NN models have showed a mediocre performance. In general, k-NN are less suited for high-dimensional data since with increasing dimensions the difference between the nearest and farthest distance is decreasing.

$\rightarrow$ Notebook 3.4 (https://github.com/felixriese/hyperspectral-regression) contains the implementation of the k-NN model for regression tasks. In the following, **best practices** are listed:

- Set the **number of considered neighbors** k either automatically with cross-validation or choose k as an odd number which is not too large or too small. One rule of thumb is $k \approx \sqrt{n}$ with the number of datapoints n. With increasing k, the bias increases while the variance decreases.
- **Re-weight** the datapoints as described in the code example $\rightarrow$ Notebook 3.4.2 (https://github.com/felixriese/hyperspectral-regression).
- **Normalize** the input data since k-NN is a distance-based method.
- Consider a **de-correlation** (dimensionality reduction, see Sect. 7.4.1) or use the Mahalanobis distance metric.
- Apply **dimensionality reduction** as described in Sect. 7.4.1.

7.5.1.5   Artificial Neural Networks

Artificial neural networks (ANNs) based on perceptrons and backpropagation have been around since the 1960s. With increasing computing power and the increasing availability of large training datasets, ANNs significantly have grown in popularity in the 2010s. Subsequently, we give an overview of the different types of ANNs and their applications in hyperspectral regression. A comprehensive introduction to ANNs can be found, for example, in [41, 47].

Inspired by biological neural networks, ANNs are networks of artificial neurons. These neurons consist of input connections from predecessor neurons as well as an activation function depending on the weighted inputs and a defined threshold. With the activation function, the neuron output is calculated which is then forwarded to the subsequent neurons. The connections between the neurons are weighted. Through the adaptation of these weights, the ANN is adapted to a regression problem, for example, through backpropagation. The most common ANN architecture for regression tasks is a **fully connected** network. It consists of several neurons organized in consecutive and connected layers.

A **deep neural network** is a network with several hidden layers. Hidden layers are located between input and output layers. Deep ANNs are able to learn hierarchical, meaning that lower level features are learned in the first layers while higher level features are composed in the following layers. This way, a deep ANN is able to adapt to more complex tasks. Deep learning with CNNs is described in Chaps. 3, 4, 5, 11, and 14. An overview of deep learning applications in image analysis with hyperspectral data is given in [93]. The application of deep learning in hyperspectral classification tasks is illustrated in [6].

A typical challenge of applying deep learning is that with increasing number of layers and neurons, the number of trainable parameters increases. This may lead to overfitting. To prevent overfitting, **regularization** techniques such as L2 regularization, dropout, and batch normalization are introduced. L2 regularization adds an

L2 term of the weights to the loss function, dropout deactivates neurons randomly during the training iterations, and batch normalization normalizes the output of each layer per batch of datapoints.

There are several types of ANN with different characteristics. **Convolutional neural networks** (CNNs) [66] are designed to reduce the number of weights and therefore the free parameters. The main idea of CNNs is to extract local features which are translation invariant. CNNs consist of filter layers which are convolved with the input data. In most cases, the input data for CNNs consists of 2D images. Because the filters are convolved with the data instead of having one filter per input dimension, this technique is called *weight sharing*. Weight sharing significantly reduces the number of free parameters which need to be trained. These filters are learned in the training process in contrast to the hand-engineered filters in classical image processing. Many popular CNN architectures also include **pooling** layers. Pooling layers reduce the input data by a factor. A CNN often includes fully connected layers at the end.

In general, ANNs and CNNs can be trained from scratch by initializing all weights randomly and iteratively adapting the weights to the training dataset. **Transfer learning** is an alternative approach [89]. Networks are pre-trained on an existing and similar dataset and are then refined on the actual task. For example, a popular dataset for 2D images is ImageNet [33]. Pre-trained networks such as VGG16 [110] and ResNet50 [52] are freely available and can be used for own classification or regression tasks. Transfer learning can save significant amounts of time compared to training from scratch since less training iterations of the network weights are necessary.

Recurrent neural networks (RNN) [107] are another type of ANN. They learn from sequences like time series. For long sequences, the gradients can vanish. **Long short-term memory** (LSTM) networks [56] solve this issue with the help of gates (update, forget, output). In recent years, RNN and LSTM have been used mainly in natural language processing. In the context of hyperspectral remote sensing, time series such as satellite images of different dates pose possible applications of these network architectures.

ANNs are widely applied in **hyperspectral regression**. An early overview of their use and their opportunities in remote sensing is given by [5]. For example, ANNs are applied on hyperspectral spectroscopy to estimate rice nitrogen status [133]. As a finding of this study, the authors emphasize the extensive need of hyperparameter tuning of the ANN. Furthermore, the results imply that dimensionality reduction (see Sect. 7.4.1) can increase the estimation performance. A comparison of backpropagation ANNs and further regression models for the estimation of pigment content in rice leaves and panicles is shown in [26]. With hyperspectral input data, ANNs noticeably outperform the compared models. Regarding the estimation of chlorophyll *a* and soil moisture, ANNs perform strongly especially with input data normalization and dimensionality reduction [61, 62].

The primary **applications of CNNs** on hyperspectral data cover, so far, only classification tasks (e.g., [101]). An example of hyperspectral regression of soil clay content with 1D CNNs on a large European dataset is presented in [71]. Instead of applying a traditional spatial 2D CNN, the introduced CNN convolves along the spectral axis. Furthermore, this study is a good example that using transfer learning

Table 7.13 Strengths and weaknesses of ANNs and CNNs for regression

	Strengths	Weaknesses
ANN	• Strong for large datasets • Flexible architecture • Solve non-linear problems • Short prediction time • Automatic feature extraction • Transfer learning	• Weak theoretical foundation • Random architecture setup • Black box • Long training • Deep $\rightarrow$ more data needed • Large number of weights
CNN	• Outperform other models on 2D data	• Only for translation invariant features • Weight sharing

provides acceptable results. Up to now (2019), there is no relevant published study about the application of basic RNNs in hyperspectral regression. However, LSTMs are used to estimate crop yield in combination with CNNs in [134]. The results of this study emphasize the potential of LSTMs.

The strengths and weaknesses of ANN and CNNs are summarized in Table 7.13. An exemplary implementation of an ANN and a CNN architecture can be found in $\rightarrow$ Notebook 3.5 (https://github.com/felixriese/hyperspectral-regression). In addition, exemplary implementations for transfer learning with CNNs can be found in [68]. In the following, we give selected **best practices** as an excerpt from our studies and based on literature:

- Use **data augmentation** for your data to increase the size of the training dataset and to make the network more robust. In the case of 2D data, flipping and rotating the 2D images is often implemented in existing software packages.
- Apply the following **training strategy**:
  1. Train the network without regularization on a small dataset until the **estimation error on the training dataset** is ≈ 0. Start with a simple architecture and extend it if needed.
  2. Implement **regularization** and train on the training dataset while evaluating the generalization abilities on the validation dataset. We recommend applying L2 regularization, dropout (e.g., 50%) and batch normalization.
- Use the **Adam optimizer** [63] in the network for first studies before trying other algorithms.
- **Visualize the training** progress, for example, with built-in tools such as *TensorBoard* of Tensorflow [1].
- **Visualize the CNNs** during and after the training according to [135].
- Use **pre-trained** networks and implementations like **early stopping** to reduce the training time. Early stopping means ending the training of the network before the defined number of training epochs is reached. Specific metrics are applied which indicate when the network starts to overfit.

7.5.1.6 Supervised Self-organizing Maps

The traditional application of **self-organizing maps** (SOM) is unsupervised data visualization and clustering (see Sect. 7.4.1). A supervised SOM was published as the **SuSi framework** in [99, 102]. This framework combines the standardized unsupervised SOM with a supervised layer. As a result, it is able to estimate discrete (classification) or continuous (regression) parameters.

Supervised SOMs are applied in the context of hyperspectral regression in several ways. One example is the estimation of the water quality parameters CDOM, chlorophyll-*a*, diatoms, green algae, and turbidity are estimated on a dataset collected from the river Elbe in Germany in [61]. Compared to other tested ML models, the supervised SOM shows comparable results. The estimation of soil moisture on bare soil [99, 102] and on a vegetated area [62] is another example application of supervised SOMs. The results emphasize the marginal differences between the estimation performance on the training and the validation dataset. This implies that the estimation performance could be evaluated purely on the training dataset as kind of **out-of-bag estimate** (see Sect. 7.5.1.2).

The code of the SuSi framework is illustrated in [96] with implementation examples in → Notebook 3.6 (https://github.com/felixriese/hyperspectral-regression). In Table 7.14, the strengths and weaknesses of the supervised SOM are summarized. In the following, we conclude helpful **best practices** for the application of supervised SOMs:

- **Visualize** distribution of training datapoints on the SOM grid during the training. For good estimations, the whole grid is utilized.
- Start with the default **hyperparameters** in the SuSi package and start tuning the grid sizes and training iteration numbers.
- Train and evaluate the model based on the full dataset in the case of **small datasets**. In most cases, the results do not differ significantly. This aspect improves the regression performance despite the limited number of datapoints.

7.5.1.7 Comparing Supervised Models

In the following, we give a brief overview of the performance of the presented supervised ML models. We applied the ML models on the introduced dataset [99] (see Sect. 7.1) with the objective to estimate soil moisture based on hyperspectral

Table 7.14 Strengths and weaknesses of supervised SOMs for regression

	Strengths	Weaknesses
SOM	• Small *and* large datasets • Robust against overfitting • Data visualization	• Limited ability to adapt • Currently in development

Table 7.15 Regression results for soil moisture estimation with the presented ML models. The last column points out the potential of the respective model to be optimized in order to improve the regression performance. https://github.com/felixriese/hyperspectral-regression

Model	Section	Implementation	R^2 in %	MAE	RMSE	Optimization potential
Linear	7.5.1.1	3.1	80.9	1.19	1.6	–
PLS	7.5.1.1	3.1	83.5	1.16	1.5	Minor
DT	7.5.1.2	3.2	92.4	0.40	1.0	Minor
RF	7.5.1.2	3.2	93.5	0.45	0.9	Minor
ET	7.5.1.2	3.2	**96.7**	**0.33**	**0.7**	Minor
GTB	7.5.1.2	3.2	93.2	0.49	1.0	Minor
SVM	7.5.1.3	3.3	94.9	0.48	0.8	Minor
k-NN	7.5.1.4	3.4	93.3	0.43	1.0	Minor
k-NN (weighted)	7.5.1.4	3.4	94.5	0.37	0.9	Minor
ANN (sklearn)	7.5.1.5	3.5	49.5	2.09	2.6	**Major**
ANN (keras)	7.5.1.5	3.5	84.5	1.04	1.5	**Major**
CNN	7.5.1.5	3.5	75.6	1.32	1.8	**Major**
SOM	7.5.1.6	3.6	93.7	0.51	0.9	Minor

point data. The implementation and a selection of illustrating plots can be found in → Notebook 3.1 to 3.6 (https://github.com/felixriese/hyperspectral-regression) as well as further illustrations in → Notebook 3.7 (https://github.com/felixriese/hyperspectral-regression).

Table 7.15 summarizes the regression results. ET, SVM, and k-NN with weighting achieve the best regression results. We dispense with intensive hyperparameter optimization. In this account, the ML models such as ANN and CNN have further optimization potential.

7.5.2 Semi-supervised Learning for Regression

In Sect. 7.5.1, we have assumed that every datapoint x_i in our training dataset comes with an associated ground truth label y_i. In practice, this may not always be the case. For example, hyperspectral satellite images as input data cover large areas while soil moisture ground truth might be limited to several point-wise measurements. It is possible to use only the datapoints x_i with existing label y_i to apply supervised learning (see Sect. 7.5.1). **Semi-supervised learning** (SSL) is a solution that also benefits from datapoints without labels. The mathematical description can be found in Sect. 7.2 as well as in [25]. For semi-supervised learning, a certain *smoothness* of the data is assumed. That means that if two datapoints $x_1, x_2 \in \mathcal{X}$ are close in the $\mathcal{X}$ space, their corresponding labels $y_1, y_2 \in \mathcal{Y}$ are close in the $\mathcal{Y}$ space as well.

7.5.2.1 Different Types and Applications of Semi-supervised Learning

Up to now (2020), there is a lack of relevant SSL applications with respect to hyperspectral regression. In the following, we give an overview of the most important types of SSL approaches and their applications with hyperspectral input data. A review of general SSL applications is given in [86].

Generative models rely on the *cluster assumption*: The datapoints of clusters and datapoints in similar clusters share similar labels [25]. Reasonable assumptions about the dataset distributions are crucial for the success of generative models. As a result, only few ML applications with generative models exist. In [58], the authors present an example of the application of generative models. The authors present a classification of agricultural classes with multispectral data of an airborne sensor.

According to an equivalent formulation of cluster assumption, the decision boundary of an estimator should lie in a low-density region of the feature space [25]. To achieve this aim, maximum margin algorithms such as SVMs (see Sect. 7.5.1.3) can be applied. This type of SSL algorithms is called **low-density separation algorithms**. One possible implementation is transductive SVMs (TSVM). TSVM maximize the margins between unlabeled as well as labeled datapoints [11, 123]. An example of a TSVM application on Landsat 5 is shown in [20]. Six land cover classes are estimated based on a missing-label dataset as ill-posed classification task.

In **graph-based methods**, each datapoint of the dataset is represented as one node of a graph [28, 60]. The nodes are linked to each other with edges. The edges between two nodes are labeled with the distance between the respective two nodes. Graph-based methods rely on the *manifold assumption*. This means that the high-dimensional datapoints lie roughly on a low(er)-dimensional manifold [25]. Therefore, manifold regularization can be applied to the graph both on the labeled and on the unlabeled subset of the dataset [8]. This includes a term to enforce the smoothness of the dataset. As a remark, these methods imply high computational costs due to matrix inversion on large datasets despite efficient methods. In [21], the graph-based method proposed by [138] is applied to the AVIRIS *Indian Pines* land cover dataset. Another graph-based method is LapSVM. It is based on Laplacian SVMs and was introduced in [9]. LapSVM is applied by [46] on urban monitoring and cloud screening with a variety of data.

Additionally, we point to one further SSL approach which is applied on hyperspectral data. In [95], a **semi-supervised neural network** (SSNN) is introduced. The SSNN is set up to compensate shortcomings of the existing SSL approaches. The results show non-monotonic improvement of class accuracies as well as indicate the added value of ANNs in semi-supervised tasks. In [131], CNNs, and RNNs are applied in combination with label clustering to deal with the missing labels. The SuSi framework (see Sect. 7.5.1.6) also contains semi-supervised estimators for regression and classification [101]. In Table 7.16, we give an overview of all presented SSL models. To solve this issue and other shortcomings, an additional technique is proposed, called active learning. Active learning is introduced in the following section.

7.5.2.2   Active Learning

With respect to supervised learning, **active learning** is another approach to address missing labels in a dataset. This is the reason why we classify active learning as a type of SSL. The basic concept is to learn from the dataset that includes labeled and unlabeled datapoints. Iteratively, the active learning model asks (queries) a data source, for example, the user, for labels from specific unlabeled datapoints that the active learning model considers as most uncertain or most helpful. Then, the user adds these labels and the active learning model most likely is able to improve its performance.

Active learning is an useful technique in the research field of hyperspectral remote sensing, in which the **collection of labels is expensive** or time-costly. With respect to the estimation of soil moisture data from hyperspectral data, this data recording can benefit significantly from active learning, especially when measuring data in areas which are difficult to access.

An overview of active learning applications with respect to hyperspectral classification is presented in [120]. They emphasize the importance of the labeled part of the training dataset being representative for the full dataset. To solve this dependence, probabilistic elements might help to include. Similar to pure SSL approaches in Sect. 7.5.2.1, there are no relevant applications of active learning in ML regression yet. Exemplary implementations for active learning can be found in → Notebook 4 (https://github.com/felixriese/hyperspectral-regression).

Table 7.16 Overview of different semi-supervised learning approaches with references and exemplary applications

Approach/Category	Reference	Exemplary applications
Generative	[32, 85]	[58]
TSVM (Low-density sep.)	[11, 123]	[20]
Graph-based	[138]	[21]
LapSVM (Graph-based)	[9]	[46]
SSNN	[95]	
SuSi	[101]	

7.5.3 Model Selection, Optimization, and Evaluation

In Sects. 7.5.1 and 7.5.2, we have introduced a number of supervised and semi-supervised models and provided references to exemplary applications for hyperspectral regression tasks. In addition, we have reviewed their strengths and weaknesses. However, we have not yet discussed the selection of a particular ML model for a given regression problem. This selection is based on criteria which are constrained by the dataset and the respective application. In the following, we list some **important selection criteria** of an ML regression model:

- What **type of input data** are we using: 1D, 2D, 3D, time series? CNNs are good for datasets if locality and translation invariance can be assumed. LSTMs are particular useful at capturing long-term dependencies in time series data.
- What are the spectral, spatial, and temporal **dimension of the input data**? RGB, multispectral, or hyperspectral? Some ML models perform better with low-dimensional data. Therefore, dimensionality reduction (see Sect. 7.4.1) might be a good idea.
- Is the given regression problem **linear or non-linear**? We recommend applying the simplest model first. If the regression problem is expected to be linear, a linear model should be used.
- What is the **size of the training dataset**? Deep learning techniques require large amounts of data to be trained from scratch. Transfer learning can be applied to solve the shortcoming in this particular case (see Sect. 7.5.1.5).
- Is low bias or low variance more important for the estimation? Setting up ML models is often a **trade-off between bias and variance** (see Sect. 7.2). For example, in DT ensemble methods (see Sect. 7.5.1.2), bagging applies deeper trees with lower variance while boosting applies shallow trees with less bias.
- How important is it to apply ML models which are **transparent and interpretable**? Models such as k-NN and DTs are easy to interpret for humans, while ANNs and SVMs are considered as *black box* models.

After choosing the model that meets the selection criteria, it needs to be optimized and then be evaluated. The hyperparameter optimization is described in Sect. 7.5.3.1. The metrics to evaluate ML models in terms of regression performance can be found in Sect. 7.5.3.2.

7.5.3.1 Hyperparameter Optimization

ML models are defined by two sets of parameters: hyperparameters and model parameters. Hyperparameters are set before the training and model parameters are learned during the training process of the ML model. In the following, we give a brief overview of different possibilities to optimize hyperparameters. A more detailed explanation of hyperparameter optimization can be found in [13].

A simple way to tune hyperparameters is manually setting and testing them, for example, with cross-validation. Since this approach is very time-consuming, a better way is to automate the tuning process. For example, in the **grid search** approach, a pre-set hyperparameter space is automatically evaluated. The grid search operates well on small hyperparameter spaces, but it is very time-consuming for larger hyperparameter spaces. A **randomized search** speeds up the grid search approach by randomly iterating through the hyperparameter space instead of testing all combinations [12].

A more sophisticated type of hyperparameter optimization is the **Bayesian optimization**. It collects more information about the dataset and the ML model with each iteration by building hypotheses about sets of hyperparameters before the actual run. An implementation of the three types of hyperparameter optimization can be found in → Notebook 5.1 (https://github.com/felixriese/hyperspectral-regression).

7.5.3.2 Model Evaluation Metrics

To evaluate an ML regression model, different metrics are available. One metric alone does only contain a fraction of the information about the estimation performance. Therefore, it makes sense to look at more than one evaluation metric to evaluate the performance of a model. In the following, we give an overview of the most important regression metrics, also referred to as measures for the *goodness of fit*.

For all of the n input datapoints x_i with their true labels y_i, the ML model returns the estimation $\hat{y}_i$ based on x_i. The **mean absolute error** (MAE) is defined as

$$\text{MAE} = \frac{1}{n} \sum_{i=1}^{n} \left| y_i - \hat{y}_i \right|. \tag{7.7}$$

The MAE is one of the easy-to-use evaluation metrics since it sums up the absolute differences $y_i - \hat{y}_i$ of the true label value and the estimated label value.

Many ML applications require the model to have as little outliers as possible. This can be achieved by including the squared error instead of the absolute error. The **mean squared error** (MSE) is defined as

$$\text{MSE} = \frac{1}{n} \sum_{i=1}^{n} \left(y_i - \hat{y}_i \right)^2. \tag{7.8}$$

Estimation errors below 1 are less important in the MSE implementation than errors above 1. One drawback of the MSE is the unit of the error being the squared unit of the target variable to be estimated. The **root mean squared error** (RMSE) solves this issue. It is defined as

$$\text{RMSE} = \sqrt{\text{MSE}} = \sqrt{\frac{1}{n} \sum_{i=1}^{n} \left(y_i - \hat{y}_i \right)^2}. \tag{7.9}$$

All the presented metrics, MAE, MSE, and RMSE, are easy to understand. MAE and RMSE return an error measure in the unit of the target variable y. As a remark, without knowing the distribution and scale of the target variable with its minimum and maximum, these metrics are difficult to interpret.

The **coefficient of determination** R^2 is a relative measure to resolve the unit issue as stated above. It is defined as

$$R^2 = 1 - \frac{n \cdot \text{MSE}}{\sum_{i=1}^{n}(y_i - \overline{y})^2} = 1 - \frac{\sum_{i=1}^{n}(y_i - \hat{y}_i)^2}{\sum_{i=1}^{n}(y_i - \overline{y})^2} \quad \text{with } \overline{y} = \frac{1}{n}\sum_{i=1}^{n} y_i. \quad (7.10)$$

In this definition, it normally returns a value between 0 and 1; $R^2 = 1$ indicates that the ML model estimation is in perfect agreement with the data. However, negative values might occur and indicate a bad estimation performance.

Assume we evaluate the performance of an ML model in two regression examples on $n = 40$ datapoints. One regression example is based on three input features, for example, the colors RGB, and the other regression example is based on 13 hyperspectral bands. The result of the first example is $R^2 = 80\%$ and $R^2 = 85\%$ for the second example. The question is now, if the performance of the second model is better, (a) since this model is better or (b) since the input data of the second model has more input features. To answer this question, a look on the **adjusted coefficient of determination** R^2_{adj} can help. It is defined as

$$R^2_{\text{adj}} = 1 - \frac{(1 - R^2) \cdot (n - 1)}{n - m - 1}, \quad (7.11)$$

with the number of input features m and n datapoints for the evaluation. With respect to the presented examples, the first model has a $R^2_{\text{adj}} = 78.3\%$ and the second $R^2_{\text{adj}} = 77.5\%$. The result implies that the first model is the better model according to the respective metrics.

Figure 7.7 shows three exemplary distributions of an ML model which adapts to simulated data with one input feature x and the target variable y. The regres-

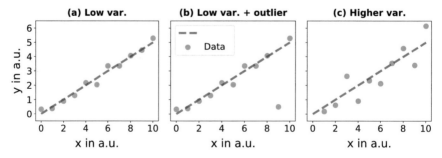

Fig. 7.7 Regression example on simulated data (**a**) with low variance (var.), (**b**) with low variance and one outlier, and (**c**) with higher variance. Both the 1D input data x and target variable y are given in arbitrary units (a.u.)

Table 7.17 Overview of the model evaluation metrics in hyperspectral regression for the three example distributions. The lowest performance with regard to the respective metric is emphasized

Example distribution	MAE in a.u.	MSE in a.u.	RMSE in a.u.	R^2 in %
Low variance	0.21	0.06	0.24	97.61
Low variance and one outlier	0.57	**1.51**	**1.23**	**39.44**
Higher variance	**0.81**	0.97	0.98	61.26

sion results are shown in Table 7.17 with different metrics. These results imply that ML regression models should to be evaluated based on an number of metrics. → Notebook 5.2 (https://github.com/felixriese/hyperspectral-regression) shows the code implementations of three exemplary distributions.

In the following, we summarize our **best practices**:

- Use the **validation** dataset (see Sect. 7.3.3) for the evaluation of the ML model performance and for choosing your hyperparameters. Use the **test** dataset only once: for the final model.
- Calculate evaluation metrics on the training dataset as well. The difference between training and validation performance is a measure for the **degree of overfitting**.
- Use the three evaluation metrics **MAE, RMSE, and R^2**. We recommended them for most hyperspectral regression applications.
- Set fixed **random seeds** of the applied ML models for reproducible results (e.g., see → Notebook 3 (https://github.com/felixriese/hyperspectral-regression)).
- Use a well understood or previously published **baseline model** to compare a new ML model with.
- Use the adjusted coefficient of determination R^2_{adj} to compare estimations based on a **different number of input features**.

7.6 Summary and Trends in Hyperspectral Regression

In the previous sections, we have presented a detailed overview of a typical **hyperspectral regression workflow**. This workflow consisting of the data level, the feature level, and the model level is illustrated in Fig. 7.1. We recommend applying this workflow for any given regression task and dataset.

The **data** level (Sect. 7.3) is divided into an overview of possible pre-processing steps, the challenge of dataset shift with proposed solutions as well as approaches for dataset splitting. The **feature** level (Sect. 7.4) consists of three parts which describe possible approaches to generate or select features. With dimensionality reduction, clustering, and feature engineering, new features are generated from the existing input features of a dataset. Feature selection describes approaches to select the best input features of a dataset according to specific metrics. On the **model** level (Sect. 7.5), an ML model is selected from several available supervised or semi-supervised models and the selection is motivated. Then, the selected model can then be optimized and evaluated, resulting in the final model for the given regression task and dataset.

New methods emerge over time from ML research which can be applied in hyperspectral regression as well. In the following sections, we give an overview of the open challenges and new methods that we consider relevant.

7.6.1 Trends at the Data Level: Generative Adversarial Networks

Generative adversarial networks (GANs) [48] are an upcoming ML model which was introduced for unsupervised data augmentation. GANs consist of two ANNs: a generator network and a discriminator network. The **generator** network learns to generate new data samples. In combination with the existing (real) training data, these new (fake) samples constitute the input data for the **discriminator** network. The discriminator network learns to differentiate between real and fake input data. During the training of a GAN, both networks learn to improve their performance regarding their respective task. Finally, a trained GAN is able to generate new training data which can be used to augment the existing training dataset (see *best practices* in Sect. 7.5.1.5).

A detailed overview on GANs is presented in [30]. In hyperspectral regression, GANs are not commonly used so far, although there are different applications in classification. For example, the implementation of GANs and their applications on open hyperspectral classification datasets is presented in [140]. A more complex approach combining GANs with semi-supervised learning is presented in [53]. Exemplary implementations of GANs are given in → Notebook 6 (https://github.com/felixriese/hyperspectral-regression).

7.6.2 Trends at the Feature Level: Domain Knowledge

In recent years, the application of (manual) feature engineering has decreased in hyperspectral regression. Deep ANNs, which are able to learn new low-level and high-level features automatically (see Sect. 7.5.1.5), are the main reason for this development. Admittedly, incorporating **domain knowledge** into data-driven ML models might still improve their performance. Especially, the estimation of physical parameters in hyperspectral regression can be improved by including domain knowledge if such knowledge is available. An overview of the domain knowledge integration into ML models is given in [104] including a review and a consistent taxonomy on previous research. The authors distinguish between four possible approaches to include prior knowledge into ML models [104]:

- Integration of the knowledge into the **training data** by feature engineering (see Sect. 7.4.3) and simulations (see Sect. 7.6.1).
- Integration of the knowledge into the **hypothesis space**, for example, by choosing an appropriate ML model such as CNNs for 2D hyperspectral data in the case of locality and translation invariance (see Sect. 7.5.1.5).
- Integration of the knowledge into the **training algorithm**, for example, by modifying the loss function.
- Integration of the knowledge into the **final hypothesis**, for example, by including physical constraints on the output variable.

7.6.3 Trends at the Model Level: Architectures and Automated ML

Innovations with respect to **ANN architectures** are continuously presented and applied in ML research. For example, hierarchical neural networks such as attention networks [132] are a promising architecture alternative to LSTMs (see Sect. 7.5.1.5) when analyzing sequential data. Further developments involve capsule networks [105], which use vectors rather than scalars to represent input features. Capsule networks might improve the estimation performance of networks, for example, in the hyperspectral image classification [139].

A further trend in ML is the automation of the ML workflow, often referred to as **automated machine learning** (AutoML). AutoML can include the automation of steps like pre-processing (Sect. 7.3.1), dimensionality reduction (Sect. 7.4.1), feature engineering and feature selection (Sect. 7.4.3), ML model selection, and optimization (Sect. 7.5.3). An overview of AutoML is given in [57]. While AutoML simplifies and speeds up the application of ML for the user, we emphasize that AutoML is not a universal solution for all hyperspectral regression problems. Two relevant implementations of AutoML are **auto-sklearn** [37, 38] and **TPOT** [87, 88]. Both implementations are based on the widely used scikit-learn [91]. Another example of AutoML is **MorphNet** [49]. MorphNet is focused on shrinking and expanding ANN structures to adapt the ANN for maximum performance with respect to constraints on computing resources.

Acknowledgements We thank Timothy D. Gebhard, Mareike Hoyer, and Raoul Gabriel for their detailed feedback on this chapter.

References

1. Abadi M, Barham P, Chen J, Chen Z, Davis A, Dean J, Devin M, Ghemawat S, Irving G, Isard M, et al. (2016) Tensorflow: a system for large-scale machine learning. In: 12th

USENIX Symposium on Operating Systems Design and Implementation (OSDI 16). USENIX Association, Savannah, GA, pp 265–283
2. Abdel-Rahman EM, Ahmed FB, Ismail R (2013) Random forest regression and spectral band selection for estimating sugarcane leaf nitrogen concentration using EO-1 Hyperion hyperspectral data. Int J Remote Sens 34(2):712–728
3. Ali I, Greifeneder F, Stamenkovic J, Neumann M, Notarnicola C (2015) Review of machine learning approaches for biomass and soil moisture retrievals from remote sensing data. Remote Sens 7(12):16398–16421
4. Altman NS (1992) An introduction to kernel and nearest-neighbor nonparametric regression. Am Stat 46(3):175–185
5. Atkinson PM, Tatnall ARL (1997) Introduction neural networks in remote sensing. Int J Remote Sens 18:699–709
6. Audebert N, Saux BL, Lefèvre S (2019) Deep learning for classification of hyperspectral data: a comparative review. In: IEEE geoscience and remote sensing magazine. pp 159–173
7. Baldeck CA, Asner GP (2013) Estimating vegetation beta diversity from airborne imaging spectroscopy and unsupervised clustering. Remote Sens 5:2057–2071
8. Belkin M, Niyogi P (2004) Semi-supervised learning on riemannian manifolds. Mach Learn 56(1–3):209–239
9. Belkin M, Niyogi P, Sindhwani V (2006) Manifold regularization: a geometric framework for learning from labeled and unlabeled examples. J Mach Learn Res 7:2399–2434
10. Bellman R, Collection KMR (1961) Adaptive control processes: a guided tour, Princeton legacy library, vol 2045. Princeton University Press, Princeton
11. Bennett KP, Demiriz A (1999) Semi-supervised support vector machines. In: Proceedings of the 1998 conference on advances in neural information processing systems II. MIT Press, Cambridge, MA, pp 368–374
12. Bergstra J, Bengio Y (2012) Random search for hyper-parameter optimization. J Mach Learn Res 13:281–305
13. Bergstra JS, Bardenet R, Bengio Y, Kégl B (2011) Algorithms for hyper-parameter optimization. In: Advances in neural information processing systems. pp 2546–2554
14. Bioucas-Dias JM, Plaza A, Camps-Valls G, Scheunders P, Nasrabadi N, Chanussot J (2013) Hyperspectral remote sensing data analysis and future challenges. IEEE Geosci Remote Sens Mag 1(2):6–36
15. Blum AL, Langley P (1997) Selection of relevant features and examples in machine learning. Artif Intell 97(1):245–271
16. Breiman L (1997) Arcing the edge. Technical Report 486, Statistics Department, University of California, Berkeley
17. Breiman L (2001) Random forests. Mach Learn 45(1):5–32
18. Breiman L, Friedman J, Olshen RA, Stone CJ (1984) Classification and regression trees. Routledge, Abingdon
19. Brown DJ, Shepherd KD, Walsh MG, Mays MD, Reinsch TG (2006) Global soil characterization with VNIR diffuse reflectance spectroscopy. Geoderma 132(3–4):273–290
20. Bruzzone L, Chi M, Marconcini M (2006) A novel transductive svm for semisupervised classification of remote-sensing images. IEEE Trans Geosci Remote Sens 44(11):3363–3373
21. Camps-Valls G, Bandos Marsheva TV, Zhou D (2007) Semi-supervised graph-based hyperspectral image classification. IEEE Trans Geosci Remote Sens 45(10):3044–3054
22. Camps-Valls G, Bruzzone L, Rojo-Alvarez JL, Melgani F (2006) Robust support vector regression for biophysical variable estimation from remotely sensed images. IEEE Geosci Remote Sens Lett 3:339–343
23. Camps-Valls G, Tuia D, Gómez-Chova L, Jiménez S, Malo J (2011) Remote sensing image processing. Synth Lect Image Video Multimed Process 5(1):1–192
24. Chang C (2018) A review of virtual dimensionality for hyperspectral imagery. IEEE J Sel Top Appl Earth Obs Remote Sens 11(4):1285–1305
25. Chapelle O, Schölkopf B, Zien A (2006) Semi-supervised learning. adaptive computation and machine learning. MIT Press, Cambridge, MA

26. Chen L, Huang JF, Wang FM, Tang YL (2007) Comparison between back propagation neural network and regression models for the estimation of pigment content in rice leaves and panicles using hyperspectral data. Int J Remote Sens 28(16):3457–3478
27. Cho MA, Skidmore A, Corsi F, van Wieren SE, Sobhan I (2007) Estimation of green grass/herb biomass from airborne hyperspectral imagery using spectral indices and partial least squares regression. Int J Appl Earth Obs Geoinformation 9(4):414–424
28. Chung FR, Graham FC (1997) Spectral graph theory. Am Math Soc 92:212
29. Colini L, Spinetti C, Amici S, Buongiorno M, Caltabiano T, Doumaz F, Favalli M, Giammanco S, Isola I, La Spina A, et al. (2014) Hyperspectral spaceborne, airborne and ground measurements campaign on Mt. Etna: multi data acquisitions in the frame of Prisma Mission (ASI-AGI Project n. I/016/11/0). Quaderni di Geofisica 119:1–51
30. Creswell A, White T, Dumoulin V, Arulkumaran K, Sengupta B, Bharath AA (2018) Generative adversarial networks: an overview. IEEE Signal Process Mag 35(1):53–65
31. Datta A, Ghosh S, Ghosh A (2012) Clustering based band selection for hyperspectral images. In: 2012 international conference on communications, devices and intelligent systems (CODIS). pp 101–104
32. Dempster AP, Laird NM, Rubin DB (1977) Maximum likelihood from incomplete data via the EM algorithm. J R Stat Soc 39(1):1–38
33. Deng J, Dong W, Socher R, Li LJ, Li K, Fei-Fei L (2009) ImageNet: a large-scale hierarchical image database. In: 2009 IEEE conference on computer vision and pattern recognition. IEEE, Piscataway, pp 248–255
34. Drucker H, Cortes C (1996) Boosting decision trees. In: Advances in neural information processing systems, pp. 479–485
35. Ester M, Kriegel HP, Sander J, Xu X (1996) A density-based algorithm for discovering clusters a density-based algorithm for discovering clusters in large spatial databases with noise. In: Proceedings of the second international conference on knowledge discovery and data mining, pp. 226–231. AAAI Press, Palo Alto, CA
36. Fawcett T, Flach PA (2005) A response to webb and ting's on the application of ROC analysis to predict classification performance under varying class distributions. Mach Learn 58(1):33–38
37. Feurer M, Klein A, Eggensperger K, Springenberg J, Blum M, Hutter F (2015) Efficient and robust automated machine learning. In: Cortes C, Lawrence ND, Lee DD, Sugiyama M, Garnett R (eds) Advances in neural information processing systems, vol 28. Curran Associates, Inc., Red Hook, NY, pp 2962–2970
38. Feurer M, Klein A, Eggensperger K, Springenberg JT, Blum M, Hutter F (2019) Auto-sklearn: efficient and robust automated machine learning. In: Hutter F, Kotthoff L, Vanschoren J (eds) Automated machine learning: methods, systems, challenges. Springer, Cham, pp 113–134
39. Fielding AH, Bell JF (1997) A review of methods for the assessment of prediction errors in conservation presence/absence models. Environ Conserv 24(1):38–49
40. Fonville JM, Carter CL, Pizarro L, Steven RT, Palmer AD, Griffiths RL, Lalor PF, Lindon JC, Nicholson JK, Holmes E, Bunch J (2013) Hyperspectral visualization of mass spectrometry imaging data. Anal Chem 85(3):1415–1423
41. Friedman J, Hastie T, Tibshirani R (2001) The elements of statistical learning, vol 1. Springer, New York
42. Friedman JH (2002) Stochastic gradient boosting. Comput Stat Data Anal 38(4):367–378
43. Geman S, Bienenstock E, Doursat R (1992) Neural networks and the bias/variance dilemma. Neural Comput 4(1):1–58
44. Geurts P, Ernst D, Wehenkel L (2006) Extremely randomized trees. Mach Learn 63(1):3–42
45. Gewali UB, Monteiro ST, Saber E (2018) Machine learning based hyperspectral image analysis: a survey. arXiv:1802.08701
46. Gomez-Chova L, Camps-Valls G, Munoz-Mari J, Calpe J (2008) Semisupervised image classification with laplacian support vector machines. IEEE Geosci Remote Sens Lett 5(3):336–340
47. Goodfellow I, Bengio Y, Courville A (2016) Deep learning. MIT Press, Cambridge

48. Goodfellow I, Pouget-Abadie J, Mirza M, Xu B, Warde-Farley D, Ozair S, Courville A, Bengio Y (2014) Generative adversarial nets. In: Advances in neural information processing systems, pp 2672–2680
49. Gordon A, Eban E, Nachum O, Chen B, Wu H, Yang TJ, Choi E (2018) MorphNet: fast & simple resource-constrained structure learning of deep networks. In: Proceedings of the IEEE conference on computer vision and pattern recognition, pp 1586–1595
50. Green AA, Berman M, Switzer P, Craig MD (1988) A transformation for ordering multispectral data in terms of image quality with implications for noise removal. IEEE Trans Geosci Remote Sens 26(1):65–74
51. Guyon I, Elisseeff A (2003) An introduction to variable and feature selection. J Mach Learn Res 3:1157–1182
52. He K, Zhang X, Ren S, Sun J (2016) Deep residual learning for image recognition. In: 2016 IEEE conference on computer vision and pattern recognition (CVPR), pp 770–778
53. He Z, Liu H, Wang Y, Hu J (2017) Generative adversarial networks-based semi-supervised learning for hyperspectral image classification. Remote Sens 9(10):1042
54. Hinton GE, Salakhutdinov RR (2006) Reducing the dimensionality of data with neural networks. Science 313(5786):504–507
55. Hinton GE, Zemel RS (1994) Autoencoders, minimum description length and helmholtz free energy. In: Cowan JD, Tesauro G, Alspector J (eds) Advances in neural information processing systems vol 6. Morgan-Kaufmann, Burlington, pp 3–10
56. Hochreiter S, Schmidhuber J (1997) Long short-term memory. Neural Comput 9(8):1735–1780
57. Hutter F, Kotthoff L, Vanschoren J (2019) Automated machine learning: methods, systems. Springer International Publishing, Challenges, Berlin
58. Jackson Q, Landgrebe DA (2001) An adaptive classifier design for high-dimensional data analysis with a limited training data set. IEEE Trans Geosci Remote Sens 39(12):2664–2679
59. Jia X, Kuo BC, Crawford MM (2013) Feature mining for hyperspectral image classification. Proc IEEE 101:676–697
60. Jordan MI (1998) Learning in graphical models, vol 89. Springer Science & Business Media, Berlin
61. Keller S, Maier PM, Riese FM, Norra S, Holbach A, Börsig N, Wilhelms A, Moldaenke C, Zaake A, Hinz S (2018) Hyperspectral data and machine learning for estimating CDOM, chlorophyll a, diatoms, green algae, and turbidity. Int J Environ Res Public Health 15(9):1881
62. Keller S, Riese FM, Stötzer J, Maier PM, Hinz S (2018) Developing a machine learning framework for estimating soil moisture with VNIR hyperspectral data. ISPRS Ann Photogramm Remote Sens Spat Inf Sci IV-1:101–108
63. Kingma DP, Ba J (2015) Adam: a method for stochastic optimization. In: 3rd International conference on learning representations, ICLR. San Diego, CA
64. Kohavi R, John GH (1997) Wrappers for feature subset selection. Artif Intell 97(1–2):273–324
65. Kohonen T (1990) The self-organizing map. Proc IEEE 78(9):1464–1480
66. LeCun Y, Boser B, Denker JS, Henderson D, Howard RE, Hubbard W, Jackel LD (1989) Backpropagation applied to handwritten zip code recognition. Neural Comput 1(4):541–551
67. Lee JB, Woodyatt AS, Berman M (1990) Enhancement of high spectral resolution remote-sensing data by a noise-adjusted principal components transform. IEEE Trans Geosci Remote Sens 28(3):295–304
68. Leitloff J, Riese FM (2018) Examples for CNN training and classification on Sentinel-2 data. https://doi.org/10.5281/zenodo.3268451
69. Li F, Mistele B, Hu Y, Chen X, Schmidhalter U (2014) Reflectance estimation of canopy nitrogen content in winter wheat using optimised hyperspectral spectral indices and partial least squares regression. Eur J Agron 52:198–209
70. Liu L, Ji M, Buchroithner M (2017) Combining partial least squares and the gradient-boosting method for soil property retrieval using visible near-infrared shortwave infrared spectra. Remote Sens 9:1299

71. Liu L, Ji M, Buchroithner M (2018) Transfer learning for soil spectroscopy based on convolutional neural networks and its application in soil clay content mapping using hyperspectral imagery. Sensors 18(9):3169

72. Liu Y, Heer J (2018) Somewhere over the rainbow: an empirical assessment of quantitative colormaps. In: Proceedings of the 2018 CHI conference on human factors in computing systems. ACM, New York, p 598

73. van der Maaten L, Hinton G (2008) Visualizing data using t-SNE. J Mach Learn Res 9:2579–2605

74. MacQueen J (1967) Some methods for classification and analysis of multivariate observations. In: Proceedings of the fifth berkeley symposium on mathematical statistics and probability, volume 1: statistics. University of California Press, Berkeley, pp 281–297

75. Maier PM, Keller S (2018) Machine learning regression on hyperspectral data to estimate multiple water parameters. In: 2018 9th workshop on hyperspectral image and signal processing: evolution in remote sensing (WHISPERS). Amsterdam, pp 1–5

76. Maier PM, Keller S (2019) Application of different simulated spectral data and machine learning to estimate the chlorophyll a concentration of several inland waters. In: 2019 10th Workshop on Hyperspectral Imaging and Signal Processing: Evolution in Remote Sensing (WHISPERS). IEEE, Amsterdam, Netherlands, pp 1–5. https://doi.org/10.1109/WHISPERS.2019.8921073

77. Maier PM, Keller S (2019) Estimating chlorophyll a concentrations of several inland waters with hyperspectral data and machine learning models. ISPRS Ann Photogramm Remote Sens Spat Inf Sci IV-2/W5:609–614

78. Matejka J, Fitzmaurice G (2017) Same stats, different graphs: generating datasets with varied appearance and identical statistics through simulated annealing. In: Proceedings of the 2017 CHI conference on human factors in computing systems. ACM, New York, NY, pp 1290–1294

79. McInnes L, Healy J, Saul N, Grossberger L (2018) UMAP: uniform manifold approximation and projection. J Open Source Softw 3(29):861

80. McKinney W (2010) Data structures for statistical computing in python. In: van der Walt S, Millman, J (eds) Proceedings of the 9th Python in science conference, pp 51–56

81. Merentitis A, Debes C, Heremans R (2014) Ensemble learning in hyperspectral image classification: toward selecting a favorable bias-variance tradeoff. IEEE J Sel Top Appl Earth Obs Remote Sens 7(4):1089–1102

82. Morellos A, Pantazi XE, Moshou D, Alexandridis T, Whetton R, Tziotzios G, Wiebensohn J, Bill R, Mouazen AM (2016) Machine learning based prediction of soil total nitrogen, organic carbon and moisture content by using VIS-NIR spectroscopy. Biosyst Eng 152:104–116

83. Moreno-Torres JG, Raeder T, Alaiz-Rodríguez R, Chawla NV, Herrera F (2012) A unifying view on dataset shift in classification. Pattern Recognit 45(1):521–530

84. Mutanga O, Adam E, Cho MA (2012) High density biomass estimation for wetland vegetation using WorldView-2 imagery and random forest regression algorithm. Int J Appl Earth Obs Geoinformation 18:399–406

85. Nigam K, McCallum AK, Thrun S, Mitchell T (2000) Text classification from labeled and unlabeled documents using EM. Mach Learn 39(2–3):103–134

86. Oliver A, Odena A, Raffel CA, Cubuk ED, Goodfellow IJ (2018) Realistic evaluation of deep semi-supervised learning algorithms. In: Bengio S, Wallach H, Larochelle H, Grauman K, Cesa-Bianchi N, Garnett R (eds) Advances in neural information processing systems, vol 31. Curran Associates, Inc., Red Hook, NY, pp 3235–3246

87. Olson RS, Bartley N, Urbanowicz RJ, Moore JH (2016) Evaluation of a tree-based pipeline optimization tool for automating data science. In: Proceedings of the genetic and evolutionary computation conference 2016. ACM, New York, NY, pp 485–492

88. Olson RS, Moore JH (2019) TPOT: a tree-based pipeline optimization tool for automating machine learning. In: Hutter F, Kotthoff L, Vanschoren J (eds) Automated machine learning: methods, systems, challenges. Springer, Cham, pp 151–160

89. Pan SJ, Yang Q (2009) A survey on transfer learning. IEEE Trans Knowl Data Eng 22(10):1345–1359

90. Pearson K (1901) On lines and planes of closest fit to systems of points in space. Lond Edinb Dublin Philos Mag J Sci 2(11):559–572
91. Pedregosa F, Varoquaux G, Gramfort A, Michel V, Thirion B, Grisel O, Blondel M, Pretten-hofer P, Weiss R, Dubourg V, Vanderplas J, Passos A, Cournapeau D, Brucher M, Perrot M, Duchesnay E (2011) Scikit-learn: machine learning in Python. J Mach Learn Res 12:2825–2830
92. Peng X, Shi T, Song A, Chen Y, Gao W (2014) Estimating soil organic carbon using VIS/NIR spectroscopy with SVMR and SPA methods. Remote Sens 6:2699–2717
93. Petersson H, Gustafsson D, Bergström D (2016) Hyperspectral image analysis using deep learning - a review. In: 2016 sixth international conference on image processing theory, tools and applications (IPTA), pp 1–6
94. Quionero-Candela J, Sugiyama M, Schwaighofer A, Lawrence ND (2009) Dataset shift in machine learning. The MIT Press, Cambridge
95. Ratle F, Camps-Valls G, Weston J (2010) Semisupervised neural networks for efficient hyper-spectral image classification. IEEE Trans Geosci Remote Sens 48(5):2271–2282
96. Riese FM (2019) SUSI: supervised self-organizing maps in Python. https://doi.org/10.5281/zenodo.2609130
97. Riese FM, Keller S (2018) Fusion of hyperspectral and ground penetrating radar data to estimate soil moisture. In: 2018 9th workshop on hyperspectral image and signal processing: evolution in remote sensing (WHISPERS). Amsterdam, pp 1–5
98. Riese FM, Keller S (2018) Hyperspectral benchmark dataset on soil moisture. https://doi.org/10.5281/zenodo.1227836
99. Riese FM, Keller S (2018) Introducing a framework of self-organizing maps for regression of soil moisture with hyperspectral data. In: IGARSS 2018 - 2018 IEEE international geoscience and remote sensing symposium. Valencia, Spain, pp 6151–6154
100. Riese FM, Keller S (2019) Hyperspectral regression: code examples. https://doi.org/10.5281/zenodo.3450676
101. Riese FM, Keller S, Hinz S (2020) Supervised and semi-supervised self-organizing maps for regression and classification focusing on hyperspectral data. Remote Sens 12(1):7. https://doi.org/10.3390/rs12010007
102. Riese FM, Keller S (2019) Susi: supervised self-organizing maps for regression and classifi-cation in python. arXiv:1903.11114
103. Rouse Jr JW, Haas R, Schell J, Deering D (1974) Monitoring vegetation systems in the great plains with ERTS. In: Third earth resources technology satellite-1 symposium. Greenbelt, pp 309–317
104. von Rueden L, Mayer S, Garcke J, Bauckhage C, Schuecker J (2019) Informed machine learning-towards a taxonomy of explicit integration of knowledge into machine learning. arXiv:1903.12394
105. Sabour S, Frosst N, Hinton GE (2017) Dynamic routing between capsules. In: Advances in neural information processing systems, pp 3856–3866
106. Schapire RE (1999) A brief introduction to boosting
107. Servan-Schreiber D, Cleeremans A, McClelland JL (1989) Learning sequential structure in simple recurrent networks. In: Advances in neural information processing systems, pp 643–652
108. Shalev-Shwartz S, Ben-David S (2014) Understanding machine learning: from theory to algorithms. Cambridge University Press, Cambridge
109. Shimodaira H (2000) Improving predictive inference under covariate shift by weighting the log-likelihood function. J Stat Plan Inference 90(2):227–244
110. Simonyan K, Zisserman A (2015) Very deep convolutional networks for large-scale image recognition. In: 3rd international conference on learning representations, ICLR. San Diego, CA
111. Smets T, Verbeeck N, Claesen M, Asperger A, Griffioen G, Tousseyn T, Waelput W, Waelkens E, De Moor B (2019) Evaluation of distance metrics and spatial autocorrelation in uniform manifold approximation and projection applied to mass spectrometry imaging data. Anal Chem 91(9):5706–5714

112. Smola AJ, Schölkopf B (2004) A tutorial on support vector regression. Stat Comput 14(3):199–222
113. Stamenkovic J, Tuia D, de Morsier F, Borgeaud M, Thiran J (2013) Estimation of soil moisture from airborne hyperspectral imagery with support vector regression. In: 2013 5th workshop on hyperspectral image and signal processing: evolution in remote sensing (WHISPERS). pp 1–4
114. Stehman SV (1999) Basic probability sampling designs for thematic map accuracy assessment. Int J Remote Sens 20(12):2423–2441
115. Storkey A (2009) When training and test sets are different: characterizing learning transfer. In: Dataset shift in machine learning, pp. 3–28
116. Su H, Yang H, Du Q, Sheng Y (2011) Semisupervised band clustering for dimensionality reduction of hyperspectral imagery. IEEE Geosci Remote Sens Lett 8(6):1135–1139
117. Tao C, Pan H, Li Y, Zou Z (2015) Unsupervised spectral-spatial feature learning with stacked sparse autoencoder for hyperspectral imagery classification. IEEE Geosci Remote Sens Lett 12(12):2438–2442
118. Theiler J, Wohlberg B (2013) Regression framework for background estimation in remote sensing imagery. In: 2013 5th workshop on hyperspectral image and signal processing: evolution in remote sensing (WHISPERS). IEEE, pp 1–4
119. Treitz PM, Howarth PJ (1999) Hyperspectral remote sensing for estimating biophysical parameters of forest ecosystems. Prog Phys Geogr: Earth Environ 23(3):359–390
120. Tuia D, Volpi M, Copa L, Kanevski M, Munoz-Mari J (2011) A survey of active learning algorithms for supervised remote sensing image classification. IEEE J Sel Top Signal Process 5(3):606–617
121. Van Der Maaten L, Postma E, Van den Herik J (2009) Dimensionality reduction: a comparative. J Mach Learn Res 10(66–71):13
122. Vapnik VN (1995) The nature of statistical learning theory. Springer, New York Inc., New York, NY
123. Vapnik VN (1998) Statistical learning theory. Wiley, Hoboken
124. Vidal M, Amigo JM (2012) Pre-processing of hyperspectral images. Essential steps before image analysis. Chemom Intell Lab Syst 117:138–148
125. Villa A, Chanussot J, Benediktsson JA, Jutten C, Dambreville R (2013) Unsupervised methods for the classification of hyperspectral images with low spatial resolution. Pattern Recognit 46(6):1556–1568
126. Virtanen P, Gommers R, Oliphant TE, Haberland M, Reddy T, Cournapeau D, Burovski E, Peterson P, Weckesser W, Bright J, et al (2020) SciPy 1.0: fundamental algorithms for scientific computing in Python. Nat Methods 1–12
127. Widmer G, Kubat M (1996) Learning in the presence of concept drift and hidden contexts. Mach Learn 23(1):69–101
128. Windrim L, Ramakrishnan R, Melkumyan A, Murphy RJ, Chlingaryan A (2019) Unsupervised feature-learning for hyperspectral data with autoencoders. Remote Sens 11(7):864
129. Wold H (1966) Estimation of principal components and related models by iterative least squares. In: Multivariate analysis. Academic, New York, pp 391–420
130. Wold S, Sjöström M, Eriksson L (2001) PLS-regression: a basic tool of chemometrics. Chemom Intell Lab Syst 58(2):109–130
131. Wu H, Prasad S (2018) Semi-supervised deep learning using pseudo labels for hyperspectral image classification. IEEE Trans Image Process 27(3):1259–1270
132. Xu K, Ba J, Kiros R, Cho K, Courville A, Salakhudinov R, Zemel R, Bengio Y (2015) Show, attend and tell: neural image caption generation with visual attention. In: International conference on machine learning, pp 2048–2057
133. Yi QX, Huang JF, Wang FM, Wang XZ, Liu ZY (2007) Monitoring rice nitrogen status using hyperspectral reflectance and artificial neural network. Environ Sci Technol 41(19):6770–6775
134. You J, Li X, Low M, Lobell D, Ermon S (2017) Deep gaussian process for crop yield prediction based on remote sensing data. In: Thirty-First AAAI conference on artificial intelligence, pp 4559–4566

135. Zeiler MD, Fergus R (2014) Visualizing and understanding convolutional networks. In: Fleet D, Pajdla T, Schiele B, Tuytelaars T (eds) Computer vision - ECCV 2014. Springer International Publishing, Cham, pp 818–833
136. Zhang J, Chen L, Zhuo L, Liang X, Li J (2018) An efficient hyperspectral image retrieval method: deep spectral-spatial feature extraction with DCGAN and dimensionality reduction using t-SNE-based NM hashing. Remote Sens 10(2):271
137. Zhong Y, Zhang L, Huang B, Li P (2006) An unsupervised artificial immune classifier for multi/hyperspectral remote sensing imagery. IEEE Trans Geosci Remote Sens 44:420–431
138. Zhou D, Bousquet O, Lal TN, Weston J, Schölkopf B (2004) Learning with local and global consistency. In: Advances in neural information processing systems, pp 321–328
139. Zhu K, Chen Y, Ghamisi P, Jia X, Benediktsson JA (2019) Deep convolutional capsule network for hyperspectral image spectral and spectral-spatial classification. Remote Sens 11(3):223
140. Zhu L, Chen Y, Ghamisi P, Benediktsson JA (2018) Generative adversarial networks for hyperspectral image classification. IEEE Trans Geosci Remote Sens 56(9):5046–5063

Chapter 8
Sparsity-Based Methods for Classification

Zebin Wu, Yang Xu and Jianjun Liu

Abstract Sparsity is an important prior for various signals, and sparsity-based methods have been widely used in hyperspectral image classification. This chapter introduces the sparse representation methodology and its related techniques for hyperspectral image classification. To start with, we provide a brief review on the mechanism, models, and algorithms of sparse representation classification (SRC). We then introduce several advanced SRC methods that can improve hyperspectral image classification accuracy by incorporating spatial–spectral information into SRC models. As a case study, a hyperspectral image SRC method based on adaptive spatial context is discussed in detail to demonstrate the performance of SRC methods in hyperspectral image classification.

8.1 Introduction

In the last few decades, sparsity has become one of the most important concepts in the field of signal processing. Sparsity concept has been widely employed in a variety of fields, e.g., source separation, restoration, and compression. Sparse representation was originally derived from compressed sensing [1–3], suggesting that if a signal is sparse or compressive, the original signal can be reconstructed with a few number of samplings. By introducing sparsity in sampling, compressed sensing has achieved great success in information theory, image acquisition, image processing, medical imaging, remote sensing, etc. Compressed sensing has also motivated many researches on sparse representation. As a matter of fact, signals in real world may not be sparse in the original space, but they can be sparse in an appropriate basis.

Z. Wu (✉) · Y. Xu
Nanjing University of Science and Technology, Nanjing, China
e-mail: wuzb@njust.edu.cn

J. Liu
Jiangnan University, Wuxi, China

© Springer Nature Switzerland AG 2020
S. Prasad and J. Chanussot (eds.), *Hyperspectral Image Analysis*,
Advances in Computer Vision and Pattern Recognition,
https://doi.org/10.1007/978-3-030-38617-7_8

Hyperspectral imaging sensors record reflected light in hundreds of narrow frequencies covering the visible, near-infrared, and shortwave infrared bands. This abundant spectral information yields more precise measures and makes it possible to gain insight into the material at each pixel in the image. Supervised classification plays a central role in hyperspectral image (HSI) analysis, such as land-use or land-cover mapping, forest inventory, or urban-area monitoring [4]. Many methods have been proposed for solving the HSI classification problem, such as logistic regression [5], support vector machines (SVM) [6], artificial neural networks [7], and k-nearest neighbor (KNN) classifier [8]. These methods can serve the purpose of generating acceptable classification results. However, the high dimensionality of hyperspectral data remains a challenge for HSI classification.

To address this problem, sparse representation [9, 10] has been employed for classifying high-dimensional signals. A sparse representation classification (SRC) method [10] has been first proposed for face recognition. A test signal is sparsely represented by an over-complete dictionary composed of labeled training samples. At the decision level, the label of each test sample is set as the class whose corresponding atoms maximally represent the original test sample. Since then, SRC has been widely used in face recognition [10, 11], speech recognition [12], and image super-resolution [13]. Chen et al. [14] proposed an SRC framework for solving the HSI classification problem, in which each sample is a pixel's spectral responses. Inspired by this work, many improved SRC methods have been proposed for HSI classification.

In this chapter, we investigate the SRC methods and present several advanced models of sparse representation for HSI classification. More specifically, we will give a case study of SRC method that improves the classification accuracy by incorporating the spectral–spatial information of HSI into the SRC framework.

8.2 Sparse Representation-Based HSI Classification

In the theory of sparse representation, given a dictionary, each signal can be linearly represented by a set of atoms in the dictionary. Designing an over-complete dictionary and obtaining the sparse representation vector through sparse coding are the two main goals of sparse representation.

In HSI classification, SRC assumes that the features belonging to the same class approximately lie in the same low-dimensional subspace spanned by dictionary atoms from the same class. Suppose we have M distinct classes and $N_i (i = 1, 2, \ldots, M)$ training samples for each class. Each class has a sub-dictionary $\mathbf{D}_i = [\mathbf{d}_{i,1}, \mathbf{d}_{i,2}, \ldots, \mathbf{d}_{i,N_i}] \in \mathbb{R}^{B \times N_i}$ in which the columns represent training samples and B is the number of spectral bands. A test pixel $\mathbf{x} \in \mathbb{R}^B$ can be represented by a sparse linear combination of the training pixels as

$$\mathbf{x} = \mathbf{D} \alpha \qquad (8.1)$$

where $\mathbf{D} = [\mathbf{D}_1 \, \mathbf{D}_2 \ldots \mathbf{D}_M] \in \mathbb{R}^{B \times N}$ with $N = \sum_{i=1}^{M} N_i$ is the dictionary constructed by combining all sub-dictionaries $\{\mathbf{D}_i\}_{i=1,\ldots,M}$. $\boldsymbol{\alpha} \in \mathbb{R}^N$ is an unknown sparse vector with K nonzero entries. Here, we denote $K = \|\boldsymbol{\alpha}\|_0$. The sparse coefficient vector $\boldsymbol{\alpha}$ is obtained by solving the following problem

$$\min_{\alpha} \|\mathbf{x} - \mathbf{D}\boldsymbol{\alpha}\|_2 \quad \text{s.t } \|\boldsymbol{\alpha}\|_0 \leq K_0 \tag{8.2}$$

where K_0 is a pre-specified upper bound of K. The class label of $\mathbf{x}$ is determined by the minimal residual between $\mathbf{x}$ and its approximation from each class sub-dictionary, i.e.,

$$\text{class}(\mathbf{x}) = \arg \min_{i=1,2,\ldots,M} \|\mathbf{x} - \mathbf{D}_i \boldsymbol{\alpha}_i\|_2 \tag{8.3}$$

where $\boldsymbol{\alpha}_i$ is the sub-vector corresponding to the i-th class, and $\mathbf{D}_i$ denotes the sub-dictionary.

Problem (2) is NP-hard, and can be approximately solved by greedy algorithms, such as orthogonal match pursuit (OMP) and subspace pursuit (SP).

In OMP algorithm, we select one atom from the dictionary that is most correlated with the residual. The algorithmic flow of the OMP algorithm is described in Algorithm 8.1.

Algorithm 1 Orthogonal Matching Pursuit

Input: Dictionary $\mathbf{D} = [\mathbf{d}_1 \ \mathbf{d}_2 \ \ldots \ \mathbf{d}_N]$, test samples $\mathbf{x}$, normalize the dictionary $\mathbf{D}$ and $\mathbf{x}$.

Initialize: Set the residual $\mathbf{r}_0 = \mathbf{x}$, set the index set $\Lambda_0 = \varnothing$, set iteration count $k = 1$;

While termination criterion not not satisfied **do**

1) Compute $\lambda_k = \arg \max_{i=1,\ldots,N} \mathbf{r}_{k-1}^T \mathbf{d}_i$, and find the atom that matches with residual most;

2) Update the index set $\Lambda_k = \Lambda_{k-1} \cup \lambda_k$;

3) Compute $\mathbf{P}_k = (\mathbf{D}_{\Lambda_k}^T \mathbf{D}_{\Lambda_k})^{-1} \mathbf{D}_{\Lambda_k}^T \mathbf{x} \in \mathbb{R}^k$, where $\mathbf{D}_{\Lambda_k}$ is the sub-dictionary composed of the atoms from the index set;

4) Compute the residual $\mathbf{r}_k = \mathbf{x} - \mathbf{D}_{\Lambda_k} \mathbf{P}_k$;

5) $k = k + 1$;

Output: the index set $\Lambda = \Lambda_{k+1}$, the sparse vector $\boldsymbol{\alpha}$, where the non-zero elements are $(\mathbf{D}_{\Lambda_k}^T \mathbf{D}_{\Lambda_k})^{-1} \mathbf{D}_{\Lambda_k}^T \mathbf{x}$, and the support is the determined by the index set.

The procedure of SP algorithm is similar to that of OMP algorithm. The difference is that SP finds all the K atoms that satisfy (8.2) during one iteration. The complete procedure of SP algorithm is provided in Algorithm 8.2.

Algorithm 2 Subspace Pursuit

Input: Dictionary $\mathbf{D} = [\mathbf{d}_1 \ \mathbf{d}_2 \ ... \ \mathbf{d}_N]$, test samples $\mathbf{x}$, sparsity K, normalize the dictionary $\mathbf{D}$ and $\mathbf{x}$.

Initialize: Set the index set Λ_0, where the element of Λ_0 are determined by the K largest elements in $\mathbf{x}^T \mathbf{D}$, set the residual $\mathbf{r}_0 = \mathbf{x} - \mathbf{D}_{\Lambda_0} (\mathbf{D}_{\Lambda_0}^T \mathbf{D}_{\Lambda_0})^{-1} \mathbf{D}_{\Lambda_0}^T \mathbf{x}$, set iteration count $k = 1$;

While stopping criterion not satisfied **do**

    1) Find the indices of K atoms according to the K largest elements in $\mathbf{r}_{k-1}^T \mathbf{d}_i$, denoted as l

    2) Update the index set $\Lambda_k = \Lambda_{k-1} \cup l$;

    3) Compute $\mathbf{P}_k = (\mathbf{D}_{\Lambda_k}^T \mathbf{D}_{\Lambda_k})^{-1} \mathbf{D}_{\Lambda_k}^T \mathbf{x} \in \mathbb{R}^{2K}$;

    4) Fin the K largest elements in $\mathbf{P}_k$, and update the index set Λ_k;

    5) Compute the residual $\mathbf{r}_k = \mathbf{x} - \mathbf{D}_{\Lambda_k} (\mathbf{D}_{\Lambda_k}^T \mathbf{D}_{\Lambda_k})^{-1} \mathbf{D}_{\Lambda_k}^T \mathbf{x}$;

    6) $k = k + 1$;

Output: the index set $\Lambda = \Lambda_{k+1}$, the sparse vector $\boldsymbol{\alpha}$, where the non-zero elements are $(\mathbf{D}_{\Lambda_k}^T \mathbf{D}_{\Lambda_k})^{-1} \mathbf{D}_{\Lambda_k}^T \mathbf{x}$, and the support is the determined by the index set.

8.3 Advanced Models of Sparse Representation for Hyperspectral Image Classification

Many advanced methods based on SRC have been proposed for HSI classification.

In HSI, pixels within a small neighborhood usually consist of similar materials. Therefore, these pixels tend to have high spatial correlation [14]. The corresponding sparse coefficient vectors share a common sparsity pattern as follows.

Let $\{\mathbf{x}_t\}_{t=1,...,T}$ be T pixels in a fixed window centered at $\mathbf{x}_1$. These pixels can be represented by

$$\begin{aligned} \mathbf{X} = [\mathbf{x}_1 \mathbf{x}_2 \ldots \mathbf{x}_T] &= [\mathbf{D}\,\boldsymbol{\alpha}_1 \ \mathbf{D}\,\boldsymbol{\alpha}_2 \ldots \mathbf{D}\,\boldsymbol{\alpha}_T] \\ &= \mathbf{D}\underbrace{[\boldsymbol{\alpha}_1 \ \boldsymbol{\alpha}_2 \ldots \boldsymbol{\alpha}_T]}_{\mathbf{S}} = \mathbf{DS} \end{aligned} \tag{8.4}$$

In the joint sparsity model (JSM), the sparse vectors $\{\boldsymbol{\alpha}_t\}_{t=1,...,T}$ share the same support Ω. $\mathbf{S}$ is a sparse matrix with $|\Omega|$ nonzero rows, which can be obtained by solving the following optimization problem,

$$\min_{\mathbf{S}} \|\mathbf{X} - \mathbf{DS}\|_F \quad \text{s.t} \quad \|\mathbf{S}\|_{\text{row},0} \leq K_0 \tag{8.5}$$

where $\|\mathbf{S}\|_{\text{row},0}$ denotes the number of nonzero rows of $\mathbf{S}$, and $\|\cdot\|_F$ denotes the Frobenius norm. The problem in (8.5) can be approximately solved by the simultaneous version of OMP (SOMP). The label of the central pixel $\mathbf{x}_1$ can be determined minimizing the total residual

$$\text{class}(\mathbf{x}_1) = \arg \min_{i=1,\ldots,M} \|\mathbf{X} - \mathbf{D}_i \mathbf{S}_i\|_F \tag{8.6}$$

where $\mathbf{S}_i$ is the sub-sparse coefficient matrix corresponding to the i-th class.

Note that, the optimization models (8.2) and (8.5) are non-convex, and can be converted into convex versions by relaxing the norm constraints:

$$\min_{\boldsymbol{\alpha}} \frac{1}{2}\|\mathbf{x} - \mathbf{D}\boldsymbol{\alpha}\|_2^2 + \lambda\|\boldsymbol{\alpha}\|_1 \tag{8.7}$$

$$\min_{\mathbf{S}} \frac{1}{2}\|\mathbf{X} - \mathbf{DS}\|_F^2 + \lambda\|\mathbf{S}\|_{1,2} \tag{8.8}$$

where $\|\boldsymbol{\alpha}\|_1 = \sum_{i=1}^{N} |\alpha_i|$ is the ℓ_1 norm, $\|\mathbf{S}\|_{1,2} = \sum_{i=1}^{N} \|\mathbf{s}^i\|_2$ is the $\ell_{1,2}$ norm, and $\mathbf{s}^i$ represents the i-th row of $\mathbf{S}$.

The JSM model enforces that the pixels in the neighborhood of the test sample are represented by the same atoms. However, if the neighboring pixels are on the boundary of several homogeneous regions, they would be classified into different classes. In this scenario, different sub-dictionaries should be used. Laplacian sparsity promotes sparse coefficients of neighboring pixels belonging to different clusters to be different from each other. For this reason, a weight matrix $\mathbf{W}$ is introduced, where w_{ij} represents the similarity between a pair of pixels $\mathbf{x}_i$ and $\mathbf{x}_j$ in the neighborhood of the text sample. As reported in [15], the optimization problem with additional Laplacian sparsity prior can be described as

$$\min_{\mathbf{S}} \frac{1}{2}\|\mathbf{X} - \mathbf{DS}\|_F^2 + \lambda_1\|\mathbf{S}\|_1 + \lambda_2 \sum_{i,j} w_{ij} \|\mathbf{s}_i - \mathbf{s}_j\|_2^2 \tag{8.9}$$

where λ_1 and λ_2 are regularization parameters. $\mathbf{s}_i$ is the i-th column of matrix $\mathbf{S}$. Weight matrix $\mathbf{W}$ can characterize the similarity among neighboring pixels in the spectral space. If two pixels are similar, the weight value will be large. As a result, their corresponding sparse codes will be similar. On the other hand, if two pixels are less similar, the weight value will be small, allowing a large difference between their sparse codes. Laplacian sparsity prior is more flexible than the joint sparsity prior. In fundamental, the joint sparsity prior can be regarded as a special case of Laplacian sparsity. Laplacian sparsity prior can well characterize more pixels in the image, since the sparse codes of the neighboring pixels are not limited to have the same supports. Suppose $\mathbf{L} = \mathbf{I} - \mathbf{H}^{-1/2}\mathbf{W}\mathbf{H}^{-1/2}$ is the normalized symmetric Laplacian matrix and, $\mathbf{H}$ is the degree matrix computed from $\mathbf{W}$. We can have the following new optimization problem:

$$\min_{\mathbf{S}} \frac{1}{2}\|\mathbf{X} - \mathbf{DS}\|_F^2 + \lambda_1\|\mathbf{S}\|_1 + \lambda_2 tr(\mathbf{SLS}^T) \tag{8.10}$$

In JSM model, each pixel is represented by the atoms in the dictionary, and is classified according to the residual between the sparse codes multiplying the sub-dictionary. It is a reasonable assumption that each pixel can only be represented by one sub-dictionary. This condition can be achieved by enforcing the sparse codes corresponding to one sub-dictionary to be active and other ones to be inactive. Group Lasso sums up the Euclidean norm of the sparse codes corresponding to all sub-dictionaries as the sparsity prior. In [15], group Lasso is introduced as the new regularization in the optimization problem, i.e.,

$$\min_{\alpha} \frac{1}{2}\|\mathbf{x} - \mathbf{D}\alpha\|_2^2 + \lambda \sum_{g\in G} \omega_g \|\alpha_g\|_2 \tag{8.11}$$

where $g \subset \{G_1, G_2, \cdots G_M\}$, $\sum_{g\in G} \|\alpha_g\|_2$ represents the group sparse prior defined in terms of M groups, ω_g is the weight and is set to the square root of the cardinality of the corresponding group. Note here that α_g represents the coefficients of different groups. In a similar way, the group sparsity [15] can be employed in the JSM model as follows:

$$\min_{\mathbf{S}} \frac{1}{2}\|\mathbf{X} - \mathbf{DS}\|_F^2 + \lambda \sum_{g\in G} \omega_g \|\mathbf{S}_g\|_F \tag{8.12}$$

where $\sum_{g\in G} \|\mathbf{S}_g\|_F$ refers to the collaborative group Lasso regularization defined in terms of groups, and $\mathbf{S}_g$ is the sub-matrix corresponding to the g-th sub-dictionary.

In models (8.11) and (8.12) only group sparsity is introduced, and the sparsity of the sparse code corresponding to sub-dictionary is not taken into consideration. When the sub-dictionary is over-complete, it is important to introduce the sparsity within each group [15]. The ℓ_1-norm regularization can be incorporated into the objective function of (8.11) as follows:

$$\min_{\alpha} \frac{1}{2}\|\mathbf{x} - \mathbf{D}\alpha\|_2^2 + \lambda_1 \sum_{g\in G} \omega_g \|\alpha_g\|_2 + \lambda_1\|\alpha\|_1 \tag{8.13}$$

Similarly, the problem in (8.13) can be extended to JSM as follows:

$$\min_{\mathbf{S}} \frac{1}{2}\|\mathbf{X} - \mathbf{DS}\|_F^2 + \lambda_1 \sum_{g\in G} \omega_g \|\mathbf{S}_g\|_F + \lambda_1 \sum_{g\in G} \omega_g \|\mathbf{S}_g\|_1 \tag{8.14}$$

Another effective method is to introduce the correlation coefficient (CC) [16]. Traditionally, CC value is used to measure the correlation between different variables. In HSI classification, we can use CCs to determine whether pixels represent the same class. In general, CC can be calculated as follows:

$$\rho = \frac{\text{cov}(\mathbf{x}_i, \mathbf{x}_j)}{\sqrt{\text{var}(\mathbf{x}_i)} \cdot \sqrt{\text{var}(\mathbf{x}_j)}} = \frac{\sum_{z=1}^{B} (\mathbf{x}_{iz} - u_{\mathbf{x}_i})(\mathbf{x}_{jz} - u_{\mathbf{x}_j})}{\sqrt{\sum_{z=1}^{B} (\mathbf{x}_{iz} - u_{\mathbf{x}_i})^2} \cdot \sqrt{\sum_{z=1}^{B} (\mathbf{x}_{jz} - u_{\mathbf{x}_j})^2}} \quad (8.15)$$

where $\text{var}(\mathbf{x}_i)$ and $\text{var}(\mathbf{x}_j)$ are the variance of $\mathbf{x}_i$ and $\mathbf{x}_j$, respectively. $\mathbf{x}_{iz}$ refers to the z-th element in $\mathbf{x}_i . u_{\mathbf{x}_i} = (1/B) \sum_{z=1}^{B} \mathbf{x}_{iz}$, and $u_{\mathbf{x}_j} = (1/B) \sum_{z=1}^{B} \mathbf{x}_{jz}$ represents the mean values of the corresponding vectors. According to the definition of CC, we have $|\rho| \leq 1$. Stronger correlation indicates that ρ is close to 1.

Following the method in [16], CCs among the training samples and test samples are first calculated. Given a test sample $\mathbf{x}$ and any training sample $\mathbf{d}_j^i$, where $\mathbf{d}_j^i$ represents the j-th atom in the i-th sub-dictionary. The CC between $\mathbf{x}$ and $\mathbf{d}_j^i$ can be calculated as follows:

$$\rho_j^i = \frac{\text{cov}(\mathbf{d}_j^i, \mathbf{x})}{\sqrt{\text{var}(\mathbf{d}_j^i)} \cdot \sqrt{\text{var}(\mathbf{x})}} = \frac{\sum_{z=1}^{B} [(\mathbf{d}_j^i)_z - u_{\mathbf{d}_j^i}][(\mathbf{x})_z - u_{\mathbf{x}}]}{\sqrt{\sum_{z=1}^{B} [(\mathbf{d}_j^i)_z - u_{\mathbf{d}_j^i}]^2} \cdot \sqrt{\sum_{z=1}^{B} [(\mathbf{x})_z - u_{\mathbf{x}}]^2}}.$$
$$(8.16)$$

We define a matrix $\boldsymbol{\rho}^i = \{\rho_1^i, \rho_2^i, \ldots, \rho_{N_i}^i\}$. This matrix is sorted in descending order according to CCs among different training samples. Subsequently, the mean of L largest $\boldsymbol{\rho}^i$ is calculated as the CC cor^i. Assuming that the L largest $\boldsymbol{\rho}^i$ consists of $\{\rho_1^i, \rho_2^i, \ldots, \rho_L^i\}$, the CC cor^i can be calculated as

$$cor^i = \frac{1}{L}(\rho_1^i + \rho_2^i + \cdots + \rho_L^i). \quad (8.17)$$

Finally, the CC is combined with the JSM at the decision level to exploit the CCs among training and test samples as well as the representation residuals.

$$\text{class}(\mathbf{x}_1) = \arg \min_{i=1,\ldots,M} \|\mathbf{X} - \mathbf{D}_i \mathbf{S}_i\|_F + \lambda(1 - cor^i(\mathbf{x}_1)) \quad (8.18)$$

where $cor^i \in [0, 1]$ represents the CCs among pixels, and λ is the regularization parameter.

One more approach to improve SRC is kernel trick. As an extension of SRC, kernel SRC (KSRC) uses the kernel trick to project data into a feature space, in which the projected data are linearly separable.

Suppose the feature mapping function $\phi : \mathbb{R}^B \to \mathbb{R}^K, (B \leq K)$ maps the features and also the dictionary to a high-dimensional feature space, $\mathbf{x} \to \phi(\mathbf{x})$, $\mathbf{D} = [\mathbf{d}_1, \mathbf{d}_2, \ldots, \mathbf{d}_N] \to \phi(\mathbf{D}) = [\phi(\mathbf{d}_1), \phi(\mathbf{d}_2), \ldots, \phi(\mathbf{d}_N)]$. By replacing the mapped features and dictionary in (8.7), we have the KSRC model,

$$\min_{\alpha} \frac{1}{2} \|\phi(\mathbf{x}) - \phi(\mathbf{D})\alpha\|_2 + \lambda \|\alpha\|_1. \tag{8.19}$$

Similarly, the class label of **x** is determined as

$$\text{class}(\mathbf{x}) = \arg \min_{i=1,2,\ldots,M} \|\phi(\mathbf{x}) - \phi(\mathbf{D}_i)\alpha_i\|_2. \tag{8.20}$$

It is worth mentioning that all ϕ mappings used in KSRC occur in the form of inner products, allowing us to define a kernel function **k** for any samples $\mathbf{x}_i \in \mathbb{R}^B$.

$$\mathbf{k}(\mathbf{x}_i, \mathbf{x}_j) = \langle \phi(\mathbf{x}_i), \phi(\mathbf{x}_j) \rangle \tag{8.21}$$

In this way, KSRC can be constructed using only the kernel function, without considering the mapping ϕ explicitly. Then, the optimization problem can be rewritten as

$$\min_{\alpha} \frac{1}{2} \alpha^T \mathbf{Q} \alpha - \alpha \mathbf{p} + \lambda \|\alpha\|_1 + C \tag{8.22}$$

where $C = \frac{1}{2}\mathbf{k}(\mathbf{x}_i, \mathbf{x}_j)$ is a constant, **Q** is a $B \times B$ matrix with $\mathbf{Q}_{ij} = \mathbf{k}(\mathbf{d}_i, \mathbf{d}_j)$, and **p** is a $B \times 1$ vector with $\mathbf{p}_i = \mathbf{k}(\mathbf{d}_i, \mathbf{x})$. Analogously, the classification criterion can be rewritten as

$$\text{class}(\mathbf{x}) = \arg \min_{i=1,2,\ldots,M} \delta_i^T(\alpha)\mathbf{Q}\delta(\alpha) - 2\delta_i^T(\alpha)\mathbf{p} \tag{8.23}$$

where $\delta_i(\cdot)$ is the characteristic function that selects coefficients within the i-th class and sets all other coefficients to zero.

Valid kernels are only those satisfying the Mercer's condition [17, 18]. Some commonly used kernels in kernel methods include linear kernel, polynomial kernel, and Gaussian radial basis function kernel. Assuming $\mathbf{k}_1$ and $\mathbf{k}_2$ are two valid Mercer's kernels over $\mathcal{X} \times \mathcal{X}$ with $\mathbf{x}_i \in \mathcal{X} \subseteq \mathbb{R}^B$ and $z > 0$, the direct sum $\mathbf{k}(\mathbf{x}_i, \mathbf{x}_j) = \mathbf{k}_1(\mathbf{x}_i, \mathbf{x}_j) + \mathbf{k}_2(\mathbf{x}_i, \mathbf{x}_j)$, tensor product $\mathbf{k}(\mathbf{x}_i, \mathbf{x}_j) = \mathbf{k}_1(\mathbf{x}_i, \mathbf{x}_j) \cdot \mathbf{k}_2(\mathbf{x}_i, \mathbf{x}_j)$, or scaling $\mathbf{k}(\mathbf{x}_i, \mathbf{x}_j) = z\mathbf{k}_1(\mathbf{x}_i, \mathbf{x}_j)$ are valid Mercer's kernels [19].

A suitable kernel is a kernel whose structure reflects data relations. To properly define such a kernel, unlabeled information and geometrical relationships between labeled and unlabeled samples are very useful. The spatial–spectral kernel sparse representation is proposed [20], in which the neighboring filtering kernel is presented and the corresponding optimization algorithm is developed.

A full family of composite kernels (CKs) for the combination of spectral and spatial contextual information have been presented in SVM [21, 22]. These kernels are valid and are all suitable for KSRC. Although one can improve the performance of KSRC by CK, it is worth noting that the kernel should learn all high-order similarities between neighboring samples directly, and should reflect the data lying in complex manifolds. For these purposes, the neighbor filtering (NF) kernel would be a good choice, which computes the spatial similarity between neighboring samples in the feature space.

Given $\mathbf{x}^m \in \Omega, m = 1, 2, \ldots, \omega^2$, with Ω being the spatial window ω around pixel. Let $\phi(\mathbf{x}^m)$ be the image of $\mathbf{x}^m$ under the mapping ϕ. In order to describe $\phi(\mathbf{x})$, a straightforward way is to use the average of spatially neighboring pixels in the kernel space. This method is similar to the mean filtering. The estimated vector is given by

$$MF(\phi(\mathbf{x})) = \frac{1}{\omega^2} \sum_{m=1}^{\omega^2} \phi(\mathbf{x}^m). \qquad (8.24)$$

However, the mean filtering rarely reflects relative contributions (which treats every neighboring pixel equally). To address this issue, the neighboring filtering is defined as

$$NF(\phi(\mathbf{x})) = \frac{1}{\sum_m \mathbf{w}^m} \sum_{m=1}^{\omega^2} \mathbf{w}^m \phi(\mathbf{x}^m) \qquad (8.25)$$

where $\mathbf{w}^m = \exp(-\gamma_0 ||\mathbf{x} - \mathbf{x}^m||_2^2)$ and parameter $\gamma_0 > 0$ acts as a degree of filtering.

Let us consider two different pixels $\mathbf{x}_i$ and $\mathbf{x}_j$. We are interested in defining a similarity function that estimates the proximity between them in a sufficiently rich feature space. A straightforward kernel function reflecting the similarity between them is obtained by evaluating the kernel function between the estimated vectors

$$
\begin{aligned}
\mathbf{k}_{NF}(\mathbf{x}_i, \mathbf{x}_j) &= \langle NF(\phi(\mathbf{x}_i)), NF(\phi(\mathbf{x}_j)) \rangle \\
&= \left\langle \frac{\sum_{m=1}^{\omega^2} \mathbf{w}_i^m \phi(\mathbf{x}_i^m)}{\sum_m \mathbf{w}_i^m}, \frac{\sum_{n=1}^{\omega^2} \mathbf{w}_j^n \phi(\mathbf{x}_j^n)}{\sum_n \mathbf{w}_i^n} \right\rangle \\
&= \frac{\sum_{m=1}^{\omega^2} \sum_{n=1}^{\omega^2} \mathbf{w}_i^m \mathbf{w}_j^n \mathbf{k}(\mathbf{x}_i^m, \mathbf{x}_j^n)}{\sum_m \mathbf{w}_i^m \sum_n \mathbf{w}_i^n},
\end{aligned} \qquad (8.26)
$$

which is referred to as neighbor filtering (NF) kernel. Similarly, we can define mean filtering (MF) kernel as follows:

$$
\begin{aligned}
\mathbf{k}_{MF}(\mathbf{x}_i, \mathbf{x}_j) &= \langle MF(\phi(\mathbf{x}_i)), MF(\phi(\mathbf{x}_j)) \rangle \\
&= \left\langle \frac{1}{\omega^2} \sum_{m=1}^{\omega^2} \phi(\mathbf{x}_j^m), \frac{1}{\omega^2} \sum_{n=1}^{\omega^2} \phi(\mathbf{x}_j^n) \right\rangle \\
&= \frac{1}{\omega^4} \sum_{m=1}^{\omega^2} \sum_{n=1}^{\omega^2} \mathbf{k}(\mathbf{x}_i^m, \mathbf{x}_j^n),
\end{aligned} \qquad (8.27)
$$

which computes the spatial similarity between neighboring samples, whereas the cluster similarity is computed in the mean map kernel.

Since $\mathbf{Q}$ is a valid kernel, the objective function of (8.22) is convex, which is the same as the objective function of (8.19) except for the definition of $\mathbf{Q}$ and $\mathbf{p}$. Therefore, alternating direction method of multipliers (ADMM) [23] can be used to solve this problem. By introducing a new variable $\mathbf{u} \in \mathbb{R}^B$, the objective function can be rewritten as

$$
\min_{\alpha} \frac{1}{2} \alpha^T \mathbf{Q} \alpha - \alpha^T \mathbf{p} + \lambda \|\alpha\|_1
$$

$$
\text{s.t.} \quad \mathbf{u} = \alpha. \tag{8.28}
$$

ADMM imposes the constraint $\mathbf{u} = \mathbf{a}$ which can be defined as

$$
\begin{cases}
(\alpha^{(t+1)}, \mathbf{u}^{(t+1)}) = \underset{\alpha, \mathbf{u}}{\arg\min} \frac{1}{2} \alpha^T \mathbf{Q} \alpha - \alpha^T \mathbf{p} + \lambda \|\alpha\|_1 + \frac{\mu}{2} \|\alpha - \mathbf{u} - \mathbf{d}^{(t)}\|_2^2, \\
\mathbf{d}^{(t+1)} = \mathbf{d}^{(t)} - (\alpha^{(t+1)} - \mathbf{u}^{(t+1)})
\end{cases}
$$

$$\tag{8.29}$$

where $t \geq 0$ and $\mu > 0$. The minimizing solution $\alpha^{(t+1)}$ is simply determined as

$$
\alpha^{(t+1)} \leftarrow (\mathbf{Q} + \mu \mathbf{I})^{-1} (\mathbf{p} + \mu (\mathbf{u}^{(t)} + \mathbf{d}^{(t)})), \tag{8.30}
$$

where $\mathbf{I}$ is the identity matrix. The minimizing solution $\mathbf{u}^{(t+1)}$ is the soft threshold [24],

$$
\mathbf{u}^{(t+1)} \leftarrow \text{soft}(\alpha^{(t+1)} - \mathbf{d}^{(t)}, \lambda/\mu), \tag{8.31}
$$

where $\text{soft}(\cdot, \tau)$ denotes the component-wise application of the soft-threshold function $y \leftarrow \text{sign}(y) \max\{|y| - \tau, 0\}$.

The optimization algorithm for KSRC is summarized in Algorithm 8.3.

Algorithm 3 Spatial-Spectral Kernel Sparse Representation Classification

---

Input: A training dictionary $\mathbf{D} \in \mathbb{R}^{B \times N}$, and a test sample $\mathbf{x} \in \mathbb{R}^B$

    1) Select the Mercer kernel $\mathbf{k}_{NF}$ (or others) and its parameters.

    2) Compute the matrix $\mathbf{Q}$, and the vector $\mathbf{p}$.

    3) Set $t = 0$, choose $\mu > 0$, $\mathbf{s}^{(0)}$, $\mathbf{u}^{(0)}$, $\mathbf{d}^{(0)}$.

    4) repeat.

    5) Compute $\mathbf{s}^{(t+1)}$, $\mathbf{u}^{(t+1)}$ and $\mathbf{d}^{(t+1)}$ using (29)

    6) $t \leftarrow t + 1$

    7) until some stopping criterion is satisfied.

    8) compute the M residuals $r_i(\mathbf{x}) = \delta_i^T(\mathbf{s}) \mathbf{Q} \delta_i(\mathbf{s}) - 2\delta_i^T(\mathbf{s}) \mathbf{p}, i = 1, 2, ..., M$.

Output: The estimated label of $\mathbf{x}$ according to (23)

8.4 A Case Study of Hyperspectral Image Sparse Representation Classification Based on Adaptive Spatial Context

8.4.1 Model and Algorithm

In model (8.5), pixels in a fixed window centered at the test pixel are selected to be simultaneously sparse represented. All pixels in the fixed window have the same correlation with the center pixel. However, this condition does not always hold, especially for pixels located on the edge which can be seen as class boundary. It is obvious that pixels on the same side of the edge will have stronger correlation. Since different pixels have different spatial context, the definition of local structure for the adaptive spatial context is essential to HSI classification.

In the field of image recovery, steering kernel (SK) [25] is a popular local method, which can effectively express the adaptive local structure. This method starts with making an initial estimate of the image gradients using a gradient estimator, and then uses the estimate to measure the dominant orientation of the local gradients in the image [26]. The obtained orientation information is then used to adaptively "steer" the local kernel, resulting in elongated, elliptical contours spread along the directions of the local edge structure.

Taking into consideration that HSI generally contains hundreds of sub-images, a high-dimensional steering kernel (HDSK) [27] is defined where the gradient estimator contains every sub-image's gradients. The gradients in vertical and horizontal directions are written as follows:

$$(\nabla \mathbf{x}_i^v, \nabla \mathbf{x}_i^h) = (\frac{\left\| \mathbf{x}_i - \mathbf{x}_{i+1}^v \right\|_1}{B}, \frac{\left\| \mathbf{x}_i - \mathbf{x}_{i+1}^h \right\|_1}{B}) \tag{8.32}$$

where $\mathbf{x}_{i+1}^v$ and $\mathbf{x}_{i+1}^h$ represent the neighboring pixels of $\mathbf{x}_i$ in vertical and horizontal directions. HDSK for pixel $\mathbf{x}_i$ is defined as

$$w_{ij} = \frac{\sqrt{\det(\mathbf{C}_i)}}{2\pi h^2} \exp(-\frac{(\mathbf{e}_i - \mathbf{e}_j)^T \mathbf{C}_i (\mathbf{e}_i - \mathbf{e}_j)}{2h^2}) \tag{8.33}$$

where $\mathbf{e}_i$ and $\mathbf{e}_j$ represent the coordinates of pixel $\mathbf{x}_i$ and pixel $\mathbf{x}_j$, respectively, h is the smoothing parameter used for controlling the supporting range of the steering kernel, and $\mathbf{C}_i$ is the symmetric gradient covariance in vertical and horizontal directions in a $M \times M$ window centered at $\mathbf{x}_i$. A naïve estimate of this covariance matrix can be obtained by $\mathbf{C}_i = \mathbf{J}_i^T \mathbf{J}_i$, where

$$\mathbf{J}_i = \begin{bmatrix} \nabla \mathbf{x}_1^v & \nabla \mathbf{x}_1^h \\ \vdots & \vdots \\ \nabla \mathbf{x}_{M \times M}^v & \nabla \mathbf{x}_{M \times M}^h \end{bmatrix} \tag{8.34}$$

Here, $\mathbf{x}_1, \cdots, \mathbf{x}_{M \times M}$ are the $M \times M$ neighboring pixels in the local window centered at $\mathbf{x}_i$. The resulting w_{ij} can be explained as the correlation between pixels $\mathbf{x}_i$ and $\mathbf{x}_j$. Since a large weight in steering kernel mean two pixels have strong correlation, HDSK could be an effective way to represent the local structure. For example, Fig. 8.1 shows the 10-th band image in the University of Pavia HSI and the calculated HDSKs for different pixels. It can be observed that when pixels are in a homogeneous region, the shape of HDSK is cycles without any directional preference. When the pixels are in the intersection or the boundary of different classes, the shape of HDSKs is oval and exhibits clear directional preference. The direction of the long axis of the oval indicates that similar pixels may appear in this direction.

Once having determined the local structure of a test pixel x_i using (8.20), we select P pixels whose weights are larger than the others. These pixels can be stacked as $\mathbf{X}^P = [\mathbf{x}_{i1} \, \mathbf{x}_{i2} \ldots \mathbf{x}_{iP}] \in \mathbb{R}^{B \times P}$, and $\mathbf{w}^P = [w_1 \, w_2 \ldots w_P]^T$ is the corresponding

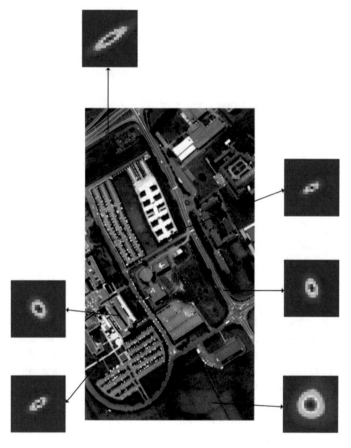

Fig. 8.1 Examples of HDSKs

weight vector. It is believed that these selected P pixels have more compact inner patterns than those in a fixed window do. The adaptive spatial contextual information is introduced by the following problem:

$$\mathbf{S}^P = \arg\min_{\mathbf{S}^P} \left\| \mathbf{X}^P - \mathbf{DS}^P \right\|_F$$
$$\text{s.t } \left\| \mathbf{S}^P \right\|_{\text{row},0} \le K_0 \tag{8.35}$$

Algorithm 4 ASC-SOMP Algorithm

Input: Dictionary $\mathbf{D} = [\mathbf{d}_1 \ \mathbf{d}_2 \ ... \ \mathbf{d}_N]$, test samples $\{\mathbf{x}_i\}_{i=1,2,...,L}$, window size M, and the number of selected pixels P

1) **Pre-calculate**: First compute gradients as in (34), then compute the gradient covariance $\mathbf{C}_i, \{i = 1, 2, ..., L\}$.

While $i \le L$ **do**

2) Compute steering kernel of x_i according to (33)

3) Sort the pixels in the window as their weights from large to small, select the first P pixels and stack them as   and record their weights

4) Initialization, residual $\mathbf{R}_0 = \mathbf{X}^P$, index set $\mathbf{\Lambda}_0 = \varnothing$, iteration counter $k = 1$

   While stopping criterion has not been met **do**

   a) Find the index of the atom that best approximates all residuals, $\lambda_k = \arg\max_{i=1,...,N} \| \mathbf{R}_{k-1}^T \mathbf{d}_i \|_\infty$

   b) Update the index set $\mathbf{\Lambda}_k = \mathbf{\Lambda}_{k-1} \cup \{\lambda_k\}$

   c) Compute $\mathbf{M}_k = (\mathbf{D}_{\Lambda_k}^T \mathbf{D}_{\Lambda_k})^{-1} \mathbf{D}_{\Lambda_k}^T \mathbf{X}^P \in \mathbb{R}^{k \times P}$, $\mathbf{D}_{\Lambda_k} \in \mathbb{R}^{B \times k}$ consists of the k atoms in $\mathbf{D}$ indexed in $\mathbf{\Lambda}_k$

   d) Determine the residual

   e) $k \leftarrow k + 1$

   Output: the sparse representation   , its nonzero rows indexed by $\mathbf{\Lambda}$ which are the K rows of the matrix $(\mathbf{D}_{\Lambda}^T \mathbf{D}_{\Lambda})^{-1} \mathbf{D}_{\Lambda}^T \mathbf{X}^P$ where $\mathbf{\Lambda} = \mathbf{\Lambda}_{k-1}$

5) Determine the label of $\mathbf{X}_i$ according to (36)

end while

Once the coefficient matrix $\mathbf{S}^P$ is obtained, a new classifier is designed based on the HDSK. As the weights in the HDSK reflect the influence of neighboring pixels on the test pixel, the original decision rule (8.6) is replaced by

$$class(\mathbf{x}_i) = \arg\min_{j=1,...,M} \left\| (\mathbf{X}^P - \mathbf{D}_j \mathbf{S}_j^P) \mathbf{w}^P \right\|_2 \tag{8.36}$$

The joint sparse HSI classification method based on adaptive spatial context is named adaptive spatial context SOMP (ASC-SOMP), of which the general flow is summarized in Algorithm 8.4.

8.4.2 *Experimental Results and Discussion*

This section uses two real hyperspectral datasets to verify the effectiveness of ASC-SOMP algorithm. For each image, the pixel-wise SVM, SVM with composite kernel (SVM-CK) [19], OMP [14], SOMP [14] are compared with ASC-SOMP both visually and quantitatively. We select Gaussian radial basis function (RBF) for the pixel-wise SVM and SVM-CK methods, since RBF has proved its capability handling complex nonlinear class distributions. The parameters in SVM-based methods are obtained by fivefold cross-validation. For methods involved with composite kernels, the spatial kernels were built by using the mean and standard deviation of the neighboring pixels in a window per spectral channel. Each value of the results is obtained after performing ten Monte Carlo runs.

The training and test samples are randomly selected from the available ground truth map. The classification accuracy is evaluated by the overall accuracy (OA) which is defined as the ratio of the number of accurately classified samples to the number of test samples, the coefficient of agreement (κ) which is the ratio of the amount of corrected agreement to the amount of expected agreement, and the average accuracy (AA). To be specific, OA is calculated by

$$OA = \sum_{i=1}^{C} \mathbf{E}_{ij}/N \tag{8.37}$$

where N is the total number of samples, and $\mathbf{E}_{ij}$ represents the number of samples in class i which are miss-classified to class j.

AA is calculated by

$$AA = \left(\sum_{i=1}^{C} \left(\mathbf{E}_{ij} \bigg/ \sum_{j=1}^{C} \mathbf{E}_{ij} \right) \right) \bigg/ C \tag{8.38}$$

The κ statistic is calculated by weighting the measured accuracies. This metric incorporates the diagonal and off-diagonal entries of the confusion matrix and is given by

$$\kappa = \left(N \left(\sum_{i=1}^{C} \mathbf{E}_{ii} \right) - \sum_{i=1}^{C} \left(\sum_{j=1}^{C} \mathbf{E}_{ij} \sum_{j=1}^{C} \mathbf{E}_{ji} \right) \right) \bigg/ \left(N^2 - \sum_{i=1}^{C} \left(\sum_{j=1}^{C} \mathbf{E}_{ij} \sum_{j=1}^{C} \mathbf{E}_{ji} \right) \right) \tag{8.39}$$

8.4.2.1   Hyperspectral Dataset of AVIRIS Indian Pines

The Indian Pines image contains 145×145 pixels and 200 spectral reflectance bands, among which 24 water absorption bands have been removed. The ground truth contains 16 land cover classes and a total of 10366 labeled pixels. We randomly choose 10% of labeled samples for training, and use the rest 90% for testing. The false color image and ground truth are shown in Fig. 8.2a, b.

The parameters for ASC-SOMP algorithm are set to $P = 120$, $K_0 = 25$, $h = 25$, and $M = 21$. The window size of SOMP algorithm is empirically set to 9×9. The classification results, in terms of overall accuracy (OA), average accuracy (AA), κ

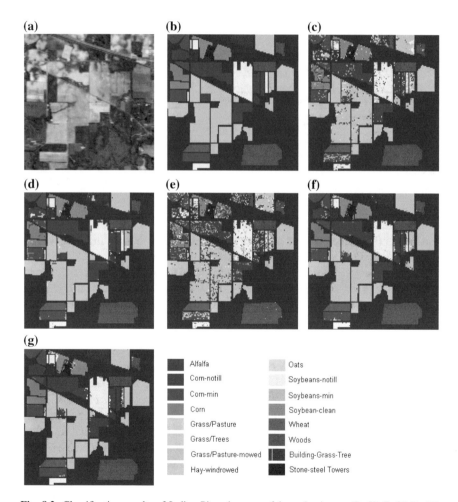

Fig. 8.2 Classification results of Indian Pines image, **a** false color image (R, 57 G, 27 B, 17), **b** ground truth, **c** SVM (OA, 85.24%), **d** SVM-CK (OA, 93.60%), **e** OMP (OA, 75.67%), **f** SOMP (OA, 95.28%), **g** ASC-SOMP (96.79%)

Table 8.1 Classification accuracy (%) For the Indian Pines image on the test set

Class	#train samples	#test samples	SVM	SVM-CK	OMP	SOMP	ASC-SOMP
Alfalfa	6	48	31.25	62.08	65.62	85.42	**91.67**
Corn-no till	144	1290	82.80	92.71	64.58	94.88	**95.74**
Corn-min till	84	750	75.01	91.29	61.36	94.93	**96.27**
Corn	24	210	64.42	79.71	44.80	91.43	**95.24**
Grass/Pasture	50	447	93.08	**95.59**	91.09	89.49	93.96
Grass/Trees	75	672	95.46	98.09	94.04	98.51	**99.70**
Grass/Pasture-mowed	3	23	4.35	49.56	84.78	**91.30**	56.20
Hay-windrowed	49	440	98.81	98.47	97.97	95.55	**100**
Oats	2	18	0.00	0.00	**43.33**	0.00	22.22
Soybeans-no till	97	871	76.76	89.97	70.76	89.44	**92.31**
Soybeans-min till	247	2221	87.76	96.13	76.22	97.34	**98.42**
Soybean-clean till	62	552	85.25	89.49	57.91	88.22	**92.39**
Wheat	22	190	98.53	96.63	97.73	**100**	99.47
Woods	130	1164	97.62	98.04	94.09	99.14	**100**
Building-Grass-Trees-Drives	38	342	56.11	89.29	44.26	99.12	**100**
Stone-steel Towers	10	85	81.17	88.11	90.47	**96.47**	95.29
OA (%)			85.24	93.60	75.67	95.28	**96.79**
AA (%)			70.52	**92.70**	72.22	88.45	89.33
κ			83.11	82.20	73.69	94.60	**96.34**

statistic, and class individual accuracies, are shown in Table 8.1. The final maps are illustrated in Fig. 8.2c–g. It can be observed that ASC-SOMP algorithm achieves the highest OA of 96.79%, which is 1.5% higher than the second-highest OA. Classification results using different percentages of labeled samples for training are shown in Fig. 8.3. In this figure and the following, error bars indicate the standard deviation by random sampling. From Fig. 8.3, both numerical and statistical differences can be observed.

Next, we demonstrate the impact of the number of selected neighboring pixels P upon the performance of ASC-SOMP algorithm. We use 10% of data in each class as training samples. The number of selected pixels P ranges from $P = 80$ to $P = 140$, and the sparsity level K_0 ranges from $K_0 = 5$ to $K_0 = 45$. The plots of overall accuracy evaluated on the entire test set are shown in Fig. 8.4. When $K_0 \geq 25$ and $P \geq 110$, a relatively high classification accuracy can be achieved. Compared with SOMP algorithm, ASC-SOMP leads to the same optimal K_0 value, but the optimal P value is significantly larger. As pixels are selected according to their spatial correlation to the center pixel, it is reasonable to select more pixels that can be sparsely represented simultaneously.

To investigate the effect of the introduced adaptive spatial context, we compare ASC-SOMP with traditional joint sparsity method in detail. It is obvious that SOMP is not able to identify any samples belonging to oats class. This observation is because oat pixels cover a very narrow region of size 10×2 located in the middle-left of

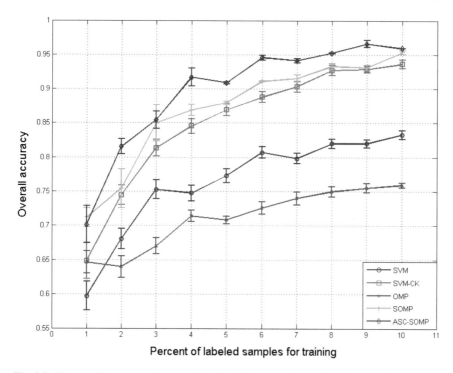

Fig. 8.3 The overall accuracy of Indian Pines for different numbers of training samples

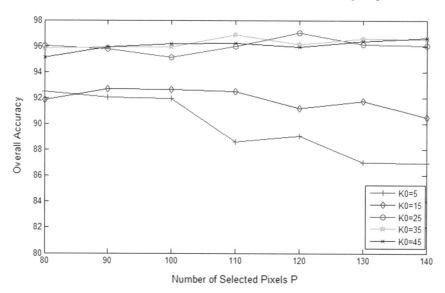

Fig. 8.4 Effects of the sparsity level K_0 and number of selected pixels P for Indian Pines

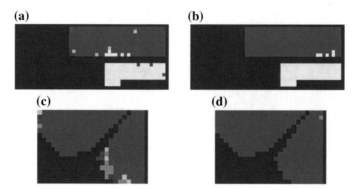

Fig. 8.5 Amplified map in two regions, **a** and **c** are results of SOMP, **b** and **d** are results of ASC-SOMP

the image. In SOMP, the optimal 9 × 9 local window centered at each oat pixel is dominated by pixels belonging to the other two adjacent classes. In contrast, ASC-SOMP achieves a 22.22% classification accuracy for oat class. By introducing adaptive spatial context, pixels distributed along the direction of the narrow region are selected as they have large correlation with the test pixel. On the other hand, pixels belonging to the other two classes whose weights are small have less impact upon our decision rule. Thus, better results can be obtained. However, the classification accuracy for oat class is still very low, because the total number of oat class is much less than the selected pixels to be sparsely represented simultaneously, and most of the selected pixels do not belong to oat class oat.

Taking into consideration that the effect of adaptive spatial context is clearer in the class boundary, more attention should be paid on the edge. We amplify the region of SOMP result and the region of ASC-SOMP result to verify the effect of adaptive spatial context. Figure 8.5 shows that our classification result has less wrong-classified pixels in the class boundary, demonstrating the advantages of the adaptive spatial context.

8.4.2.2 Hyperspectral Dataset of ROSIS Pavia University

The second hyperspectral data set was collected by the ROSIS optical sensor over the urban area of the Pavia University, Italy. The image size in pixels is 610 × 340, with a very high spatial resolution of 1.3 m per pixel. The number of data channels in the acquired image is 103 (with the spectral range from 0.43 to 0.86 μm). Nine classes of interest were considered, including tree, asphalt, bitumen, gravel, metal sheet, shadow, bricks, meadow, and soil. Figure 8.6a, b shows the three-band false color image and the ground truth map, respectively. We randomly sampled 60 pixels for each class as the training samples and use the remainder as test samples. The optimal parameter settings for the ASC-SOMP method are $P = 100$ and $K_0 = 5$. In SOMP, the window size was set to 9 × 9, and the sparsity level was set to $K_0 =$

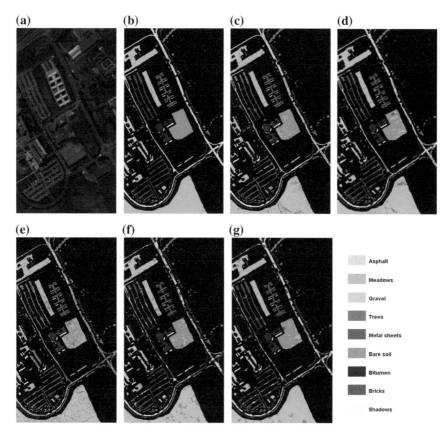

Fig. 8.6 Classification results of University of Pavia image, **a** false color image (R, 57 G, 27 B, 17), **b** ground truth, **c** SVM (OA, 84.26%), **d** SVM-CK (OA, 91.60%), **e** OMP (OA, 71.12%), **f** SOMP (OA, 83.60%), **g** ASC-SOMP (85.07%)

15. We set $h = 25$ and $M = 21$ as in the previous set of experiments. The final classification maps are illustrated in Fig. 8.6c–g. The classification results, in term of overall accuracy (OA), average accuracy (AA), k statistic, and class individual accuracies, are provided in Table 8.2. The ASC-SOMP method outperforms other methods except for SVM-CK. SVM-CK achieves the best results since it is a spectral–spatial nonlinear kernel method. Figure 8.7 illustrates the classification accuracies by using different number of training samples. This result justifies the robustness of ASC-SOMP method. Figure 8.8 shows the performance in terms of overall accuracy with different numbers of selected pixels P at sparsity level $K_0 = 5$ and $K_0 = 10$, respectively. The number of selected pixels P ranges from 50 to 110. Figure 8.8 also shows that the overall accuracy improves as P value increases. This conclusion isconsistent with the conclusion drawn on the dataset of AVIRIS Indian Pines.

Table 8.2 Classification accuracy (%) for University of Pavia on the test set

Class	#train samples	#test samples	SVM	SVM-CK	OMP	SOMP	ASC-SOMP
Asphalt	60	6571	77.92	**88.98**	57.62	47.87	52.01
Bare soil	60	18589	81.67	**93.09**	71.96	91.59	91.36
Bitumen	60	2039	82.13	87.65	65.85	92.15	**93.52**
Bricks	60	3004	95.33	**97.52**	89.83	89.34	95.97
Gravel	60	1285	99.15	99.47	99.75	**100**	99.24
Meadows	60	4969	87.92	**89.66**	63.38	87.74	86.76
Metal sheets	60	1270	93.59	94.55	85.85	95.98	**97.92**
Shadows	60	3622	83.70	83.03	68.30	84.40	**87.00**
Trees	60	887	**99.96**	99.14	94.61	73.95	85.49
OA (%)			84.26	**91.60**	71.12	83.60	85.07
AA (%)			79.75	**88.95**	63.17	78.56	80.50
κ			89.04	**92.57**	77.46	84.78	87.70

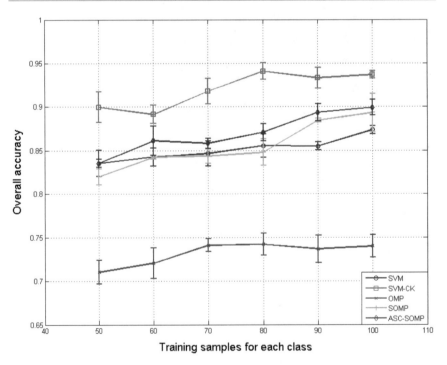

Fig. 8.7 The overall accuracy of University of Pavia for different numbers of training samples

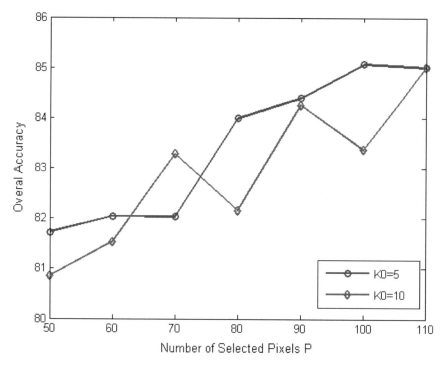

Fig. 8.8 Effect of different numbers of selected pixels P for University of Pavia

8.4.2.3 Discussion

The ASC-SOMP method and the nonlocal-weighted version of SOMP (NLW-JSRC) [28] both were developed for improving the original SOMP method. The weights for the neighboring pixels are calculated in both methods. We compared our method with NLW-JSRC. All experiments were performed using the same experimental setup as in the work of NLW-JSRC, where 9% of the labeled data are randomly sampled as the training samples, and the remainder of the data are used as test samples. Tables 8.3 and 8.4 present the comparisons of results by both methods. We can observe that the ASC-SOMP method outperforms the NLW-JSRC method, indicating that the steering kernel can better describe the spatial context than the nonlocal weights can.

h and M are two important parameters that control the supporting range of the steering kernel and determine the contributions of the selected pixels to the classification of test pixel. We further evaluate the classification accuracy on the two images for different h and M values. We use the same training samples as in previous experiments. h ranges from 1 to 45, and the window size M ranges from 13×13 to 29×29. Figure 8.9a indicates that the classification accuracy is relatively high when h is between 10 and 35. If h is too small, the variance of the weights is large, resulting in the outcome that a few pixels with large weights dominate the classification decision. If h is too large, on the other hand, the gap between different pixels' weights

Table 8.3 Numerical comparison with NLW-JSRC for Indian Pines

Class	#train samples	#test samples	NLW-JSRC	ASC-SOMP
1	6	46	**95.00**	81.25
2	129	1299	92.99	**94.25**
3	83	747	87.82	**95.33**
4	24	213	85.45	**96.66**
5	48	435	93.33	**93.76**
6	73	657	**100**	99.25
7	5	23	73.91	**95.23**
8	48	430	**100**	**100**
9	4	16	31.25	**50.00**
10	97	875	90.51	**92.30**
11	196	2259	96.90	**98.85**
12	59	534	**96.82**	86.12
13	21	184	**100**	**100**
14	114	1151	**99.91**	**99.91**
15	39	347	96.25	**98.82**
16	12	81	97.53	100
OA (%)			95.19	**96.35**
κ			94.50	**95.83**

Table 8.4 Numerical comparison with NLW-JSRC for University of Pavia

Class	#train samples	#test samples	NLW-JSRC	ASC-SOMP
1	579	6034	87.67	**96.56**
2	932	17717	98.91	**99.90**
3	189	1910	79.42	**98.69**
4	276	2788	92.90	**96.77**
5	269	1076	**100**	**100**
6	453	4576	77.69	**99.08**
7	266	1064	96.43	**99.62**
8	331	3351	85.68	**97.01**
9	189	758	**98.81**	88.39
OA (%)			92.98	**98.54**
κ			90.46	**98.03**

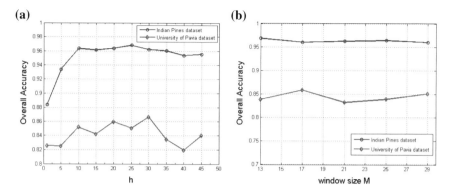

Fig. 8.9 a The classification accuracy for different h. **b** The classification accuracy for different window size M

is not clear enough, as the adaptive spatial context information is not used as much as possible. We can also observe from Fig. 8.9b that the classification accuracy is robust to the window size M as long as there are enough pixels to be selected.

8.5 Conclusions

Sparsity-based methods play an important role in HSI classification. Taking into consideration that the spectrum of a pixel lies in the low-dimensional subspace spanned by the training samples of the same class, sparse representation classification (SRC) is widely employed in HSI classification. Many advanced SRC models are presented to improve the classification accuracy, based on the structural sparsity priors, spectral–spatial information, kernel tricks, etc. This chapter reviews the structural sparsity priors and explains how the spectral–spatial information of HSI is incorporated into the SRC method. More specifically, a case study of HSI sparse representation classification based on adaptive spatial context is presented in detail. Experimental results demonstrate that, by combining SRC and adaptive spectral–spatial information, the performances of SRC can be significantly improved. Future work can be directed toward tensor sparse representation which can take full advantage of the high-order correlation in HSI and can preserve the spectral–spatial structure of HSI.

References

1. Donoho DL (2006) Compressed sensing. IEEE Trans Inf Theory 52(4):1289–1306
2. Baraniuk RG (2007) Compressive sensing. IEEE Signal Process Mag 24(4):118–121
3. Candès EJ, Romberg J, Tao T (2006) Robust uncertainty principles, Exact signal reconstruction from highly incomplete frequency information. IEEE Trans Inf Theory 52(2):489–509

4. Sun L, Zebin W, L Jianjun, X Liang, Wei Z (2015) Supervised spectral-spatial hyperspectral image classification with weighted markov random fields. IEEE Trans Geosci Remote Sens 53(3):1490–1503
5. Li J, Bioucas-Dias JM, Plaza A (2010) Semisupervised hyperspectral image segmentation using multinomial logistic regression with active learning. IEEE Trans Geosci Remote Sens 48(11):4085–4098
6. Melgani F, Bruzzone L (2004) Classification of hyperspectral remote sensing images with support vector machines. IEEE Trans Geosci Remote Sens 42(8):1778–1790
7. Pan B, Shi Z, Xu X (2018) MugNet, deep learning for hyperspectral image classification using limited samples. ISPRS J Photogramm Remote Sens 145:108–119
8. Ma L, Crawford MM, Tian J (2010) Local manifold learning-based k-nearest-neighbor for hyperspectral image classification. IEEE Trans Geosci Remote Sens 48(11):4099–4109
9. Aharon M, Elad M, Bruckstein A (2006) K-SVD, an algorithm for designing overcomplete dictionaries for sparse representation. IEEE Trans Signal Process 54(11):4311
10. Wright J, Yang AY, Ganesh A et al (2009) Robust face recognition via sparse representation. IEEE Trans Pattern Anal Mach Intell 31(2):210–227
11. Wagner A, Wright J, Ganesh A et al (2012) Toward a practical face recognition system, robust alignment and illumination by sparse representation. IEEE Trans Pattern Anal Mach Intell 34(2):372–386
12. Gemmeke JF, Virtanen T, Hurmalainen A (2011) Exemplar-based sparse representations for noise robust automatic speech recognition. IEEE Trans Audio Speech Lang Process 19(7):2067–2080
13. Yang J, Wright J, Huang TS et al (2010) Image super-resolution via sparse representation. IEEE Trans Image Process 19(11):2861–2873
14. Chen Y, Nasrabadi NM, Tran TD (2011) Hyperspectral image classification using dictionary-based sparse representation. IEEE Trans Geosci Remote Sens 49(10):3973–3985
15. Sun X, Qu Q, Nasrabadi NM et al (2014) Structured priors for sparse-representation-based hyperspectral image classification. IEEE Geosci Remote Sens Lett 11(7):1235–1239
16. Tu B, Zhang X, Kang X et al (2018) Hyperspectral image classification via fusing correlation coefficient and joint sparse representation. IEEE Geosci Remote Sens Lett 15(3):340–344
17. Liu J, Wu Z, Xiao Z, Yang J (2017) Hyperspectral image classification via kernel fully constrained least squares. In: 2017 IEEE international geoscience and remote sensing symposium, Fort Worth, 23–28 July 2017, pp 2219–2222
18. Aizerman A, Braverman E, Rozoner L (1964) Theoretical foundations of the potential function method in pattern recognition learning. Autom Remote Control 25:821–837
19. Camps-Valls G, Gomez-Chova L, Muñoz-Mari' J, Vila-Francés J, Calpe-Maravilla J (2006) Composite kernels for hyperspectral image classification. IEEE Geosci Remote Sens Lett 3(1):93–97
20. Liu J, Wu Z, Wei Z et al (2013) Spatial-spectral kernel sparse representation for hyperspectral image classification. IEEE J Sel Top Appl Earth Obs Remote Sens 6(6):2462–2471
21. Tuia D, Camps-Valls G (2001) Urban image classification with semisupervised multiscale cluster kernels. IEEE J Sel Top Appl Earth Obs Remote Sens 4(1):65–74
22. Gomez-Chova L, Camps-Valls G, Bruzzone L, Calpe-Maravilla J (2010) Mean map kernel methods for semisupervised cloud classification. IEEE Trans Geosci Remote Sens 48(1):207–220
23. Bioucas-Dias J, Figueiredo M (2010) Alternating direction algorithms for constrained sparse regression, Application to hyperspectral unmixing. In: Proceedings of WHISPERS, Reykjavik, Iceland, June 2010, pp 1–4. IEEE
24. Combettes P et al (2006) Signal recovery by proximal forward-backward splitting. Multiscale Model Simul 4(4):1168–1200
25. Takeda H, Farsiu S, Milanfar P (2007) Kernel regression for image processing and reconstruction. IEEE Trans Image Process 16(2):349–366
26. Feng X, Milanfar P (2002) Multiscale principal components analysis for image local orientation estimation. In: 36th Asilomar conference signals, systems and computers, Pacific Grove, CA

27. Xu Y, Wu Z, Wei ZH (2014) Joint sparse hyperspectral image classification based on adaptive spatial context. J Appl Remote Sens 8(1):083552
28. Zhang H et al (2013) A nonlocal weighted joint sparse representation classification method for hyperspectral imagery. IEEE J Sel Top Appl Earth Obs Remote Sens 7(6):2056–2065
29. Takeda H, Farsiu S, Milanfar P (2007) Kernel regression for image processing and reconstruction. IEEE Trans Image Process 16:349–366

Chapter 9
Multiple Kernel Learning for Hyperspectral Image Classification

Tianzhu Liu and Yanfeng Gu

Abstract With the rapid development of spectral imaging techniques, classification of hyperspectral images (HSIs) has attracted great attention in various applications such as land survey and resource monitoring in the field of remote sensing. A key challenge in HSI classification is how to explore effective approaches to fully use the spatial–spectral information provided by the data cube. Multiple Kernel Learning (MKL) has been successfully applied to HSI classification due to its capacity to handle heterogeneous fusion of both spectral and spatial features. This approach can generate an adaptive kernel as an optimally weighted sum of a few fixed kernels to model a nonlinear data structure. In this way, the difficulty of kernel selection and the limitation of a fixed kernel can be alleviated. Various MKL algorithms have been developed in recent years, such as the general MKL, the subspace MKL, the nonlinear MKL, the sparse MKL, and the ensemble MKL. The goal of this chapter is to provide a systematic review of MKL methods, which have been applied to HSI classification. We also analyze and evaluate different MKL algorithms and their respective characteristics in different cases of HSI classification cases. Finally, we discuss the future direction and trends of research in this area.

Keywords Remote sensing · Hyperspectral images · Multiple kernel learning (MKL) · Heterogeneous features · Classification

9.1 Introduction

A wide range of pixel-level processing techniques for the classification of HSIs has been developed; the illustration of HSI supervised classification is shown in Fig. 9.1. Kernel methods have been successfully applied to HSI classification [1] while providing an elegant way to deal with nonlinear problems [2]. The main idea of kernel methods is to map the input data from the original space to a convenient feature space by a nonlinear mapping function. Inner products in the feature space can be computed

T. Liu · Y. Gu (✉)
Harbin Institute of Technology, Harbin, China
e-mail: guyf@hit.edu.cn

© Springer Nature Switzerland AG 2020
S. Prasad and J. Chanussot (eds.), *Hyperspectral Image Analysis*,
Advances in Computer Vision and Pattern Recognition,
https://doi.org/10.1007/978-3-030-38617-7_9

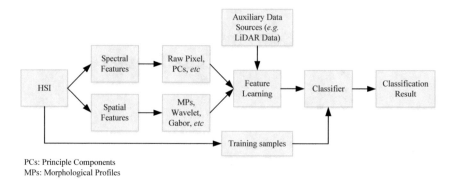

PCs: Principle Components
MPs: Morphological Profiles

Fig. 9.1 Illustration of HSI supervised classification

by a kernel function without knowing the nonlinear mapping function explicitly. Then, the nonlinear problems in the input space can be processed by building linear algorithms in the feature space [3]. The kernel support vector machine (SVM) is the most popular approach applied to HSI classification among various kernel methods [3–7]. SVM is based on the margin maximization principle, which does not require an estimation of the statistical distributions of classes. To address the limitation of the curse of dimensionality for HSI classification, some improved methods based on SVM have been proposed, such as multiple classifiers system based on Adaptive Boosting (AdaBoost) [8], rotation-based SVM ensemble [9], particle swarm optimization (PSO) SVM [10], subspace-based SVM [11]. To enhance the ability of similarity measurements using the kernel trick, a region-kernel-based support vector machine (RKSVM) was proposed [12]. Considering the tensor data structure of HSI, multiclass support tensor machine (STM) was specifically developed for HSI classification [13]. However, the standard SVM classifier can only use the labeled samples to provide predicted classes for new samples. In order to consider the data structure during the classification process, some clustering algorithms have been used [14], such as the hierarchical semisupervised SVM [15] and spatial–spectral Laplacian support vector machine (SS-LapSVM) [16].

There are some other families of kernel methods for HSI classification, such as Gaussian processes (GPs) and kernel-based representation. GPs provide a Bayesian nonparametric approach of the considered classification problem [17–19]. GPs assume that the probability of belonging to a class label for an input sample is monotonically related to the value of some latent function at that sample. In GP, the covariance kernel represents the prior assumption, which characterizes correlation between samples in the training data. Kernel-based representation was derived from representation-based learning (RL) to solve nonlinear problems in HSI, which assumes that a test pixel can be linearly represented by training samples in the feature space. RL has already been applied to HSI classification [20–39], which includes sparse representation-based classification (SRC) [40, 41] collaborative representation-based classification (CRC) [42], and their extensions [22, 32, 33, 38].

For example, to exploit spatial contexts of HSI, Chen et al. [20] proposed a joint sparse representation classification (JSRC) method under the assumption of a joint sparsity model (JSM) [43]. These RL methods can be kernelized as kernel SRC (KSRC) [22], kernelized JSRC (KJSRC) [44], kernel nonlocal joint CRC [32], and kernel CRC (KCRC) [36, 37] etc.

Furthermore, Multiple Kernel Learning (MKL) methods have been proposed for HSI classification, as there is a very limited selection of a single kernel, which is able to fit complex data structures. MKL methods aim at constructing a composite kernel by combining a set of predefined base kernels [45]. A framework of composite kernel machines was presented to enhance classification of HSIs [46], which opens a wide field of subsequent developments for integrating spatial and spectral information [47, 48], such as the spatial–spectral composite kernel of superpixel [49, 50], the extreme learning machine with spatial–spectral composite kernel [51], spatial–spectral composite kernels discriminant analysis [52], and the locality preserving composite kernel [53]. In addition, MKL methods generally focus on determining key kernels to be preserved and their significance in optimal kernel combination. Some typical MKL methods have been gradually proposed for HSI classification, such as subspace MKL methods [54–57], SimpleMKL [58], class-specific sparse MKL (CS-SMKL) [59], and nonlinear MKL [60, 61].

In the following, we will present a survey of the existing work related to MKL with special emphasis on remote sensing image classification. First, general MKL framework will be discussed. Then, several MKL methods are introduced which have been divided into six categories: subspace MKL methods and nonlinear MKL method for spatial–spectral joint classification of HSI, sparse MKL methods for feature interpretation in HSI classification, MK-Boosting for ensemble learning, heterogeneous feature fusion with MKL and MKL with superpixel. Next, several examples with MKL for HSI classification are demonstrated, followed by the drawn conclusions. For easy reference, Table 9.1 lists the notations of all the symbols used in this chapter.

9.2 Learning from Multiple Kernels

Given a labeled training data set with N samples $\mathbf{X} = \{\mathbf{x}_i | i = 1, 2, \ldots, N\}$, $\mathbf{x}_i \in \mathbb{R}^D$, $\mathbf{Y} = \{y_i | i = 1, 2, \ldots, N\}$, where $\mathbf{x}_i$ is a pixel vector with D-dimension, y_i is the class label, and D is the number of hyperspectral bands. The classes in the original feature space are often linearly inseparable as shown in Fig. 9.2. Then the kernel method maps these classes to a higher dimensional feature space via nonlinear mapping function Φ. The mapped higher dimensional feature space is denoted as $\mathbb{Q}$, i.e.:

$$\Phi : \mathbb{R}^D \to \mathbb{Q}, \mathbf{X} \to \Phi(\mathbf{X}) \tag{9.1}$$

Table 9.1 Summary of the notations

Relational data

Symbol	Meaning	Symbol	Meaning
N	Number of training samples	$y_i \in \{-1, +1\}$	The ith sample label
D	Number of HSI bands	$\mathbb{Q}$	Feature space
$\mathbf{X}$	Training data matrix with samples as rows	Φ	Nonlinear mapping function
		$\mathbf{x}_i \in \mathbb{R}^d$	The ith sample

Kernel methods

Symbol	Meaning	Symbol	Meaning
$\mathbf{K}$	Kernel matrix/kernel function	$\mathbf{K}_m$	The mth base kernel matrix
$\mathbf{k}_m$	The vector stacking all columns of mth base kernel matrix	M	Number of candidate base kernels for combination in MKL
η_m	The weight of the mth base kernel	$\boldsymbol{\eta}$	The vector of base kernels weights
$\mathbf{Q}$	Kernel matrix vectorization	$\mathbf{D}$	Projection matrix
$\mathbf{I}$	Identify matrix	$\mathbf{S}_t$	Within-class scatter matrix
$\mathbf{S}_b$	Between-class scatter matrix	ν	Constraint term
μ	Nonnegative constant	ρ	A parameter controlling sparsity
S	Kinds of SEs	λ	Scales of attribute filters (AFs)
T	Number of boosting tails	γ	Measures the misclassification performance of the weak classifiers
W_t	Samples probability distribution in tth boosting rail		

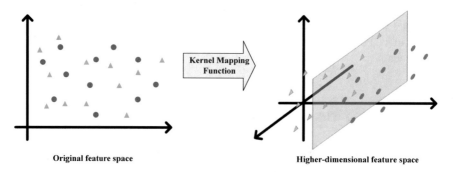

Original feature space   Higher-dimensional feature space

Fig. 9.2 Illustration of nonlinear kernel mapping

9.2.1 General MKL

MKL provides a more flexible framework so as to more effectively mine information, compared with using a single kernel. In MKL, a flexible combined kernel is generated by a linear or nonlinear combination of a series of base kernels and is used to replace the single kernel in a learning model to achieve better ability to learn. Each base

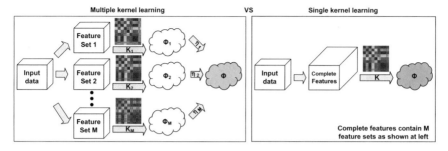

Fig. 9.3 Comparison of the multiple kernel trick and the single kernel method

kernel may exploit the full set of features or a subset of features [58]. Figure 9.3 provides an illustration of the comparison of multiple kernel trick and single kernel case. The dual problem of general linear combined MKL is expressed as follows:

$$\min_{\eta} \max_{\alpha} \left\{ \sum_{i=1}^{N} \alpha_i - \frac{1}{2} \sum_{i,j=1}^{N} \alpha_i \alpha_j y_i y_j \sum_{m=1}^{M} \eta_m \mathbf{K}_m(\mathbf{x}_i, \mathbf{x}_j) \right\}$$

$$\text{s.t. } \eta_m \geq 0, \text{ and } \sum_{m=1}^{M} \eta_m = 1 \tag{9.2}$$

where M is the number of candidate base kernels for combination, η_m is the weight of the mth base kernel.

All the weighting coefficients are nonnegative and sum to one in order to ensure that the combined kernel fulfills the positive semi-definite (PSD) condition and retains normalization as base kernels. The MKL problem is designed to optimize both the combining weights η_m and the solutions to the original learning problem, i.e., the solutions of α_i and α_j for SVM in (9.2).

Learning from multiple kernels can provide better similarity measuring ability, for example, multiscale kernels, which are RBF kernels with multiple scale parameters σ (i.e., bandwidth) [54]. Figure 9.4 shows the multiscale kernel matrices. According to the visual display of kernel matrices in Fig. 9.4, the kernelized similarity measuring appears with multiscale characteristics. The kernel with a small scale is sensitive to variation of similarities, but may result in a highly diagonal kernel matrix, which loses

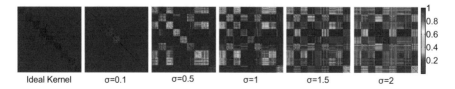

Fig. 9.4 Multiscale kernel matrices

generalization capability. On the contrary, with large scale, the kernel becomes insensitive to small variations of similarities. Therefore, by learning multiscale kernels, an optimal kernel with the best discriminative ability can be achieved.

For various applications in real world, there are plenty of heterogeneous data or features [62]. In terms of remote sensing, the features could be spectra, spatial distribution, digital elevation model (DEM) or height, and temporal information, which need to be learned with not only a single kernel but multiple kernels where each base kernel corresponds to one type of features.

9.2.2 Strategies for MKL

The strategies for determining the kernel combination can be basically divided into three major categories [45, 63].

(a) **Criterion-based approaches**. They use a criterion function to obtain the kernel or the kernel weights. For example, kernel alignment selects the most similar kernel to the ideal kernel. Representative MKL (RMKL) obtains the kernel weights by performing principal component analysis (PCA) on the base kernels [54]. Sparse MKL acquires the kernel by robust sparse PCA [64]. Nonnegative matrix factorization (NMF) and kernel NMF (KNMF) MKL [55] find the kernel weights by NMF and KNMF. Rule-based multiple kernel learning (RBMKL) generates the kernel via summation or multiplication of the base kernels. The spatial–spectral composite kernel assigns fixed values as the kernel weights [46, 49, 51–53].

(b) **Optimization approaches**. They obtain the base kernel weights and the decision function of classification simultaneously by solving the optimization problem. For instance, class-specific MKL (CS-SMKL) [59], SimpleMKL [58], and discriminative MKL (DMKL) [57] are determined using the optimization approach.

(c) **Ensemble approaches**. They use the idea of ensemble learning. The new base kernel is added iteratively until the minimum of cost function or the optimal classification performance, for example, MK-Boosting [65], which adopts boosting to determine base kernel and corresponding weights. Besides, in the ensemble MKL-Active Learning (AL) approach [66], an ensemble of probabilistic multiple kernel classifiers is embedded into a maximum disagreement-based AL system, which adaptively optimizes the kernel for each source during the AL process.

9.2.3 Basic Training for MKL

In terms of training manners for MKL, the existing algorithms can be partitioned into two categories:

(a) **One-stage methods**: solve both classifier parameters and base kernel weights by simultaneously optimizing a target function based on the risk function of classifier. The algorithms of one-stage MKL can be further split into the two sub-categories of *direct* and *wrapper* methods according to the order of solution of classifier parameters and base kernel weights. The *direct* methods simultaneously solve the base kernel weights and the parameters [45]. The *wrapper* methods solve the two kinds of parameters separately and alternately at a given iteration. First, they optimize the base kernel weights by fixing the classifier parameters, and then optimize the classifier parameters by fixing the base kernel weights [58, 59, 66].

(b) **Two-stage methods**: solve the base kernel weights independently from the classifier [54, 55, 57]. Usually, they solve the base kernel weights first, and then take the base kernel weights as the known conditions to solve the parameters of the classifier.

The computational time of one-stage and two-stage MKL depends on two factors, which are the number of considered kernels and the number of available training samples. The one-stage algorithms are usually faster than the two-stage algorithms when both the number and size of the base kernels are small. The two-stage algorithms are generally faster than the one-stage algorithms when the number of base kernels is high or the number of training samples used for kernel construction is large.

9.3 MKL Algorithms

9.3.1 Subspace MKL

Recently, some effective MKL algorithms have been proposed for HSI classification, called subspace MKL, which use subspace method to obtain the weights of base kernels in the linear combination. These algorithms include RMKL [54], NMF-MKL, KNMF-MKL [55], and DMKL [57]. Given M base kernel matrices $\{\mathbf{K}_m, m = 1, 2, \ldots, M, \mathbf{K}_m \in \mathbb{R}^{N \times N}\}$, which are composed of a 3-D data cube of size $N \times N \times M$. In order to facilitate the subsequent operations, the 3-D data cube of the kernel matrices is converted into a 2-D matrix with the help of a vectorization operator, where all kernel matrices are separately converted into column vectors $\mathbf{k}_m = vec(\mathbf{K}_m)$. After the vectorization, a new form of the base kernels is denoted as $\mathbf{Q} = [\mathbf{k}_1, \mathbf{k}_2, \ldots, \mathbf{k}_M]^T \in \mathbb{R}^{M \times N^2}$. Subspace MKL algorithms build a loss function as follows:

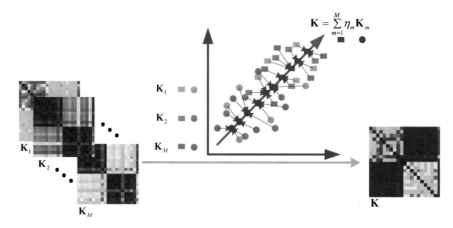

Fig. 9.5 Illustration of subspace MKL methods. The square and circle, respectively, denote training samples from two classes. The combination weights of subspace MKL methods can be obtained by base kernels projection with a few projection directions

$$\Gamma(\mathbf{K}, \boldsymbol{\eta}) = \|\mathbf{Q} - \mathbf{DK}\|_F^2 \qquad (9.3)$$

where $\mathbf{D} \in \mathbb{R}^{M \times l}$ is the projection matrix whose columns $\{\boldsymbol{\eta}_r\}_{r=1}^l$ are the bases of l-dimensional linear subspace, $\mathbf{K} \in \mathbb{R}^{l \times N^2}$ is the projected matrix onto the linear subspace spanned by $\mathbf{D}$, and $\|\bullet\|_F$ is Frobenius norm of matrix. Adopting different optimization criteria to solve $\mathbf{D}$ and $\mathbf{K}$ forms different subspace MKL methods.

The visual illustration of subspace MKL methods is shown in Fig. 9.5. Table 9.2 summarizes the three subspace MKL methods with different ways to solve the combination weights. RMKL is to determine optimal kernel combination weights by projecting onto the max-variance direction. In NMF-MKL and KNMF-MKL, NMF and KNMF are used to solve the problem of weights and the optimal combined kernel due to the nonnegativity of both matrix and combination weights. Moreover, the core idea of DMKL is to learn an optimally combined kernel from predefined base kernels by maximizing separability in reproduction kernel Hilbert space, which leads to the minimum within-class scatter and maximum between-class scatter.

9.3.2 Nonlinear MKL

Nonlinear MKL (NMKL) is motivated by the justifiable assumption that the nonlinear combination of different linear kernels can improve classification performance [45]. In [61], a nonlinear MKL (NMKL) is introduced to learn an optimally combined kernel from the predefined base kernels for HSI classification. The NMKL method can fully exploit the mutual discriminability of the inter-base-kernels corresponding

Table 9.2 Summary of subspace MKL methods

Methods	Solving strategy	Characteristics or significance
RMKL [54]	$\arg\max_{\mathbf{D}} \left\| \mathbf{D}\Sigma_{\mathbf{Q}}\mathbf{D} \right\|_F =$ $\arg\max_{\mathbf{D}} \left\| \mathbf{D}^T\mathbf{Q} \right\|_F$ s.t. $\mathbf{D}^T\mathbf{D} = \mathbf{I}_l$	Singular value decomposition
NMF/KNMF-MKL [55]	$k_{ij}^{t+1} = k_{ij}^t \dfrac{(\mathbf{QD})_{ij}}{(\mathbf{KD}^T\mathbf{D})_{ij}} \quad \eta_{ij}^{t+1} = \eta_{ij}^t \dfrac{(\mathbf{D}^T\mathbf{K})_{ij}}{(\mathbf{DK}^T\mathbf{K})_{ij}}$	NMF is used for optimization
	$k_{ij}^{t+1} = k_{ij}^t \dfrac{\left(\widehat{\mathbf{K}}\mathbf{D}\right)_{ij}}{(\mathbf{KD}^T\mathbf{D})_{ij}} \quad \eta_{ij}^{t+1} = \eta_{ij}^t \dfrac{(\mathbf{D}^T\mathbf{K})_{ij}}{(\mathbf{DK}^T\mathbf{K})_{ij}}$	Kernel NMF is used for optimization, where $\widehat{\mathbf{K}} = \Phi(\mathbf{Q})^T\Phi(\mathbf{Q})$
DMKL [57]	$\mathbf{D}^* =$ $\arg\max_{\mathbf{D}} \left\{ \mathrm{trace}\left(\left(\mathbf{D}^T(\mathbf{S}_t+\nu\mathbf{I})\mathbf{D}\right)^{-1}\mathbf{D}^T\mathbf{S}_b\mathbf{D} \right) \right\}$	Maximizing separability by Fisher criterion (FC)
	$\mathbf{D}^* = \arg\max_{\mathbf{D}} \left\{ \mathbf{D}^T(\mathbf{S}_b - \mu\mathbf{S}_t)\mathbf{D} \right\}$	Maximizing separability by maximum margin criterion (MMC)

to spatial–spectral features. Then the corresponding improvement in classification performance can be expected.

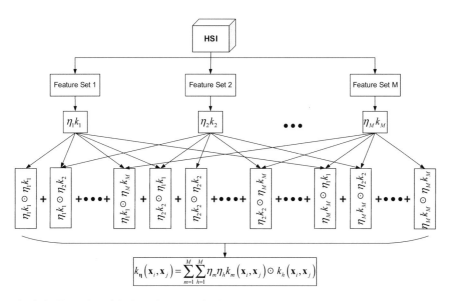

Fig. 9.6 Illustration of the kernel construction in NMKL

The framework of NMKL is shown in Fig. 9.6. First, M spatial–spectral feature sets are extracted from the HSI data cube. Each feature set is associated with one base kernel, which is defined as $\mathbf{K}_m(\mathbf{x}_i, \mathbf{x}_j) = \eta_m(\mathbf{x}_i, \mathbf{x}_j), m = 1, 2, \ldots, M$. Therefore, $\boldsymbol{\eta} = [\eta_1, \eta_2, \ldots, \eta_M]$ is the vector of kernel weights associated with the base kernels as shown in Fig. 9.6. Then, nonlinear combined kernel is computed from original kernels. M^2 new kernel matrices are given by the Hadamard product of any two base kernels, and the final kernel matrix is the weighted sum of these new kernel matrices. The final kernel matrix is shown as follows:

$$\mathbf{K}_{\boldsymbol{\eta}}(\mathbf{x}_i, \mathbf{x}_j) = \sum_{m=1}^{M} \sum_{h=1}^{M} \eta_m \eta_h \mathbf{K}_m(\mathbf{x}_i, \mathbf{x}_j) \odot \mathbf{K}_h(\mathbf{x}_i, \mathbf{x}_j) \tag{9.4}$$

Applying $\mathbf{K}_{\boldsymbol{\eta}}(\mathbf{x}_i, \mathbf{x}_j)$ to SVM, the related problem of learning the kernel $\mathbf{K}_{\boldsymbol{\eta}}$ can be concomitantly formulated as the following min-max optimization problem:

$$\min_{\boldsymbol{\eta} \in \Omega} \max_{\boldsymbol{\alpha} \in \mathbb{R}^N} \sum_{i=1}^{N} \alpha_i - \frac{1}{2} \sum_{i=1}^{N} \sum_{j=1}^{N} \alpha_i \alpha_j y_i y_j \mathbf{K}_{\boldsymbol{\eta}}(\mathbf{x}_i, \mathbf{x}_j) \tag{9.5}$$

where $\Omega = \{\boldsymbol{\eta} | \boldsymbol{\eta} \geq 0 \wedge \|\boldsymbol{\eta} - \boldsymbol{\eta}_0\|_2 \leq \Lambda\}$ is a positive, bounded, and convex set. A positive $\boldsymbol{\eta}$ ensures that the combined kernel function is positive semi-definite (PSD), and the regularization of the boundary controls the norm of $\boldsymbol{\eta}$. The definition includes an offset parameter $\boldsymbol{\eta}_0$ for the weight $\boldsymbol{\eta}$. Natural choices for $\boldsymbol{\eta}_0$ are $\boldsymbol{\eta}_0 = 0$ or $\boldsymbol{\eta}_0 / \|\boldsymbol{\eta}_0\| = 1$.

A projection-based gradient-descent algorithm can be used to solve this min-max optimization problem. At each iteration, α is obtained by solving a kernel ridge regression (KRR) problem with the current kernel matrix and $\boldsymbol{\eta}$ is updated with the gradients calculated using α while considering the bound constraints on $\boldsymbol{\eta}$ due to Ω.

9.3.3 Sparsity-Constrained MKL

(a) Sparse MKL

There is redundancy among the multiple base kernels, especially the kernels with similar scales (shown in Fig. 9.7). In [64], a sparse MKL framework was proposed to achieve a good classification performance by using a linear combination of only a few kernels from multiple base kernels. In sparse MKL, learning with multiple base kernels from hyperspectral data is carried out by two stages. The first stage is to learn an optimally sparse combined kernel from all base kernels, and the second stage is to perform the standard SVM optimization with the optimal kernel. In the first step, a sparsity constraint is introduced to control the number of nonzero weights and improve the interpretability of base kernels in classification. The learning model in the first step can be written as the following optimization problem:

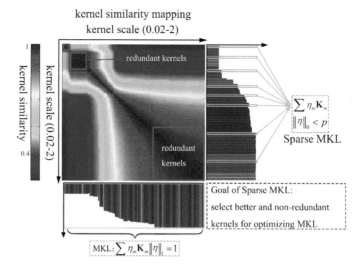

Fig. 9.7 Illustration of sparse multiple kernel learning

$$\max_{\boldsymbol{\eta}} \boldsymbol{\eta}^T \Sigma \boldsymbol{\eta} - \rho \text{Card}(\boldsymbol{\eta}) \quad \text{s.t.} \quad \boldsymbol{\eta}^T \boldsymbol{\eta} = 1 \tag{9.6}$$

where $\text{Card}(\boldsymbol{\eta})$ is the cardinality of $\boldsymbol{\eta}$ and corresponds to the number of nonzero weights, and ρ is a parameter to control sparsity.

Maximization in (9.6) can be interpreted as a robust maximum eigenvalue problem and solved with a first-order algorithm given as

$$\max \text{Tr}(\Sigma \mathbf{Z}) - \rho \mathbf{1}^T |\mathbf{Z}| \mathbf{1} \quad \text{s.t.} \quad \text{Tr}(\mathbf{Z}) = 1, \ \mathbf{Z} \geq 0 \tag{9.7}$$

(b) **Class-Specific MKL**

A class-specific sparse multiple kernel learning (CS-SMKL) framework has been proposed for spatial–spectral classification of HSIs, which can effectively utilize the multiple features with multiple scales [59]. CS-SMKL classifies the HSIs by simultaneously learning class-specific significant features and selecting class-specific weights.

The framework of CS-SMKL is illustrated in Fig. 9.8. First, feature extraction is performed on the original data set, and M feature sets are obtained. Then, M base kernels associated with M feature sets were constructed. At the kernel learning stage, a class-specific way via the one-vs-one learning strategy is used to select the class-specific weights for different feature sets and remove the redundancy of those features when classifying any two categories. As shown in Fig. 9.8, when classifying one class-pair (take, e.g., class 2 and class 5), first we find their position coordinates according to the label of training samples, then the associate class-specific kernel κ_m, $m = 1, 2, \ldots, M$, is extracted from the base kernels via the corresponding location.

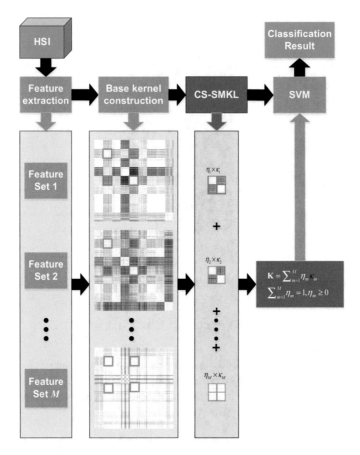

Fig. 9.8 Illustration of the class-specific kernel learning (taking class 2 and 5 as examples)

After that, the optimal kernel is obtained by the linear combination of these class-specific kernels. The weights of the linear combination are constrained by the criteria $\sum_{m=1}^{M} \eta_m = 1, \eta_m \geq 0$. The criteria can enforce the sparsity at the group/feature level and automatically learn a compact feature set for classification purposes. The combined kernel was embedded into SVM to complete the final classification.

In CS-SMKL approach, an efficient optimization method has been adopted by using the equivalence between MKL and group lasso [67]. The MKL optimization problem is equivalent to the optimization problem:

$$\min_{\eta \in \Omega} \min_{\{f_m \in \mathrm{H}_m\}_{m=1}^M} \left[\frac{1}{2} \sum_{m=1}^{M} \eta_m \| f_m \|_{\mathrm{H}_m}^2 + \max_{\alpha \in [0,\mathbf{C}]^N} \sum_{i=1}^{N} \alpha_i \left(1 - \sum_{m=1}^{M} y_i \eta_m f_m(\mathbf{x}_i) \right) \right] \quad (9.8)$$

The main differences among the three sparse MKL methods are summarized in Table 9.3.

Table 9.3 Summary of sparse MKL methods

Classifier	Solving strategy	Characteristics or significance
Sparse MKL [64]	$\max_{\boldsymbol{\eta}} \boldsymbol{\eta}^T \boldsymbol{\Sigma} \boldsymbol{\eta} - \rho \mathrm{Card}(\boldsymbol{\eta}), \quad \text{s.t. } \boldsymbol{\eta}^T \boldsymbol{\eta} = \mathbf{I}$	Robust sparse PCA is used for optimization
CS-SMKL [59]	$\min_{\boldsymbol{\eta} \in \Omega} \max_{\boldsymbol{\alpha} \in [\mathbf{0}, \mathbf{C}]^N} \left[\mathbf{1}^T \boldsymbol{\alpha} - \frac{1}{2} \sum_{m=1}^{M} \eta_m (\boldsymbol{\alpha} \circ \mathbf{y})^T \kappa_m (\boldsymbol{\alpha} \circ \mathbf{y}) \right]$	Learn class-specific significant features and select class-specific weights for each class-pair simultaneously
SimpleMKL [58]	$\min_{\boldsymbol{\eta} \in \Omega} \max_{\boldsymbol{\alpha} \in [\mathbf{0}, \mathbf{C}]^N} \left[\mathbf{1}^T \boldsymbol{\alpha} - \frac{1}{2} \sum_{m=1}^{M} \eta_m (\boldsymbol{\alpha} \circ \mathbf{y})^T \mathbf{K}_m (\boldsymbol{\alpha} \circ \mathbf{y}) \right]$	Solve both classifier parameters and base kernel weights simultaneously

9.3.4 Ensemble MKL

Ensemble learning strategy can be applied to the MKL framework to select more effective training samples. As being a main way to ensemble learning, Boosting was proposed [68] and improved in [69]. The idea is based on the way to iteratively select training samples, which sequentially pays more attention to these easily misclassified samples to train base classifiers. The idea of using boosting techniques to learn kernel-based classifiers was introduced in [70]. Recently Boosting has been integrated to the MKL with extended morphological profiles (EMP) features in [65] for HSI classification.

Let T be the number of boosting tails. The base classifiers are constructed by SVM classifiers with the input of the complete set of multiple features. The method screens samples by probability distribution $W_t \subset W$, $t = 1, 2, \ldots T$, which indicates the importance of the training samples for designing a classifier. The incorrectly classified samples have much higher probability to be chosen as screened samples in the next iteration. In this way, MK-Boosting provides a strategy to select more effective training samples for HSI classification. SVM classifier is used as a weak classifier in this case. In each iteration, the base classifier f_t is obtained from M weak classifiers:

$$f_t = \underset{f_t^m, j=\{1,\ldots,M\}}{\arg\min} \gamma_t^m = \underset{f_t^m, j=\{1,\ldots,M\}}{\arg\min} \gamma(f_t^m) \tag{9.9}$$

where γ measures the misclassification performance of the weak classifiers.

In each iteration, the weights of the distribution are adjusted by increasing the values of incorrectly classified samples and decreasing the values of correctly classified samples in order to make the classifier focus on the "hard" samples in the training set, as shown in Fig. 9.9.

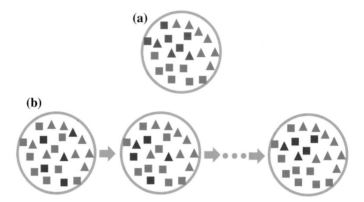

Fig. 9.9 Illustration of the sample screening process during boosting trails by taking two classes as a simple example. **a** Training samples set: the triangle and square, respectively, denote training samples from two classes and samples marked in red mean "hard" samples, which are easily misclassified. **b** Sequent screened samples: the screened samples (sample ratio = 0.2) marked in purple color during boosting trails, and the screened samples focus on "hard samples" as shown in **a**

Taking the morphological profile as an example, the architecture of this method is shown in Fig. 9.10. The features respectively are the input to SVM, and then the best classifier with the best performance will be selected as a base classifier, and the last T base classifiers are combined as the final classifier. Furthermore, the coefficients are determined by the classification accuracy of the base classifiers during the boosting trails.

9.3.5 Heterogeneous Feature Fusion with MKL

This subsection introduces a heterogeneous feature fusion framework with MKL, as shown in Fig. 9.11. It can be found that there are two levels of MKL in column and row, respectively. First, different kernel functions are used to measure the similarity of samples on each feature subset. This is the "column" MKL, $\mathbf{K}_{Col}^{(m)}(\mathbf{x}_i^{(m)}, \mathbf{x}_j^{(m)}) = \sum_{s=1}^{S} h_s^{(m)} \mathbf{K}_s^{(m)}(\mathbf{x}_i^{(m)}, \mathbf{x}_j^{(m)})$. In this way, the discriminative ability of each feature subset is exploited at different kernels and is integrated to generate an optimally combined kernel for each feature subset. Then, the multiple combined kernels resulted by MKL on each feature subset are integrated using a linear combination. This is the "row" MKL $\mathbf{K}_{Row}(\mathbf{x}_i, \mathbf{x}_j) = \sum_{m=1}^{M} d_m \mathbf{K}_{Col}^{(m)}(\mathbf{x}_i^{(m)}, \mathbf{x}_j^{(m)})$. As a result, the information contained in different feature subsets is mined and integrated into the final classification kernel. In this framework, the weights of the base kernels can be determined by any MKL algorithm, such as RMKL, NMF-MKL, and DMKL. It is worth noting that sparse MKL can be carried out on both each feature subset

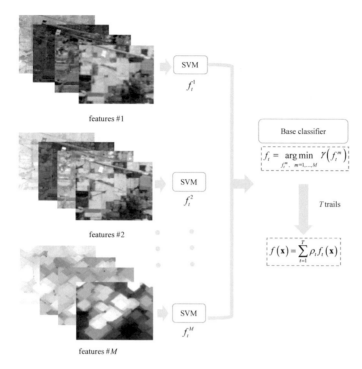

Fig. 9.10 The architecture of MK-boosting method

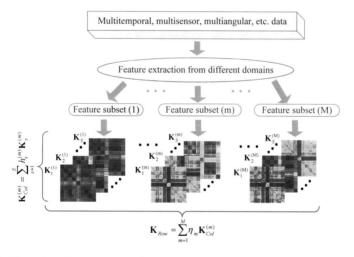

Fig. 9.11 Illustration of heterogeneous feature fusion with MKL

level and between feature subsets level for base kernels and features interpretation, respectively.

9.3.6 MKL with Superpixel

MKL provides a very effective means of learning, and can conveniently be embedded in a variety of characteristics. Therefore, it is critical to apply MKL to effective features. Recently, a superpixel approach has been applied to HSI classification as an effective spatial feature extraction means. Each superpixel is a local region, whose size and shape can be adaptively adjusted according to local structures. And the pixels in the same superpixel are assumed to have very similar spectral characteristics, which mean that superpixel can provide more accurate spatial information. Utilizing the feature explored by superpixel, the salt and pepper phenomenon appearing in the classification result will be reduced. In consequence, superpixel MKL will lead to a better classification performance.

(a) MKL with Multi-morphological Superpixel (MMSP)

This MMSP model for HSI classification consists of four steps [71]. The flowchart of the proposed framework is shown in Fig. 9.12. The first step is MMSP generation using SLIC method performed on the principle components (PCs) extracted from original spectral feature and each morphological filtered image after obtaining the multi-morphological features. Note that multi-morphological features are multi-SE EMPs or multi-AF extended multi-attribute profiles (EMAPs). The second step is the merging of MMSPs from the same class according to a uniformity constraint. The third step is the spatial feature extraction inner- and inter- the MMSPs by applying a mean filter on the MMSPs and merged MMSPs. The last step is HSI classification using MKL methods where base kernels are calculated, respectively, from the original spectral feature, spatial features inner- and inter- MMSPs.

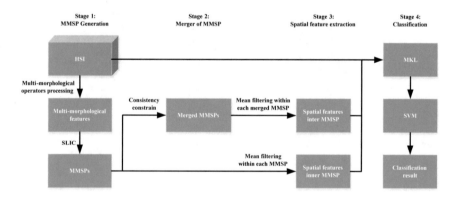

Fig. 9.12 Flowchart of the proposed MMSP model

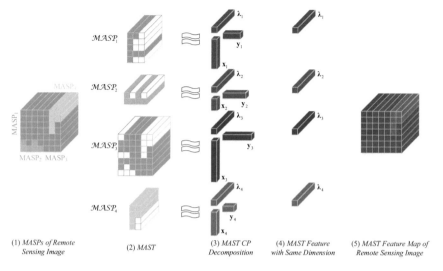

(1) *MASPs of Remote Sensing Image*   (2) *MAST*   (3) *MAST CP Decomposition*   (4) *MAST Feature with Same Dimension*   (5) *MAST Feature Map of Remote Sensing Image*

Fig. 9.13 Detailed procedure of tensor representation of MASP and integrated feature extraction by CP decomposition

(b)   MKL with Multi-attribute Super-Tensor (MAST)

Based on the multi-attribute MASP, a super-tensor model which treats each superpixel as a tensor, exploits the third-order nature of HSI. The first step is the super-tensor representation of MASPs. Then, MAST feature is extracted by applying CP decomposition. Finally, HSI classification is achieved by MKL methods where base kernels are calculated, respectively, from the original spectral feature, EMAP features, and MAST features. The illustration of the main procedure of the proposed STM model is shown in Fig. 9.13.

9.4   MKL for HSI Classification

9.4.1   Hyperspectral Data Sets

Five data sets are used in this chapter. Three of them are HSIs, which were used to validate classification performance. The 4th and 5th data sets consist of two parts, i.e., MSI and LiDAR, which are used to perform multisource classification. The first two HSIs are from cropland scenes acquired by the Airborne Visible/Infrared Imaging Spectrometer (AVIRIS) sensor. The AVIRIS sensor acquires 224 bands of 10 nm width with center wavelengths from 400 to 2500 nm. The third HSI was acquired with the Reflective Optics System Imaging Spectrometer (ROSIS-03) optical sensor over an urban area [72]. The flight over the city of Pavia, Italy, was operated by the Deutschen Zentrum für Luft- und Raumfahrt (DLR, German Aerospace Agency) within the context of the HySens project, managed and sponsored by the European Union. The ROSIS-03 sensor provides 115 bands with a spectral coverage ranging from 430 to 860 nm. The spatial resolution is 1.3 m per pixel.

(a) **Indian Pine Data set**: This HSI was acquired over the agricultural Indian Pine test site in Northwestern Indiana. It has a spatial size of 145 × 145 pixels with a spatial resolution of 20 m per pixel. Twenty water absorption bands were removed, and a 200-band image was used for the experiments. The data set contains 10,366 labeled pixels and 16 ground reference classes, most of which are different types of crops. A false color image and the reference map are presented in Fig. 9.14a.

(b) **Salinas data set**: This hyperspectral image was acquired in Southern California [73]. It has a spatial size of 512 × 217 pixels with a spatial resolution of 3.7 m per pixel. Twenty water absorption bands were removed, and a 200-band image was used for the experiments. The ground reference map was composed of 54,129 pixels and 16 land-cover classes. Figure 9.14b shows a false color image and information of the labeled classes.

(c) **Pavia University Area**: This HSI with 610 × 340 pixels was collected near the Engineering School, University of Pavia, Pavia, Italy. Twelve channels were removed due to noise [46]. The remaining 103 spectral channels were processed. There are 43,923 labeled samples in total, and nine classes of interest. Figure 9.14c presents false color images of this data set.

(d) **Bayview Park**: The data set is from 2012 IEEE GRSS Data Fusion Contest and is one of subregions of a whole scene around downtown area of San Francisco, USA. This data set contains multispectral images with eight bands acquired by WorldView2 on October 9, 2011 and corresponding LiDAR data acquired in June 2010. It has a spatial size of 300 × 200 pixels with a spatial resolution of 1.8 m per pixel. There are 19,537 labeled pixels and 7 classes. The false color image and ground reference map are shown in Fig. 9.14d.

(e) **Recology**: The source of this data set is the same as Bayview Park, which is another subregion of whole scene. It has 200 × 250 pixels with 11,811 labeled pixels and 11 classes. Figure 9.14e shows the false color image and ground reference map.

More details about these data sets are listed in Table 9.4.

9.4.2 Experimental Settings and Evaluation

To evaluate the performance of the various MKL methods for the classification task, MKL methods and typical comparison methods are shown in Table 9.5. The single kernel method represents the best performance by standard SVM, which can be used as a standard to evaluate whether a MKL method is effective or not. The number of training samples per class was varied (n = {1%, 2%, 3%} or n = {10, 20, 30}). The overall accuracy (OA [%]) and computation time were measured. Average

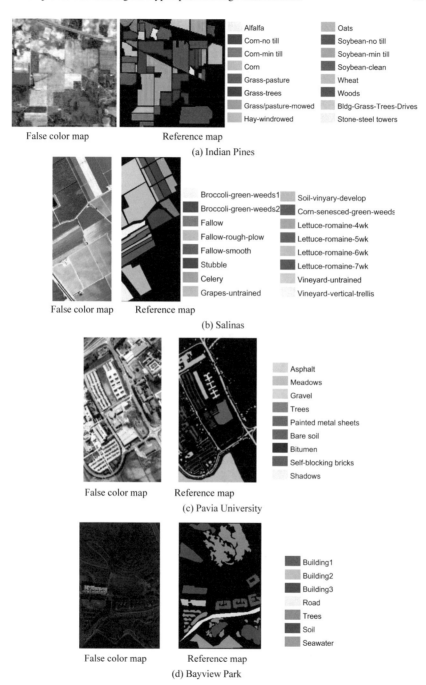

Fig. 9.14 Ground reference maps for the five data sets

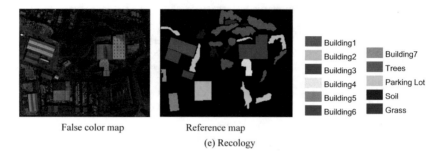

False color map   Reference map
(e) Recology

Fig. 9.14 (continued)

results for a number of ten realizations are shown. To guarantee the generality, all
the experiments were conducted on typical HSI data sets.

In the first experiment of spectral classification, all spectral bands are stacked into
a feature vector as input features. The feature vector was input into a Gaussian kernel
with different scales. For all of the classifiers, the range of the scale of Gaussian kernel
was set to [0.05, 2], and uniform sampling that selects scales from the interval with
a fixed step size of 0.05 was used to select 40 scales within the given range.

In the second experiment of spatial and spectral classification, all the data sets
were processed first by PCA and then by mathematical morphology (MM). The
eigenvalues were arranged in descending order. The first p PCs that account for 99%
of the total variation in terms of eigenvalues were reserved. Hence, the construction
of the morphological profile (MP) was based on the PCs, and a stacked vector was
built with the MP on each PC. Here, three kinds of SEs were used to obtain the MP
features, including diamond, square, and disk SEs. For each kind of SE, a step size
of an increment of 1 was used, and ten closings and ten openings were computed for
each PC. Each structure of MPs with ten closings and ten openings and the original
spectral features were, respectively, stacked as the input vector of each base kernel
for MKL algorithms. The base kernels were 4 Gaussian kernels, i.e., the values {0.1,
1, 1.5, 2}, which corresponds to three kinds of structures of MPs and original spectral
features, respectively, namely 20 base kernels for MKL methods, except for NMKL,
which is with 3 Gaussian kernels, i.e., the values {1, 1.5, 2} for NMKL-Gaussian,
and 4 linear base kernels function for NMKL-Linear.

Heterogeneous features were used in the third experiment, including spectral features, elevation features, normalized digital surface model (nDSM) from LiDAR
data, and spatial features of MPs. MPs features are extracted from original multispectral bands and nDSM uses the diamond structure element with the sizes [3, 5,
7, 9, 11, 13, 15, 17, 19, 21]. Heterogeneous features are stacked as a single vector of
features to be the input of fusion methods.

Superpixel-based spatial–spectral features were used in the fourth experiment. The
Multiple SEs and multiple AFs were carried out on the extracted p PCs, respectively.
Three kinds of SEs including line, square, and disk with three scales [3, 6, 9] are used.
Four kinds of AFs are adopted, including (1) area of the region (related to the size

Table 9.4 Information of all the data sets

Data set	No.	Categories	Samples	No.	Categories	Samples
Indian Pine	C1	Alfalfa	54	C9	Oats	20
	C2	Corn-no till	1434	C10	Soybean-no till	968
	C3	Corn-min till	834	C11	Soybean-min till	2468
	C4	Corn	234	C12	Soybean-clean	614
	C5	Grass-pasture	497	C13	Wheat	212
	C6	Grass-trees	747	C14	Woods	1294
	C7	Grass/pasture-mowed	26	C15	Bldg-Grass-Trees-Drives	380
	C8	Hay-windrowed	489	C16	Stone-steel towers	95
	Total			10,366		
Salinas	C1	Broccoli-green-weeds_1	2009	C9	Soil-vinyary-develop	6203
	C2	Broccoli-green-weeds_2	3726	C10	Corn-senesced-green-weeds	3278
	C3	Fallow	1976	C11	Lettuce-romaine-4wk	1068
	C4	Fallow-rough-plow	1394	C12	Lettuce-romaine-5wk	1927
	C5	Fallow-smooth	2678	C13	Lettuce-romaine-6wk	916
	C6	Stubble	3959	C14	Lettuce-romaine-7wk	1070
	C7	Celery	3579	C15	Vineyard-untrained	7268
	C8	Grapes-untrained	11,271	C16	Vineyard-vertical-trellis	1807
	Total			54,129		
Pavia University	C1	Asphalt	6852	C6	Bare soil	5104
	C2	Meadows	18,686	C7	Bitumen	1356
	C3	Gravel	2207	C8	Self-blocking bricks	3878
	C4	Trees	3436	C9	Shadows	1026
	C5	Painted metal sheets	1378			
	Total			43,923		
Bayview Park	C1	Building 1	2282	C5	Trees	7684
	C2	Building 2	719	C6	Soil	4283
	C3	Building 3	995	C7	Seawater	2008
	C4	Road	1566			
	Total			19,537		
Recology	C1	Building 1	1080	C7	Building 7	167
	C2	Building 2	1136	C8	Trees	3321
	C3	Building 3	1849	C9	Parking Lot	1783
	C4	Building 4	431	C10	Soil	561
	C5	Building 5	549	C11	Grass	149
	C6	Building 6	785			
	Total			11,811		

Table 9.5 Experimental methods and setting

Category	Methods	Setting
Single kernel	Standard SVM	A single kernel whose scale parameter was optimized by kernel alignment (KA) [75], denoted as Spe-SK (spectral features as the input) or MPs + Spe-SK (MPs and spectral features as the input)
General MKL	Mean MKL [45] (ruled-based method)	
	CKL [46]	$\mu = 0.4$ is used to weigh the spectral kernel and spatial kernel. In experiment for spatial–spectral classification, and the weights of spectral, spatial, and elevation kernels were set to 0.5, 0.1, and 0.4, separately for heterogeneous features classification
Subspace MKL	RMKL [54], NMF MKL, KNMF MKL [55], DMKL [57]	
Nonlinear MKL	NMKL-Linear [61]	The base kernels are linear kernels
	NMKL-Gaussian	The base kernels are Gaussian kernels
Sparse methods	SimpleMKL [58], CS-SMKL [59], Sparse MKL [64]	
	SRC [76], CRC [73]	All the original spectral features are used, and the regularization parameter is set as the optimal parameter
Ensemble MKL	MK-Boosting [65]	The sampling ratio was 0.2 and the boosting trails T was 200
MFL	Multiple feature learning (MFL) [77]	All the MP features are used for this method

of the regions), (2) diagonal of the box bounding the regions, (3) moment of inertia (as an index for measuring the elongation of the regions), (4) standard deviation (as an index for showing the homogeneity of the regions), and the setting is the same as which is presented in [59].

The summary of the experimental setup is listed in Table 9.6.

Table 9.6 Summary of the experimental setup for Sect. 9.4

Experiment#	Kernel type	Features	Base kernel construction	# kernels
1	Gaussian	Spectral features	Single scale of kernel in each Gaussian kernel with the same input features	40
2	Gaussian and linear	Multiple structures of MPs and spectral features	Multiple scales of kernel in each Gaussian kernel or linear kernel with the same input features	4 for NMKL-Linear 15 for NMKL-Gaussian 20 for others
3	Gaussian	Heterogonous features	A single scale of kernel in each Gaussian kernel with the same input features	40 (40 × **D** for HF-MKL)

9.4.3 Spectral Classification

The numerical classification results of different MKL methods for different data sets are given in Table 9.7. The performance of MKL methods is mainly determined by the ways of constructing base kernel and the solutions of weights for base kernels. The resulting base kernel matrices from the different ways of constructing base kernel contain all the information that will be used for the subsequent classification task. The weights of base kernels learned by different MKL methods represent how to combine this information with the objective of strengthening information extraction and curbing useless information for classification.

Observing the results on the three data sets, some conclusions can be drawn as follows. (1) There is a situation that the classification performance of some MKL methods is not as good in terms of classification accuracies as for that of the single kernel method. This reveals that MKL methods need good learning algorithms to ensure the performance. (2) In the three benchmark HSI data sets, the best classification performance in terms of accuracies is derived from the MKL methods. This proves that using multiple kernels instead of a single one can improve performances for HSI classification and the key is to choose the suitable learning algorithm. (3) In most cases, the subspace MKL methods are superior to the comparative MKL methods and single kernel method in terms of OA.

Table 9.7 OA (%) of MKL methods under multiple scale base kernel construction

Data sets	Indian Pines			Pavia University			Salinas		
Classifiers	Percentage of training samples			Percentage of training samples			Percentage of training samples		
	1%	3%	5%	1%	3%	5%	1%	3%	5%
Spe-SK	63.44	76.32	82.03	88.39	91.43	92.27	89.34	91.00	91.96
SimpleMKL	57.39	67.12	73.26	81.89	87.11	89.41	86.89	90.33	91.91
Mean MKL	61.50	72.99	78.70	86.82	91.45	92.90	88.71	91.40	92.86
RMKL	63.30	75.29	81.47	88.44	91.97	93.15	89.47	91.75	93.12
NMF-MKL	64.38	76.36	82.31	88.77	92.035	93.12	89.63	91.88	93.31
KNMF-MKL	66.19	77.56	82.70	88.47	91.63	92.60	90.23	92.81	93.77
DMKL	65.56	77.67	83.17	88.78	92.01	93.08	90.39	91.51	92.86

9.4.4 Spatial–Spectral Classification

The classification results of all these compared methods on three data sets are shown in Table 9.8. And the overall time of training and test process of Pavia University data set with 1% training samples is shown in Fig. 9.15. Several conclusions can be derived. First, as the number of training samples increases, accuracy increases. Second, the MK-Boosting method has the best classification accuracy with the cost of computation time. It is also important to note that there is not a large difference between the methods in terms of classification accuracy. It can be explained that MPs can mine well, information for classification by the way of MKL and, then, the difference among MKL algorithms mainly concentrate on complexity and sparsity of the solution. The conclusion is consistent with [45]. SimpleMKL shows the worst classification performance in terms of accuracies under multiple-scale constructions in the first experiment, but is comparable to the other methods in terms of classi- fication accuracy in this experiment. The example of SimpleMKL illustrates that a MKL method is difficult to guarantee the best classification performance in terms of accuracies in all cases. Feature extraction and classification are both important steps for classification. If the information extraction via features is successful for classification, the classifier design can be easy in terms of complexity and sparsity, and vice versa. The subspace MKL algorithms as two-stage methods have a lower complexity than one-step methods such as SimpleMKL, CS-SMKL.

It can be noted that the NMKL with the linear kernels demonstrates a little lower accuracy than subspace MKL algorithms with the Gaussian kernel. NMKL with the Gaussian kernels obtains comparable classification accuracy compared with NMKL with linear kernels in the Pavia University data set and the Salinas data set, but with a lower accuracy in the Indian data set. In general, using a linear combination of Gaussian kernels is more promising than a nonlinear combination of linear kernels. However, the nonlinear combinations of Gaussian kernels need to be researched further. Feature combination and the scale of the Gaussian kernels have a big influence on the accuracy of NMKL with a Gaussian kernel. And the NMKL method also demonstrates a different performance trend for different data sets. In this experiment, some tries were attempted and the results show relatively better results compared to other approaches in some situations. More work of theoretical analysis needs to be done in this area.

It can be found that among all the sparse methods, CS-SMKL demonstrated com- parable classification accuracies for the Indian Pines and Salinas data sets. And for Pavia data set, as the number of training samples grows, the classification perfor- mance of CS-SMKL increased significantly and reached a comparable accuracy, too. In order to visualize the contribution of each feature type and these corresponding base kernels in these MKL methods, we plot the kernel weights of the base kernels for RMKL, DMKL, SimpleMKL, Sparse MKL, and CS-SMKL in Fig. 9.16. For simplicity, here only three one against one classifiers of Pavia University data set (*Painted metal sheets vs. Bare soil, Painted metal sheets vs. Bitumen, Painted metal sheets vs. Self-blocking bricks*) are listed. RMKL, DMKL, SimpleMKL and Sparse

Table 9.8 OA (%) of MKL methods under MPs base kernel construction

Data sets		Indian Pines			Pavia University			Salinas		
Classifiers		Percentage of training samples			Percentage of training samples			Percentage of training samples		
		1%	3%	5%	1%	3%	5%	1%	3%	5%
Comparison methods	Spe-SK	64.32	77.19	81.49	88.43	91.65	92.90	89.37	92.28	93.17
	MPs + Spe-SK	75.09	85.63	90.05	92.95	96.16	97.21	92.00	94.29	94.71
	MFL	51.98	72.70	80.86	86.33	94.89	95.87	85.00	90.34	92.02
	Mean MKL	76.37	89.07	92.74	96.25	98.31	98.99	94.39	96.61	97.48
	CKL	74.43	88.02	91.92	95.68	98.08	98.83	94.05	96.45	97.32
	SimpleMKL	75.89	89.30	93.41	95.88	98.17	98.84	93.79	96.32	97.22
	CRC	57.67	65.14	68.13	70.95	72.81	73.08	85.14	87.15	87.81
	SRC	64.90	72.46	74.24	76.66	80.47	80.78	83.30	86.32	88.37
Subspace and nonlinear MKL	RMKL	77.27	89.50	93.17	96.65	98.50	**99.11**	94.67	96.73	**97.59**
	NMF-MKL	78.59	90.13	**93.53**	96.73	98.53	**99.15**	94.85	96.80	**97.54**
	KNMF-MKL	78.33	89.41	93.17	96.05	97.98	98.67	95.21	96.84	**97.53**
	DMKL	77.56	89.82	**93.51**	96.84	98.55	**99.12**	94.49	96.69	**97.55**
	NMKL-Linear	76.17	87.34	90.98	95.23	97.39	98.19	92.02	94.03	94.88
	NMKL- Gaussian	69.32	81.98	85.90	95.22	97.41	98.21	91.58	94.35	94.39
Sparse MKL	CS-SMKL	77.61	86.76	91.75	79.33	89.97	98.88	93.37	96.01	96.65
	Sparse MKL	76.68	88.11	92.23	95.81	98.10	98.63	91.99	94.72	95.75
Ensemble	MK-Boosting	79.80	90.48	**94.65**	96.87	98.63	**99.15**	95.37	97.37	**97.98**

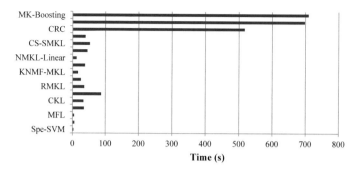

Fig. 9.15 The overall time of training and testing process in all the methods

MKL used the same kernel weights as shown in Fig. 9.16a–d for all the class-pairs. From Fig. 9.16e, it is easy to find that CS-SMKL selected different sparse base kernel sets for different class-pairs, and the spectral features are important for these three class-pair. For the CS-SMKL, it only selected very few base kernels for classification purposes, while the kernel weight for the spectral features is very high. However, these corresponding kernel weights in RMKL, DMKL are much lower, and Sparse MKL did not select any kernel related to the spectral features; SimpleMKL selects the first three kernels related to the spectral features, but obviously, the corresponding kernel weights are lower than that related to the EMP feature obtained by the square SE. This is an example showing that CS-SMKL provides more flexibility in selecting kernels (features) for improving classification.

9.4.5 Classification with Heterogeneous Features

This subsection shows the performance of the fusion framework of heterogeneous features with MKL (denoted as HF-MKL) under realistic ill-posed situations, and the results compared with other MKL methods. In fusion framework of HF-MKL, RMKL was adopted to determine the weights of the base kernels on both levels of MKL in column and row. Joint classification with the spectral features, elevation features, and spatial features was carried out, and the results of classification for two data sets are shown in Table 9.9. SK represents a natural and simple strategy to fuse heterogeneous features, and it can be used as a standard to evaluate the effectiveness of different fusion strategies for heterogeneous features. With this standard, CKL is poor. The performance of CKL is affected by the weights of spectral, spatial, and elevation kernels. All the MKL methods outperform the stacked-vector approach strategy. This reveals that features from different sources obviously have different meanings and statistical significance. Therefore, they may play different roles in classification. Consequently, the stacked-vector approach is not a good choice for the joint classification. However, MKL is an effective fusion strategy for heterogeneous features, and the further HF-MKL framework is a good choice.

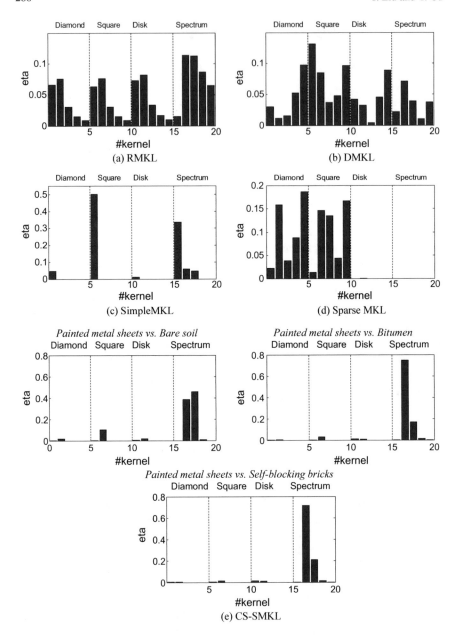

Fig. 9.16 Weights η determined for each base kernel and the corresponding feature type. **a–d** A fixed set of kernel weights selected by RMKL. **e** The kernel weights selected for three different class-pairs by CS-SMKL

Table 9.9   OA (%) of different MKL methods on two data sets

Data sets	Bayview Park			Recology		
Classifiers	Number of training samples			Number of training samples		
	10	20	30	10	20	30
SK	92.16	96.02	95.77	84.76	91.40	92.84
SimpleMKL	92.62	96.37	96.16	85.30	91.52	92.95
Mean MKL	93.53	96.73	96.56	85.49	91.47	93.21
CKL	91.89	94.77	95.49	82.23	89.88	92.20
RMKL	93.59	96.78	**96.71**	85.96	91.95	93.68
NMF-MKL	93.48	96.76	**96.68**	85.99	92.06	93.82
KNMF-MKL	93.02	96.51	96.26	85.87	92.21	**94.16**
DMKL	93.21	96.59	96.42	85.82	92.13	**93.93**
HF-MKL	94.50	96.93	*97.07*	89.06	93.81	*95.49*

9.4.6   Superpixel-Based Classification

The OA with the standard deviation for two data sets were shown in Table 9.10. The best results were given in bold. It is clear that the classification accuracy of EMP-SP-SVM is higher than EMP-SVM and the proposed framework can achieve the highest classification accuracy for both data sets, which demonstrate the effectiveness of the MMSP model. For Pavia University data set, the best results were obtained from MPSP-DMKL method, the maximum increment is 5.22% when the number of training samples was 50 per class. As the number of training samples increased, the

Table 9.10   OA (%) of two data sets

Data sets	Pavia University			Salinas		
Classifier	Number of training samples			Number of training samples		
	50	100	150	50	100	150
Spe-SVM	76.38	79.90	81.42	89.43	91.04	91.82
SCMK [49]	93.21	96.60	97.57	94.58	96.22	97.09
EMP-SVM	90.95	93.65	94.77	91.07	92.84	93.77
MPSP-SVM	91.64	95.46	96.70	94.01	95.95	97.00
EMAP-SVM	91.72	95.32	96.66	94.49	96.59	97.46
MPSP-RMKL	96.81	98.77	99.12	94.52	96.24	97.12
MPSP-DMKL	95.91	98.22	98.65	92.44	94.19	95.08
MPSP-CKL	95.41	98.02	98.49	93.58	95.54	96.39
MASP-RMKL	97.94	98.91	99.18	96.64	98.11	98.72
MASP-DMKL	**98.43**	**99.24**	**99.42**	**97.08**	**98.30**	**98.74**
MASP-CKL	98.26	99.04	99.26	96.46	97.99	98.67

Table 9.11 OAS (%) of two data sets

Data sets		Indian Pines			Pavia University		
Classifier		Number of training samples			Number of training samples		
		1%	2%	3%	1%	2%	3%
Spe-SVM		65.36	73.18	78.00	89.19	91.18	92.30
SCMK [49]		71.92	80.80	86.87	95.16	96.97	98.08
EMAP-SVM		77.69	85.69	89.08	97.90	98.74	99.07
MPCA-SVM		68.34	74.17	78.40	87.12	89.81	91.24
3D-Gabor-SVM		58.08	67.26	74.77	88.17	92.80	94.72
3D-Gabor-DMKL		66.67	76.26	82.57	92.89	95.96	97.28
MAST-DMKL	Missing	80.21	87.20	91.02	97.91	98.79	99.15
	0 vector	80.79	87.17	91.12	98.08	98.85	99.18
	Mean vector	80.48	87.66	**91.59**	98.07	98.80	99.15
	Original pixels	**81.03**	**87.97**	91.46	**98.41**	**98.95**	**99.25**

increment decreased to 1.95%. The relative low proposed method was MPSP-CKL whose increment was between 0.92 and 2.21%. Note that in [74], for Pavia University data set, the OA of SCMK was 99.22% with 200 training samples per class. While in our proposed methods, the OA of MASP-DMKL can achieve 99.24% with only 100 training samples per class. For Salinas data set, not all the proposed methods achieved satisfactory classification results. Only these EMAP-based frameworks can outperform the other approaches. The reason might be that the geometry structure in agriculture scene is simple and mostly polygon, disk SE cannot detect the size and shape of the object exactly, but introduce wrong edges because of erosion and dilation operations, leading to imprecise spatial information. The method which showed the best classification performance is MASP-DMKL with an increment between 1.65 and 2.50%.

The OAs for all the data sets are presented in Table 9.11. It is clear that on both data sets, the proposed MAST-DMKL framework on four different tensor construction means outperforms the other methods, exhibiting the availability of the MAST model. In addition, the MAST-DMKL method where MASTs are filled up with original pixels can accomplish the highest classification accuracy for the other three data sets. For Indian Pines data set, when the number of training sample is 1% per class, MAST-DMKL where MAST is filled up with original pixels achieves the best classification effect with an increment of 3.34% (compared with EMAP-SVM). When the number is 2%, MAST-DMKL in which MAST is filled up with original pixels achieves the second-highest OA with an increment of 2.28% (compared with EMAP-SVM). When the number is 3%, the highest OA is obtained by MAST-DMKL of mean vectors with an increment of 2.51%. For Pavia University data set, the four kinds of MAST frameworks achieve similar classification results. With the increasing number of training samples, the increment of the OA becomes smaller, i.e., from 0.51 to 0.18% (compared with EMAP-SVM).

9.5   Conclusion

In general, the MKL methods can improve the classification performance in most cases compared with single kernel method. For classification of spectral information of HSI, Subspace MKL methods using a trained, weighted combination on the average outperform the untrained, unweighted sum, namely, RBMKL (Mean MKL), and have significant superiority of accuracy and computational efficiency compared with the SimpleMKL method. Ensemble MKL method (MK-Boosting) has higher classification performance in terms of classification accuracy but an additional cost of computation time. It is also important to note that there is not a large difference in classification accuracy among different MKL methods. If we can extract effective spatial–spectral features for HSI classification, the choice of MKL algorithms mainly concentrates on complexity and sparsity of the solution. In general, using the linear combination of kernels with Gaussian kernels is effective compared to a nonlinear combination of linear kernels. However, more research needs to be carried out to fully develop the nonlinear combinations of Gaussian kernels. This is still an open problem, which is affected by many factors such as the manner in which features are combined, as well as the scale of Gaussian kernels.

Currently, with the improvement of the quality of HSI, we can extract more and more accurate features for classification task. These features could be multiscale, multi-attribute, multi-dimension and multi-components. Since MKL provides a very effective means of learning, it is natural considering to utilize these features by MKL framework. Expanding the feature spaces with a number of information diversities, these multiple features provide excellent ability to improve the classification performance. However, there exists a high redundancy of information among these multiple features, and each kind of them has different contribution to classification task. As a solution, sparse MKL methods are developed. The sparse MKL framework allows to embed a variety of characteristics in the classifier, it removes the redundancy of multiple features effectively to learn a compact set of features and selects the weights of corresponding base kernels, leading to a remarkable discriminability. The experimental results on three different hyperspectral data sets, corresponding to different contexts (urban, agricultural) and different spectral and spatial resolutions, demonstrate that the sparse methods offer good performance.

Heterogeneous features from different sources have different meanings, dimension units, and statistical significance. Therefore, they may play different roles in classification and should be treated differently. MKL performs heterogeneous features fusion in implicit high-dimensional feature representation. Utilizing different heterogeneous features to construct different base kernels can distinguish those different roles and fuse the complementary information contained in heterogeneous features. Consequently, MKL is a more reasonable choice than stacked-vector approach, and our experimental results also demonstrated this point. Furthermore, the two-stage MKL framework is a good choice in terms of OA.

References

1. Demir B, Erturk S (2010) Empirical mode decomposition of hyperspectral images for support vector machine classification. IEEE Trans Geosci Remote Sens 48(11):4071–4084
2. Hughes GF (1968) On the mean accuracy of statistical pattern recognizers. IEEE Trans Inf Theory 14(1):55–63
3. Kuo B, Li C, Yang J (2009) Kernel nonparametric weighted feature extraction for hyperspectral image classification. IEEE Trans Geosci Remote Sens 47(4):1139–1155
4. Melgani F, Bruzzone L (2004) Classification of hyperspectral remote sensing images with support vector machines. IEEE Trans Geosci Remote Sens 42(8):1778–1790
5. Camps-Valls G, Bruzzone L (2005) Kernel-based methods for hyperspectral image classification. IEEE Trans Geosci Remote Sens 43(6):1351–1362
6. Kuo B, Ho H, Li C et al (2014) A kernel-based feature selection method for SVM With RBF kernel for hyperspectral image classification. IEEE J Sel Topics Appl Earth Observ Remote Sens 7(1):317–326
7. Gehler PV, Schölkopf B (2009) An introduction to kernel learning algorithms. In: Camps-Valls G, Bruzzone L (eds) Kernel methods for remote sensing data analysis. Wiley, Chichester, UK, pp 25–48
8. Ramzi P, Samadzadegan F, Reinartz P (2014) Classification of hyperspectral data using an AdaBoostSVM technique applied on band clusters. IEEE J Sel Topics Appl Earth Observ Remote Sens 7(6):2066–2079
9. Xia J, Chanussot J, Du P, He X (2016) Rotation-based support vector machine ensemble in classification of hyperspectral data with limited training samples. IEEE Trans Geosci Remote Sens 54(3):1519–1531
10. Xue Z, Du P, Su H (2014) Harmonic analysis for hyperspectral image classification integrated with PSO optimized SVM. IEEE J Sel Topics Appl Earth Observ Remote Sens 7(6):2131–2146
11. Gao L, Li J, Khodadadzadeh M et al (2015) Subspace-based support vector machines for hyperspectral image classification. IEEE Geosci Remote Sens Lett 12(2):349–353
12. Peng J, Zhou Y, Chen CLP (2015) Region-kernel-based support vector machines for hyperspectral image classification. IEEE Trans Geosci Remote Sens 53(9):4810–4824
13. Guo X, Huang X, Zhang L et al (2016) Support tensor machines for classification of hyperspectral remote sensing imagery. IEEE Trans Geosci Remote Sens 54(6):3248–3264
14. Stork CL, Keenan MR (2010) Advantages of clustering in the phase classification of hyperspectral materials images. Microsc Microanal 16(6):810–820
15. Shao Z, Zhang L, Zhou X et al (2014) A novel hierarchical semisupervised SVM for classification of hyperspectral images. IEEE Geosci Remote Sens Lett 11(9):1609–1613
16. Yang L, Yang S, Jin P et al (2014) Semi-supervised hyperspectral image classification using spatio-spectral Laplacian support vector machine. IEEE Geosci Remote Sens Lett 11(3):651–655
17. Bazi Y, Melgani F (2008) Classification of hyperspectral remote sensing images using gaussian processes. IEEE Int Geosci Remote Sens Symp
18. Bazi Y, Melgani F (2010) Gaussian process approach to remote sensing image classification. IEEE Trans Geosci Remote Sens 48(1):186–197
19. Liao W, Tang J, Rosenhahn B, Yang MY (2015) Integration of Gaussian process and MRF for hyperspectral image classification. IEEE Urban Remote Sens Event
20. Chen Y, Nasrabadi NM, Tran TD (2011) Hyperspectral image classification using dictionary-based sparse representation. IEEE Trans Geosci Remote Sens 49(102):3973–3985

21. Liu J, Wu Z, Wei Z et al (2013) Spatial-spectral kernel sparse representation for hyperspectral image classification. IEEE J Sel Topics Appl Earth Observ Remote Sens 6(6):2462–2471
22. Chen Y, Nasrabadi NM, Tran TD (2013) Hyperspectral image classification via kernel sparse representation. IEEE Trans Geosci Remote Sens 51(11):217–231
23. Srinivas U, Chen Y, Monga V et al (2013) Exploiting sparsity in hyperspectral image classification via graphical models. IEEE Geosci Remote Sens Lett 10(3):505–509
24. Zhang H, Li J, Huang Y, Zhang L (2014) A nonlocal weighted joint sparse representation classification method for hyperspectral imagery. IEEE J Sel Topics Appl Earth Observ Remote Sens 7(6):2056–2065
25. Fang L, Li S, Kang X et al (2014) Spectral-spatial hyperspectral image classification via multiscale adaptive sparse representation. IEEE Trans Geosci Remote Sens 52(12):7738–7749
26. Li J, Zhang H, Zhang L (2015) Efficient superpixel-level multitask joint sparse representation for hyperspectral image classification. IEEE Trans Geosci Remote Sens 53(10):5338–5351
27. Ul Haq QS, Tao L, Sun F et al (2012) A fast and robust sparse approach for hyperspectral data classification using a few labeled samples. IEEE Trans Geosci Remote Sens 50(6):2287–2302
28. Yang S, Jin H, Wang M et al (2014) Data-driven compressive sampling and learning sparse coding for hyperspectral image classification. IEEE Geosci Remote Sens Lett 11(2):479–483
29. Qian Y, Ye M, Zhou J (2013) Hyperspectral image classification based on structured sparse logistic regression and three-dimensional wavelet texture features. IEEE Trans Geosci Remote Sens 51(42):2276–2291
30. Tang YY, Yuan H, Li L (2014) Manifold-based sparse representation for hyperspectral image classification. IEEE Trans Geosci Remote Sens 52(12):7606–7618
31. Yuan H, Tang YY, Lu Y et al (2014) Hyperspectral image classification based on regularized sparse representation. IEEE J Sel Topics Appl Earth Observ Remote Sens 7(6):2174–2182
32. Li J, Zhang H, Zhang L (2014) Column-generation kernel nonlocal joint collaborative representation for hyperspectral image classification. ISPRS J Int Soc Photo Remote Sens 94:25–36
33. Li J, Zhang H, Huang Y, Zhang L (2014) Hyperspectral image classification by non-local joint collaborative representation with a locally adaptive dictionary. IEEE Trans Geosci Remote Sens 52(6):3707–3719
34. Li J, Zhang H, Zhang L et al (2014) Joint collaborative representation with multitask learning for hyperspectral image classification. IEEE Trans Geosci Remote Sens 52(9):5923–5936
35. Li W, Du Q (2014) Joint within-class collaborative representation for hyperspectral image classification. IEEE J Sel Topics Appl Earth Observ Remote Sens 7(6):2200–2208
36. Li W, Du Q, Xiong M (2015) Kernel collaborative representation with Tikhonov regularization for hyperspectral image classification. IEEE Geosci Remote Sens Lett 12(1):48–52
37. Liu J, Wu Z, Li J, Plaza A, Yuan Y (2016) Probabilistic-kernel collaborative representation for spatial-spectral hyperspectral image classification. IEEE Trans Geosci Remote Sens 54(4):2371–2384
38. He Z, Wang Q, Shen Y, Sun M (2014) Kernel sparse multitask learning for hyper-spectral image classification with empirical mode decomposition and morphological wave-let-based features. IEEE Trans Geosci Remote Sens 52(8):5150–5163
39. Xiong M, Ran Q, Li W et al (2015) Hyperspectral image classification using weighted joint collaborative representation. IEEE Geosci Remote Sens Lett 12(6):1209–1213
40. Wright J, Ma Y, Mairal J et al (2010) Sparse representation for computer vision and pattern recognition. Proc IEEE 98(6):1031–1044
41. Wright J, Yang AY, Ganesh A et al (2009) Robust face recognition via sparse representation. IEEE Trans Pattern Anal Mach Intell 31(2):210–227
42. Zhang L, Yang M, Feng X, Ma Y, Zhang D (2012) Collaborative representation based classification for face recognition. Comput Sci
43. Baron D, Duarte MF, Wakin MB, Sarvotham S, Baraniuk RG (2009) Distributed compressive sensing. arXiv:0901.3403 [cs.IT]
44. Gu Y, Wang Q, Xie B (2017) Multiple kernel sparse representation for airborne Li-DAR data classification. IEEE Trans Geosci Remote Sens 5(2):1085–1105

45. Gonen M, Alpaydin E (2011) Multiple kernel learning algorithms. J Mach Learn Res 12:2211–2268
46. Camps-Valls G, Gomez-Chova L, Munoz-Mari J et al (2006) Composite Kernels for hyperspectral image classification. IEEE Geosci Remote Sens Lett 3(1):93–97
47. Wang J, Jiao L, Wang S et al (2016) Adaptive nonlocal spatial–spectral kernel for hyperspectral imagery classification. IEEE J Sel Topics Appl Earth Observ Remote Sens 9(9):4086–4101
48. Camps-Valls G, Gomez-Chova L, Munoz-Mari J et al (2008) Kernel-based framework for multitemporal and multisource remote sensing data classification and change detection. IEEE Trans Geosci Remote Sens 46(6):1822–1835
49. Fang L, Li S, Duan W et al (2015) Classification of hyperspectral images by exploiting spectral-spatial information of superpixel via multiple kernels. IEEE Trans Geosci Remote Sens 53(12):6663–6674
50. Valero S, Salembier P, Chanussot J (2013) Hyperspectral image representation and processing with binary partition trees. IEEE Trans Image Process 22(4):1430–1443
51. Zhou Y, Peng J, Chen CLP (2015) Extreme learning machine with composite kernels for hyperspectral image classification. IEEE J Sel Topics Appl Earth Observ Remote Sens 8(6):2351–2360
52. Li H, Ye Z, Xiao G (2015) Hyperspectral image classification using spectral-spatial composite kernels discriminant analysis. IEEE J Sel Topics Appl Earth Observ Remote Sens 8(6):2341–2350
53. Zhang Y, Prasad S (2015) Locality preserving composite kernel feature extraction for multi-source geospatial image analysis. IEEE J Sel Topics Appl Earth Observ Remote Sens 8(3):1385–1392
54. Gu Y, Wang C, You D et al (2012) Representative multiple kernel learning for classification in hyperspectral imagery. IEEE Trans Geosci Remote Sens 50(72):2852–2865
55. Gu Y, Wang Q, Wang H et al (2015) Multiple kernel learning via low-rank nonnegative matrix factorization for classification of hyperspectral imagery. IEEE J Sel Topics Appl Earth Observ Remote Sens 8(6):2739–2751
56. Gu Y, Wang Q, Jia X et al (2015) A novel MKL model of integrating LiDAR data and MSI for urban area classification. IEEE Trans Geosci Remote Sens 53(10):5312–5326
57. Wang Q, Gu Y, Tuia D (2016) Discriminative multiple kernel learning for hyperspectral image classification. IEEE Trans Geosci Remote Sens 54(7):3912–3927
58. Rakotomamonjy A, Bach FR, Canu SE et al (2008) SimpleMKL. J Mach Learn Res 9(11):2491–2521
59. Liu T, Gu Y, Jia X et al (2016) Class-specific sparse multiple kernel learning for spectral-spatial hyperspectral image classification. IEEE Trans Geosci Remote Sens 54(12):7351–7365
60. Wang L, Hao S, Wang Q, Atkinson PM (2015) A multiple-mapping kernel for hyper-spectral image classification. IEEE Geosci Remote Sens Letters 12(5):978–982
61. Gu Y, Liu T, Jia X, Benediktsson JA, Chanussot J (2016) Nonlinear multiple kernel learning with multiple-structure-element extended morphological profiles for hyperspectral image classification. IEEE Trans Geosci Remote Sens 54(6):3235–3247
62. Do TTH (2012) A unified framework for support vector machines, multiple kernel learning and metric learning. In: A unified framework for support vector machines, multiple kernel learning and metric learning, vol. Docteur, Series A unified framework for support vector machines, multiple kernel learning and metric learning. Universite De Geneve
63. Cristianini N, Shawe-Taylor J (1999) An introduction to support vector machines and other kernel-based learning methods. Cambridge University Press, New York, NY, USA
64. Gu Y, Gao G, Zuo D, You D (2014) Model selection and classification with multiple kernel learning for hyperspectral images via sparsity. IEEE J Sel Topics Appl Earth Observ Remote Sens 7(6):2119–2130
65. Gu Y, Liu H (2016) Sample-screening MKL method via boosting strategy for hyperspectral image classification. Neurocomputing 173:1630–1639
66. Zhang Y, Yang HL, Prasad S et al (2015) Ensemble multiple kernel active learning for classification of multisource remote sensing data. IEEE J Sel Topics Appl Earth Observ Remote Sens 8(2):845–858

67. Cortes C, Mohri M, Rostamizadeh A (2009) Learning non-linear combinations of kernels. In: Proceedings of advances in neural information processing systems, pp 396–404
68. Xu Z, Jin R, Yang H et al (2010) Simple and efficient multiple kernel learning by group Lasso. In: Proceedings of the 27th international conference on machine learning
69. Schapire RE (1990) The strength of weak learnability. Mach Learn 5(2):197–227
70. Freund Y, Schapire RE (1997) A decision-theoretic generalization of on-line learning and an application to boosting. J Comput System Sci 55(1):119–139
71. Li J, Huang X, Gamba P et al (2015) Multiple feature learning for hyperspectral image classification. IEEE Trans Geosci Remote Sens 53(3):1592–1606
72. Xia H, Hoi SCH (2013) MKBoost: a framework of multiple kernel boosting. IEEE Trans Knowl Data Eng 25(7):1574–1586
73. Gamba P (2004) A collection of data for urban area characterization. IEEE Int Geosci Remote Sens Symp (IGARSS), pp 69–72
74. Liu T, Gu Y, Chanussot J et al (2017) Multimorphological superpixel model for hyperspectral image classification. IEEE Trans Geosci Remote Sens 55(12):6950–6963
75. Jia S, Zhu Z, Shen L, Li Q (2014) A two-stage feature selection framework for hyperspectral image classification using few labeled samples. IEEE J Sel Topics Appl Earth Observ Remote Sens 7(4):1023–1035
76. Cristianini N, Shawe-Taylor J, Elisseeff A, Kandola J (2002) On kernel-target alignment. In: Dietterich TG, Becker S, Ghahramani Z (eds) Advance in neural information processing systems, vol 14, pp 367–373
77. Li W, Du Q (2016) A survey on representation-based classification and detection in hyperspectral remote sensing imagery. Pattern Recogn Lett 83:115–123

Chapter 10
Low Dimensional Manifold Model in Hyperspectral Image Reconstruction

Wei Zhu, Zuoqiang Shi and Stanley Osher

Abstract In this chapter, we present a low dimensional manifold model (LDMM) for hyperspectral image reconstruction. This model is based on the observation that the spatial–spectral blocks of hyperspectral images typically lie close to a collection of low dimensional manifolds. To emphasize this, we directly use the dimension of the manifold as a regularization term in a variational functional, which can be solved efficiently by alternating direction of minimization and advanced numerical discretization. Experiments on the reconstruction of hyperspectral images from sparse and noisy sampling demonstrate the superiority of LDMM in terms of both speed and accuracy.

10.1   Introduction

Hyperspectral imagery is an important domain in the field of remote sensing with numerous applications in agriculture, environmental science, and surveillance [6]. When capturing a hyperspectral image (HSI), the sensors detect the intensity of reflection at a wide range of continuous wavelengths, from the infrared to ultraviolet, to form a 3D data cube with up to thousands of spectral bands. When such data of high dimensionality are collected, the observed images are very likely degraded due to various reasons. For instance, the collected images might be extremely noisy because of limited exposure time, or some of the voxels can be missing due to the

W. Zhu (✉)
Department of Mathematics, Duke University, Durham, NC, USA
e-mail: zhu@math.duke.edu

Z. Shi
Department of Mathematical Sciences & Yau Mathematical Sciences Center,
Tsinghua University, Beijing, China
e-mail: zqshi@tsinghua.edu.cn

S. Osher
Department of Mathematics, University of California, Los Angeles, CA, USA
e-mail: sjo@math.ucla.edu

© Springer Nature Switzerland AG 2020
S. Prasad and J. Chanussot (eds.), *Hyperspectral Image Analysis*,
Advances in Computer Vision and Pattern Recognition,
https://doi.org/10.1007/978-3-030-38617-7_10

malfunctions of the hyperspectral cameras. Thus an important task in HSI analysis is to recover the original image from its noisy incomplete observation. This is an ill-posed inverse problem, and some prior knowledge of the original data must be exploited.

One widely used prior information of HSI is that the 3D data cube has a low-rank structure under the linear mixing model (LMM) [3]. More specifically, the spectral signature of each pixel is assumed to be a linear combination of a few constituent end-members. Under such an assumption, low-rank matrix completion and sparse representation techniques have been used for HSI reconstruction [7, 16, 28]. Despite the simplicity of LMM, the linear mixing assumption has been shown to be physically inaccurate in certain situations [9].

Partial differential equation (PDEs) and graph-based image processing techniques have also been applied to HSI reconstruction. The total variation (TV) method [23] has been widely used as a regularization in hyperspectral image processing [1, 13, 15, 29]. The nonlocal total variation (NLTV) [11], which computes the gradient in a nonlocal graph-based manner, has also been applied to the analysis of hyperspectral images [14, 18, 30]. However, such methods typically fail to produce satisfactory results when there is a significant number of missing voxels in the degraded HSI.

Over the past decade, patch-based manifold models have achieved great success in image processing. The key assumption in the manifold model is that image patches typically concentrate around a low dimensional smooth manifold [5, 17, 21, 22]. Based on such assumption, a low dimensional manifold model (LDMM) has been proposed for general image processing problems [20, 24], in which the dimension of the patch manifold is directly used as a regularization term in a variational functional. LDMM achieved excellent results, especially in image inpainting problems from very sparse sampling.

In this chapter, we will illustrate how LDMM can be used in HSI reconstruction. The direct extension of LDMM to higher dimensional data reconstruction has been considered in [32], but such generalization typically has poor scalability and requires huge memory storage. A considerable amount of computational burden can be reduced, however, if the special structure of hyperspectral images is utilized. Because an HSI is a collection of 2D images of the same spatial location, a single spatial similarity matrix can be shared across all spectral bands [31]. The resulting algorithm is considerably faster than its 3D counterpart: it typically takes less than two minutes given a proper initialization as compared to hours in [32].

10.2   Low Dimensional Manifold Model

We provide a detailed explanation of LDMM in HSI reconstruction, which includes the definition of the patch manifold of an HSI, the variational functional with the manifold dimension as a regularizer, and how to compute the dimension of a manifold sampled from a point cloud.

10.2.1   Patch Manifold

Let $u \in \mathbb{R}^{m \times n \times B}$ be a hyperspectral image, where $m \times n$ and B are the spatial and spectral dimensions of the image. For any $x \in \bar{\Omega} = [m] \times [n]$, where $[m] = \{1, 2, \ldots, m\}$, we define a patch $\mathcal{P}u(x)$ as a 3D block of size $s_1 \times s_2 \times B$ of the original data cube u, and the pixel x is the top-left corner of the rectangle of size $s_1 \times s_2$. The *patch set* $\mathcal{P}(u)$ is defined as the collection of all patches:

$$\mathcal{P}(u) = \{\mathcal{P}u(x) : x \in \bar{\Omega}\} \subset \mathbb{R}^d, \quad d = s_1 \times s_2 \times B. \tag{10.1}$$

It has been shown in [20, 32] that the point cloud $\mathcal{P}(u)$ is typically close to a collection of low dimensional smooth manifolds $\mathcal{M} = \cup_{l=1}^{L} \mathcal{M}_l$ embedded in $\mathbb{R}^d$. This collection of manifolds is called the *patch manifold* of u.

Remark 10.1 We sometimes regard $\mathcal{P} : \mathbb{R}^{m \times n \times B} \to \mathbb{R}^{d \times |\bar{\Omega}|}$ as an operator that maps an HSI $u \in \mathbb{R}^{m \times n \times B}$ to its patch set $\mathcal{P}(u) \in \mathbb{R}^{d \times |\bar{\Omega}|}$. This point of view will be assumed throughout Sect. 10.3.

10.2.2   Model Formulation and Calculating the Manifold Dimension

Our objective is to reconstruct the unknown HSI u from its noisy and incomplete observation $b \in \mathbb{R}^{m \times n \times B}$. Assume that for each spectral band $t \in [B]$, b is only known on a random subset $\Omega^t \subset \bar{\Omega} = [m] \times [n]$, with a sampling rate r (in our experiments $r = 5\%$ or 10%.) We also use the notation

$$\bar{\Omega}_{\text{all}} = [m] \times [n] \times [B] \quad \text{and} \quad \Omega_{\text{all}} = \cup_{t=1}^{B} \Omega^t \tag{10.2}$$

to denote the domain of the entire 3D data cube and its sampled subset. According to the analysis in Sect. 10.2.1, we can use the dimension of the patch manifold as a regularizer to reconstruct u from b:

$$\min_{\substack{u \in \mathbb{R}^{m \times n \times B} \\ \mathcal{M} \subset \mathbb{R}^d}} \int_{\mathcal{M}} \dim(\mathcal{M}(p)) \mathrm{d}p + \lambda \sum_{t=1}^{B} \|u^t - b^t\|_{L^2(\Omega^t)}^2$$

$$\text{subject to:} \quad \mathcal{P}(u) \subset \mathcal{M}, \tag{10.3}$$

where $u^t \in \mathbb{R}^{m \times n}$ is the t-th spectral band of the HSI $u \in \mathbb{R}^{m \times n \times B}$, $\mathcal{M}(p)$ denotes the smooth manifold $\mathcal{M}_l$ to which p belongs, $\mathcal{M} = \cup_{l=1}^{L} \mathcal{M}_l$, and

$$\int_{\mathcal{M}} \dim(\mathcal{M}(p)) \mathrm{d}p = \sum_{l=1}^{L} |\mathcal{M}_l| \dim(\mathcal{M}_l)$$

is the L^1 norm of the local dimension. Note that (10.3) is not mathematically well defined, since we do not know how to compute the manifold dimension given only a point cloud sampling the manifold. Fortunately, the following formula from differential geometry provides a simple way of calculating the dimension of a smooth manifold [20]:

Proposition 10.1 *Let $\mathcal{M}$ be a smooth submanifold isometrically embedded in $\mathbb{R}^d$. For any $p \in \mathcal{M}$, we have*

$$\dim(\mathcal{M}) = \sum_{j=1}^{d} \|\nabla_{\mathcal{M}} \alpha_j(p)\|^2,$$

where $\alpha_i, i = 1, \ldots, d$ are the coordinate functions on $\mathcal{M}$, i.e.,

$$\alpha_i(p) = p_i, \quad \forall p = (p_1, \ldots, p_d) \in \mathcal{M}.$$

Proof Since $\mathcal{M}$ is a smooth submanifold isometrically embedded in $\mathbb{R}^d$, it can be locally parametrized as

$$p = \psi(\gamma) : U \subset \mathbb{R}^k \to \mathcal{M} \subset \mathbb{R}^d, \tag{10.4}$$

where $k = \dim(\mathcal{M})$, $\gamma = (\gamma^1, \ldots, \gamma^k)^T \in \mathbb{R}^k$, and $p = (p^1, \ldots, p^d)^T \in \mathcal{M}$. With the induced metric from $\mathbb{R}^d$, we have $\partial_{i'} = (\partial_{i'}\psi^1, \ldots, \partial_{i'}\psi^d)$, and the metric tensor is

$$g_{i'j'} = <\partial_{i'}, \partial_{j'}> = \sum_{l=1}^{d} \partial_{i'}\psi^l \partial_{j'}\psi^l. \tag{10.5}$$

Let $g^{i'j'}$ denote the inverse of $g_{i'j'}$, i.e.,

$$\sum_{l'=1}^{k} g_{i'l'} g^{l'j'} = \delta_{i'j'} = \begin{cases} 1, & i' = j', \\ 0, & i' \neq j'. \end{cases} \tag{10.6}$$

For any function u on $\mathcal{M}$, its gradient $\nabla_{\mathcal{M}} u$ is defined as

$$\nabla_{\mathcal{M}} u = \sum_{i',j'=1}^{k} g^{i'j'} \partial_{j'} u \, \partial_{i'}. \tag{10.7}$$

When viewed as a vector in the ambient space $\mathbb{R}^d$, the j-th component of $\nabla_{\mathcal{M}} u$ in the ambient coordinates can be written as

$$\nabla_{\mathcal{M}}^j u = \sum_{i',j'=1}^{k} \partial_{i'}\psi^j g^{i'j'} \partial_{j'} u, \quad j = 1, \ldots, d. \tag{10.8}$$

Following the definition of $\nabla_{\mathcal{M}}$ in (10.8), we have

$$\sum_{j=1}^{d} \|\nabla_{\mathcal{M}}\alpha_j\|^2 = \sum_{i,j=1}^{d} \nabla_{\mathcal{M}}^i \alpha_j \nabla_{\mathcal{M}}^i \alpha_j$$

$$= \sum_{i,j=1}^{d} \left(\sum_{i',j'=1}^{k} \partial_{i'}\psi^j g^{i'j'} \partial_{j'}\alpha_j \right) \left(\sum_{i'',j''=1}^{k} \partial_{i''}\psi^j g^{i''j''} \partial_{j''}\alpha_j \right)$$

$$= \sum_{j=1}^{d} \sum_{i',j',i'',j''=1}^{k} \left(\sum_{i=1}^{d} \partial_{i'}\psi^i \partial_{i''}\psi^i \right) g^{i'j'} g^{i''j''} \partial_{j'}\alpha_j \partial_{j''}\alpha_j$$

$$= \sum_{j=1}^{d} \sum_{j',i'',j''=1}^{k} \left(\sum_{i'=1}^{k} g_{i'i''} g^{i'j'} \right) g^{i''j''} \partial_{j'}\alpha_j \partial_{j''}\alpha_j$$

$$= \sum_{j=1}^{d} \sum_{j',i'',j''=1}^{k} \delta_{i''j'} g^{i''j''} \partial_{j'}\alpha_j \partial_{j''}\alpha_j$$

$$= \sum_{j=1}^{d} \sum_{j',j''=1}^{k} g^{j'j''} \partial_{j'}\alpha_j \partial_{j''}\alpha_j.$$

Notice that $\partial_{j'}\alpha_j = \frac{\partial}{\partial \gamma^{j'}}\alpha_j(\psi(\boldsymbol{\gamma})) = \partial_{j'}\psi^j$. We thus have

$$\sum_{j=1}^{d} \|\nabla_{\mathcal{M}}\alpha_j\|^2 = \sum_{j=1}^{d} \sum_{j',j''=1}^{k} g^{j'j''} \partial_{j'}\psi^j \partial_{j''}\psi^j$$

$$= \sum_{j',j''=1}^{k} g^{j'j''} \left(\sum_{j=1}^{d} \partial_{j'}\psi^j \partial_{j''}\psi^j \right)$$

$$= \sum_{j',j''=1}^{k} g^{j'j''} g_{j'j''}$$

$$= \sum_{j'=1}^{k} \delta_{j'j'} = k = \dim(\mathcal{M}) \qquad (10.9)$$

This concludes the proof.

Based on Proposition 10.1, we can rewrite (10.3) as

$$\min_{\substack{u \in \mathbb{R}^{m \times n \times B} \\ \mathcal{M} \subset \mathbb{R}^d}} \sum_{i=1}^{d_s} \sum_{t=1}^{B} \|\nabla_{\mathcal{M}}\alpha_i^t\|_{L^2(\mathcal{M})}^2 + \lambda \sum_{t=1}^{B} \|u^t - b^t\|_{L^2(\Omega^t)}^2$$

$$\text{subject to:} \quad \mathcal{P}(\boldsymbol{u}) \subset \mathcal{M}, \qquad (10.10)$$

where $d_s = s_1 \times s_2$ is the spatial dimension, α_i^t is the coordinate function that maps a point $\boldsymbol{p} = \left(p_i^t\right)_{i\in[d_s],t\in[B]} \in \mathcal{M} \subset \mathbb{R}^d$ into its (i, t)-th coordinate p_i^t. Note that (10.10) is a constrained nonconvex optimization problem with respect to $\mathcal{M}$ and $\boldsymbol{u}$, the solution of which will be explained in detail in the next section.

10.3 Two Numerical Approaches of Solving the LDMM Model

Because (10.10) is nonconvex, we attempt to solve it by alternating the direction of minimization with respect to $\boldsymbol{u}$ and $\mathcal{M}$. More specifically, given $\mathcal{M}^{(k)}$ and $\boldsymbol{u}^{(k)}$ at step k satisfying $\mathcal{P}(\boldsymbol{u}^{(k)}) \subset \mathcal{M}^{(k)}$

- With fixed $\mathcal{M}^{(k)}$, update the data $\boldsymbol{u}^{(k+1)}$ and the perturbed coordinate functions $\boldsymbol{\alpha}^{(k+1)} = \left\{[\alpha_i^t]^{(k+1)}\right\}_{i,t}$ by solving:

$$(\boldsymbol{u}^{(k+1)}, \boldsymbol{\alpha}^{(k+1)}) = \arg\min_{\boldsymbol{u},\boldsymbol{\alpha}} \sum_{i,t} \|\nabla_{\mathcal{M}^{(k)}} \alpha_i^t\|_{L^2(\mathcal{M}^{(k)})}^2 + \lambda \sum_{t=1}^{B} \|\boldsymbol{u}^t - \boldsymbol{b}^t\|_{L^2(\Omega^t)}^2.$$

$$\text{subject to: } \boldsymbol{\alpha}(\mathcal{P}\boldsymbol{u}^{(k)}) = \mathcal{P}(\boldsymbol{u}) \qquad (10.11)$$

- Update the manifold $\mathcal{M}^{(k+1)}$ as the image of $\mathcal{M}^{(k)}$ under the perturbed coordinate function $\boldsymbol{\alpha}^{(k+1)}$

$$\mathcal{M}^{(k+1)} = \boldsymbol{\alpha}^{(k+1)}(\mathcal{M}^{(k)}). \qquad (10.12)$$

Remark 10.2 Note that $\mathcal{P}(\boldsymbol{u}^{(k+1)}) \subset \mathcal{M}^{(k+1)}$ holds because they are the images of $\mathcal{P}(\boldsymbol{u}^{(k)})$ and $\mathcal{M}^{(k)}$ under the same perturbed coordinate functions $\boldsymbol{\alpha}^{(k+1)}$. Moreover, if the above iterative procedure converges, then the adjacent iterates $\mathcal{M}^{(k)} \approx \mathcal{M}^{(k+1)}$ when k is large enough. Therefore, the perturbed coordinate function $\boldsymbol{\alpha}^{(k+1)}$ is indeed very close to identity as defined in Proposition 10.1.

Notice that (10.12) is easy to implement, whereas (10.11) is a constrained optimization problem whose numerical implementation involves the discretization of the manifold gradient operator $\nabla_{\mathcal{M}}$ over an unstructured point cloud $\mathcal{P}(\boldsymbol{u}^{(k)})$. In what follows, we provide two numerical procedures to solve the LDMM model. In the first method, problem (10.11) is further split into two subproblems, whose Euler–Lagrange equation, a Laplace–Beltrami equation over the manifold $\mathcal{M}^{(k)}$, is solved by the point integral method (PIM) [19]. In the second approach, we directly discretize the Dirichlet energy $\|\nabla_{\mathcal{M}} \alpha_i^t\|_{L^2(\mathcal{M})}^2$ in (10.11) using the weighted nonlocal

Laplacian (WNLL) [25], a practical way of finding a smooth interpolation function on a point cloud.

10.3.1 The First Approach

Notice that (10.11) is a convex optimization problem subject to a linear constraint, which can be further split into two simpler subproblems using the split Bregman iteration [12]. More specifically, given the l-th iterates $\boldsymbol{\alpha}^{(k+1),l}$, $\boldsymbol{u}^{(k+1),l}$, and $\boldsymbol{z}^l$:

- Update $\boldsymbol{\alpha}^{(k+1),l+1} = \left\{[\alpha_i^t]^{(k+1),l+1}\right\}_{i,t}$ with fixed $\boldsymbol{u}^{(k+1),l}$ and $\boldsymbol{z}^l$,

$$\boldsymbol{\alpha}^{(k+1),l+1} = \min_{\boldsymbol{\alpha}} \sum_{i,t} \|\nabla_{\mathcal{M}^{(k)}} \alpha_i^t\|_{L^2(\mathcal{M}^{(k)})}^2 + \mu \|\boldsymbol{\alpha}(\mathcal{P}(\boldsymbol{u}^{(k)})) - \mathcal{P}(\boldsymbol{u}^{(k+1),l}) + \boldsymbol{z}^l\|_F^2,$$
(10.13)

  where both the patch set $\mathcal{P}(\boldsymbol{u}^{(k+1),l})$ and the image of the patch set under the perturbed coordinate functions $\boldsymbol{\alpha}(\mathcal{P}(\boldsymbol{u}^{(k)}))$ are treated as matrices in $\mathbb{R}^{d \times |\bar{\Omega}|}$.
- Update $\boldsymbol{u}^{(k+1),l+1}$ with fixed $\boldsymbol{\alpha}^{(k+1),l+1}$ and $\boldsymbol{z}^l$,

$$\boldsymbol{u}^{(k+1),l+1} = \min_{\boldsymbol{u}} \lambda \sum_{t=1}^{B} \|\boldsymbol{u}^t - \boldsymbol{b}^t\|_{L^2(\Omega^t)}^2 + \mu \|\boldsymbol{\alpha}^{(k+1),l+1}(\mathcal{P}(\boldsymbol{u}^{(k)})) - \mathcal{P}(\boldsymbol{u}) + \boldsymbol{z}^l\|_F^2$$
$$= \min_{\boldsymbol{u}} \lambda \|I_{\Omega_{\mathrm{all}}} \boldsymbol{u} - \boldsymbol{b}\|_{L^2(\bar{\Omega}_{\mathrm{all}})}^2 + \mu \|\boldsymbol{\alpha}^{(k+1),l+1}(\mathcal{P}(\boldsymbol{u}^{(k)})) - \mathcal{P}(\boldsymbol{u}) + \boldsymbol{z}^l\|_F^2,$$
(10.14)

  where $\Omega_{\mathrm{all}} \subset \bar{\Omega}_{\mathrm{all}}$ are defined in (10.2), and $I_{\Omega_{\mathrm{all}}} : \mathbb{R}^{m \times n \times B} \to \mathbb{R}^{m \times n \times B}$ is the projection operator that sets $\boldsymbol{u}(\boldsymbol{x}, t)$ to zero for $(\boldsymbol{x}, t) \notin \Omega_{\mathrm{all}}$, i.e.,

$$I_{\Omega_{\mathrm{all}}} \boldsymbol{u}(\boldsymbol{x}, t) = \begin{cases} \boldsymbol{u}(\boldsymbol{x}, t) & , (\boldsymbol{x}, t) \in \Omega_{\mathrm{all}}, \\ 0 & , (\boldsymbol{x}, t) \notin \Omega_{\mathrm{all}}, \end{cases}$$
(10.15)

- Update $\boldsymbol{z}^{l+1}$,

$$\boldsymbol{z}^{l+1} = \boldsymbol{z}^l + \boldsymbol{\alpha}^{(k+1),l+1}(\mathcal{P}(\boldsymbol{u}^{(k)})) - \mathcal{P}(\boldsymbol{u}^{(k+1),l+1}).$$
(10.16)

Note that among (10.13), (10.14), and (10.16), the dual variable update (10.16) is trivial to implement, and the $\boldsymbol{u}$ update (10.14) has the following closed form solution:

$$\boldsymbol{u}^{(k+1),l+1} = \left(\lambda I_{\Omega_{\mathrm{all}}}^* I_{\Omega_{\mathrm{all}}} + \mu \mathcal{P}^* \mathcal{P}\right)^{-1} \left[\lambda I_{\Omega_{\mathrm{all}}}^* \boldsymbol{b} + \mu \mathcal{P}^* \left(\boldsymbol{z}^l + \boldsymbol{\alpha}^{(k+1),l+1}(\mathcal{P}(\boldsymbol{u}^{(k)}))\right)\right],$$
(10.17)

where $I^*_{\Omega_{\text{all}}} : \mathbb{R}^{m \times n \times B} \to \mathbb{R}^{m \times n \times B}$ and $\mathcal{P}^* : \mathbb{R}^{d \times |\bar{\Omega}|} \to \mathbb{R}^{m \times n \times B}$ are the adjoint operators of $I_{\Omega_{\text{all}}}$ and $\mathcal{P}$. It is worth mentioning that $\left(\lambda I^*_{\Omega_{\text{all}}} I_{\Omega_{\text{all}}} + \mu \mathcal{P}^* \mathcal{P} \right)$ is a diagonal operator, and hence (10.17) can be solved efficiently. As for the $\boldsymbol{\alpha}$ update (10.13), one can easily check that the coordinate functions $\left\{ [\alpha^t_i]^{(k+1),l+1} \right\}_{i,t}$ are decoupled, and thus (10.13) can be solved separately,

$$[\alpha^t_i]^{(k+1),l+1} = \min_{\alpha^t_i} \|\nabla_{\mathcal{M}^{(k)}} \alpha^t_i\|^2_{L^2(\mathcal{M}^{(k)})} + \mu \|\alpha^t_i (\mathcal{P}(\boldsymbol{u}^{(k)})) - \mathcal{P}^t_i(\boldsymbol{u}^{(k+1),l}) + (z^l)^t_i\|^2,$$

(10.18)

where

$$\mathcal{P}^t_i(\boldsymbol{u}) = \left(\mathcal{P}^t_i \boldsymbol{u}(\boldsymbol{x}) \right)_{x \in \bar{\Omega}} \in \mathbb{R}^{|\bar{\Omega}|},$$

(10.19)

and $\mathcal{P}^t_i \boldsymbol{u}(\boldsymbol{x})$ is the (i, t)-th element in the patch $\mathcal{P}\boldsymbol{u}(\boldsymbol{x})$. We next explain how to solve problem (10.18) using the point integral method.

10.3.1.1   Discretization with the Point Integral Method

Note that problem (10.18) for each individual $[\alpha^t_i]^{(k+1),l+1}$ can be cast into the following canonical form,

$$\min_{u \in H^1(\mathcal{M})} \|\nabla_{\mathcal{M}} u\|^2_{L^2(\mathcal{M})} + \mu \sum_{\boldsymbol{p} \in P} |u(\boldsymbol{p}) - v(\boldsymbol{p})|^2,$$

(10.20)

where u can be any α^t_i, $\mathcal{M} = \mathcal{M}^{(k)}$, $\mathcal{M} \supset P = \mathcal{P}(\boldsymbol{u}^{(k)})$, and $v(\boldsymbol{p})$ is a given function on P. It can easily be checked by standard variational methods that (10.20) is equivalent to the following Euler–Lagrange equation:

$$\begin{cases} -\Delta_{\mathcal{M}} u(\boldsymbol{p}) + \mu \sum_{\boldsymbol{q} \in P} \delta(\boldsymbol{p} - \boldsymbol{q})(u(\boldsymbol{q}) - v(\boldsymbol{q})) = 0, & \boldsymbol{p} \in \mathcal{M} \\[2mm] \dfrac{\partial u}{\partial \mathbf{n}} = 0, & \boldsymbol{p} \in \partial \mathcal{M}. \end{cases}$$

(10.21)

This is a Laplace–Beltrami equation over the manifold $\mathcal{M}$ sampled by the point cloud P, which can be solved efficiently using the point integral method (PIM) [19]. The key observation in PIM is the following integral approximation.

Theorem 10.1 *If $u \in C^3(\mathcal{M})$ is a function on $\mathcal{M}$, then*

$$\left\| \int_{\mathcal{M}} \Delta_{\mathcal{M}} u(\boldsymbol{q}) R_t(\boldsymbol{p}, \boldsymbol{q}) d\boldsymbol{q} - 2 \int_{\partial \mathcal{M}} \frac{\partial u(\boldsymbol{q})}{\partial \mathbf{n}} R_t(\boldsymbol{p}, \boldsymbol{q}) d\tau_{\boldsymbol{q}} \right.$$
$$\left. + \frac{1}{t} \int_{\mathcal{M}} (u(\boldsymbol{p}) - u(\boldsymbol{q})) R_t(\boldsymbol{p}, \boldsymbol{q}) d\boldsymbol{q} \right\|_{L^2(\mathcal{M})} = O(t^{1/4}),$$

(10.22)

where R_t is the normalized heat kernel:

$$R_t(\boldsymbol{p}, \boldsymbol{q}) = C_t \exp\left(-\frac{|\boldsymbol{p} - \boldsymbol{q}|^2}{4t}\right). \tag{10.23}$$

Theorem 10.1 suggests that the solution u of (10.21) should approximately satisfy the following integral equation:

$$\int_{\mathcal{M}} (u(\boldsymbol{p}) - u(\boldsymbol{q}))\, R_t(\boldsymbol{p}, \boldsymbol{q})\mathrm{d}\boldsymbol{q} + \mu t \sum_{\boldsymbol{q} \in P} R_t(\boldsymbol{p}, \boldsymbol{q})\, (u(\boldsymbol{q}) - v(\boldsymbol{q})) = 0. \tag{10.24}$$

If we further assume that $P = \mathcal{P}(\boldsymbol{u}^{(k)}) = \{\boldsymbol{p}_1, \ldots, \boldsymbol{p}_{|\bar{\Omega}|}\} \subset \mathbb{R}^d$ samples the manifold $\mathcal{M}$ uniformly at random, then (10.24) can be discretized into the following linear system:

$$\frac{|\mathcal{M}|}{|\bar{\Omega}|} \sum_{j=1}^{|\bar{\Omega}|} R_{t,ij}(u_i - u_j) + \mu t \sum_{j=1}^{|\bar{\Omega}|} R_{t,ij}(u_j - v_j) = 0, \tag{10.25}$$

where $u_i = u(\boldsymbol{p}_i)$, $v_i = v(\boldsymbol{p}_i)$, and $R_{t,ij} = R_t(\boldsymbol{p}_i, \boldsymbol{p}_j)$. Equation (10.25) can be written in the matrix form

$$(\boldsymbol{L} + \bar{\mu}\boldsymbol{W})\boldsymbol{u} = \bar{\mu}\boldsymbol{W}\boldsymbol{v}, \tag{10.26}$$

where $\boldsymbol{u} = (u_1, \ldots, u_{|\bar{\Omega}|})^T$, $\boldsymbol{v} = (v_1, \ldots, v_{|\bar{\Omega}|})^T$, $\bar{\mu} = \frac{\mu t |\bar{\Omega}|}{|\mathcal{M}|}$, $\boldsymbol{W} = (w_{ij})_{i, j \in \{1, \ldots, |\bar{\Omega}|\}}$ is the weight matrix with $w_{ij} = R_{t,ij}$, and $\boldsymbol{L}$ is the difference between $\boldsymbol{D}$ and $\boldsymbol{W}$,

$$\boldsymbol{L} = \boldsymbol{D} - \boldsymbol{W}, \tag{10.27}$$

where $\boldsymbol{D} = diag(d_i)$ is the degree matrix with $d_i = \sum_{j=1}^{|\bar{\Omega}|} w_{ij}$. In the numerical experiments, we always truncate the weight matrix $\boldsymbol{W}$ to 20 nearest neighbors, searched efficiently using the k-d tree data structure [10]. Thus (10.26) is a sparse symmetric linear system, which can be solved by the conjugate gradient method. We summarize the numerical procedures of solving LDMM with the PIM discretization in Algorithm 1.

10.3.2   The Second Approach

We now explain our second approach of solving (10.11). This involves directly discretizing the Dirichlet energy $\|\nabla_{\mathcal{M}^{(k)}} \alpha_i^t\|_{L^2(\mathcal{M}^{(k)})}^2$ in (10.11) using the weighted nonlocal Laplacian (WNLL) [25] without splitting the update of $\boldsymbol{\alpha}$ and $\boldsymbol{u}$ [31].

Algorithm 1 LDMM for HSI reconstruction with the PIM discretization

---

Input: A noisy and incomplete observation b of an unknown hyperspectral image $u \in \mathbb{R}^{m \times n \times B}$.
For every spectral band $t \in [B]$, u is only partially observed on a random subset Ω^t of $\bar{\Omega} = [m] \times [n]$.

Output: Reconstructed HSI u.
  Initial guess $u^{(0)}$, $z^0 = 0$.
  while not converge **do**
    1. Compute the weight matrix $W = (w_{ij})_{1 \le i, j \le |\bar{\Omega}|}$ and L from $\mathcal{P}(u^{(k)})$,

$$w_{ij} = R_{t,ij} = R_t(p_i, p_j), \quad p_i, p_j \in \mathcal{P}(u^{(k)}), \quad i, j = 1, \dots, |\bar{\Omega}|, \quad L = D - W.$$

  while not converge **do**
    1. Solve the linear systems for $U^{l+1} = \alpha^{(k+1),l+1}(\mathcal{P}(u^{(k)})) \in \mathbb{R}^{d \times |\bar{\Omega}|}$

$$(L + \bar{\mu} W)(U^{l+1})^T = \bar{\mu} W V^l,$$

  where $V^l = \left(\mathcal{P}(u^{(k+1),l}) - z^l\right)^T$.
    2. Update $u^{(k+1),l+1}$

$$u^{(k+1),l+1} = \left(\lambda I_{\Omega_{\text{all}}}^* I_{\Omega_{\text{all}}} + \mu \mathcal{P}^* \mathcal{P}\right)^{-1} \left[\lambda I_{\Omega_{\text{all}}}^* b + \mu \mathcal{P}^* \left(z^l + U^{l+1}\right)\right],$$

    3. Update z^{l+1}

$$z^{l+1} = z^l + U^{l+1} - \mathcal{P}(u^{(k+1),l+1}).$$

    4. $l \leftarrow l + 1$.
  end while
  1. $u^{(k+1)} \leftarrow u^{(k+1),l}$.
  2. $k \leftarrow k + 1$.
  end while
  $u \leftarrow u^{(k)}$.

10.3.2.1 Weighted Nonlocal Laplacian

The weighted nonlocal Laplacian was proposed in [25] to find a smooth interpolation of a function on a point cloud. Suppose that $C = \{c_1, c_2, \dots, c_n\}$ is a set of points in $\mathbb{R}^d$, and let g be a function defined on a subset $S = \{s_1, s_2, \dots, s_n\} \subset C$. The objective is to extend g to C by finding a smooth function u on $\mathcal{M}$ that agrees with g when restricted to S.

A widely used method to solve the above interpolation problem is the harmonic extension model [8, 33], which seeks to minimize the following energy:

$$\mathcal{J}(u) = \|\nabla_{\mathcal{M}} u\|_{L^2(\mathcal{M})}^2, \quad \text{subject to: } u(p) = g(p) \text{ on } S. \tag{10.28}$$

A common way of discretizing the manifold gradient $\nabla_{\mathcal{M}} u$ is to use its nonlocal approximation:

$$\nabla_{\mathcal{M}} u(\boldsymbol{p})(\boldsymbol{q}) \approx \sqrt{w(\boldsymbol{p}, \boldsymbol{q})} \left(u(\boldsymbol{p}) - u(\boldsymbol{q}) \right),$$

where w is a positive weight function, e.g., $w(\boldsymbol{p}, \boldsymbol{q}) = \exp\left(-\frac{\|\boldsymbol{p}-\boldsymbol{q}\|^2}{\sigma^2} \right)$. With this approximation, we have

$$\mathcal{J}(u) \approx \sum_{\boldsymbol{p}, \boldsymbol{q} \in P} w(\boldsymbol{p}, \boldsymbol{q}) \left(u(\boldsymbol{p}) - u(\boldsymbol{q}) \right)^2. \qquad (10.29)$$

Such discretization of solving the harmonic extension model leads to the well-known graph Laplacian method [2, 4, 33]. However, a closer look into the energy $\mathcal{J}$ in (10.29) reveals that the model fails to achieve satisfactory results when the sample rate $|S|/|C|$ is low [20, 25]. More specifically, after splitting the sum in (10.29) into two terms, we have

$$\mathcal{J}(u) = \sum_{\boldsymbol{p} \in S} \sum_{\boldsymbol{q} \in C} w(\boldsymbol{p}, \boldsymbol{q}) \left(u(\boldsymbol{p}) - u(\boldsymbol{q}) \right)^2 + \sum_{\boldsymbol{p} \in C \backslash S} \sum_{\boldsymbol{q} \in C} w(\boldsymbol{p}, \boldsymbol{q}) \left(u(\boldsymbol{p}) - u(\boldsymbol{q}) \right)^2.$$

$$(10.30)$$

Note that the first term in (10.30) is much smaller than the second term when $|S| \ll |C|$. Therefore, the second term will be prioritized when minimizing (10.30), and the continuity of u on the sampled set S will be sacrifice. An easy remedy is to add a large weight $\mu = |C|/|S|$ in front of the first term in (10.30) to balance the two terms:

$$\mathcal{J}_{\mathrm{WNLL}}(u) = \mu \sum_{\boldsymbol{p} \in S} \sum_{\boldsymbol{q} \in C} w(\boldsymbol{p}, \boldsymbol{q}) \left(u(\boldsymbol{p}) - u(\boldsymbol{q}) \right)^2 + \sum_{\boldsymbol{p} \in C \backslash S} \sum_{\boldsymbol{q} \in C} w(\boldsymbol{p}, \boldsymbol{q}) \left(u(\boldsymbol{p}) - u(\boldsymbol{q}) \right)^2.$$

$$(10.31)$$

It is readily checked that $\mathcal{J}_{\mathrm{WNLL}}$ generalizes the graph Laplacian $\mathcal{J}$ in the sense that $\mathcal{J}_{\mathrm{WNLL}} = \mathcal{J}$ when $|S| = |C|$. The generalized energy functional $\mathcal{J}_{\mathrm{WNLL}}$ is called the weighted nonlocal Laplacian.

   We point out that such intuition can be made precise by deriving (10.31) through the point integral method [19]. The interested reader can refer to [25] for the details.

10.3.2.2   Numerical Discretization

We now explain how to solve the optimization problem (10.11) using the weighted nonlocal Laplacian (10.31). Using the terminology introduced in Sect. 10.3.2.1, the functions to be interpolated in (10.11) are α_i^t, the point cloud C is $\mathcal{P}(\boldsymbol{u}^{(k)})$, and the sampled set for α_i^t is

$$S_i^t = \left\{ \mathcal{P}\boldsymbol{u}^{(k)}(\boldsymbol{x}) : \mathcal{P}_i \boldsymbol{u}^{(k)}(\boldsymbol{x}) \text{ is sampled} \right\} \subset C.$$

Based on the discussion of WNLL in Sect. 10.3.2.1, we can discretize the Dirichlet energy $\|\nabla_{\mathcal{M}^{(k)}} \alpha_i^t\|_{L^2(\mathcal{M}^{(k)})}^2$ as

$$
\|\nabla_{\mathcal{M}^{(k)}} \alpha_i^t\|_{L^2(\mathcal{M}^{(k)})}^2 = \frac{|\bar{\Omega}|}{|\Omega_i^t|} \sum_{x \in \Omega_i^t} \sum_{y \in \bar{\Omega}} \bar{w}(x, y) \left(\alpha_i^t(\mathcal{P}u^{(k)}(x)) - \alpha_i^t(\mathcal{P}u^{(k)}(y))\right)^2
$$
$$
+ \sum_{x \in \bar{\Omega} \setminus \Omega_i^t} \sum_{y \in \bar{\Omega}} \bar{w}(x, y) \left(\alpha_i^t(\mathcal{P}u^{(k)}(x)) - \alpha_i^t(\mathcal{P}u^{(k)}(y))\right)^2, \quad (10.32)
$$

where

$$
\Omega_i^t = \left\{x \in \bar{\Omega} : \mathcal{P}_i^t u^{(k)}(x) \text{ is sampled}\right\}
$$

is a spatially translated version of Ω^t, $|\bar{\Omega}|/|\Omega_i^t| = 1/r$ is the inverse of the sampling rate, and $\bar{w}(x, y) = w(\mathcal{P}u^{(k)}(x), \mathcal{P}u^{(k)}(y))$ is the similarity between the patches, with

$$
w(p, q) = \exp\left(-\frac{\|p - q\|^2}{\sigma(p)\sigma(q)}\right), \quad (10.33)
$$

where $\sigma(p)$ is the normalizing factor. Combining the WNLL discretization (10.32) and the constraint in (10.11), the update of u in (10.11) can be discretized as

$$
\min_{u} \ \lambda \sum_{t=1}^{B} \|u^t - b^t\|_{L^2(\Omega^t)}^2 + \sum_{i,t} \left[\sum_{x \in \bar{\Omega} \setminus \Omega_i^t} \sum_{y \in \bar{\Omega}} \bar{w}(x, y) \left(\mathcal{P}_i^t u(x) - \mathcal{P}_i^t u(y)\right)^2 \right.
$$
$$
\left. + \frac{1}{r} \sum_{x \in \Omega_i^t} \sum_{y \in \bar{\Omega}} \bar{w}(x, y) \left(\mathcal{P}_i^t u(x) - \mathcal{P}_i^t u(y)\right)^2 \right]. \quad (10.34)
$$

Remark 10.3 Unlike our first approach of solving (10.11) detailed in Sect. 10.3.1, we do not explicitly update the perturbed coordinate function α. The reason is that the value of α on the point cloud $\mathcal{P}(u^{(k)})$ is already implicitly determined for a given u, and this is enough to discretize the Dirichlet energy $\|\nabla_{\mathcal{M}^{(k)}} \alpha_i^t\|_{L^2(\mathcal{M}^{(k)})}^2$ on the manifold $\mathcal{M}^{(k)}$.

Note that (10.34) is decoupled with respect to the spectral coordinate t, and for any given $t \in [B]$, we only need to solve the following problem:

$$\min_{\boldsymbol{u}^t} \ \lambda \|\boldsymbol{u}^t - \boldsymbol{b}^t\|_{L^2(\Omega^t)}^2 + \sum_{i=1}^{d_s} \left[\sum_{\boldsymbol{x} \in \bar{\Omega} \setminus \Omega_i^t} \sum_{\boldsymbol{y} \in \bar{\Omega}} \bar{w}(\boldsymbol{x}, \boldsymbol{y}) \left(\mathcal{P}_i \boldsymbol{u}^t(\boldsymbol{x}) - \mathcal{P}_i \boldsymbol{u}^t(\boldsymbol{y}) \right)^2 \right.$$

$$\left. + \frac{1}{r} \sum_{\boldsymbol{x} \in \Omega_i^t} \sum_{\boldsymbol{y} \in \bar{\Omega}} \bar{w}(\boldsymbol{x}, \boldsymbol{y}) \left(\mathcal{P}_i \boldsymbol{u}^t(\boldsymbol{x}) - \mathcal{P}_i \boldsymbol{u}^t(\boldsymbol{y}) \right)^2 \right], \tag{10.35}$$

where $\mathcal{P}_i : \mathbb{R}^{m \times n} \to \mathbb{R}^{m \times n}$ satisfies $\mathcal{P}_i \boldsymbol{u}^t(\boldsymbol{x}) = \mathcal{P}_i^t \boldsymbol{u}(\boldsymbol{x})$. A standard variational technique shows that (10.35) is equivalent to the following Euler–Lagrange equation:

$$\left[\mu \sum_{i=1}^{d_s} \mathcal{P}_i^* I_{\Omega_i^t} \boldsymbol{h}_i^t(\boldsymbol{x}) + \sum_{i=1}^{d_s} \mathcal{P}_i^* \boldsymbol{g}_i^t(\boldsymbol{x}) + \lambda I_{\Omega^t} \left(\boldsymbol{u}^t - \boldsymbol{b}^t \right) \right](\boldsymbol{x}) = 0, \quad \forall \boldsymbol{x} \in \bar{\Omega} \ (10.36)$$

where $\mu = 1/r - 1$, $\mathcal{P}_i^*$ is the adjoint operator of $\mathcal{P}_i$, I_{Ω^t} is the projection operator that sets $\boldsymbol{u}^t(\boldsymbol{x})$ to zero for $\boldsymbol{x} \notin \Omega^t$, i.e.,

$$I_{\Omega^t} \boldsymbol{u}^t(\boldsymbol{x}) = \begin{cases} \boldsymbol{u}^t(\boldsymbol{x}) & , \boldsymbol{x} \in \Omega^t, \\ 0 & , \boldsymbol{x} \notin \Omega^t, \end{cases} \tag{10.37}$$

and

$$\begin{cases} \boldsymbol{h}_i^t(\boldsymbol{x}) = \displaystyle\sum_{\boldsymbol{y} \in \bar{\Omega}} \bar{w}(\boldsymbol{x}, \boldsymbol{y}) \left(\mathcal{P}_i \boldsymbol{u}^t(\boldsymbol{x}) - \mathcal{P}_i \boldsymbol{u}^t(\boldsymbol{y}) \right) \\ \boldsymbol{g}_i^t(\boldsymbol{x}) = \displaystyle\sum_{\boldsymbol{y} \in \bar{\Omega}} 2\bar{w}(\boldsymbol{x}, \boldsymbol{y}) \left(\mathcal{P}_i \boldsymbol{u}^t(\boldsymbol{x}) - \mathcal{P}_i \boldsymbol{u}^t(\boldsymbol{y}) \right) + \mu \displaystyle\sum_{\boldsymbol{y} \in \Omega_i^t} \bar{w}(\boldsymbol{x}, \boldsymbol{y}) \left(\mathcal{P}_i \boldsymbol{u}^t(\boldsymbol{x}) - \mathcal{P}_i \boldsymbol{u}^t(\boldsymbol{y}) \right) \end{cases}$$

We use the notation $\boldsymbol{x}_{\widehat{j}}$ to denote the j-th component (in the spatial domain) after $\boldsymbol{x}$ in a patch. Assuming a periodic padding is used when patches exceed the spatial domain of HSI, one can easily verify that

$$\begin{cases} \mathcal{P}_i \boldsymbol{u}^t(\boldsymbol{x}) = \boldsymbol{u}^t(\boldsymbol{x}_{\widehat{i-1}}), \\ \mathcal{P}_i^* \boldsymbol{u}^t(\boldsymbol{x}) = \boldsymbol{u}^t(\boldsymbol{x}_{\widehat{1-i}}). \end{cases}$$

With such notations, it follows that

$$\mathcal{P}_i^* I_{\Omega_i^t} \boldsymbol{h}_i^t(\boldsymbol{x}) = \left[I_{\Omega_i^t} \boldsymbol{h}_i^t \right] (\boldsymbol{x}_{\widehat{1-i}})$$
$$= I_{\Omega^t} \left[\boldsymbol{h}_i^t(\boldsymbol{x}_{\widehat{1-i}}) \right]$$
$$= I_{\Omega^t} \left[\sum_{\boldsymbol{y} \in \bar{\Omega}} \bar{w}(\boldsymbol{x}_{\widehat{1-i}}, \boldsymbol{y}_{\widehat{1-i}}) \left(\mathcal{P}_i \boldsymbol{u}^t(\boldsymbol{x}_{\widehat{1-i}}) - \mathcal{P}_i \boldsymbol{u}^t(\boldsymbol{y}_{\widehat{1-i}}) \right) \right]$$

$$= I_{\Omega^t} \left[\sum_{y \in \bar{\Omega}} \bar{w}(x_{\widehat{1-i}}, y_{\widehat{1-i}}) \left(u^t(x) - u^t(y) \right) \right], \tag{10.38}$$

and

$$
\begin{aligned}
\mathcal{P}_i^* g_i^t(x) &= g_i^t(x_{\widehat{1-i}}) \\
&= \sum_{y \in \bar{\Omega}} 2\bar{w}(x_{\widehat{1-i}}, y) \left(\mathcal{P}_i u^t(x_{\widehat{1-i}}) - \mathcal{P}_i u^t(y) \right) \\
&\quad + \mu \sum_{y \in \Omega_i^t} \bar{w}(x_{\widehat{1-i}}, y) \left(\mathcal{P}_i u^t(x_{\widehat{1-i}}) - \mathcal{P}_i u^t(y) \right) \\
&= \sum_{y \in \bar{\Omega}} 2\bar{w}(x_{\widehat{1-i}}, y_{\widehat{1-i}}) \left(\mathcal{P}_i u^t(x_{\widehat{1-i}}) - \mathcal{P}_i u^t(y_{\widehat{1-i}}) \right) \\
&\quad + \mu \sum_{y \in \Omega^t} \bar{w}(x_{\widehat{1-i}}, y_{\widehat{1-i}}) \left(\mathcal{P}_i u^t(x_{\widehat{1-i}}) - \mathcal{P}_i u^t(y_{\widehat{1-i}}) \right) \\
&= \sum_{y \in \bar{\Omega}} 2\bar{w}(x_{\widehat{1-i}}, y_{\widehat{1-i}}) \left(u^t(x) - u^t(y) \right) \\
&\quad + \mu \sum_{y \in \Omega^t} \bar{w}(x_{\widehat{1-i}}, y_{\widehat{1-i}}) \left(u^t(x) - u^t(y) \right). \tag{10.39}
\end{aligned}
$$

Therefore, we can rewrite (10.36) as

$$
\sum_{i=1}^{d_s} \left[\sum_{y \in \bar{\Omega}} 2\bar{w}(x_{\widehat{1-i}}, y_{\widehat{1-i}}) \left(u^t(x) - u^t(y) \right) + \mu \sum_{y \in \Omega^t} \bar{w}(x_{\widehat{1-i}}, y_{\widehat{1-i}}) \left(u^t(x) - u^t(y) \right) \right]
$$
$$
+ \mu I_{\Omega^t} \left[\sum_{y \in \bar{\Omega}} \sum_{i=1}^{d_s} \bar{w}(x_{\widehat{1-i}}, y_{\widehat{1-i}}) \left(u^t(x) - u^t(y) \right) \right] + \lambda I_{\Omega^t} \left(u^t - b^t \right) = 0, \ \forall x \in \bar{\Omega}. \tag{10.40}
$$

After assembling the weight matrices $\bar{w}(x_{\widehat{1-i}}, y_{\widehat{1-i}})$ into

$$\tilde{w}(x, y) = \sum_{i=1}^{d_s} \bar{w}(x_{\widehat{1-i}}, y_{\widehat{1-i}}), \tag{10.41}$$

it follows that (10.40) is equivalent to

$$2 \sum_{y \in \bar{\Omega}} \tilde{w}(x, y) \left(u^t(x) - u^t(y) \right) + \mu \sum_{y \in \Omega^t} \tilde{w}(x, y) \left(u^t(x) - u^t(y) \right)$$

$$+ \mu I_{\Omega^t} \left[\sum_{y \in \bar{\Omega}} \tilde{w}(x, y) \left(u^t(x) - u^t(y) \right) \right] + \lambda I_{\Omega^t} \left(u^t - b^t \right) = 0, \quad \forall x \in \bar{\Omega}$$

$$(10.42)$$

Note that (10.42) is a sparse linear system for u^t in $\mathbb{R}^{mn}$, but unlike (10.26), the coefficient matrix is not symmetric because of the projection operator I_{Ω^t}. We thus use the generalized minimal residual method (GMRES) to solve the systems (10.42). The numerical procedures of solving LDMM with the WNLL discretization is summarized in Algorithm 2.

Algorithm 2 LDMM for HSI reconstruction with the WNLL discretization

---

Input: A noisy and incomplete observation b of an unknown hyperspectral image $u \in \mathbb{R}^{m \times n \times B}$. For every spectral band $t \in [B]$, u is only partially observed on a random subset Ω^t of $\bar{\Omega} = [m] \times [n]$.

Output: Reconstructed HSI u.

Initial guess $u^{(0)}$.

while not converge **do**

    1. Extract the patch set $\mathcal{P}u^{(k)}$ from $u^{(k)}$.

    2. Compute the similarity matrix on the spatial domain

$$\bar{w}(x, y) = w(\mathcal{P}u^{(k)}(x), \mathcal{P}u^{(k)}(y)), \quad x, y \in \bar{\Omega}.$$

    3. Assemble the new similarity matrix

$$\tilde{w}(x, y) = \sum_{i=1}^{d_s} \bar{w}(x_{\widehat{1-i}}, y_{\widehat{1-i}})$$

    4. For every spectral band t, Update $(u^t)^{(k+1)}$ as the solution of (10.42) using GMRES.

    5. $k \leftarrow k + 1$.

end while

$u = u^{(k)}$.

10.3.3 A Comparison of the Two Approaches

We first compare the computational cost of the two approaches. The most time-consuming part of both algorithms is solving the $|\bar{\Omega}|$-dimensional sparse linear systems. For each iteration in the inner loop of Algorithm 1, one needs to solve $d = s_1 \times s_2 \times B$ linear systems. On the other hand, one only needs to solve B linear systems in each iteration of Algorithm 2. Moreover, unlike Algorithm 1, Algorithm 2

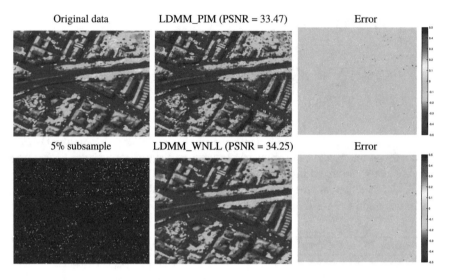

Fig. 10.1 Reconstruction of the Pavia Center dataset from its 5% noise-free subsample. The first column displays the original image and its 5% random subsample at one spectral band. The remaining two columns display the reconstructed images and the error (the difference between the original data and the reconstruction) using LDMM with the PIM (first row) and the WNLL (second row) discretization

does not have an inner loop. The reason is that the weight assembly step (10.41) in Algorithm 2 combines $s_1 \times s_2$ equations in the spatial patch domain into only one equation, and the WNLL discretization enforces the constraint in (10.11) directly without a further splitting. Therefore, Algorithm 2 is much more computationally efficient as compared to Algorithm 1.

We also compare the numerical accuracy of the two algorithms in the reconstruction of hyperspectral images from their 5% noise-free random subsamples. In the experiments, we set the spatial patch size $s_1 \times s_2$ to 2×2, and Figs. 10.1 and 10.2 present the performance of the two algorithms on the Pavia Center and Pavia University datasets. The peak signal-to-noise ratio,

$$\text{PSNR} = 10 \log_{10} \left(\frac{\|\boldsymbol{u}^*\|_\infty}{\text{MSE}} \right), \tag{10.43}$$

is used to evaluate the reconstruction, where $\boldsymbol{u}^*$ is the ground truth, and MSE is the mean squared error. As can be seen from Figs. 10.1 and 10.2, even though both algorithms lead to remarkable results of HSI reconstruction, LDMM with the WNLL discretization has a slight edge over PIM in terms of accuracy as well. Due to the advantage of WNLL in both computational efficiency and reconstruction accuracy, we will report the results of LDMM with WNLL discretization only in the following experiments, even though PIM has more theoretical guarantee on the consistency of the discretization (c.f. Theorem 10.1).

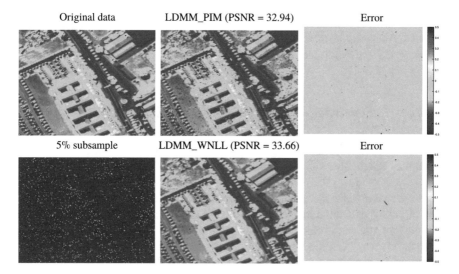

Fig. 10.2 Reconstruction of the Pavia University dataset from its 5% noise-free subsample. The first column displays the original image and its 5% random subsample at one spectral band. The remaining two columns display the reconstructed images and the error (the difference between the original data and the reconstruction) using LDMM with the PIM (first row) and the WNLL (second row) discretization

10.4   Numerical Experiments

In this section, we present the numerical results on the following datasets: Pavia University, Pavia Center, Indian Pine, and San Diego Airport. All images have been cropped in the spatial dimension to 200×200 for easy comparison. The objective of the experiment is to reconstruct the original HSI from 5% random subsample (10% random subsample for noisy data). As discussed in Sect. 10.3.3, we choose LDMM with the WNLL discretization (Algorithm 2) as the default method because of its computational efficiency and superior numerical accuracy.

10.4.1   Experimental Setup

Empirically, we discovered that it is easier for LDMM to converge if we use a reasonable initialization. In our experiments, we always use the result of the low-rank matrix completion algorithm APG [26] as an initialization, and run three iterations of manifold update for LDMM. The peak signal-to-noise ratio (PSNR) defined in (10.43) is used to evaluate the reconstruction accuracy. All experiments were run on a Linux machine with 8 Intel core i7-7820X 3.6 GHz CPUs and 64 GB of RAM. Codes and datasets are available for download at http://services.math.duke.edu/~zhu/software/HSI_LDMM_public.tar.gz.

10.4.2 Reconstruction from Noise-Free Subsample

We first present the results of the reconstruction of hyperspectral images from their 5% noise-free random subsamples. Table 10.1 displays the computational time and accuracy of the low-rank matrix completion initialization (APG) and LDMM with different spatial patch sizes. Unlike high-resolution RGB images, the spatial resolution of the hyperspectral images considered in this chapter is typically much lower, and thus we limited the choice of spatial patch size to only 1×1 and 2×2. It is clear from the table that LDMM significantly improves the accuracy of the APG initialization with comparable extra computational time. Figures 10.3 and 10.4 provide a visual illustration of the results. It can be observed from the figures that the reconstructions of the subsampled hyperspectral images by LDMM are spatially much smoother than the APG initialization because of the low-dimensionality regularization on the patch manifold. What is also quite interesting from the error map of LDMM in Fig. 10.3 is the relatively poorer reconstruction of the several "rare" objects in the scene, e.g., the two airplanes on the upper right corner of the image. The poor reconstruction of these "anomalies" in the scene is due to the fact that the patch manifolds of these rare objects are not well-resolved with only limited samples. This observation does mean that LDMM is less robust when reconstructing images with many sparsely sampled distinct objects. However, on the other hand, it also suggests that LDMM might be used as an "anomaly detection" algorithm by purposefully subsampling the original image and identifying the anomalies as the objects that are least well reconstructed by LDMM. This usage of LDMM as an hyperspectral anomaly detection algorithm has been studied recently in [27].

Table 10.1 Reconstruction of the HSIs from their noise-free 5% subsamples. LDMM (1×1) and LDMM (2×2) stand for LDMM with spatial patch size of 1×1 and 2×2. The reported time of LDMM does not include that of the AGP initialization

	APG		LDMM (1×1)		LDMM (2×2)	
	PSNR	Time (s)	PSNR	Time (s)	PSNR	Time (s)
Indian Pine	26.80	13	32.09	8	**34.08**	22
Pavia Center	32.61	17	**34.54**	11	34.25	31
Pavia University	31.51	13	33.38	11	**33.66**	29
San Diego Airport	32.43	23	40.33	16	**44.21**	46

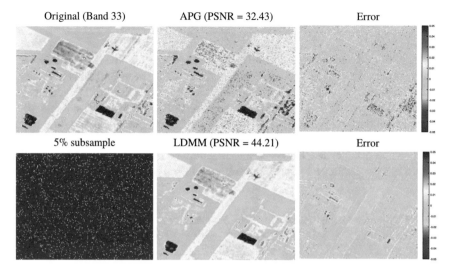

Fig. 10.3 Reconstruction of the San Diego Airport dataset from its 5% noise-free subsample. Note that the error is displayed with a scale 1/20 of the original data to visually amplify the difference

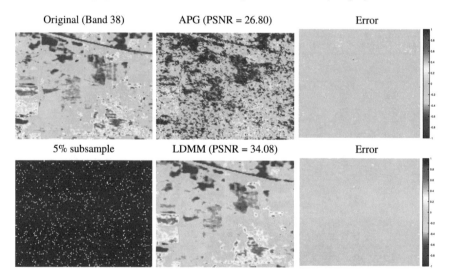

Fig. 10.4 Reconstruction of the Indian Pine dataset from its 5% noise-free subsample

10.4.3 Reconstruction from Noisy Subsample

We next show the results of the reconstruction of hyperspectral images from their 10% noisy subsample. A Gaussian noise with a standard deviation of 0.05 is first added to the original image, and 90% of the voxels are later removed from the data cube. We report the accuracy and computational time of the experiments in Table 10.2. Note that when noise is present, LDMM with 2×2 patches typically produce better results than that with 1×1 patches. This is due to the stronger spatial regularization by choosing a larger patch size. Visual illustrations of the reconstruction are displayed in Figs. 10.5 and 10.6. It can be observed from the figures that, even with the

Table 10.2 Reconstruction of the noisy HSIs from their 10% subsamples. LDMM (1×1) and LDMM (2×2) stand for LDMM with spatial patch size of 1×1 and 2×2. The reported time of LDMM does not include that of the AGP initialization

	APG		LDMM (1×1)		LDMM (2×2)	
	PSNR	Time (s)	PSNR	Time (s)	PSNR	Time (s)
Indian Pine	31.56	18	**34.03**	54	34.02	56
Pavia Center	30.22	47	30.55	82	**31.61**	82
Pavia University	29.88	38	30.26	77	**31.40**	86
San Diego Airport	33.90	69	39.17	186	**41.31**	231

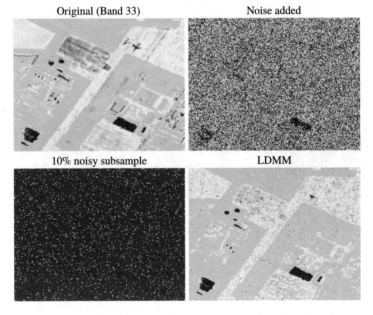

Original (Band 33)　　　　　　Noise added

10% noisy subsample　　　　　　LDMM

Fig. 10.5 Reconstruction of the San Diego Airport dataset from its 10% noisy subsample

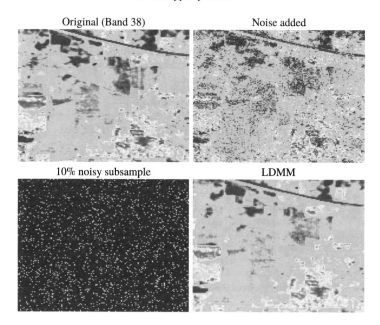

Fig. 10.6 Reconstruction of the Indian Pine dataset from its 10% noisy subsample

presence of significant noise in the remaining subsampled voxels, LDMM is able to achieve a reasonable reconstruction of the original HSI, especially for objects that are abundantly represented in the image.

10.5   Conclusion

We explained in this chapter the low dimensional manifold model for the reconstruction of hyperspectral images from noisy and incomplete observations with a significant number of missing voxels. LDMM is based on the assumption that the 3D patches in a hyperspectral image tend to sample a collection of low dimensional manifolds. As a result, we directly use the dimension of the patch manifold as a regularizer in a variational functional, which can be solved using either the point integral method or the weighted nonlocal Laplacian. Because of the special data structure of hyperspectral images, the same similarity matrix can be shared across all spectral bands, which significantly reduces the computational burden. Numerical experiments show that the proposed algorithm is both accurate and efficient for HSI reconstruction from its noisy and incomplete observation.

Acknowledgements This work was supported by STROBE: A National Science Foundation Science & Technology Center, under Grant No. DMR 1548924 as well as DOE-DE-SC0013838 and NSFC 11671005.

References

1. Aggarwal HK, Majumdar A (2016) Hyperspectral image denoising using spatio-spectral total variation. IEEE Geosci Remote Sens Lett 13(3):442–446
2. Bertozzi AL, Flenner A (2012) Diffuse interface models on graphs for classification of high dimensional data. Multiscale Model Simul 10(3):1090–1118
3. Bioucas-Dias J, Plaza A, Dobigeon N, Parente M, Du Q, Gader P, Chanussot J (2012) Hyperspectral unmixing overview: geometrical, statistical, and sparse regression-based approaches. IEEE J Sel Top Appl Earth Obs Remote Sens 5(2):354–379
4. Bühler T, Hein M (2009) Spectral clustering based on the graph p-Laplacian. In: Proceedings of the 26th annual international conference on machine learning. ACM, pp 81–88
5. Carlsson G, Ishkhanov T, de Silva V, Zomorodian A (2008) On the local behavior of spaces of natural images. Int J Comput Vis 76(1):1–12
6. Chang C-I (2003) Hyperspectral imaging: techniques for spectral detection and classification, vol 1. Springer Science & Business Media
7. Charles AS, Olshausen BA, Rozell CJ (2011) Learning sparse codes for hyperspectral imagery. IEEE J Sel Top Signal Process 5(5):963–978
8. Chung FR, Graham FC (1997) Spectral graph theory, vol 92. American Mathematical Society
9. Dobigeon N, Tourneret J-Y, Richard C, Bermudez J, McLaughlin S, Hero A (2014) Nonlinear unmixing of hyperspectral images: models and algorithms. IEEE Signal Process Mag 31(1):82–94
10. Friedman JH, Bentley JL, Finkel RA (1976) An algorithm for finding best matches in logarithmic time. ACM Trans Math Softw, 3(SLAC-PUB-1549-REV. 2):209–226
11. Gilboa G, Osher S (2009) Nonlocal operators with applications to image processing. Multiscale Model Simul 7(3):1005–1028
12. Goldstein T, Osher S (2009) The split bregman method for l1-regularized problems. SIAM J Imaging Sci 2(2):323–343
13. He W, Zhang H, Zhang L, Shen H (2016) Total-variation-regularized low-rank matrix factorization for hyperspectral image restoration. IEEE Trans Geosci Remote Sens 54(1):178–188
14. Hu H, Sunu J, Bertozzi AL (2015) Multi-class graph Mumford-Shah model for plume detection using the MBO scheme. In: Energy minimization methods in computer vision and pattern recognition: 10th international conference, EMMCVPR 2015, Hong Kong, China, January 13–16, 2015. Proceedings. Springer International Publishing, Cham, pp 209–222
15. Iordache MD, Bioucas-Dias JM, Plaza A (2012) Total variation spatial regularization for sparse hyperspectral unmixing. IEEE Trans Geosci Remote Sens 50(11):4484–4502
16. Kawakami R, Matsushita Y, Wright J, Ben-Ezra M, Tai YW, Ikeuchi K (2011) High-resolution hyperspectral imaging via matrix factorization. In: 2011 IEEE conference on computer vision and pattern recognition (CVPR), pp 2329–2336
17. Lee AB, Pedersen KS, Mumford D (2003) The nonlinear statistics of high-contrast patches in natural images. Int J Comput Vis 54(1–3):83–103
18. Li J, Yuan Q, Shen H, Zhang L (2015) Hyperspectral image recovery employing a multidimensional nonlocal total variation model. Signal Process 111:230–248
19. Li Z, Shi Z, Sun J (2017) Point integral method for solving poisson-type equations on manifolds from point clouds with convergence guarantees. Commun Comput Phys 22(1):228–258
20. Osher S, Shi Z, Zhu W (2017) Low dimensional manifold model for image processing. SIAM J Imaging Sci 10(4):1669–1690
21. Peyré G (2008) Image processing with nonlocal spectral bases. Multiscale Model Simul 7(2):703–730
22. Peyré G (2011) A review of adaptive image representations. IEEE J Sel Top Signal Process 5(5):896–911
23. Rudin LI, Osher S, Fatemi E (1992) Nonlinear total variation based noise removal algorithms. Phys D 60:259–268
24. Shi Z, Osher S, Zhu W (2017) Generalization of the weighted nonlocal Laplacian in low dimensional manifold model. J Sci Comput

25. Shi Z, Osher S, Zhu W (2017) Weighted nonlocal Laplacian on interpolation from sparse data. J Sci Comput 73(2):1164–1177
26. Toh K-C, Yun S (2010) An accelerated proximal gradient algorithm for nuclear norm regularized linear least squares problems. Pac J Optim 6(615–640):15
27. Wu Z, Zhu W, Chanussot J, Xu Y, Osher S (2019) Hyperspectral anomaly detection via global and local joint modeling of background. IEEE Trans Signal Process 67(14):3858–3869
28. Xing Z, Zhou M, Castrodad A, Sapiro G, Carin L (2012) Dictionary learning for noisy and incomplete hyperspectral images. SIAM J Imaging Sci 5(1):33–56
29. Yuan Q, Zhang L, Shen H (2012) Hyperspectral image denoising employing a spectral-spatial adaptive total variation model. IEEE Trans Geosci Remote Sens 50(10):3660–3677
30. Zhu W, Chayes V, Tiard A, Sanchez S, Dahlberg D, Bertozzi AL, Osher S, Zosso D, Kuang D (2017) Unsupervised classification in hyperspectral imagery with nonlocal total variation and primal-dual hybrid gradient algorithm. IEEE Trans Geosci Remote Sens 55(5):2786–2798
31. Zhu W, Shi Z, Osher S (2018) Scalable low dimensional manifold model in the reconstruction of noisy and incomplete hyperspectral images. In: 9th workshop on hyperspectral image and signal processing: evolution in remote sensing (WHISPERS)
32. Zhu W, Wang B, Barnard R, Hauck CD, Jenko F, Osher S (2018) Scientific data interpolation with low dimensional manifold model. J Comput Phys 352:213–245
33. Zhu X, Ghahramani Z, Lafferty JD (2003) Semi-supervised learning using Gaussian fields and harmonic functions. In: Proceedings of the 20th international conference on machine learning (ICML-03), pp 912–919

Chapter 11
Deep Sparse Band Selection for Hyperspectral Face Recognition

Fariborz Taherkhani, Jeremy Dawson and Nasser M. Nasrabadi

Abstract Hyperspectral imaging systems collect and process information from specific wavelengths across the electromagnetic spectrum. The fusion of multispectral bands in the visible spectrum has been exploited to improve face recognition performance over all the conventional broadband face images. In this chapter, we propose a new Convolutional Neural Network (CNN) framework which adopts a structural sparsity learning technique to select the optimal spectral bands to obtain the best face recognition performance over all of the spectral bands. Specifically, in this method, images from all bands are fed to a CNN, and the convolutional filters in the first layer of the CNN are then regularized by employing a group Lasso algorithm to zero out the redundant bands during the training of the network. Contrary to other methods which usually select the useful bands manually or in a greedy fashion, our method selects the optimal spectral bands automatically to achieve the best face recognition performance over all spectral bands. Moreover, experimental results demonstrate that our method outperforms state-of-the-art band selection methods for face recognition on several publicly available hyperspectral face image datasets.

11.1 Introduction

In recent years, hyperspectral imaging has attracted much attention due to the decreasing cost of hyperspectral cameras used for image accusation [1]. A hyperspectral image consists of many narrow spectral bands within the visible spectrum and beyond. This data is structured as a hyperspectral "cube", with x- and y-coordinates making up the imaging pixels and the z-coordinate the imaging wavelength, which,

F. Taherkhani · J. Dawson · N. M. Nasrabadi (✉)
West Virginia University, Morgantown, USA
e-mail: nasser.nasrabadi@mail.wvu.edu

F. Taherkhani
e-mail: ft0009@mix.wvu.edu

J. Dawson
e-mail: jeremy.dawson@mail.wvu.edu

© Springer Nature Switzerland AG 2020
S. Prasad and J. Chanussot (eds.), *Hyperspectral Image Analysis*,
Advances in Computer Vision and Pattern Recognition,
https://doi.org/10.1007/978-3-030-38617-7_11

in the case of facial imaging, results in several co-registered face images captured at varying wavelengths. Hyperspectral imaging has provided new opportunities for improving the performance of different imaging tasks, such as face recognition in biometrics, that exploits the spectral characteristics of facial tissues to increase the inter-subject differences [2]. It has been demonstrated that, by adding the extra spectral dimension, the size of the feature space representing a face image is increased which results in a larger inter-class feature difference between subjects for face recognition. Beyond the surface appearance, spectral measurements in the infrared (i.e., 700–1000 nm) can penetrate the subsurface tissue which can notably produce different biometric features for each subject [3].

A hyperspectral imaging camera simultaneously measures hundreds of adjacent spectral bands with a small spectral resolution (e.g., 10 nm). For example, AVIRIS hyperspectral imaging includes 224 spectral bands from 400 to 2500 nm [4]. Such a large number of bands implies high-dimensional data which remarkably influences the performance of face recognition. This is because, a redundancy exists between spectral bands, and some bands may hold less discriminative information than others. Therefore, it is advantageous to discard bands which carry little or no discriminative information during the recognition task. To deal with this problem, many band selection approaches have been proposed in order to choose the optimal and informative bands for face recognition. Most of these methods, such as those presented in [5], are based on dimensionality reduction, but in an ad hoc fashion. These methods, however, suffer from a lack of comprehensive and consolidated evaluation due to (a) the small number of subjects used during the testing of the methods and (b) lack of publicly available datasets for comparison. Moreover, these studies do not compare the performance of their algorithms comprehensively with other face recognition approaches that can be used for this challenge with some modifications [3].

The development of hyperspectral cameras has introduced many useful techniques that merge spectral and spatial information. Since hyperspectral cameras have become more readily available, computational approaches introduced initially for remote sensing challenges have been leveraged to other applications such as biomedical applications. Considering the vast person-to-person spectral variability for different types of tissue, hyperspectral imaging has the power to enhance the capability of automated systems for human re-identification. Recent face recognition protocols essentially apply spatial discriminants that are based on geometric facial features [4]. Many of these protocols have provided promising results on databases captured under controlled conditions. However, these methods often indicate significant performance drop in the presence of variation in face orientation [2, 6].

The work in [7], for instance, indicated that there is significant drop in the performance of recognition for images of faces which are rotated more than 32° from a frontal image that is used to train the model. Furthermore, [8], which uses a light-field model for pose-invariant face recognition, provided well recognition results for probe faces which are rotated more than 60° from a gallery face. The method, however, requires the manual determination of the 3D transformation to register face images. The methods that use geometric features can also perform poorly if subjects are imaged across varying spans of time. For instance, recognition performance can decrease by a maximum of 20% if imaging sessions are separated by a 2-week interval [7]. Partial face occlusion also usually results in poor performance. An approach [9] that divided the face images into regions for isolated analysis can tolerate up to 1/6 face occlusion without a decrease in matching accuracy. Thermal infrared imaging provides an alternative imaging modality that has been leveraged for face recognition [10]. However, algorithms based on thermal images utilize spatial features and have difficulty recognizing faces when presented with images containing pose variation.

A 3D morphable face approach has been introduced for face recognition across variant poses [11]. This method has provided a good performance on a 68-subject dataset. However, this method is currently computationally intensive and requires significant manual intervention. Many of the limitations of current face recognition methods can be overcome by leveraging spectral information. The interaction of light with human tissue has been explored comprehensively by many works [12] which consider the spectral properties of tissue. The epidermal and dermal layers of human skin are essentially a scattering medium that consists of several pigments such as hemoglobin, melanin, bilirubin, and carotene. Small changes in the distribution of these pigments cause considerable changes in the skin's spectral reflectance [13]. For instance, the impacts are large enough to enable algorithms for the automated separation of melanin and hemoglobin from RGB images [14]. Recent work [15] has calculated skin reflectance spectra over the visible wavelengths and introduced algorithms for the spectra.

11.2   Related Work

11.2.1   Hyperspectral Imaging Techniques

There are three common techniques used to construct a hyperspectral image: spatial scanning, spectral scanning, or snapshot imaging. These techniques will be described in detail in the following sections.

Spatial scan systems capture each spectral band along a single dimension as a scanned composite image of the object or area being viewed. The scanning aspect of these systems describes the narrow imaging field of view (e.g., a $1 \times N$ pixel array) of the system. The system creates images using an optical slit to allow only a thin strip of the image to pass through a prism or grating that then projects the diffracted scene onto an imaging sensor. By limiting the amount of scene (i.e., spatial) information into the system at any given instance, most of the imaging sensor area can be utilized to capture spectral information. This reduction in spatial resolution allows for simultaneous capture of data at a higher spectral resolution. This data capture technique is a practical solution for applications where a scanning operation is possible, specifically for airborne mounted systems that image the ground surface as an aircraft flies overhead. Food quality inspection is another successful application of these systems, as they can rapidly detect defective or unhealthy produce on a pro-

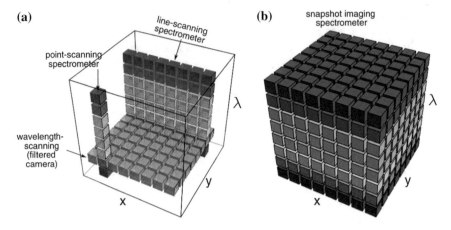

Fig. 11.1 Building the spectral data cube in both line scan and snapshot systems

duction or sorting line. While this technique provides both high spatial and spectral resolution, line scan Hyperspectral Imaging Systems (HSIs) are highly susceptible to the changes of the morphology of the target. This means the system must be fixed to a steady structure as the subject passes through its linear field of view or that the subject remains stationary as the imaging scan is conducted.

HSIs, such as those employing an Acousto-Optical Tunable Filter (AOTF) or a Liquid Crystal Tunable Filter (LCTF), use tunable optical devices that allow specific wavelengths of electromagnetic radiation to pass through to a broadband camera sensor. While the fundamental technology behind these tunable filters is different, their application achieves the same goal in a similar fashion by iteratively selecting the spectral bands of a subject that fall on the imaging sensor. Depending on the type of filter used, the integration time between the capture of each band can vary based on the driving frequency of the tunable optics and the integration time of the imaging plane. One limitation of scanning HSIs is that all bands in the data cube cannot be captured simultaneously. Figure 11.1a [16] illustrates a diagram which depicts the creation of the hyperspectral data cube by spatial and spectral scanning.

In contrast to scanning methods, a snapshot hyperspectral camera can capture hyperspectral image data in which all wavelengths are captured instantly to create the hypercube, as shown in Fig. 11.1b [16]. Snapshot hyperspectral technology is designed and built in configurations different from line scan imaging systems, often employing a prism to break up the light and causing the diffracted, spatially separated wavelengths to fall on different portions of the imaging sensor dedicated to collecting light energy from a specific wavelength. Software is used to sort the varying wavelengths of light falling onto different pixels into wavelength-specific groups. While conventional line scan hyperspectral cameras build the data cube by scanning through various filtered wavelengths or spatial dimensions, the snapshot HSI acquires an image and the spectral signature at each pixel simultaneously. Snapshot systems have an advantage of faster measurement and higher sensitivity. However, one drawback is that the resolution is limited by downsampling the light falling onto the imaging array into a smaller number of spectral channels.

11.2.2 Spectral Face Recognition

Most hyperspectral face recognition approaches are an extension of typical face recognition methods which have been adjusted to this challenge. For example, each band of a hyperspectral image can be considered as a separate image, and as a result, grayscale face recognition approaches can be applied to them.

Considering a hyperspectral cube as a set of images, image-set classification approaches can be leveraged for this problem without using a dimensionality reduction algorithm [3]. For example, Pan et al. [2] used 31 spectral band signatures at manually chosen landmarks on face images which were captured within the near-infrared spectrum. Their method provided high recognition accuracy under pose variations on a dataset which contains 1400 hyperspectral images from 200 people. However, the method does not achieve the same promising results on the public hyperspectral datasets used in [6].

Later on, Pan et al. [5] incorporated spatial and spectral information to improve the recognition results on the same dataset. Robila [17] distinguished spectral signatures of different face locations by leveraging spectral angle measurements. Their experiments are restricted to a very small dataset which consists of only eight subjects. Di et al. [18] projected the cube of hyperspectral images to a lower dimensional space by using a two-dimensional PCA method, and then Euclidean distance was calculated for face recognition. Shen and Zheng [19] used Gabor wavelets on hyperspectral data cubes to generate 52 new cubes from each given cube. Then, they used an ad hoc sub-sampling algorithm to reduce the large amount of data for face recognition.

A wide variety of approaches have been used to address the challenge of band selection for hyperspectral face recognition. Some of these methods are information-based methods [20], transform-based methods [21], search-based methods [22], and other techniques which include maximization of a spectral angle mapper [23], high-order moments [24], wavelet analysis [25], and a scheme trading spectral for spatial resolution [26]. Nevertheless, there are still some challenges with these approaches due to the presence of local-minima problems, difficulties for real-time implementation, and high computational cost. Hyperspectral imaging techniques for face recognition have provided promising results in the field of biometrics, overcoming challenges such as pose variations, lighting variations, presentation attacks, and facial expression variations [27]. The fusion of narrowband spectral images in the visible spectrum has been explored to enhance face recognition performance [28]. For example, Chang et al. [21] have demonstrated that the fusion of 25 spectral bands can surpass the performance of conventional broad band images for face recognition, mainly in cases where the training and testing images are collected under different types of illumination.

Despite the new opportunities provided by hyperspectral imaging, challenges still exist due to low signal-to-noise ratios, high dimensionality, and difficulty in data acquisition [29]. For example, hyperspectral images are usually stacked sequentially; hence, subject movements, specifically blinking of the eyes, can lead to band misalignment. This misalignment causes intra-class variations which cannot be compensated for by adding spectral dimension. Moreover, adding a spectral dimension makes the recognition task challenging due to the difficulty of choosing the required discriminative information. Furthermore, the spectral dimension causes a curse of dimensionality concern, because the ratio between the dimension of the data and the number of training data becomes very large [3].

Sparse dictionary learning has only been extended to the hyperspectral image classification [30]. Sparse-based hyperspectral image classification methods usually rank the contribution of each band in the classification task, such that each band is approximated by a linear combination of a dictionary, which contains other band images. The sparse coefficients represent the contribution of each dictionary atom to the target band image, where the large coefficient shows that the band has significant contribution for classification, while the small coefficient indicates that the band has negligible contribution for classification.

In recent years, deep learning methods have shown impressive learning ability in image retrieval [31–33], generating images [34–36], security purposes [37, 38], image classification [39–41], object detection [42, 43], face recognition [44–50], and many other computer vision and biometrics tasks. In addition to improving performance in computer vision and biometrics tasks, deep learning in combination with reinforcement learning methods was able to defeat the human champion in challenging games such as Go [51]. CNN-based models have also been applied to hyperspectral image classification [52], band selection [53, 54], and hyperspectral face recognition [55]. However, few of these methods have provided promising results for hyperspectral image classification due to a sub-optimal learning process caused by an insufficient amount of training data and the use of comparatively small-scale CNNs [56].

11.2.3 Spectral Band Selection

Previous research on band selection for face recognition usually works in an ad hoc fashion where the combination of different bands is evaluated to determine the best recognition performance. For instance, Di et al. [18] manually choose two disjoint subsets of bands which are centered at 540 and 580 nm to examine their discrimination power. However, selecting the optimal bands manually may not be appropriate because of the huge search space of many spectral bands.

In another case, Guo et al. [57] select the optimal bands by using an exhaustive search in such a way that the bands are first evaluated individually for face recognition, and a combination of the results are then selected by using a score-level fusion method. However, evaluating each band individually may not consider the complementary relationships between different bands. As a result, the selected subset of bands may not provide an optimal solution. To address this problem, Uzair et al. [3] leverage a sequential backward selection algorithm to search for a set of most discriminative bands. Sharma et al. [55] adopt a CNN-based model for band selection which uses a CNN to obtain the features from each spectral band independently, and then they use Adaboost in a greedy fashion (similar to other methods in the literature) for feature selection to determine the best bands. This method selects one band at a time, which ignores the complementary relationships between different bands for face recognition.

In this chapter, we propose a CNN-based model which adopts a Structural Sparsity Learning (SSL) technique to select the optimal bands to obtain the best recognition performance over *all* broadband images. We employ a group Lasso regularization algorithm [58] to sparsify the redundant spectral bands for face recognition. The group Lasso puts a constraint on the structure of the filters in the first layer of our CNN during the training process. This constraint is a loss term augmented to the total loss function used for face recognition to zero out the redundant bands during the training of the CNN. To summarize, the main contributions of this chapter include the following:

1. Joint face recognition and spectral band selection: We propose an end-to-end deep framework which jointly recognizes hyperspectral face images and selects the optimal spectral bands for the face recognition task.
2. Using group sparsity to automatically select the optimal bands: We adopt a group sparsity technique to reduce the depth of convolutional filters in the first layer of our CNN network. This is done to zero out the redundant bands during face recognition. Contrary to most of the existing methods which select the optimal bands in a greedy fashion or manually, our group sparsity technique selects the optimal bands automatically to obtain the best face recognition performance over all the spectral bands.
3. Comprehensive evaluation and obtaining the best recognition accuracy: We evaluate our algorithm comprehensively on three standard publicly available hyperspectral face image datasets. The results indicate that our method outperforms state-of-the-art spectral band selection methods for face recognition.

11.3 Sparsity

The sparsity of signals has been a powerful tool in many classical signal processing applications, such as denoising and compression. This is because most natural signals can be represented compactly by only a few coefficients that carry the most principal information in a certain dictionary or basis. Currently, applications in sparse data representation have also been leveraged to the field of pattern recognition and computer vision by the development of compressed sensing (CS) framework and sparse modeling of signals and images. These applications are essentially based on the fact that, when contrasted to the high dimensionality of natural signals, the signals in the same category usually exist in a low-dimensional subspace. Thus, for each sample, there is a sparse representation with respect to some proper basis which encodes the important information. The CS concepts guarantee that a sparse signal can be recovered from its incomplete but incoherent projections with a high probability. This enables the recovery of the sparse representation by decomposing the sample over an often over-complete dictionary constructed by or learned from the representative samples. Once the sparse representation vector is constructed, the important information can be obtained directly from the recovered vector.

Sparsity was also introduced to enhance the accuracy of prediction and interpretability of regression models by altering the model fitting process to choose only a subset of provided covariates for use in the final model rather than using all of them. Sparsity is important for many reasons as follows:

(a) It is necessary to have the smallest possible number of neurons in neural network firing at a given time when a stimulus is presented. This means that a sparse model is faster as it is possible to make use of that sparsity to construct faster specialized algorithms. For instance, in structure from motion, the obtained data matrix is sparse when applying bundle adjustments of many methods that have been proposed to take advantage of the sparseness and speedup things. Sparse models are normally very scalable but they are compact. Recently, large-scale deep learning models can easily have larger than 200k nodes. But why are they not very functional? This is because they are not sparse.

(b) Sparse models can allow more functionalities to be compressed into a neural network. Therefore, it is essential to have sparsity at the neural activity level in deep learning and exploring a way to keep more neurons inactive at any given time through neural region specialization. Neurological studies of biological brains indicate this region specialization is similar to face region's firing if a face is presented, while other regions remain mainly inactive. This means finding ways to channel the stimuli to the right regions of the deep model and prevent computations that end up resulting in no response. This can help in making deep model not only more efficient but more functional as well.

(c) In a deep neural network architecture, the main characteristic that matters is sparsity of connections; each unit should often be connected to comparatively few other units. In the human brain, estimates of the number of neurons are around 10^{10}–10^{11} neurons. However, each neuron is only connected to about 10^4 other neurons on average. In deep learning, we see this in convolutional network architectures. Each neuron receives information only from a very small patch in the lower layer.

(d) Sparsity of connections can be considered as resembling sparsity of weights. This is because it is equivalent to having a fully connected network that has zero weights in most places. However, sparsity of connections is better, because we do not spend the computational cost of explicitly multiplying each input by zero and augmenting all those zeros.

Statisticians usually learn sparse models to understand which variables are most critical. However, it is an analysis strategy, not a strategy for making better predictions. The process of learning activations that are sparse does not really seem to matter as well. Previously, researchers thought that part of the reason that the Rectified Linear Unit (ReLU) worked well was that they were sparse. However, it was shown that all that matters is that they are piece-wise linear.

11.4 Compression Approaches for Neural Networks

Our algorithm is closely related to a compression technique based on sparsity. Here, we also provide a brief overview of other two popular methods: quantization and decomposition.

11.4.1 Network Pruning

Initial research on neural network compression concentrates on removing useless connections by using weight decay. Hanson and Pratt [59] propose hyperbolic and exponential biases to the cost objective function. Optimal Brain Damage and Optimal Brain Surgeon [60] prune the networks by using second-order derivatives of the objectives. Recent research by Han et al. [61] alternates between pruning near-zero weights, which are encouraged by ℓ_1 or ℓ_2 regularization and retraining the pruned networks. More complex regularizers have also been introduced. Wen et al. [62] and Li et al. [63] place structured sparsity regularizers on the weights, while Murray and Chiang [64] place them on the hidden units. Feng and Darrell [65] propose a nonparametric prior based on the Indian buffet processes [66] on the network layers. Hu et al. [67] prune neurons were based on the analysis of their outputs on a large dataset. Anwar et al. [68] use particular sparsity patterns: channel-wise (deleting a channel from a layer or feature map), kernel-wise (deleting all connections between two feature maps in successive layers), and intra-kernel-strided (deleting connections between two features with special stride and offset). They also introduce the use of a particle filter to point out the necessity of the connections and paths over the course of training. Another line of research introduces fixed network architectures with some subsets of connections deleted. For instance, LeCun et al. [69] delete connections between the first two convolutional feature maps in an entirely uniform fashion. This approach, however, only considers a pre-defined pattern in which the same number of input feature map is assigned to each output feature map. Moreover, this method does not investigate how sparse connections influence the performance compared to dense networks.

Likewise, Ciresan et al. [70] delete random connections in their MNIST experiments. However, they do not aim to preserve the spatial convolutional density and it may be challenging to harvest the savings on existing hardware. Ioannou et al. [71]

investigate three kinds of hierarchical arrangements of filter groups for CNNs, which depend on different assumptions about co-dependency of filters within each layer. These arrangements contain columnar topologies which are inspired by AlexNet [40], tree-like topologies have been previously used by Ioannou et al. [71], and root-like topologies. Finally, [72] introduces the depth multiplier technique to scale down the number of filters in each convolutional layer by using a scalar. In this case, the depth multiplier can be considered as a channel-wise pruning method, which has been introduced in [68]. However, the depth multiplier changes the network architectures before the training phase and deletes feature maps of each layer in a uniform fashion. With the exception of [68] and the depth multiplier [72], the above previous work performs connection pruning that causes nonuniform network architectures. Therefore, these approaches need additional efforts to represent network connections and may or may not lead to a reduction in computational cost.

11.4.2   Quantization

Decreasing the degree of redundancy of the parameters of the model can be performed in the form of quantization of the network parameters. Arora et al. [73] propose to train CNNs with binary and ternary weights, accordingly. Gong et al. [74] leverage vector quantization for parameters in fully connected layers. Anwar et al. [75] quantize a network with the squared error minimization. Chen et al. [76] group network parameters randomly by using a hash function. Note that this method can be complementary to the network pruning method. For instance, Han et al. [77] merge connection pruning in Han et al. [61] with quantization and Huffman coding.

11.4.3   Decomposition

Decomposition is another method which is based on low-rank decomposition of the parameters. Decomposition approaches include truncated Singular Value Decomposition (SVD) [78], decomposition to rank-1 bases [79], Canonical Polyadic Decomposition (CPD) [80], sparse dictionary learning, asymmetric (3D) decomposition by using reconstruction loss of nonlinear responses which is integrated with a rank selection method based on Principal Component Analysis (PCA) [81], and Tucker decomposition by applying a kernel tensor reconstruction loss which is integrated with a rank selection approach based on global analytic variational Bayesian matrix factorization [82].

11.5   Regularization of Neural Network

Alex et al. [40] proposed dropout to regularize fully connected layers in the neural networks layers by randomly setting a subset of activations to zero over the course of training. Later, Wan et al. [83] introduced DropConnect, a generalization of Dropout that instead randomly zero out a subset of weights or connections. Recently, Han et al. [77] and Jin et al. [84] propose a kind of regularization where dropped connections are unfrozen and the network is retrained. This method can be thought of as an incremental training approach.

11.6   Neural Network Architectures

Network architectures and compression are closely related. The purpose of compression is to eliminate redundancy in network parameters. Therefore, the knowledge about traits that indicate the success of architecture success is advantageous. Other than the discovery that depth is an essential factor, little is known regarding such traits. Some previous research performs architecture search but without the main purpose of performing compression. Recent work introduces skip connections or shortcut to convolutional networks such as residual networks [39].

11.7   Convolutional Neural Network

CNN is a well-known used deep learning framework which was inspired by the visual cortex of animals. First, it was widely applied for object recognition but now it is used in other areas as well like object tracking [85], pose estimation [86], visual saliency detection [87], action recognition [88], and object detection [89]. CNNs are similar to traditional neural network in such a way that they consist of neurons that self-optimize through learning. Each neuron receives an input and then performs an operation (such as a product of scalar followed by a nonlinear function) on the basis of countless neural networks. From the given input image to the final output of the class score, the entire network still represents a single perceptive score function. The last layer consists of loss functions associated with the classes, and all of the regular methodologies and techniques introduced for traditional neural network still can be used. The only important difference between CNNs and traditional neural network is that CNNs are essentially used in the field of pattern recognition within images. This allows us to encode image-specific features into the architecture, making the network more suitable for image-focused tasks, while further reducing the parameters which are required to set up the model. One of the largest limitations of traditional forms of neural network is that they aim to challenge with the computational complexity needed to compute image data. Common machine learning datasets such as the MNIST database of handwritten digits are appropriate for most types of neural network, because of its relatively small image dimensionality of just 28×28. With this dataset, a single neuron in the first hidden layer will consist of 784 weights ($28 \times 28 \times 1$ where one considers that MNIST is normalized to just black and white values), which can be controlled for most types of neural networks. Here, we used a CNN for our hyperspectral band selection for face recognition. We used the VGG-19 [41] as our baseline CNN [90].

11.7.1 Convolutional Layer

The convolutional layer constructs the basic unit of a CNN where most of the computation is conducted. It is basically a set of feature maps with neurons organized in it. The weights of the convolutional layer are a set of filters or kernels which are learned during the training. These filters are convolved by the feature maps to create a separate two-dimensional activation map stacked together alongside the depth dimension, providing the output volume. Neurons that exist in the same feature map share the weight whereby decreasing the complexity of the network by keeping the number of weights low. The spatial extension of sparse connectivity between the neurons of two layers is a hyperparameter named the receptive field. The hyperparameters that manage the size of the output volume are the depth (number of filters at a layer), stride (for moving the filter), and zero-padding (to manage spatial size of the output). The CNNs are trained by backpropagation and the backward pass as well performs a convolution operation, but with spatially flipped filters. Figure 11.2 shows the basic convolution operation of a CNN.

One of the traditional versions of a CNN is "Network In Network" (NIN), introduced by Lin et al. [91], where the 1×1 convolution filter leveraged is a Multilayer Perceptron (MLP) instead of the typical linear filters and the fully connected layers are replaced by a Global Average Pooling (GAP) layer. The output structure is named the MLP-Conv layer because the micro-network contains stack of MLP-Conv layers. Dissimilar to a regular CNN, NIN can improve the abstraction ability of the latent concepts. They work very well in providing for justification that the last MLP-Conv layers of NIN were confidence maps of the classes leading to the possibility of conducting object recognition using NIN. The GAP layer within the architecture is used to reduce the parameters of our framework. Indeed, reducing the dimension of the CNN output by the GAP layer prevents our model from becoming over-parametrized and having a large dimension. Therefore, the chance of overfitting in model is potentially reduced.

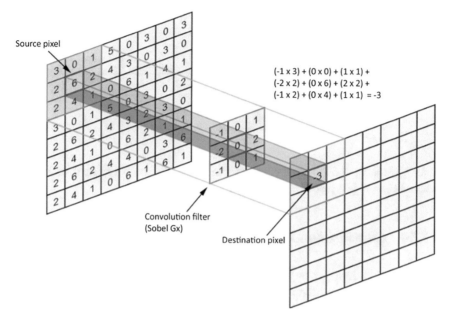

$(-1 \times 3) + (0 \times 0) + (1 \times 1) +$
$(-2 \times 2) + (0 \times 6) + (2 \times 2) +$
$(-1 \times 2) + (0 \times 4) + (1 \times 1) = -3$

Source pixel

Convolution filter
(Sobel Gx)

Destination pixel

Fig. 11.2 Convolution operation

11.7.2   Pooling Layer

Basic CNN architectures have alternating convolutional and pooling layers and the latter functions to reduce the spatial dimension of the activation maps (without loss of information) and the number of parameters in the network and therefore decreasing the overall computational complexity. This manages the problem of overfitting. Some of the common pooling operations are max pooling, average pooling, stochastic pooling [92], spectral pooling [93], spatial pyramid pooling [94], and multiscale orderless pooling [95]. The work by Springenberg et al. [96] evaluates the functionality of different components of a CNN and has found that max pooling layers can be replaced with convolutional layers with stride of two. This essentially can be applied for simple networks which have proven to beat many existing intricate architectures. We used max pooling in our deep model. Figure 11.3 shows the operation of max pooling.

11.7.3   Fully Connected Layer

Neurons in this layer are Fully Connected (FC) to all neurons in the previous layer, as in a regular neural network. High level reasoning is performed here. The neurons are not spatially arranged so there cannot be a convolution layer after a fully connected

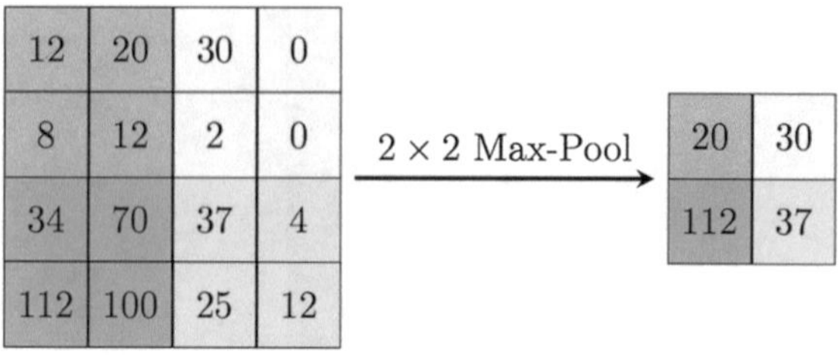

Fig. 11.3 Max pooling
operation

12	20	30	0
8	12	2	0
34	70	37	4
112	100	25	12

2×2 Max-Pool $\longrightarrow$

20	30
112	37

layer. Currently, some deep architecture have their FC layer replaced, as in NIN, where FC layer is replaced by a GAP layer.

11.7.4 Classification Layer

The last FC layer serves as the classification layer that calculates the loss or error which is a penalty for discrepancy between actual output and desired. For predicting a single class out of k mutually exclusive classes, we use Softmax loss. It is the commonly and widely used loss function. Specifically, it is multinomial logistic regression. It maps the predictions to non-negative values and is normalized to achieve probability distribution over classes. Large margin classifier, SVM, is trained by computing a Hinge loss. For regressing to real-valued labels, Euclidean loss can be calculated. We used Softmax loss to train our deep model. The Softmax loss is formulated as follows:

$$\mathcal{L}(w) = -\sum_{i=1}^{n}\sum_{j=1}^{k} y_i^{(j)} \log(p_i^{(j)}), \qquad (11.1)$$

where n is the number of training samples, $y^{(i)}$ is the one-hot encoding label for the i-th sample, and $y_i^{(j)}$ is the j-th element in the label vector y_i. The variable p_i is the probability vector and $p_i^{(j)}$ is the j-th element in the label vector p_i which indicate the probability that CNN assigns to class j. The variable w is the parameter of the CNN.

11.7.5 Activation Function: ReLU

ReLU is the regular activation function that is used in CNN models. It is a linear activation function which has thresholding at zero as shown in Eq. 11.2. It has been shown that the convergence of gradient descent is accelerated by applying ReLU. The ReLU activation function is shown in Fig. 11.4

$$f(x) = \max\{0, x\}. \qquad (11.2)$$

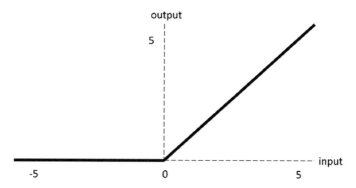

Fig. 11.4 ReLu activation function

11.7.6 VGG-19 Architecture

Our band selection algorithm can be used for any other deep architecture including ResNet [97] and AlexNet [98], and there is no restriction on choosing a specific deep model during the process of band selection in the first convolutional layer of these networks using our algorithm. We used VGG-19 network since (a) it is easy to implement in Tensorflow and it is more popular than other deep models and (b) it achieved excellent results on the ILSVRC-2014 dataset (i.e., ImageNet competition). The input to our VGG-19-based CNN is a fixed-size 224 × 224 hyperspectral image. The only pre-processing that we perform is to subtract the mean spectral value, calculated on the training set, from each pixel. The image is sent through a stack of convolutional operation, where we use filters with a very small receptive field of 3 × 3. This filter size is the smallest size that captures the notion of left and right, up and down, and center. In one of the configurations, we also can use 1 × 1 convolutional filters, which can be considered as a linear transformation of the input channels. The convolutional stride is set to 1 pixel. The spatial padding of the convolutional layer input is such that the spatial resolution is preserved after convolution, which means that the padding is 1 pixel for 3 × 3 convolutional layers. Spatial pooling is performed by five max pooling layers, which follow some of the convolutional layers. Note that not all of the convolutional layers are followed by max pooling. In VGG-19 network, max pooling is carried out on a 2 × 2 pixel window, with stride of 2. A stack of convolutional layers is followed by two FC layers as follows: the first has 4096 nodes and the second performs k nodes (i.e., one for each class). The second layer is basically the Softmax layer. The hidden layer is followed by rectification ReLU nonlinearity. The overall architecture of VGG-19 is shown in Fig. 11.5.

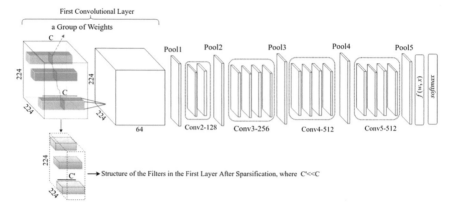

Fig. 11.5 Block diagram of hyperspectral band selection for face recognition based on structurally sparsified CNN

11.8 SSL Framework for Band Selection

We propose a regularization scheme which uses an SSL technique to specify the optimal spectral bands to obtain the best face recognition performance over all the spectral bands. Our regularization method is based on a group Lasso algorithm [58] which shrinks a set of groups of weights during the training of our CNN architecture. By using this regularization method, our algorithm recognizes face images with high accuracy, and simultaneously forces some groups of weights corresponding to redundant bands to become zero. In our framework, the goal is achieved by adding the ℓ_{12} norm of the groups as a sparsity constraint term to the total loss function of the network for face recognition. Depending on how much sparsity that we want to impose on our model, we scale the sparsity term by a hyperparameter. The hyperparameter creates a balance between face recognition loss and the sparsity constraint during the training step. It can be shown that if we enlarge the hyperparameter value, we impose more sparsity on our model, and if the hyperparameter is set to a value close to zero, we add less sparsity constraint to our model.

11.8.1 Proposed Structured Sparsity Learning for Generic Structures

In our regularization framework, the hyperspectral images are directly fed to the CNN. Therefore, the depth of each convolutional filter in the first layer of the CNN is equal to the number of spectral bands, and all the weights belonging to the same channel for all the convolutional filters in the first layer construct a group of weights. This results in the number of groups in our regularization scheme being equal to the number of spectral bands. The group Lasso regularization algorithm attempts to zero out the groups of weights that are related to the redundant bands during the training of our CNN.

11.8.2   Total Loss Function of the Framework

Suppose that w is all the weights for the convolutional filters of our CNN and w_1 denotes all the weights in the first convolutional layer of our CNN. Therefore, each weight in a given layer is identified by a 4-D tensor (i.e., $\mathbb{R}^{L \times C \times P \times Q}$, where L, C, P, and Q are the dimensions of the weight in the tensor space along the axes of the filter, channel, spatial height, and width, respectively). The proposed loss function which uses SSL to train our CNN is formulated as follows:

$$\mathcal{L}(w) = \mathcal{L}_r(w) + \lambda_g . \mathcal{R}_g(w_1), \tag{11.3}$$

neural network (it is ℓ_1 norm in our case) where $\mathcal{L}_r(.)$ is loss function used for face recognition and $\mathcal{R}_g(.)$ is SSL loss term applied on the convolutional filters in the first layer. The variable λ_g is a hyperparameter used to balance the two loss terms in (11.3). Since group Lasso can effectively zero out all of the weights in some groups [58], we leverage it in our total loss function to zero out groups of weights corresponding to the redundant spectral bands in the band selection process. Indeed the total loss function in (11.3) consists of two terms in which the first term performs face recognition, while the second term performs band selection based on the SSL. These two terms are optimized jointly during the training of the network.

11.8.3   Face Recognition Loss Function

In this section, we describe the loss function, $\mathcal{L}_r(w)$, that we have used for face recognition. We use the center loss [99] to learn a set of discriminative features for hyperspectral face images. The Softmax classifier loss which is typically used in a CNN only forces the CNN features of different classes to stay apart. However, the center loss not only does this, but also efficiently brings the CNN features of the same class close to each other. Therefore, by considering the center loss during the training of the network, not only are the inter-class feature differences enlarged, but also the intra-class feature variations are reduced. The center loss function for face recognition is formulated as follows:

$$\mathcal{L}_r(w) = -\sum_{i=1}^{n} \sum_{j=1}^{k} y_i^{(j)} log(p_i^{(j)}) + \frac{\gamma}{2} \sum_{i=1}^{n} ||f(w, x_i) - c_{yi}||_2^2, \tag{11.4}$$

where n is the number of training data, $f(w, x_i)$ is the output of the CNN, and x_i is the ith image in the training batch. The variable y_i is one-hot encoding label corresponding to the sample x_i, $y_i^{(j)}$ is the jth element in vector y_i, k is the number of classes, and p_i is the output of the Softmax applied only on the output of the CNN (i.e., $f(w, x_i)$). The variable c_{yi} indicates the center of the features corresponding to the ith class. The variable γ is a hyperparameter used to balance the two terms in the center loss.

11.8.4 Band Selection via Group Lasso

Assume that each hyperspectral image has C number of spectral bands. Since, in our regularization scheme, hyperspectral images are directly fed to the CNN, the depth of each convolutional filter in the first layer of our CNN is equal to C. Here, we adopt a group Lasso to regularize the depth of each convolutional filter in the first layer of our CNN. We use the group Lasso because it can effectively zero out all of the weights in some groups [58]. Therefore, the group Lasso can zero out groups of weights which correspond to redundant spectral bands. In the setup of our group Lasso regularization, weights belonging to the same channel for all the convolutional filters in the first layer form a group (red squares in Fig. 11.5) which can be removed during the training step by using $\mathcal{R}_g(w_1)$ function as defined in (11.3). Therefore, there are C number of groups in our regularization framework. The group Lasso regularization on the parameters of w_1 is an ℓ_{12} norm which can be expressed as follows:

$$\mathcal{R}_g(w_1) = \sum_{g=1}^{C} ||w_1^{(g)}||_2, \qquad (11.5)$$

where $w_1^{(g)}$ is the subset of weights (i.e., a group of weights) from w_1 and C is the total number of groups. Generally, different groups may overlap in the group Lasso regularization. However, this does not happen in our case. The notation $||.||_2$ represents an ℓ_2 norm on the parameters of the group $w_1^{(g)}$. Therefore, the group Lasso regularization as a sparsity constraint for band selection can be expressed as follows:

$$\mathcal{R}_g(w_1) = \sum_{c=1}^{C} \sqrt{\sum_{l=1}^{L} \sum_{p=1}^{P} \sum_{q=1}^{Q} (w_1(l, c, p, q))^2}, \qquad (11.6)$$

where $w_1(l, c, p, q)$ denotes a weight located in lth convolutional filter, cth channel, and (p, q) spatial position. In this formulation, all of the weights $w_1(:, c, :, :)$ (i.e., the weights which have the same index c) belong to the same group $w_1^{(c)}$. Therefore, $\mathcal{R}_g(w_1)$ is an ℓ_{12} regularization term in which ℓ_1 is performed on the ℓ_2 norm of each group.

11.8.5   Sparsification Procedure

The proposed framework automatically selects the optimal bands from all spectral bands for face recognition during the training phase. For clarification, we can assume that in a typical RGB image, we have three bands and the depth of each filter in the first convolutional layer is three. However, here, there are C spectral bands and as a consequence, the depth of each filter in the first layer is C. As shown in Fig. 11.5, hyperspectral images are fed into the CNN directly. The group Lasso efficiently removes redundant weight groups (associated with different spectral bands) to improve the recognition accuracy during the training phase. In the beginning of the training, the depth of the filters is C, and once we start to sparsify the depth of the convolutional filters, the depth of each filter will be reduced (i.e., $C' \ll C$).

It should be noted that the dashed cube in Fig. 11.5 is not part of our CNN architecture. This is the structure of the convolutional filters in the first layer after several epochs training the network using the network loss function defined in (11.3).

11.9   Experimental Setup and Results

11.9.1   CNN Architecture

We use the VGG-19 [90] architecture as shown in Fig. 11.5 with the same filter size, pooling operation, and convolutional layers. However, the depth of the filters in the first convolutional layer of our CNN is set to the number of the hyperspectral bands. The network uses filters with a receptive field of 3×3. We set the convolution stride to 1 pixel. To preserve spatial resolution after convolution, the spatial padding of the convolutional layer is fixed to 1 pixel for all the 3×3 convolutional layers. In this framework, each hidden layer is followed by a ReLU activation function. We apply batch normalization (i.e., shifting inputs to zero mean and unit variance) after each convolutional and fully connected layer and before performing the ReLU activation function. Batch normalization potentially helps to achieve faster learning as well as higher overall accuracy. Furthermore, batch normalization allows us to use a higher learning rate, which potentially provides another boost in speed (Fig. 11.6).

11.9.2   Initializing Parameters of the Network

In this section, we describe how we initialize the parameters of our network for the training phase. Thousands of images are needed to train such a deep model. For this reason, we initialize the parameters of our network by a VGG-19 network pre-trained on the ImageNet dataset and then we fine-tune it as a classifier by using the CASIA-Web Face dataset [100]. CASIA-Web Face contains 10,575 subjects and

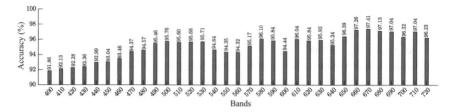

Fig. 11.6 Face recognition accuracy of each individual band on the UWA-HSFD

494,414 images. To the best of our knowledge, this is the largest publicly available face image dataset, second only to the private Facebook dataset. In our case, however, since the depth of the filters in the first layer is the number of spectral bands, we initialize these filters by duplicating the filters of the pre-trained VGG-19 network in the first convolutional layer. For example, assume that the depth of the filters in the first layer is $3n$ (we have $3n$ spectral bands). Then, in such a case, we duplicate filters of the first layer n times as an initialization point for training the network.

11.9.3 Training the Network

We use the Adam optimizer [101] with the default hyperparameter values ($\epsilon = 10^{-3}$, $\beta_1 = 0.9$, $\beta_2 = 0.999$) to minimize the total loss function of our network defined in (11.3). The Adam optimizer is a robust and well-adapted optimizer that can be applied to a variety of non-convex optimization problems in the field of deep neural networks. We set the learning rate to 0.001 to minimize loss function (11.3) during the training process. The hyperparameter λ_g is selected by cross-validation in our experiments. We ran the CNN model through 100 epochs, although the model nearly converged after 30 epochs. The batch size in all experiments is fixed to 32. We implemented our algorithm in TensorFlow, and all experiments are conducted on two GeForce GTX TITAN X 12GB GPUs (Fig. 11.7).

11.9.4 Hyperspectral Face Datasets

We performed our experiments on three standard and publicly available hyperspectral face image datasets including CMU [102], HK PolyU [18], and UWA [103]. Descriptions of these datasets are as follows:

CMU-HSFD: The face cubes in this dataset have been obtained by a spectropolarimetric camera. The spectral wavelength range during the image acquisition is from 450 to 1100 nm with a step size of 10 nm. The images of this dataset have been collected in multiple sessions from 48 subjects.

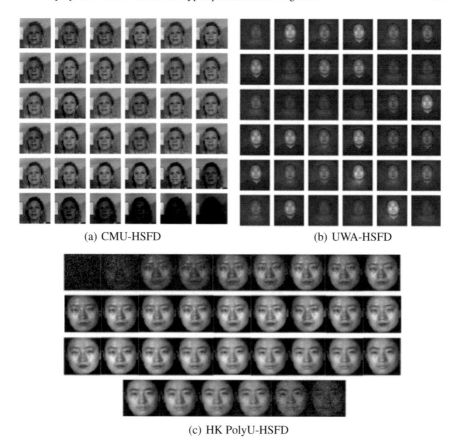

(a) CMU-HSFD (b) UWA-HSFD

(c) HK PolyU-HSFD

Fig. 11.7 Samples of hyperspectral images

HK PolyU-HSFD: The face images in this dataset have been obtained by using an indoor system made up of CRI's VariSpec Liquid Crystal Tunable Filter with a halogen light source. The spectral wavelength range during the image acquisition is from 400 to 720 nm with a step size of 10 nm, which creates 33 bands in total. There are 300 hyperspectral face cubes captured from 24 subjects. For each subject, the hyperspectral face cubes have been collected from multiple sessions in an average span of 5 months.

UWA-HSFD: Similar to the HK PolyU dataset, the face images in this dataset have been acquired by using an indoor imaging system made up of CRI's VariSpec Liquid Crystal Tunable Filter integrated with a photon focus camera. However, the camera exposure time is set and altered based on the signal-to-noise ratio for different bands. Therefore, this dataset has the advantage of having lower noise levels in comparison to other two datasets. There are 70 subjects in this dataset and the spectral wavelength range during the image acquisition is from 400 to 720 nm with a step size of 10 nm.

Table 11.1 A summary of hyperspectral face datasets

Dataset	Subjects	HS cubes	Bands	Spectral range (nm)
CMU	48	147	65	450–1090
HK PolyU	24	113	33	400–720
UWA	70	120	33	400–720

Table 11.1 indicates a summary of the datasets that we have used in our experiments.

11.9.5 Parameter Sensitivity

We explore the influence of the hyperparameter λ_g defined in (11.3) on face recognition performance. Figure 11.8 shows the CMC curves for CMU, HK PolyU, and UWA HSFD with different values of $\{10, 1, 10^{-1}, 10^{-2}, 10^{-3}\}$, respectively. We can see that our network total loss defined in (11.3) is not significantly sensitive to λ_g if we set these parameters within $[10^{-3}, 10]$ interval.

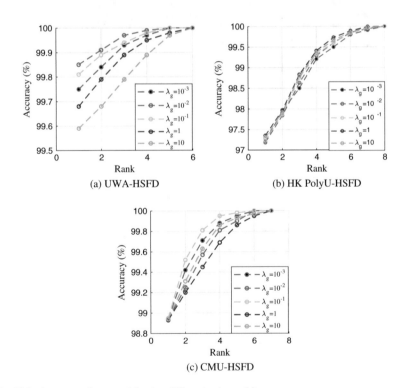

Fig. 11.8 Accuracy of our model using different values of λ_g

11.9.6   Updating Centers in Center Loss

We used the strategy presented in [99] to update the center of each class (i.e., c_{yi} in (11.4)). In this strategy, first, instead of updating the centers with respect to the entire training set, we update the centers based on a mini-batch such that, in each iteration, the centers are obtained by averaging the features of the corresponding classes. Second, to prevent the large perturbations made by a few mislabeled samples, we scale it by a small number of 0.001 to control the learning rate of the centers, as suggested in [99].

11.9.7   Band Selection

RGB cameras produce three bands over the whole visible spectrum. However, hyper-spectral imaging camera divides this range into many narrow bands (e.g., 10 nm). Both of these types of imaging cameras are the extreme cases of spectral resolution. Even though RGB cameras divide the visible spectrum into three bands, they are wide and the center of the wavelengths in these bands is selected to approximate the human visual system instead of maximizing the performance of the face recognition task.

In this work, we conducted experiments to find the optimal number of bands and their center wavelengths that maximize face recognition accuracy. Our method adopts the SSL technique during the training of our CNN to automatically select spectral bands which provide the maximum recognition accuracy. The results indicate that maximum discrimination power can be achieved by using a small number of bands rather than all the spectral bands but more than three bands in RGB for the CMU dataset. Specifically, the results demonstrate that the most discriminative spectral wavelengths for face recognition are obtained by a subset of red and green wavelengths (Figs. 11.9 and 11.10).

In addition to the improvement in face recognition accuracy, other advantages of the band selection include a reduction in computational complexity, a reduction in the cost and time during image acquisition for hyperspectral cameras, and reduction in redundancy of the data. This is because one can capture the bands which are more discriminative for a face recognition task instead of capturing images from the entire visible spectrum. Table 11.2 indicates the optimal spectral bands from all of the bands selected by our method. Our algorithm selects four bands including {750, 810 , 920, 990} for the CMU dataset, three bands including {580, 640, 700} for PolyU, and three bands including {570, 650, 680} for the UWA dataset. The results show that SSL selects the optimal bands from the green and red spectra and ignores bands within the blue spectrum. Figures 11.11 and 11.12 demonstrate some of the face images from the bands which are selected by our algorithm. The experimental results indicate that the blue wavelength bands are discarded earlier during the sparsification procedure because they are less discriminative and they are less useful compared to the green, red, and IR ranges for the task of face recognition.

Table 11.2 Center wavelengths of the selected bands for different hyperspectral datasets

Dataset	Bands (nm)
CMU	{750, 810, 920, 990}
HK PolyU	{580, 640, 700}
UWA	{570, 650, 680, 710}

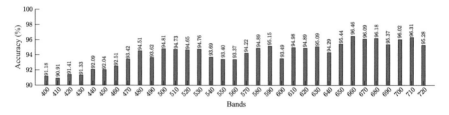

Fig. 11.9 Face recognition accuracy of each individual band on the HK PolyU-HSFD

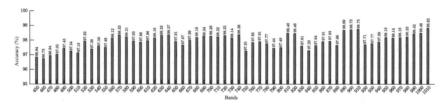

Fig. 11.10 Face recognition accuracy of each individual band on the CMU-HSFD

The group sparsity technique used in our algorithm automatically selects the optimal bands by combining the informative bands so that the selected bands have the most discriminative information for the task of face recognition.

11.9.8 Effectiveness of SSL

Figures 11.6, 11.9, and 11.10 indicate the face recognition accuracy for each individual band on the UWA, CMU, and PolyU datasets, respectively. In Table 11.3, we reported the maximum and minimum accuracy obtained from each spectral band when we use each band individually during the training. We also reported the case where we use all bands without using the SSL technique for face recognition. Finally, we provided the results of our framework in the case where we use SSL during the training. The results show that using SSL not only removes the redundant spectral bands for the face recognition task, but it can also improve the recognition performance in comparison to the case where all the spectral bands are used for face recognition. These improvements are around 0.59 %, 0.36%, and 0.32% on the CMU, HK PolyU, and UWA datasets, respectively.

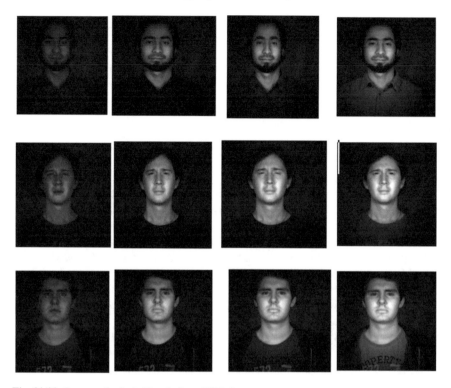

Fig. 11.11   Images of selected bands from UWA dataset

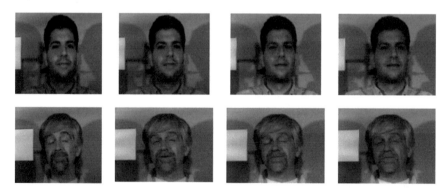

Fig. 11.12   Images of selected bands from CMU dataset

Table 11.3 Accuracy (%) of our band selection algorithm in different cases

Dataset	Min	Max	All the bands	SSL
CMU	96.73	98.82	99.34	99.93
HK PolyU	90.91	96.46	99.52	99.88
UWA	91.86	97.41	99.63	99.95

11.9.9 Comparison

We compared our proposed algorithm with several existing face recognition techniques that are extended to the hyperspectral face recognition methods. We categorize these methods into four groups including four existing hyperspectral face recognition methods [5, 17–19], eight image-set classification methods [3, 104–109] three RGB/grayscale face recognition algorithms [110–112], and one existing CNN-based model for hyperspectral face recognition [55]. For a fair comparison, we have been consistent with other compared methods in experimental setup including the number of images in the gallery and probe data. Specifically, for the PolyU-HSFD dataset, we use the first 24 subjects which contain 113 hyperspectral image cubes. For each subject, we randomly select two cubes for the gallery and we use the remaining 63 cubes for probes. For the CMU-HSFD dataset, the dataset includes 48 subjects, each subject has 4 to 20 cubes obtained from different sessions and different lighting conditions. We use only the cubes which are obtained in a condition that all lights are turned on. Thus, there are 147 hyperspectral cubes of 48 subjects such that each subject has 1 to 5 cubes. We construct the gallery randomly by selecting one cube per subject, and we use the remaining 99 cubes for probes. For the UWA-HSFD dataset, we randomly select one cube for each of 70 subjects to construct a gallery and we use the remaining 50 cubes for probes.

Table 11.4 indicates the average accuracy of the compared methods when all bands are available for different algorithms during the face recognition. The Deep-Baseline is the case where we use all the bands in our CNN framework for face recognition. Therefore, in this case, we turn off the SSL regularization term in (11.3), while Deep-SSL is the case that we perform face recognition using the SSL regularization term. We reported the face recognition accuracy of Deep-SSL in Table 11.4 to compare it with the best recognition results reported in the literature. The results show that Deep-SSL outperforms the state-of-the-art methods including PLS* and S-CNN+SVM* methods. The symbol * represents the case that the algorithms perform face recognition when they use their optimal hyperspectral bands.

Please email us[1] if you want to receive the data and the source code of our proposed algorithm presented in this chapter.

---

[1] ft0009@mix.wvu.edu.

Table 11.4 Comparing accuracy (%) of different band selection methods with our proposed method

Methods	CMU	PolyU	UWA
Hyperspectral			
Spectral angle [17]	38.1	25.4	37.9
Spectral eigenface [5]	84.5	70.3	91.5
2D PCA [18]	72.1	71.1	83.8
3D Gabor wavelets [19]	91.6	90.1	91.5
Image-set classification			
DCC [104]	87.5	76.0	91.5
MMD [105]	90.0	83.8	82.8
MDA [106]	90.6	87.9	91.0
AHISD [107]	90.6	89.9	92.5
SHIDS [107]	91.1	90.3	92.5
SANP [108]	90.9	90.5	92.5
CDL[109]	92.7	89.3	93.1
PLS [3]	99.1	95.2	98.2
PLS* [3]	99.1	95.2	98.2
Grayscale and RGB			
SRC [110]	91.0	85.6	96.2
CRC [111]	93.8	86.1	96.2
LCVBP+RLDA [112]	87.3	80.3	97.0
CNN-based models			
S-CNN [55]	98.8	97.2	–
S-CNN+SVM [55]	99.2	99.3	–
S-CNN+SVM * [55]	99.4	99.6	–
Deep-baseline	99.3	99.5	99.6
Deep-SSL	**99.9**	**99.8**	**99.9**

11.10 Conclusion

In this work, we proposed a CNN-based model which uses an SSL technique to select the optimal spectral bands to obtain the best face recognition performance from all the spectral bands. In this method, convolutional filters in the first layer of our CNN are regularized by using a group Lasso algorithm to remove the redundant bands during the training. Experimental results indicate that our method automatically selects the optimal bands to obtain the best face recognition performance over that achieved using conventional broadband (RGB) face images. Moreover, the results indicate that our model outperforms existing methods which also perform band selection for face recognition.

References

1. Allen DW (2016) An overview of spectral imaging of human skin toward face recognition. Face recognition across the imaging spectrum. Springer, pp 1–19
2. Pan Z, Healey G, Prasad M, Tromberg B (2003) Face recognition in hyperspectral images. IEEE Trans Pattern Anal Mach Intell 25(12):1552–1560
3. Uzair M, Mahmood A, Mian A (2015) Hyperspectral face recognition with spatiospectral information fusion and PLS regression. IEEE Trans Image Process 24(3):1127–1137
4. Kruse FA et al (2002) Comparison of AVIRIS and hyperion for hyperspectral mineral mapping. In: 11th JPL airborne geoscience workshop, vol 4
5. Pan Z, Healey G, Tromberg B (2009) Comparison of spectral-only and spectral/spatial face recognition for personal identity verification. EURASIP J Adv Signal Process 2009:8
6. Ryer DM, Bihl TJ, Bauer KW, Rogers SK (2012) Quest hierarchy for hyperspectral face recognition. Adv Artif Intell 2012:1
7. Gross R, Shi J, Cohn JF (2001) Quo vadis face recognition? Carnegie Mellon University, The Robotics Institute
8. Gross R, Matthews I, Baker S (2004) Appearance-based face recognition and light-fields. IEEE Trans Pattern Anal Mach Intell 26(4):449–465
9. Martínez AM (2002) Recognizing imprecisely localized, partially occluded, and expression variant faces from a single sample per class. IEEE Trans Pattern Anal Mach Intell 6:748–763
10. Wilder J, Phillips PJ, Jiang C, Wiener S (1996) Comparison of visible and infra-red imagery for face recognition. In: Proceedings of the second international conference on automatic face and gesture recognition, pp 182–187. IEEE
11. Blanz V, Romdhani S, Vetter T (2002) Face identification across different poses and illuminations with a 3D morphable model. In: Proceedings of fifth IEEE international conference on automatic face gesture recognition, pp 202–207. IEEE
12. Anderson RR, Parrish JA (1981) The optics of human skin. J Investig Derm 77(1):13–19
13. Edwards EA, Duntley SQ (1939) The pigments and color of living human skin. Am J Anat 65(1):1–33
14. Tsumura N, Haneishi H, Miyake Y (1999) Independent-component analysis of skin color image. JOSA A 16(9):2169–2176
15. Angelopoulo E, Molana R, Daniilidis K (2001) Multispectral skin color modeling. In: Proceedings of the 2001 IEEE computer society conference on computer vision and pattern recognition, CVPR 2001, vol 2, pp II–II. IEEE
16. Hagen NA, Kudenov MW (2013) Review of snapshot spectral imaging technologies. Opt Eng 52(9):090901
17. Robila SA (2008) Toward hyperspectral face recognition. In: Proceedings of image processing: algorithms and systems VI, vol 6812, p 68120X. International Society for Optics and Photonics
18. Di W, Zhang L, Zhang D, Pan Q (2010) Studies on hyperspectral face recognition in visible spectrum with feature band selection. IEEE Trans Syst Man Cybern Part A Syst Hum 40(6):1354–1361
19. Shen L, Zheng S (2012) Hyperspectral face recognition using 3D Gabor wavelets. In: 2012 21st international conference on pattern recognition (ICPR), pp 1574–1577. IEEE
20. Bajcsy P, Groves P (2004) Methodology for hyperspectral band selection. Photogramm Eng Remote Sens 70(7):793–802
21. Chang C-I, Du Q, Sun T-L, Althouse ML (1999) A joint band prioritization and band-decorrelation approach to band selection for hyperspectral image classification. IEEE Trans Geosci Remote Sens 37(6):2631–2641
22. Melgani F, Bruzzone L (2004) Classification of hyperspectral remote sensing images with support vector machines. IEEE Trans Geosci Remote Sens 42(8):1778–1790
23. Keshava N (2001) Best bands selection for detection in hyperspectral processing. In: Proceedings of 2001 IEEE international conference on acoustics, speech, and signal processing, ICASSP'01, 2001, vol 5, pp 3149–3152. IEEE

24. Du Q (2003) Band selection and its impact on target detection and classification in hyperspectral image analysis. In: 2003 IEEE workshop on advances in techniques for analysis of remotely sensed data, pp 374–377. IEEE
25. Kaewpijit S, Le Moigne J, El-Ghazawi T (2003) Automatic reduction of hyperspectral imagery using wavelet spectral analysis. IEEE Trans Geosci Remote Sens 41(4):863–871
26. Price JC (1997) Spectral band selection for visible-near infrared remote sensing: spectral-spatial resolution tradeoffs. IEEE Trans Geosci Remote Sens 35(5):1277–1285
27. Steiner H, Kolb A, Jung N (2016) Reliable face anti-spoofing using multispectral SWIR imaging. In: 2016 international conference on biometrics (ICB), pp 1–8. IEEE
28. Bouchech HJ, Foufou S, Abidi M (2014) Dynamic best spectral bands selection for face recognition. In: 2014 48th annual conference on information sciences and systems (CISS), pp 1–6. IEEE
29. Taherkhani F, Jamzad M (2017) Restoring highly corrupted images by impulse noise using radial basis functions interpolation. IET Image Process 12(1):20–30
30. Chen Y, Nasrabadi NM, Tran TD (2013) Hyperspectral image classification via kernel sparse representation. IEEE Trans Geosci Remote Sens 51(1):217–231
31. Talreja V, Taherkhani F, Valenti MC, Nasrabadi NM (2019) Attribute-guided coupled GAN for cross-resolution face recognition. arXiv:1908.01790
32. Taherkhani F, Talreja V, Kazemi H, Nasrabadi N (2018) Facial attribute guided deep cross-modal hashing for face image retrieval. In: 2018 international conference of the biometrics special interest group (BIOSIG), pp 1–6. IEEE
33. Talreja V, Taherkhani F, Valenti MC, Nasrabadi NM (2018) Using deep cross modal hashing and error correcting codes for improving the efficiency of attribute guided facial image retrieval. In: 2018 IEEE global conference on signal and information processing (GlobalSIP), pp 564–568. IEEE
34. Taherkhani F, Kazemi H, Nasrabadi NM (2019) Matrix completion for graph-based deep semi-supervised learning. In: Thirty-third AAAI conference on artificial intelligence
35. Kazemi H, Soleymani S, Taherkhani F, Iranmanesh S, Nasrabadi N (2018) Unsupervised image-to-image translation using domain-specific variational information bound. Advances in neural information processing systems, pp 10369–10379
36. Kazemi H, Taherkhani F, Nasrabadi NM (2018) Unsupervised facial geometry learning for sketch to photo synthesis. In: 2018 international conference of the biometrics special interest group (BIOSIG), pp 1–5. IEEE
37. Talreja V, Valenti MC, Nasrabadi NM (2017) Multibiometric secure system based on deep learning. In: 2017 IEEE global conference on signal and information processing (globalSIP), pp 298–302. IEEE
38. Talreja V, Ferrett T, Valenti MC, Ross A (2018) Biometrics-as-a-service: a framework to promote innovative biometric recognition in the cloud. In: 2018 IEEE international conference on consumer electronics (ICCE), pp 1–6. IEEE
39. He K, Zhang X, Ren S, Sun J (2016) Deep residual learning for image recognition. In: Proceedings of IEEE conference on computer vision and pattern recognition, pp 770–778
40. Krizhevsky A, Sutskever I, Hinton GE (2012) Imagenet classification with deep convolutional neural networks. In: Proceedings of advances in neural information processing systems, pp 1097–1105, Dec 2012
41. Simonyan K, Zisserman A (2014) Very deep convolutional networks for large-scale image recognition. CoRR abs/1409.1556
42. Erhan D, Szegedy C, Toshev A, Anguelov D (2014) Scalable object detection using deep neural networks. In: Proceedings of IEEE conference on computer vision and pattern recognition, June 2014
43. Ren S, He K, Girshick R, Sun J (2015) Faster R-CNN: towards real-time object detection with region proposal networks. In: Proceedings of advances in neural information processing systems, pp 91–99, Dec 2015
44. Soleymani S, Dabouei A, Dawson J, Nasrabadi NM (2019) Defending against adversarial iris examples using wavelet decomposition. arXiv:1908.03176

45. Soleymani S, Dabouei A, Iranmanesh SM, Kazemi H, Dawson J, Nasrabadi NM (2018) Prosodic-enhanced siamese convolutional neural networks for cross-device text-independent speaker verification. In: 2018 IEEE 9th international conference on biometrics theory, applications and systems (BTAS), pp 1–7. IEEE

46. Soleymani S, Dabouei A, Dawson J, Nasrabadi NM (2019) Adversarial examples to fool iris recognition systems. arXiv:1906.09300

47. Taherkhani F, Nasrabadi NM, Dawson J (2018) A deep face identification network enhanced by facial attributes prediction. In: Proceedings of the IEEE conference on computer vision and pattern recognition workshops, pp 553–560

48. Schroff F, Kalenichenko D, Philbin J (2015) Facenet: a unified embedding for face recognition and clustering. In: Proceedings of the IEEE conference on computer vision and pattern recognition, pp 815–823

49. Taigman Y, Yang M, Ranzato M, Wolf L (2015) Web-scale training for face identification. In: Proceedings of the IEEE conference on computer vision and pattern recognition, pp 2746–2754

50. Sun Y, Chen Y, Wang X, Tang X (2014) Deep learning face representation by joint identification-verification. Advances in neural information processing systems, pp 1988–1996

51. Silver D, Huang A, Maddison CJ, Guez A, Sifre L, Van Den Driessche G, Schrittwieser J, Antonoglou I, Panneershelvam V, Lanctot M et al (2016) Mastering the game of go with deep neural networks and tree search. Nature 529:484

52. Zhong Z, Li J, Ma L, Jiang H, Zhao H (2017) Deep residual networks for hyperspectral image classification. In: 2017 IEEE international geoscience and remote sensing symposium (IGARSS), pp 1824–1827. IEEE

53. Zhan Y, Hu D, Xing H, Yu X (2017) Hyperspectral band selection based on deep convolutional neural network and distance density. IEEE Geosci Remote Sens Lett 14(12):2365–2369

54. Zhan Y, Tian H, Liu W, Yang Z, Wu K, Wang G, Chen P, Yu X (2017) A new hyperspectral band selection approach based on convolutional neural network. In: 2017 IEEE international geoscience and remote sensing symposium (IGARSS), pp 3660–3663. IEEE

55. Sharma V, Diba A, Tuytelaars T, Van Gool L (2016) Hyperspectral CNN for image classification & band selection, with application to face recognition

56. Li N, Wang C, Zhao H, Gong X, Wang D (2018) A novel deep convolutional neural network for spectral-spatial classification of hyperspectral data. Int Arch Photogramm Remote Sens Spat Inf Sci 42(3)

57. Guo Z, Zhang D, Zhang L, Liu W (2012) Feature band selection for online multispectral palmprint recognition. IEEE Trans Inf Forensics Secur 7(3):1094–1099

58. Yuan M, Lin Y (2006) Model selection and estimation in regression with grouped variables. J R Stat Soc Ser B (Stat Methodol) 68(1):49–67

59. Hanson SJ, Pratt LY (1989) Comparing biases for minimal network construction with back-propagation. Advances in neural information processing systems, pp 177–185

60. Hassibi B, Stork DG (1993) Second order derivatives for network pruning: optimal brain surgeon. Advances in neural information processing systems, pp 164–171

61. Han S, Pool J, Tran J, Dally W (2015) Learning both weights and connections for efficient neural network. Advances in neural information processing systems, pp 1135–1143

62. Wen W, Wu C, Wang Y, Chen Y, Li H (2016) Learning structured sparsity in deep neural networks. Advances in neural information processing systems, pp 2074–2082

63. Li H, Kadav A, Durdanovic I, Samet I, Graf HP (2016) Pruning filters for efficient convnets. arXiv:1608.08710

64. Murray K, Chiang D (2015) Auto-sizing neural networks: with applications to n-gram language models. arXiv:1508.05051

65. Feng J, Darrell T (2015) Learning the structure of deep convolutional networks. In: Proceedings of the IEEE international conference on computer vision, pp 2749–2757

66. Griffiths TL, Ghahramani Z (2011) The Indian buffet process: an introduction and review. J Mach Learn Res 12(Apr):1185–1224

67. Hu H, Peng R, Tai Y-W, Tang C-K (2016) Network trimming: a data-driven neuron pruning approach towards efficient deep architectures. arXiv:1607.03250
68. Anwar S, Hwang K, Sung W (2017) Structured pruning of deep convolutional neural networks. ACM J Emerg Technol Comput Syst (JETC) 13(3):32
69. LeCun Y, Denker JS, Solla SA (1990) Optimal brain damage. Advances in neural information processing systems, pp 598–605
70. Cireşan DC, Meier U, Masci J, Gambardella LM, Schmidhuber J (2011) High-performance neural networks for visual object classification. arXiv:1102.0183
71. Ioannou Y, Robertson D, Cipolla D, Criminisi A (2017) Deep roots: improving CNN efficiency with hierarchical filter groups. In: Proceedings of the IEEE conference on computer vision and pattern recognition, pp 1231–1240
72. Howard AG, Zhu M, Chen B, Kalenichenko D, Wang W, Weyand T, Andreetto M, Adam H (2017) Mobilenets: efficient convolutional neural networks for mobile vision applications. arXiv:1704.04861
73. Arora S, Bhaskara A, Ge R, Ma T (2014) Provable bounds for learning some deep representations. In: International conference on machine learning, pp 584–592
74. Gong Y, Liu L, Yang M, Bourdev L (2014) Compressing deep convolutional networks using vector quantization. arXiv:1412.6115
75. Anwar S, Hwang K, Sung W (2015) Fixed point optimization of deep convolutional neural networks for object recognition. In: 2015 IEEE international conference on acoustics, speech and signal processing (ICASSP), pp 1131–1135. IEEE
76. Chen W, Wilson J, Tyree S, Weinberger K, Chen Y (2015) Compressing neural networks with the hashing trick. In: International conference on machine learning, pp 2285–2294
77. Han S, Pool J, Narang S, Mao H, Tang S, Elsen E, Catanzaro B, Tran J, Dally WJ (2016) DSD: regularizing deep neural networks with dense-sparse-dense training flow. 3(6). arXiv:1607.04381
78. Denton EL, Zaremba W, Bruna J, LeCun Y, Fergus R (2014) Exploiting linear structure within convolutional networks for efficient evaluation. Advances in neural information processing systems, pp 1269–1277
79. Jaderberg M, Vedaldi A, Zisserman A (2014) Speeding up convolutional neural networks with low rank expansions. arXiv:1405.3866
80. Lebedev V, Ganin Y, Rakhuba M, Oseledets I, Lempitsky V (2014) Speeding-up convolutional neural networks using fine-tuned CP-decomposition. arXiv:1412.6553
81. Zhang X, Zou J, He K, Sun J (2016) Accelerating very deep convolutional networks for classification and detection. IEEE Trans Pattern Anal Mach Intell 38(10):1943–1955
82. Kim Y-D, Park E, Yoo S, Choi T, Yang L, Shin D (2015) Compression of deep convolutional neural networks for fast and low power mobile applications. arXiv:1511.06530
83. Wan L, Zeiler M, Zhang S, Le Cun Y, Fergus R (2013) Regularization of neural networks using dropconnect. In: International conference on machine learning, pp 1058–1066
84. Jin X, Yuan X, Feng J, Yan S (2016) Training skinny deep neural networks with iterative hard thresholding methods. arXiv:1607.05423
85. Fan J, Xu W, Wu Y, Gong Y (2010) Human tracking using convolutional neural networks. IEEE Trans Neural Netw 21(10):1610–1623
86. Toshev A, Szegedy C (2014) Deeppose: human pose estimation via deep neural networks. In: Proceedings of the IEEE conference on computer vision and pattern recognition, pp 1653–1660
87. Zhao R, Ouyang W, Li H, Wang X (2015) Saliency detection by multi-context deep learning. In: Proceedings of the IEEE conference on computer vision and pattern recognition, pp 1265–1274
88. Donahue J, Jia Y, Vinyals O, Hoffman J, Zhang N, Tzeng E, Darrell T (2014) Decaf: a deep convolutional activation feature for generic visual recognition. In: International conference on machine learning, pp 647–655
89. Zhao Z-Q, Zheng P, Xu S-t, Wu X (2019) Object detection with deep learning: a review. IEEE Trans Neural Netw Learn Syst

90. Simonyan K, Zisserman A (2014) Very deep convolutional networks for large-scale image recognition. arXiv:1409.1556

91. Zeiler MD, Fergus R (2014) Visualizing and understanding convolutional networks. In: European conference on computer vision, pp 818–833. Springer

92. Zeiler MD, Fergus R (2013) Stochastic pooling for regularization of deep convolutional neural networks. arXiv:1301.3557

93. Rippel O, Snoek J, Adams RP (2015) Spectral representations for convolutional neural networks. Advances in neural information processing systems, pp 2449–2457

94. Nguyen A, Yosinski J, Clune J (2015) Deep neural networks are easily fooled: high confidence predictions for unrecognizable images. In: Proceedings of the IEEE conference on computer vision and pattern recognition, pp 427–436

95. Gong Y, Wang L, Guo R, Lazebnik S (2014) Multi-scale orderless pooling of deep convolutional activation features. In: European conference on computer vision, pp 392–407. Springer

96. Springenberg JT, Dosovitskiy A, Brox T, Riedmiller M (2014) Striving for simplicity: the all convolutional net. arXiv:1412.6806

97. Zagoruyko S, Komodakis N (2016) Wide residual networks. arXiv:1605.07146

98. Krizhevsky A, Sutskever I, Hinton GE (2012) Imagenet classification with deep convolutional neural networks. Advances in neural information processing systems, pp 1097–1105

99. Wen Y, Zhang K, Li Z, Qiao Y (2016) A discriminative feature learning approach for deep face recognition. In: European conference on computer vision, pp 499–515. Springer

100. Yi D, Lei Z, Liao S, Li SZ (2014) Learning face representation from scratch. arXiv:1411.7923

101. Kingma DP, Ba J (2014) Adam: a method for stochastic optimization. arXiv:1412.6980

102. Denes LJ, Metes P, Liu Y (2002) Hyperspectral face database. Carnegie Mellon University, The Robotics Institute

103. Uzair M, Mahmood A, Mian AS (2013) Hyperspectral face recognition using 3D-DCT and partial least squares. In: Proceedings of *BMVC*

104. Kim T-K, Kittler J, Cipolla R (2007) Discriminative learning and recognition of image set classes using canonical correlations. IEEE Trans Pattern Anal Mach Intell 29(6):1005–1018

105. Wang R, Shan S, Chen X, Gao W (2008) Manifold-manifold distance with application to face recognition based on image set. In: IEEE conference on computer vision and pattern recognition, 2008, CVPR 2008, pp 1–8. IEEE

106. Wang R, Chen X (2009) Manifold discriminant analysis. In: IEEE conference on computer vision and pattern recognition, 2009, CVPR 2009, pp 429–436. IEEE

107. Cevikalp H, Triggs B (2010) Face recognition based on image sets. In: 2010 IEEE conference on computer vision and pattern recognition (CVPR), pp 2567–2573. IEEE

108. Hu Y, Mian AS, Owens R (2012) Face recognition using sparse approximated nearest points between image sets. IEEE Trans Pattern Anal Mach Intell 34(10):1992–2004

109. Wang R, Guo H, Davis LS, Dai LSQ (2012) Covariance discriminative learning: a natural and efficient approach to image set classification. In: 2012 IEEE conference on computer vision and pattern recognition (CVPR), pp 2496–2503. IEEE

110. Wright J, Yang AY, Ganesh A, Sastry SS, Ma Y (2009) Robust face recognition via sparse representation. IEEE Trans Pattern Anal Mach Intell 31(2):210–227

111. Zhang L, Yang M, Feng X (2011) Sparse representation or collaborative representation: which helps face recognition? In: 2011 IEEE international conference on computer vision (ICCV), pp 471–478. IEEE

112. Lee SH, Choi JY, Ro YM, Plataniotis KN (2012) Local color vector binary patterns from multichannel face images for face recognition. IEEE Trans Image Process 21(4):2347–2353

Chapter 12
Detection of Large-Scale and Anomalous Changes

Amanda Ziemann and Stefania Matteoli

Abstract We survey algorithms and methodologies for detecting and delineating changes of interest in remote sensing imagery. We consider both broad salient changes and rare anomalous changes, and we describe strategies for exploiting imagery containing these changes. The perennial challenge in change detection is in translating the application-dependent concept of an "interesting change" to a mathematical framework; as such, the mathematical approaches for detecting these types of changes can be quite different. In large-scale change detection (LSCD), the goal is to identify changes that have broadly occurred in the scene. The paradigm for anomalous change detection (ACD), which is grounded in concepts from anomaly detection, seeks to identify changes that are different from how everything else might have changed. This borrows from the classic anomaly detection framework, which attempts to characterize that which is "typical" and then uses that to identify deviations from what is expected or common. This chapter provides an overview of change detection, including a discussion of LSCD and ACD approaches, operational considerations, relevant datasets for testing the various algorithms, and some illustrative results.

A. Ziemann (✉)
Los Alamos National Laboratory, Los Alamos, NM 87545, USA
e-mail: ziemann@lanl.gov

S. Matteoli
National Research Council (CNR) of Italy, Institute of Electronics,
Computers and Telecommunication Engineering (IEIIT), Pisa, Italy
e-mail: stefania.matteoli@ieiit.cnr.it

S. Prasad and J. Chanussot (eds.), *Hyperspectral Image Analysis*,
Advances in Computer Vision and Pattern Recognition,
https://doi.org/10.1007/978-3-030-38617-7_12

351

12.1 Introduction

Come gather 'round people
Wherever you roam
And admit that the waters
Around you have grown
And accept it that soon
You'll be drenched to the bone.
If your time to you
Is worth savin'
Then you better start swimmin'
Or you'll sink like a stone
For the times they are a-changin'.
        —Bob Dylan, *The Times They Are A-Changin'*

To first order, the concept of change detection in spectral imagery is straightforward: given two (or more) images of a scene at different points in time, find what has changed. However, the concept of "change" is highly application-dependent and, as is true for all good signal detection algorithms, the utility of such an approach is intimately tied to the application. In other words, *not all changes are created equal*, and as a result—and to take further creative liberty with famous quotes—*there is not one change detection algorithm to rule them all*.

Consider a case in which a region experiences extreme flooding. This is a large-scale change, i.e., one that will appear broadly throughout the scene. In response to these types of events, the ability to analyze remote sensing imagery before and after a natural disaster in order to generate damage maps is critical for effective disaster response. An appropriate change detection algorithm will be one that can delineate such changes that potentially take up considerable portions of the scene.

Alternatively, consider a case in which a forested region experiences pervasive seasonal changes from spring to autumn, and during that time a building is covertly constructed in the forest. For analysts that are looking for nefarious activity, the pervasive seasonal changes will not be of interest, but the anomalous changes related to the covert building will be of great interest. In this case, an appropriate change detection algorithm will be one that can identify rare, anomalous changes and distinguish them from potentially pervasive differences due to seasonal variations, atmospheric variations, changes in sun illumination angle, etc.

We can generally categorize change detection algorithms under large-scale change detection (LSCD) or anomalous change detection (ACD). How "interesting" a particular change might be is tied to the application, and the intended application should be taken into consideration when choosing the approach. LSCD seeks to identify changes that may have broadly occurred throughout the scene. The ACD paradigm has its roots in the anomaly detection (AD) paradigm: characterize what is typical or normal (i.e., for AD, the background; for ACD, a common change), and identify deviations from that (i.e., for AD, what is different from the background; for ACD, a change that is different from a common change). Pervasive differences (or nuisance changes) are changes that are mostly uninteresting and are due to variations in, for example, calibration, atmosphere, and illumination. This chapter will start with an overview of change detection fundamentals (Sect. 12.2), continue with a discussion of large-scale change detection algorithms (Sect. 12.3), follow with a discussion of anomalous change detection algorithms (Sect. 12.4), and finish with a summary of operational considerations and relevant datasets (Sects. 12.5 and 12.6). The chapter concludes with some remarks on the current state of change detection in spectral imaging as well as open areas for future work.

12.2   Change Detection: The Fundamentals

Change detection is the process of identifying pixels in multi-temporal spectral images that have changed between subsequent acquisitions. By *change* we mean a *modification of the spectral signature* due to a transition, alteration, or temporal evolution of the land-cover or materials occupying the given pixels or regions. From a statistical perspective, we can approach the change detection problem using a binary hypothesis testing procedure [35]. For simplicity, we restrict our review to pixel-wise change detection schemes, where the determination of a change at a given pixel is done only relative to observations at the same pixel location across multi-temporal acquisitions. Pixel-based approaches assume that the images have been registered, and that corresponding pixel locations have imaged the same portion of the scene. While not explored in detail here, it is worth noting that there are also approaches that leverage spatial structure and, specifically, the correlation often existing among neighboring pixels [34, 72, 92]. In that same vein, there are also approaches that attempt to be robust to small misregistration errors [15, 47, 59, 78, 79, 85, 89, 95]. Again, for simplicity, we focus our review on bi-temporal imagery, i.e., change detection across two images, while noting that sequences of images have been explored in the literature [28, 74, 94].

To that end, let $\mathbf{X}$ and $\mathbf{Y}$ be a pair of co-registered hyperspectral images of dimension $B \times K \times L$ and acquired at two different times, with B denoting the number of spectral components (i.e., spectral bands) and $N = K \times L$ the number of pixels. The change detection binary hypothesis testing can be expressed as

$$H\left(\mathbf{x}_{k,l}, \mathbf{y}_{k,l}\right) = \begin{cases} H_0 & (k,l) \in \Omega_0 \\ H_1 & (k,l) \in \Omega_1 \end{cases}, \tag{12.1}$$

where $\mathbf{x}_{k,l} \in \mathbb{R}^B$ denotes the spectrum of the pixel at position (k,l) in image $\mathbf{X}$ and $\mathbf{y}_{k,l} \in \mathbb{R}^B$ denotes the spectrum of the corresponding pixel in image $\mathbf{Y}$; H_0 and H_1 are the competing hypotheses indicating, respectively, that *no change has occurred* at position (k,l) and that a *change has occurred* at (k,l); and Ω_0 and Ω_1 are the classes of unchanged (Ω_0) and changed (Ω_1) pixels. Because this is a binary approach, the class Ω_1 of changed pixels may contain several subclasses [6] associated with the different *types* of changes that have occurred.

Definitions

- **Pervasive differences** are changes that are mostly uninteresting, such as those due to calibration, atmospheric, and illumination variations.
- **Large-scale changes** are material changes that affect a high number of pixels in the scene, such as those due to land-cover changes like snow or floods.
- **Anomalous changes** are material changes that are rare, such as those due to new buildings or moved vehicles.

While our interest here is in material changes, there will inevitably be spectral changes that are *pervasive* and mostly uninteresting, such as those due to calibration, atmospheric, and illumination variations; these changes, sometimes termed *nuisance changes* [48], should ideally be ignored by change detection algorithms. To mitigate these effects, preprocessing steps are often applied (see Sect. 12.5) so that change detection techniques can focus on the salient changes. Among the salient changes, we make a distinction between *large-scale changes* affecting a high number of pixels in the scene (e.g., land-cover changes like snow or flooding) and rare *anomalous changes* involving a relatively small number of pixels (e.g., new buildings or vehicles). While some change detection techniques do not differentiate between these types of changes, there are approaches in the literature that give special treatment to temporal anomalies [1, 72, 76, 77, 97], and so here we give separate consideration to anomalous change detection in Sect. 12.4. We assume that the images have been registered, and for LSCD we assume the images have been radiometrically equalized; we provide more discussion on those operational considerations in Sect. 12.5.

Questions

When deciding which change detection techniques to apply to images of a scene, it is worth asking yourself the following:
- *Are there a lot of pervasive differences between the images?*
- *Am I interested in large-scale changes?*
- *Am I interested in anomalous changes?*

12.3 Approaches for Large-Scale Change Detection

The detection of large-scale changes has been dramatically advanced as an application space by the increasing availability of satellite-based spectral imagery. In particular, the extensive repetitive coverage and short revisit time of constellations like the NASA/USGS Landsat mission [81], which has been continuously imaging the Earth since 1972, and the ESA Sentinel constellation [24] as part of their Copernicus program, which has been collecting since 2014, provide the historical archive of imagery necessary to detect changes over time. LSCD has proven valuable in a variety of applications such as land use and land-cover monitoring, detecting urban development, ecosystem surveillance, management of natural resources, disaster response and post-damage analysis, crop-stress detection, and ice-melt assessment [86, 92–94, 99]. In the literature, a multitude of methodologies have been developed to detect large-scale changes in multispectral and hyperspectral images [8, 14, 28, 64]. As such, they may be grouped according to a variety of criteria. For instance, there are both *unsupervised* and *supervised* methods, depending on the availability of ground truth data. Alternatively, a taxonomy may be outlined by distinguishing between *binary* change detection methods and *multiple* change detection methods (i.e., *change classification*); this depends on whether the method is capable of just identifying the presence/absence of changes, or if it also seeks to understand the *type* of change. In the latter, the outcome is a partitioning of the Ω_1 class into multiple subclasses related to different types of changes. In this chapter, we choose to organize the change detection methods from a technique perspective, and thus group the methods in the following sections based on their methodological approaches. For the LSCD overview in this section, we are assuming that the images have been radiometrically equalized (see Sect. 12.5). It is generally advantageous to "normalize" out as many pervasive differences as possible, but worth noting that not all change detection approaches require this in advance (i.e., many ACD approaches have it explicitly built into the algorithm itself).

12.3.1 Change Vector Analysis and Related Methods

Several change detection methods are based on the Change Vector Analysis (CVA) approach, which dates back to the 1980s [43] and, with subsequent extensions and modifications, has continued to be used throughout the literature [33, 50, 51, 83, 87]. Like most other techniques, it is based on the *difference image*, which is defined as

$$\mathbf{D} = \mathbf{Y} - \mathbf{X}. \tag{12.2}$$

The associated spectrum $\mathbf{d}_{k,l} \in \mathbb{R}^B$ contains the difference between the spectra $\mathbf{x}_{k,l}$ (taken at one time) and $\mathbf{y}_{k,l}$ (taken at another time).

CVA is conceptually based on thresholding the magnitude of the difference spectra in $\mathbf{D}$ to detect the presence of changes. Here we provide a brief overview of CVA, while a rigorous theoretical framework for its formulation and interpretation can be found in [6]. In CVA, a change of coordinate system from Cartesian to hyperspherical is performed so that, for the B-dimensional spectrum $\mathbf{d}$ associated with a generic pixel in the difference image, $\mathbf{d}$ can be represented in terms of its magnitude $\rho = \|\mathbf{d}\|$ and its $B - 1$ angular coordinates $\{\theta_n\}_{n=1}^{B-1}$. In this construct, a higher magnitude of ρ corresponds to a greater difference between $\mathbf{x}$ and $\mathbf{y}$, which means (in theory) that the change affecting the given pixel is also greater. A "change" threshold for ρ allows us to identify two regions in the magnitude domain: an inner hyperspherical region of *unchanged* pixels, and an outer hyperspherical annulus of *changed* pixels (whose upper-bound comes from the maximum magnitude $\rho_{\max}$ in the difference image). In practice, the decision rule for solving the binary hypothesis test in Eq. (12.1) can be derived using the magnitude of $\rho_{k,l}$ in the (k, l) position of the difference image:

$$\rho_{k,l} \underset{H_0}{\overset{H_1}{\gtrless}} \xi \tag{12.3}$$

where ξ is a suitable change threshold.

An improved understating of the types of changes can be achieved by exploiting the $B - 1$ angular coordinates. In particular, by further thresholding the $\{\theta_n\}_{n=1}^{B-1}$ coordinates, solid sectors of the annulus can be identified that correspond to different kinds of changes. An illustration using a simplified two-dimensional domain (i.e., $\mathbf{d} = [d^{(1)}\ d^{(2)}]^\top$) is shown in Fig. 12.1a, where there is one single angular coordinate equal to $\theta = \arctan\left(d^{(2)}/d^{(1)}\right) \in [0, 2\pi]$. In [6], a theoretical analysis of the statistical distribution of changed and unchanged pixels in the hyperspherical domain is provided.

CVA can quickly become mathematically unwieldy in higher dimensions, so it has generally been applied in two-dimensional domains, and the information associated with the remaining $B - 2$ bands is not exploited [7]. With this in mind, the Compressed CVA (C²VA) approach employs a transformation from $\mathbb{R}^B \to \mathbb{R}^2$, thus compressing the hyperspherical domain into a two-dimensional domain while retaining information derived from all spectral channels [7]. Specifically, C²VA employs the magnitude ρ exactly as in CVA, but instead all directional information associated with the B bands is compressed into a single angular coordinate:

$$\theta' = \arccos\left[\frac{1}{\sqrt{B}}\left(\frac{1}{\rho}\sum_{b=1}^{B} d^{(b)}\right)\right] \in [0, \pi]. \tag{12.4}$$

The goal of this step is to "compress" the data by effectively computing the spectral angle between $\mathbf{d}$ and some reference vector; because this is an unsupervised method, the diagonal vector $\mathbb{1} = [1\ 1\ 1\ ..\ 1]$ is taken as that reference vector. This results in semicircular and semi-annular regions of unchanged and changed pixels, respectively, as well as solid sectors of the annulus that are associated with different kinds of changes, as illustrated in Fig. 12.1b.

(a)

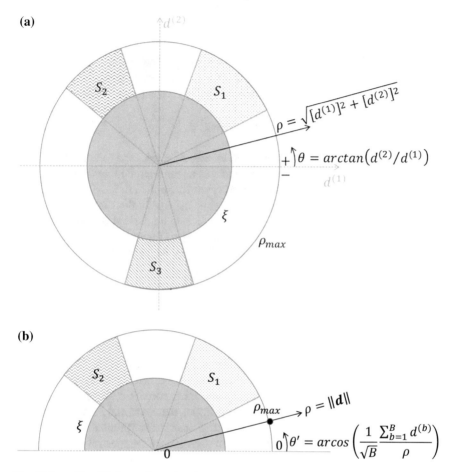

(b)

Fig. 12.1   **a** Simplified two-dimensional representation of CVA in polar coordinates. The inner darker gray circle represents the region of unchanged pixels, whereas the lighter gray outer annulus is the region of changed pixels. Annular sectors such as S_1, S_2, S_3 shown in the figure are associated with different types of changes. In the figure, $d^{(b)}$ denotes the bth spectral component of **d**. **b** Similar two-dimensional example for C^2VA in polar coordinates

The utility and simplicity of both CVA and C^2VA have led to a variety of CVA-based methods. Hierarchical Spectral CVA (HSCVA) [38], Sequential Spectral CVA (SSCVA) [40], and Multiscale Morphological C^2VA (M^2C^2VA) [41] are further variants of CVA that are more sensitive to subtle spectral variations, generally not even detectable in multispectral images, that can be found when dealing with hyperspectral imagery. In the dataset discussion in Sect. 12.6, an example of C^2VA is provided in Figs. 12.4 and 12.5.

12.3.2 Match-Based Methods

The match-based methodologies include those methods that evaluate metrics of spectral similarity or dissimilarity between spectra at corresponding pixel locations of $\mathbf{X}$ and $\mathbf{Y}$. In general, for the pixel position (k, l) (for convenience, we denote $\mathbf{x} = \mathbf{x}_{k,l}$ and $\mathbf{y} = \mathbf{y}_{k,l}$), we write:

$$\mathcal{D}\left(\mathbf{x}, \mathbf{y}\right) \gtrless_{H_0}^{H_1} \xi \tag{12.5}$$

where $\mathcal{D}(\mathbf{x}, \mathbf{y})$ is some spectral distance measure (as opposed to spatial) between the two B-dimensional pixels $\mathbf{x}$ and $\mathbf{y}$, and ξ is an appropriate threshold. There are a number of similarity or distance metrics typically applied in hyperspectral image analysis that may be employed [82]. The simplest one is the Euclidean Distance (ED) metric $\mathcal{D}_{ED}\left(\mathbf{x}, \mathbf{y}\right) = \|\mathbf{y} - \mathbf{x}\| = \rho_{k,l}$, which is equivalent to the magnitude-only portion of the CVA-based decision rule. One of the most ubiquitous distance metrics is the Spectral Angle Mapper (SAM) [36, 63, 84], which evaluates similarity in terms of the angle subtended by $\mathbf{x}$ and $\mathbf{y}$ in $\mathbb{R}^B$:

$$\mathcal{D}_{SAM}\left(\mathbf{x}, \mathbf{y}\right) = \arccos\left(\frac{\mathbf{x}^\top \mathbf{y}}{\|\mathbf{x}\| \, \|\mathbf{y}\|}\right). \tag{12.6}$$

Many other metrics can be found in the literature, such as Mahalanobis distance derived measures [16]; the Spectral Correlation Mapper (SCM) [16, 84, 92], which evaluates similarity by way of the sample correlation coefficient over the two sets of B observations given by the entries of $\mathbf{x}$ and $\mathbf{y}$; and other correlation-based indices [63]. Among the various measurements, it is worth noting that SAM and SCM are invariant to scaling factors and are thus more robust to changes in illumination within the scene. Furthermore, SCM is also invariant to variations of the spectral mean, making it robust to residual bias effects [16, 82].

12.3.3 Transformation-Based Methods

The methods in this category apply some transformation to the image data that is aimed at suppressing the effects of pervasive changes and, at the same time, maintaining or enhancing the salient changes. Some of these methods transform the data into a low-dimensional feature space that, although limited to a few components, generally retains the information related to the changes of interest. The C^2VA approach discussed above can also be considered as belonging to this category, as well as the Temporal Principal Component Analysis (TPCA) [56]. Other approaches, such as the commonly known Multivariate Alteration Detection (MAD) approach [25, 54] and its Iteratively Reweighted (IR) variant IR-MAD [25, 53, 54], perform a change of coordinate system so that the components that address the majority of change information can easily be identified. While MAD is more appropriately categorized

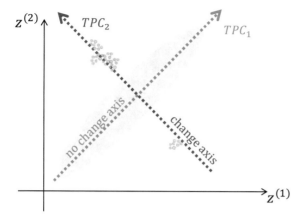

Fig. 12.2 Illustration of the idea behind TPCA for a two-dimensional (bi-temporal) domain. In this example, $z^{(t)}$ denotes the tth temporal component of $\mathbf{Z}$ (recall that $\mathbf{Z}$ is of size $2 \times BN$, so its components are $z^{(1)}$ and $z^{(2)}$). TPC_1 is the first temporal principal component, which is in the direction of unchanged pixels. TPC_2 is the second temporal principal component, along which the changed pixels are displaced

as an ACD approach, IR-MAD can be considered as an LSCD approach; the point of the iterative reweighting steps is to enable it to detect larger changes.

TPCA differs from conventional PCA in that TPCA is applied to the data in a feature space spanned by the *temporal* coordinates [56]. For a bi-temporal case, for example, this space is two-dimensional. Specifically, $\mathbf{X}$ and $\mathbf{Y}$ are rearranged by stacking their spectral components (in vector form) one over each other to obtain the $BN \times 1$ vectors Ψ and Υ, and PCA is then applied to the $2 \times BN$ matrix $\Delta = [\Psi \Upsilon]^{\top}$. In practice, $\mathbf{Z} = \mathbf{E}^{\top}\Delta$ is the $2 \times BN$ TPCA transformed data matrix, where $\mathbf{E}$ is the 2×2 matrix of eigenvectors of the covariance matrix Σ_{Δ} of Δ, which expresses the covariance between the temporal variables. The original data, which has a stronger temporal correlation, undergoes a rotation of the axes that diagonalizes $\Sigma_{\mathbf{Z}}$, thus de-correlating the temporal variables. In the two-dimensional case, the first principal component is mostly associated with the common information across the temporal features, whereas the second component tends to mostly address the "change information" [56]. Figure 12.2 provides a two-dimensional illustration of TPCA.

MAD is essentially an application of the Canonical Correlation Analysis (CCA) approach [31], and the MAD transformation is as follows [53]:

$$\mathbf{d}_{\text{MAD}} = \begin{bmatrix} \mathbf{b}_1^{\top}\mathbf{y}_{k,l} - \mathbf{a}_1^{\top}\mathbf{x}_{k,l} \\ \cdots \\ \mathbf{b}_B^{\top}\mathbf{y}_{k,l} - \mathbf{a}_B^{\top}\mathbf{x}_{k,l} \end{bmatrix} \qquad (12.7)$$

where $\{\mathbf{b}_b, \mathbf{a}_b\}_{b=1}^{B}$ are B-dimensional vectors of coefficients from a standard CCA. The MAD-transformed difference image $\mathbf{D}_{\text{MAD}}$ (for which $\mathbf{d}_{\text{MAD}} \in \mathbf{D}_{\text{MAD}}$) has B

components in the MAD feature space. By extracting MAD components starting from the last, we sequentially extract uncorrelated gray-level difference images in a way such that each extracted image exhibits maximum change information subject to the constraint of being uncorrelated to the previously extracted image [25, 54]. The MAD components retaining the highest change information can thus be retained for further processing. MAD is also invariant to linear and affine transformations [25, 54]. This holds for a number of ACD methods as well, as mentioned in Sect. 12.4. MAD is usually followed by post-processing procedures based on the Maximum Autocorrelation Factor (MAF), which allows the spatial context of neighboring pixels to be accounted for in the overall change detection analysis [54].

Another approach is a subspace-based change detection method [90] that builds on the Orthogonal Subspace Projection (OSP) target detection algorithm [27]. This method treats the generic pixel $\mathbf{y}$ as a potential target and exploits the corresponding pixel $\mathbf{x}$, together with additional spatial–spectral data, to build a background subspace; spectral dissimilarity is then evaluated against this subspace [90]. The underlying assumption is that unchanged pixels will lie in or close to the background subspace, while changed pixels will have a larger distance from the subspace. The subspace-based change detector for the pixel position (k, l) can be expressed as follows (where for convenience, we denote $\mathbf{P} = \mathbf{P}_{k,l}$ and $\mathbf{B} = \mathbf{B}_{k,l}$):

$$\mathcal{D}_{\text{SSCD}}(\mathbf{x}, \mathbf{y}) = \frac{\mathbf{y}^{\top} \mathbf{P}^{\perp} \mathbf{y}}{\mathbf{y}^{\top} \mathbf{y}} \underset{H_0}{\overset{H_1}{\gtrless}} \xi. \tag{12.8}$$

In Eq. (12.8), $\mathbf{P}^{\perp} = \mathbf{I} - \mathbf{P}$ is the $B \times B$ orthogonal projection matrix, where $\mathbf{I}$ is the $B \times B$ identity matrix and $\mathbf{P} = \mathbf{B} (\mathbf{B}^{\top} \mathbf{B})^{-1} \mathbf{B}^{\top}$ is the $B \times B$ projection matrix over the background subspace. The background subspace is spanned by the columns of the $B \times (m + 1)$ basis matrix $\mathbf{B} = \begin{bmatrix} \mathbf{x} & \mathbf{c}_1 & \mathbf{c}_2 & \cdots & \mathbf{c}_m \end{bmatrix}$, which is made up of the pixel $\mathbf{x}$ and the additional $m \geq 0$ basis vectors obtained by exploiting spatial–spectral information [90]. If $m = 0$ and the background subspace is spanned solely by $\mathbf{x}$, then the subspace-based method in Eq. (12.8) becomes

$$\mathcal{D}_{\text{SSCD\_X}}(\mathbf{x}, \mathbf{y}) = 1 - \left| \cos \left[\mathcal{D}_{\text{SAM}}(\mathbf{x}, \mathbf{y}) \right] \right|^2. \tag{12.9}$$

In [90], several strategies are proposed for learning the background subspace and, specifically, the matrix $\mathbf{C} = \begin{bmatrix} \mathbf{c}_1 & \mathbf{c}_2 & \cdots & \mathbf{c}_m \end{bmatrix}$. Radiometrically-equalized spectra of specific land covers associated with undesired changes can be used within a supervised change detection framework to suppress effects of uninteresting changes, and the spectra of pixels spatially close to $\mathbf{x}$ can be used to reduce the effects of misregistration.

12.3.4 Unmixing-Based Methods

The methods presented thus far focus on identifying *if* a pixel has undergone a change over time, which then, in the case of CVA, allows for exploration of the nature of the change (potentially including material characterization). An alternative approach, however, is to first perform a detailed subpixel material characterization of a pixel, and then to see how those material abundance values change over time. In this general unmixing-based framework, the pixel is modeled as a (generally assumed linear) mixture of pure material spectra (i.e., *endmembers*) [17, 30, 44]. Spectral changes affecting a given pixel may be due to a variety of scenarios [21, 23, 39] such as an actual material transition (and thus the appearance or disappearance of endmember), or a change in the abundance of already present endmembers (and thus a change in the mixing proportions). Several efforts have been made in the literature to develop change detection methods capable of providing subpixel-level material information. This is achieved by embedding endmember extraction [10, 88] and unmixing techniques [5] into the change detection process. Some approaches, for example, perform post-unmixing comparisons to detect subpixel changes by differencing the abundance values for each endmember [18, 21–23]. Others perform endmember extraction and unmixing on the joint $2B$-dimensional image $\mathbf{U}$ obtained by stacking $\mathbf{X}$ and $\mathbf{Y}$ along the spectral dimension, and then perform a CVA/match-based hybrid change analysis of the multi-temporal extracted endmembers [39].

12.3.5 Other Methods

This overview of LSCD methods is far from being exhaustive, and there are several methods that, although not closely fitting with the methodological categories used here, should be mentioned. As noted earlier, there are a variety of approaches that are not simply pixel-wise but also exploit the *spatial* structure of spectral images. Among these, some employ Markov random field models [34], spatial correlation [92], spatial–spectral feature vectors [85], and local *tile-based* change detection schemes [4, 96, 99]. Some methods also exploit geometrical concepts such as graph theory [4], convex hull estimation [96], and segmentation [93, 99]. Additionally, there are methods based on more recent research trends such as sparsity [11, 12], deep learning [32], and convolutional neural networks [66, 86].

12.4 Approaches for Anomalous Change Detection

The ultimate aim of change detection is to key in on changes that are interesting, but because that is so dependent on the application, there is no straightforward way to place "interesting" into a mathematical construct. In anomalous change detection, however, an *anomalous change* is one that is unusual when compared with ordinary, common changes that have occurred throughout the scene [76]. This is a definition that can be characterized mathematically, and ultimately be used to cue a human analyst to *potentially* interesting changes that are *rare*. ACD is a special case of change detection whose applications might include detecting the construction of buildings, localizing small areas of diseased crops, or cueing an analyst to nefarious activities. An example of when ACD might be used is if an analyst is interested in identifying unsanctioned construction in a particular area, and their two images are taken in the winter and in the spring; if the surrounding area goes from snow-covered to grass-covered, those are all true spectral (and material) changes, but are not interesting in this application. In this example ACD will suppress those common seasonal changes (snow to grass), and emphasize any anomalous changes (construction). It should be noted that broader "changes" can also encompass *pervasive differences* due to changes in illumination, atmosphere, etc.; all of these will also be suppressed by the ACD framework [20]. Some of the LSCD methods identified in Sect. 12.3 can also be applied to the ACD problem [54, 55], and so this section focuses on methods that have specifically been developed for ACD. Note that while our LSCD overview assumed radiometric equalization of the images, we *do not* make that same requirement here.

As noted in Sect. 12.2, we let $\mathbf{x}_{k,l} \in \mathbb{R}^{B_x}$ denote the spectrum of the pixel at position (k, l) in the first image $\mathbf{X}$, and let $\mathbf{y}_{k,l} \in \mathbb{R}^{B_y}$ denote the spectrum of the corresponding pixel in the (k, l) position in the second image $\mathbf{Y}$. However, in the LSCD overview we assumed that the spectral channels between the two images were the same (though not always required); here we are making an explicit distinction between B_x and B_y. For the ACD methods discussed here, B_x and B_y need not be equal.

12.4.1 Difference-Based Methods

The approaches here generally assume global mean subtraction, i.e., the mean is subtracted from both images so that $\langle \mathbf{x} \rangle = 0$ and $\langle \mathbf{y} \rangle = 0$. Although a detailed discussion of mean subtraction is beyond the scope of this chapter, it is often used out of mathematical convenience, and there are important considerations about what it means from a physics-based perspective that should be noted with its use [98]. We will focus on representations of the algorithms as quadratic functions of the input arguments as described in [70]:

$$\mathcal{A}(\mathbf{x}, \mathbf{y}) = \begin{bmatrix} \mathbf{x}^\top & \mathbf{y}^\top \end{bmatrix} Q \begin{bmatrix} \mathbf{x} \\ \mathbf{y} \end{bmatrix}. \tag{12.10}$$

In this formulation Q is a square matrix whose dimension is given by $B_x + B_y$, and a change will be considered anomalous when $\mathcal{A}(\mathbf{x}, \mathbf{y})$ is above a user-defined threshold ξ. For a given pair of images $\mathbf{X}$ and $\mathbf{Y}$ (each mean-subtracted), we consider

$$\Sigma_x = \langle \mathbf{xx}^\top \rangle, \tag{12.11}$$

$$\Sigma_y = \langle \mathbf{yy}^\top \rangle, \tag{12.12}$$

$$\Sigma_{yx} = \langle \mathbf{yx}^\top \rangle. \tag{12.13}$$

The algorithms in this section first identify two linear transformations, applying one to the first image $\mathbf{X}$ and one to the second image $\mathbf{Y}$. The specific transformations vary for the different approaches, but share the same overarching goal of suppressing pervasive differences and common changes. For approaches that involve a transformation step, it is not required for the original images to share the same number of bands, but it is necessary for the *transformed* images to have the same number of bands. The transformed images are then subtracted to produce a difference image, and a measure of anomalousness is computed based on the magnitude of the residuals in that difference image [1, 60, 61].

If the pervasive differences are minimal, the simplest algorithm is to just subtract each element of the two images, i.e., $\mathbf{D} = \mathbf{Y} - \mathbf{X}$. This approach does not involve any transformations, and is typically not used in practice (and requires $B = B_x = B_y$). The associated detector and quadratic function for simple difference are the following:

$$\mathcal{A}_{\mathrm{SD}}(\mathbf{x}, \mathbf{y}) = (\mathbf{y} - \mathbf{x})^\top \left[\Sigma_y - \Sigma_{yx} - \Sigma_{yx}^\top + \Sigma_x \right]^{-1} (\mathbf{y} - \mathbf{x}), \tag{12.14}$$

with quadratic coefficient matrix:

$$Q_{\mathrm{SD}} = \begin{bmatrix} -\mathbf{I} \\ \mathbf{I} \end{bmatrix} \left[\Sigma_y - \Sigma_{yx} - \Sigma_{yx}^\top + \Sigma_x \right]^{-1} [-\mathbf{I} \ \ \mathbf{I}]. \tag{12.15}$$

Two of the most commonly used difference-based ACD methods are Chronochrome and Covariance Equalization. The Chronochrome algorithm (CC) [60, 61] uses a linear transformation $\mathbf{L}$ to compute a best estimator for $\mathbf{y}$ given by $\hat{\mathbf{y}} = \mathbf{Lx}$, where $\mathbf{L} = \Sigma_{yx} \Sigma_x^{-1}$. In this formulation, anomalousness scales by the difference $\mathbf{y} - \hat{\mathbf{y}} = \mathbf{y} - \mathbf{Lx}$ (where a larger difference is more anomalous). A similar approach is taken by Covariance Equalization (CE) [61], where a transformation is applied to $\mathbf{X}$ or $\mathbf{Y}$, or to both of them, so that the transformed image cubes have equal covariance [70]. The underlying assumption for both of these methods is that the affine transformation captures the pervasive differences (e.g., illumination, atmosphere, sun angle, environmental changes) and is space invariant [47]. These methods attempt to statistically suppress these potentially spurious changes, leaving behind any "true" differences. Then, standard anomaly detection approaches are applied to the residual image after differencing the two.

The associated detector and quadratic function for Chronochrome are:

$$\mathcal{A}_{\mathrm{CC}}(\mathbf{x}, \mathbf{y}) = \left(\mathbf{y} - \Sigma_{yx} \Sigma_x \mathbf{x} \right)^\top \left[\Sigma_y - \Sigma_{yx} \Sigma_x^{-1} \Sigma_{yx}^\top \right]^{-1} \left(\mathbf{y} - \Sigma_{yx} \Sigma_x \mathbf{x} \right), \tag{12.16}$$

$$Q_{CC} = \begin{bmatrix} -\Sigma_x^{-1}\Sigma_{yx}^\top \\ \mathbf{I}_y \end{bmatrix} \left[\Sigma_y - \Sigma_{yx}\Sigma_x^{-1}\Sigma_{yx}^\top \right]^{-1} \left[-\Sigma_{yx}\Sigma_x^{-1} \ \mathbf{I}_y \right]. \qquad (12.17)$$

And for the whitened version of Chronochrome:

$$\tilde{Q}_{CC} = \begin{bmatrix} -\tilde{\Sigma}_{yx}^\top \\ \mathbf{I}_y \end{bmatrix} \left[\mathbf{I}_y - \tilde{\Sigma}_{yx}\tilde{\Sigma}_{yx}^\top \right]^{-1} \left[-\tilde{\Sigma}_{yx} \ \mathbf{I}_y \right]. \qquad (12.18)$$

We can use a similar construct for the detector and quadratic for the whitened version of standard Covariance Equalization:

$$\mathcal{A}_{CE}(\tilde{\mathbf{x}}, \tilde{\mathbf{y}}) = (\tilde{\mathbf{y}} - \tilde{\mathbf{x}})^\top \left[2\mathbf{I} - \tilde{\Sigma}_{yx} - \tilde{\Sigma}_{yx}^\top \right]^{-1} (\tilde{\mathbf{y}} - \tilde{\mathbf{x}}), \qquad (12.19)$$

$$Q_{CE} = \begin{bmatrix} -\mathbf{I} \\ \mathbf{I} \end{bmatrix} \left[2\mathbf{I} - \tilde{\Sigma}_{yx} - \tilde{\Sigma}_{yx}^\top \right]^{-1} [-\mathbf{I} \ \mathbf{I}]. \qquad (12.20)$$

Additional formulations can be extended to Optimal CE and Diagonalized CE, both of which are generalizations of Standard CE [70]. Of note is that these methods have also been extended to target detection, where CC changes are used to characterize the target-free background at any given pixel [62, 69].

12.4.2 Straight Anomaly Detection Methods

In this straightforward approach, the two images are stacked together and treated as one image with $B_x + B_y$ spectral bands. From there, a direct anomaly detection algorithm such as RX [58] can be applied to the stacked image cube [73]. The RX algorithm uses Mahalanobis distance to identify image elements (in this case, stacked pixels) that are statistically "different" from the background. This identification of deviations from a well-characterized background is the cornerstone of anomaly detection, and allows for the application of any hyperspectral anomaly detection approach [45, 71].

12.4.3 Joint-Distribution Methods

As remarked at the beginning of this section, there are some ACD methods that do not require that the images have the same number of spectral channels, i.e., B_x need not equal B_y. That is true in particular for the joint-distribution methods presented here. In the framework developed by Theiler [70, 74–78], $P(\mathbf{x}, \mathbf{y})$ is the probability density function from which data samples $(\mathbf{x}, \mathbf{y})$ are drawn. The goal is then to find a function $f(\mathbf{x}, \mathbf{y})$ such that $f(\mathbf{x}, \mathbf{y}) > 0$ identifies $(\mathbf{x}, \mathbf{y})$ as an anomalous change;

doing so reframes the ACD problem as a classification problem. These methods are invariant to linear transformations of $\mathbf{X}$ and $\mathbf{Y}$, so there is no need to apply linear transformations as with, e.g., Chronochrome or Covariance Equalization.

We can define a "background distribution" $Q(\mathbf{x}, \mathbf{y}) = P(\mathbf{x})P(\mathbf{y})$ with

$$P(\mathbf{x}) = \int P(\mathbf{x}, \mathbf{y})d\mathbf{y}, \tag{12.21}$$

$$P(\mathbf{y}) = \int P(\mathbf{x}, \mathbf{y})d\mathbf{x}. \tag{12.22}$$

Then, the anomalous changes are those with high values of mutual information. The distribution above is the one that would be exhibited if $\mathbf{x}$ and $\mathbf{y}$ were independent; by using this as background, we identify anomalies whose $(\mathbf{x}, \mathbf{y})$ dependency is unusual when compared to that encoded in $P(\mathbf{x}, \mathbf{y})$. Thus, the joint-distribution anomalous change detector is given by level curves of the ratio:

$$\frac{P(\mathbf{x}, \mathbf{y})}{P(\mathbf{x})P(\mathbf{y})}. \tag{12.23}$$

By interpreting anomalousness as mutual information as mentioned above, this leads to the following detector for Hyperbolic Anomalous Change Detection, where data distribution is assumed Gaussian:

$$\mathcal{A}_{\mathrm{HACD}}(\mathbf{x}, \mathbf{y}) = -\log \frac{P(\mathbf{x}, \mathbf{y})}{P(\mathbf{x})P(\mathbf{y})} \tag{12.24}$$

$$= \log P(\mathbf{x}) + \log P(\mathbf{y}) - \log P(\mathbf{x}, \mathbf{y}). \tag{12.25}$$

This detector is so-named due to the hyperbolic boundaries that separate the anomalous pixels. We focus here on HACD, but there are other variants such as EC-HACD, where the distributions are elliptically contoured [77], and there has been research into using parametric distributions [75] and sequences of images [74]. While not explicitly making use of spatial information, there is also an implementation of HACD using spatial windowing (i.e., local co-registration adjustment [LCRA]) that is more robust to potential misregistration issues [78].

12.4.4 Other Methods

As with the LSCD section, this section provides an overview of ACD methods but is certainly not exhaustive. Other methods that were not discussed in detail here but deserve mention include kernel-based approaches [42] and joint sparse representation approaches [91].

12.5 Operational Considerations

There are a number of operational considerations that should be taken into account when implementing any change detection pipeline. The first main step is to register the two (or more) images. We are not exploring registration in this chapter beyond noting that there are algorithms that are designed to be robust to small registration errors [15, 47, 59, 78, 79, 85, 89, 95]. Any ensuing preprocessing steps generally aim to transform the images into a similar domain such that the true interesting changes are as distinct as possible. Due to the physical nature of these images, the preprocessing steps often involve physical modeling. However, that is not always the case.

There is a subtle but important distinction between *identifying relative changes* and *quantifying material changes*. The common hyperspectral image processing pipeline involves atmospheric compensation (i.e., backing out any radiative transfer effects due to the atmosphere [46]) to convert the image from sensor-reaching radiance to approximate surface reflectance. However, when seeking to find *relative* changes between two images, it is not necessary to go through the atmospheric compensation process; the images can be appropriately preprocessed through relative *radiometric equalization* (sometimes called radiometric normalization) [19]. In contrast, for the unmixing approaches presented in Sect. 12.3.4, atmospheric compensation would be important as the methods exploit full material characterization of the scene. The emphasis here is that atmospheric compensation is not always needed depending on the goals of the change detection analysis; it is really only required when training data (e.g., a spectral library) is used [67]. In other words, relative radiometric equalization is often sufficient, and can be accomplished through a simple relative normalization between the two images [29]. For a number of the LSCD methods in Sect. 12.3, this is perfectly appropriate and advisable (and, in many cases, assumed). For the ACD methods described in Sect. 12.4, it is not required. This is for two reasons: either it is directly integrated into the method itself, as in CE and CC [60, 61], or the method does not require it, as in HACD [70, 76].

> **Tip**
> Atmospheric compensation is not always needed. Relative radiometric equalization is often sufficient, and can be accomplished through a simple radiometric normalization between the two images.

Radiometric equalization is often considered in the literature to be a *predictor*, which more broadly refers to any algorithm that seeks to transform, without loss of generality, image X to the observation conditions of image Y such that changes due to illumination, atmosphere, etc., are suppressed. These prediction algorithms can be linear [37] or nonlinear [9], and seek to suppress stationary background clutter [47]. But, as mentioned, not all change detection algorithms require this preprocessing step, although it is generally advantageous. When the amount of image data grows (e.g., due to high spatial coverage by the sensor), this can also become an operational challenge. In such cases, the scaled techniques proposed in [3] maybe used to tackle real problems as explored in Chap. 2

12.6 Overview: Datasets

In the literature, several different datasets have been used for change detection algorithm testing and benchmarking. Some of them include multi-temporal hyperspectral images acquired by satellite (spaceborne) sensors, and mostly exhibit large-scale land-cover changes or transitions [28, 99]. Other datasets include airborne (and ground-based) hyperspectral images, and mostly exhibit anomalous changes of vehicles and other man-made objects [2, 20, 26, 47, 49]. When testing an algorithm, the choice of dataset should ideally contain ground truth or reference data to enable quantitative performance evaluation. Here we report a brief overview of the major *ground truthed* multi-temporal datasets available to the scientific community, as well as some example results. An overview of both large-scale and anomalous change detection datasets can be found in Fig. 12.3, which also summarizes the main characteristics of each dataset.

12.6.1 *Details on the Data*

Two of the most widely employed datasets for large-scale changes were acquired by the spaceborne hyperspectral Hyperion sensor (launched aboard the satellite Earth Observing-1, or EO-1) [52, 57, 80], and are sometimes referred to as the **China Farmland** and **USA Irrigation** datasets [28]. They refer to agricultural areas in China and the United States, respectively, where land-cover transitions between crops, soil, and water, as well as variations in soil and vegetation water content, can be found. Bi-temporal images and regions of interest (ROIs) can be downloaded at https://rslab.ut.ac.ir/data. The ground truth (reference data) set of ROIs was constructed by visual inspection of higher resolution imagery acquired over the same area (e.g., panchromatic images acquired by EO-1's Advanced Land Imager, or ALI), as well as by looking at the outcomes of change detection analyses performed in the literature [28].

For anomalous changes there are three primary datasets with ground truth, two of which are from Rochester Institute of Technology (RIT), and one of which is from the University of Pisa. RIT's **SHARE 2012** dataset [26] was acquired using the ProSpecTIR VS hyperspectral sensor [68] over Avon, NY, USA, and is available at https://www.rit.edu/cos/share2012. RIT's **Cooke City** dataset [65] was acquired using the HyMap hyperspectral sensor [13] over Cooke City, MT, USA, and is available at http://dirsapps.cis.rit.edu/blindtest. University of Pisa's **Viareggio 2013 Trial** dataset [2] was acquired by the SIM.GA hyperspectral sensor over Viareggio, Italy,

China Farmland

| spaceborne | bi-temporal ~1 year interval | ground truth ROIs | large-scale changes |

USA Irrigation

| spaceborne | bi-temporal ~3 year interval | ground truth ROIs | large-scale changes |

SHARE 2012

| airborne | bi-temporal ~3 hour interval | ground truth coordinates context imagery | anomalous changes |

Cooke City

| airborne | bi-temporal ~20 min. interval | ground truth ROIs | anomalous changes |

Viareggio 2013 Trial

| airborne | multi-temporal ≤1 day interval | ground truth ROIs | anomalous changes |

Fig. 12.3 Synopsis of the primary datasets (with ground truth) for LSCD and ACD algorithm benchmarking

and is available at http://rsipg.dii.unipi.it. All of these images were collected by airborne sensors, and feature small, rare, and man-made changes including, e.g., vehicles, panels, and tarps. The ground truth data (reference data) were each constructed using ad hoc, in situ campaigns performed during data acquisition [2, 26, 65].

12.6.2 Illustrative Examples

Illustrative examples of large-scale change detection (Figs. 12.4 and 12.5) and anomalous change detection (Fig. 12.6) are provided as well. Note that these are *not optimized implementations* of the algorithms, and are just meant to demonstrate the techniques.

The LSCD example is shown for C^2VA applied to the USA Irrigation dataset, where Figs. 12.4a, b show the bi-temporal images and Fig. 12.4c shows the truth mask for the changes. The result of our implementation with four change classes is shown in Fig. 12.4d, where the different colors indicate different sectors; the black class represents data in the "no-change" semi-annular region, whereas the other

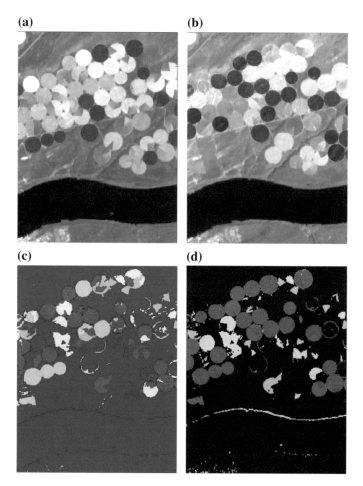

Fig. 12.4 USA Irrigation results. **a** RGB of the first image. **b** RGB of the second image. **c** Truth mask of changes, where blue corresponds to "no change." **d** C^2VA results with 4 classes

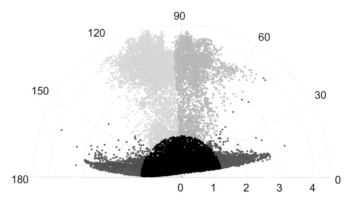

Fig. 12.5 Polar plot for C^2VA applied to the USA Irrigation dataset. For the ρ axis, note that the data has been normalized between [0,1], and that $\rho_{max} \approx 4.4$

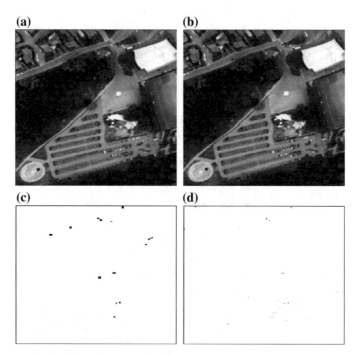

Fig. 12.6 Viareggio 2013 Trial results. **a** RGB of the first image. **b** RGB of the second image. **c** Truth mask of changes (including both insertions and deletions). **d** HACD + LCRA detection map

colors denote solid annulus sectors associated with different types of changes (after the annulus is partitioned). The corresponding polar plot is shown in Fig. 12.5, which illustrates how those categories are delineated. In this example, most of the changes are correctly differentiated, although the truth mask indicates more than four change classes. An example of HACD + LCRA (i.e., with local co-registration adjustment) applied to the Viareggio 2013 Trial dataset is shown in Fig. 12.6, with the bi-temporal images in Figs. 12.6a, b and the truth mask of changes (both deletions and insertions) in Fig. 12.6c. The HACD + LCRA detection map is shown in Fig. 12.6d, where most of the changes have been identified together with some false alarms.

12.7 Conclusions

Change detection is an area of spectral image analysis that is both well-defined and also quite nebulous. The challenge is that the utility of any given change detection algorithm is intimately tied to the application and, in turn, the *types* of changes that are considered interesting in that context. In this chapter, we make a distinction between *large-scale changes* and *anomalous changes*. Large-scale change detection aims to find broad material changes within a scene (e.g., flooding, seasonal vegetation changes). In contrast, anomalous change detection aims to find small, rare changes that have changed in a way that is different from how everything else might have changed. Within those distinctions, we presented a number of mathematical frameworks that seek to place those definitions into a rigorous mathematical construct. The approaches are typically founded in statistical inference, but also use graph theory and machine learning. Even with these rigorous mathematical frameworks, change detection continues to be a growing field in part due to the increasing accessibility of airborne and spaceborne remote sensing. Areas of ongoing research include strategies for handling misregistration and seasonal variations, leveraging cross-sensor datasets, and categorizing types of changes.

Acknowledgements This work was supported by the Los Alamos Laboratory Directed Research and Development (LDRD) program. The authors are very grateful to James Theiler for helpful suggestions that improved the chapter.

References

1. Acito N, Diani M, Corsini G, Resta S (2017) Introductory view of anomalous change detection in hyperspectral images within a theoretical gaussian framework. IEEE Aerosp Electron Syst Mag 32(7):2–27
2. Acito N, Matteoli S, Rossi A, Diani M, Corsini G (2016) Hyperspectral airborne "Viareggio 2013 Trial" data collection for detection algorithm assessment. IEEE J Sel Top Appl Earth Obs Remote Sens 9(6):2365–2376

3. Adsuara JE, Pérez-Suay A, Munoz-Mari J, Mateo-Sanchis A, Piles M, Camps-Valls G (2019) Nonlinear distribution regression for remote sensing applications. IEEE Trans Geosci Remote Sens. https://www.ieeexplore.ieee.org/document/8809360

4. Albano JA, Messinger DW, Schlamm A, Basener W (2011) Graph theoretic metrics for spectral imagery with application to change detection. In: Proceedings of SPIE, vol 8048, p 804809

5. Bioucas-Dias JM, Plaza A, Dobigeon N, Parente M, Du Q, Gader P, Chanussot J (2012) Hyperspectral unmixing overview: geometrical, statistical, and sparse regression-based approaches. IEEE J Sel Top Appl Earth Obs Remote Sens 5(2):354–379

6. Bovolo F, Bruzzone L (2007) A theoretical framework for unsupervised change detection based on change vector analysis in the polar domain. IEEE Trans Geosci Remote Sens 45(1):218–236

7. Bovolo F, Marchesi S, Bruzzone L (2012) A framework for automatic and unsupervised detection of multiple changes in multitemporal images. IEEE Trans Geosci Remote Sens 50(6):2196–2212

8. Bruzzone L, Liu S, Bovolo F, Du P (2016) Change detection in multitemporal hyperspectral images. In: Multitemporal remote sensing. Springer, pp 63–88

9. Carlotto MJ (2000) Nonlinear background estimation and change detection for wide area search. Opt Eng 39(5):1223–1229

10. Chan TH, Ma WK, Ambikapathi A, Chi CY (2011) A simplex volume maximization framework for hyperspectral endmember extraction. IEEE Trans Geosci Remote Sens 49(11):4177–4193 (2011)

11. Chen Z, Wang B (2017) Spectrally-spatially regularized low-rank and sparse decomposition: a novel method for change detection in multitemporal hyperspectral images. Remote Sens 9(1044):1–21

12. Chen Z, Yang B, Wang B, Liu G, Xia W (2017) Change detection in hyperspectral imagery based on spectrally-spatially regularized low-rank matrix decomposition. In: Proceedings of IEEE IGARSS, pp 1–4

13. Cocks T, Jenssen R, Stewart A, Wilson I, Shields T (1998) The HyMap^{TM} airborne hyperspectral sensor: the system, calibration, and performance. In: Proceedings 1st EARSeL workshop on imaging spectroscopy, pp 37–42

14. Coppin P, Jonckheere I, Nackaerts K, Muys B, Lambin E (2004) Digital change detection methods in ecosystem monitoring: a review. Int J Remote Sens 25(9):1565–1596

15. Dai X, Khorram S (1998) The effects of image misregistration on the accuracy of remotely sensed change detection. IEEE Trans Geosci Remote Sens 36(5):1566–1577

16. de Carvalho Jr OA, Meneses PR (2000) Spectral Correlation Mapper (SCM): an improvement on the Spectral Angle Mapper (SAM). In: Proceedings of 9th JPL airborne earth science workshop 9:65–74

17. Dobigeon N, Moussaoui S, Coulon M, Tourneret JY, Hero AO (2009) Joint Bayesian endmember extraction and linear unmixing for hyperspectral imagery. IEEE Trans Geosci Remote Sens 57(11):4355–4368

18. Du Q, Wasson L, King R (2005) Unsupervised linear unmixing for change detection in multitemporal airborne hyperspectral imagery. In: Proceedings of international workshop on the analysis of multi-temporal remote sensing images, pp 136–140

19. Du Y, Teillet PM, Cihlar J (2002) Radiometric normalization of multitemporal high-resolution satellite images with quality control for land cover change detection. Remote Sens Environ 82(1):123–134

20. Eismann MT, Meola J, Hardie RC (2007) Hyperspectral change detection in the presence of diurnal and seasonal variations. IEEE Trans Geosci Remote Sens 46(1):237–249

21. Ertürk A, Iordache MD, Plaza A (2015) Hyperspectral change detection by sparse unmixing with dictionary pruning. In: Proceedings of IEEE WHISPERS

22. Ertürk A, Iordache MD, Plaza A (2017) Sparse unmixing with dictionary pruning for hyperspectral change detection. IEEE J Sel Topics Appl Earth Obs Remote Sens 10(1):321–330

23. Ertürk A, Plaza A (2015) Informative change detection by unmixing for hyperspectral images. IEEE Geosci Remote Sens Lett 12(6):1252–1256

24. ESA: Sentinel online. https://sentinel.esa.int/

25. Frank M, Canty M (2003) Unsupervised change detection for hyperspectral images. In: Proceedings of 12th JPL airborne earth science workshop, pp 63–72

26. Giannandrea A, Raqueno N, Messinger DW, Faulring J, Kerekes JP, van Aardt J, Canham K, Hagstrom S, Ontiveros E, Gerace A, Kaufman J, Vongsy K, Griffith H, Bartlett BD (2013) The SHARE 2012 data campaign. In: Proceedings of SPIE, vol 8743, p 87430F

27. Harsanyi JC. Chang C (1994) Hyperspectral image classification and dimensionality reduction: an orthogonal subspace projection. IEEE Trans Geosci Remote Sens 32(4):779–785

28. Hasanlou M, Seydi ST (2018) Hyperspectral change detection: an experimental comparative study. Int J Remote Sens 39(20):7029–7083

29. Heo J, FitzHugh TW (2000) A standardized radiometric normalization method for change detection using remotely sensed imagery. Photogramm Eng Remote Sens 66(2):173–181

30. Holben BN, Shimabukuro YE (1993) Linear mixing model applied to coarse spatial resolution data from multispectral satellite sensors. Int J Remote Sens 14(11):2231–2240

31. Hotelling H (1936) Relations between two sets of variates. Biometrika 28(3–4):321–377

32. Huang F, Yu Y, Feng T (2018) Hyperspectral remote sensing image change detection based on tensor and deep learning. J Vis Commun Image R 58:233–244

33. Johnson RD, Kasischke ES (1998) Change vector analysis: a technique for multispectral monitoring of land cover and condition. Int J Remote Sens 19(3):411–426

34. Kasetkasem T, Varshney PK (2002) An image change detection algorithm based on Markov random field models. IEEE Trans Geosci Remote Sens 40(8):1815–1823

35. Kay SM (1998) Fundamentals of statistical signal processing, Volume II: Detection theory, 1 edn. Prentice Hall (1998)

36. Kruse FA, Lefkoff AB, Boardman JW, Heidebrecht KB, Shapiro AT, Barloon PJ, Goetz AFH (1993) The Spectral Image Processing System (SIPS)—interactive visualization and analysis of imaging spectrometer data. Remote Sens Environ 44:145–163

37. Lee BG, Tom VT, Carlotto MJ (1986) A signal-symbol approach to change detection. In: Proceedings of association for the advancement of artificial intelligence (AAAI), pp 1138–1143

38. Liu S, Bruzzone L, Bovolo F, Du P (2014) Hierarchical unsupervised change detection in multitemporal hyperspectral images. IEEE Trans Geosci Remote Sens 53(1):244–260

39. Liu S, Bruzzone L, Bovolo F, Du P (2016) Unsupervised multitemporal spectral unmixing for detecting multiple changes in hyperspectral images. IEEE Trans Geosci Remote Sens 54(5)

40. Liu S, Bruzzone L, Bovolo F, Zanetti M, Du P (2015) Sequential spectral change vector analysis for iteratively discovering and detecting multiple changes in hyperspectral images. IEEE Trans Geosci Remote Sens 53(8):4363–4378

41. Liu S, Du Q, Tong X, Samat A, Bruzzone L, Bovolo F (2017) Multiscale morphological compressed change vector analysis for unsupervised multiple change detection. IEEE J Sel Top Appl Earth Obs Remote Sens 10(9):4124–4137

42. Longbotham N, Camps-Valls G (2014) A family of kernel anomaly change detectors. In: Proceedings of IEEE WHISPERS, pp 1–4

43. Malila WA (1980) Change vector analysis: an approach for detecting forest changes with Landsat. In: Proceedings of IEEE symposium on machine processing of remotely sensed data, pp 326–335

44. Manolakis D, Siracusa C, Shaw G (2001) Hyperspectral subpixel target detection using the linear mixing model. IEEE Trans Geosci Remote Sens 39(7):1392–1409

45. Matteoli S, Diani M, Theiler J (2014) An overview of background modeling for detection of targets and anomalies in hyperspectral remote sensing imagery. IEEE J Sel Top Appl Earth Obs Remote Sens 7(6):2317–2336

46. Matteoli S, Ientilucci EJ, Kerekes JP (2011) Operational and performance considerations of radiative-transfer modeling in hyperspectral target detection. IEEE Trans Geosci Remote Sens 49(4):1343–1355

47. Meola J, Eismann MT (2008) Image misregistration effects on hyperspectral change detection. In: Proceedings of SPIE, vol 6966, p 69660Y

48. Meola J, Eismann MT, Moses RL, Ash JN (2011) Detecting changes in hyperspectral imagery using a model-based approach. IEEE Trans Geosci Remote Sens 49(7):2647–2661
49. Meola J, Eismann MT, Moses RL, Ash JN (2012) Application of model-based change detection to airborne VNIR/SWIR hyperspectral imagery. IEEE Trans Geosci Remote Sens 50(10):3693–3706
50. Michalek JL, Wagner TW, Luczkovich JJ, Stoffle RW (1993) Multispectral change vector analysis for monitoring coastal marine environments. Photogramm Eng & Remote Sens 59(3):381–384
51. Nackaerts K, Vaesen K, Muys B, Coppin P (2003) Comparative performance of a modified change vector analysis in forest change detection. Int J Remote Sens 26(5):839–852
52. NASA Goddard Space Flight Center: Earth Observing-1: the extended mission. https://eo1.gsfc.nasa.gov/
53. Nielsen AA (2007) The regularized iteratively reweighted MAD method for change detection in multi- and hyperspectral data. IEEE Trans Image Process 16(2):463–478
54. Nielsen AA, Conradsen K, Simpson JJ (1998) Multivariate Alteration Detection (MAD) and MAF postprocessing in multispectral, bitemporal image data: new approaches to change detection studies. Remote Sens Environ 64:1–19
55. Nielsen AA, Müller A (2003) Change detection by the MAD method in hyperspectral image data. In: Proceedings of 3rd EARSeL workshop on imaging spectroscopy, pp 1–2
56. Ortiz-Rivera V, Velez-Reyes M, Roysam B (2006) Change detection in hyperspectral imagery using temporal principal components. In: Proceedings of SPIE, vol 6233, p 623312
57. Pearlman JS, Barry PS, Segal CC, Shepanski J, Beiso D, Carman SL (2003) Hyperion, a space-based imaging spectrometer. IEEE Trans Geosci Remote Sens 41(6):1160–1173
58. Reed IS, Yu X (1990) Adaptive multiple-band CFAR detection of an optical pattern with unknown spectral distribution. IEEE Trans Acoust, Speech, Signal Process 38(10):1760–1770
59. Rotman S, Shalev H (2017) Evaluating hyperspectral imaging change detection methods. In: Proceedings of IEEE IGARSS, pp 1946–1949
60. Schaum A, Stocker A (1997) Long-interval chronochrome target detection. In: Proceedings of international symposium on spectral sensing research
61. Schaum A, Stocker A (1997) Spectrally selective target detection. In: Proceedings of the first international workshop on technology and policy for accessing spectrum
62. Schaum AP, Stocker AD (1997) Subclutter target detection using sequences of thermal infrared multispectral imagery. In: Proceedings of SPIE, vol 3071, pp 12–22
63. Seydi ST, Hasanlou M (2017) A new land-cover match-based change detection for hyperspectral imagery. Eur J Remote Sens 50(1):517–533
64. Singh A (1989) Review article: digital change detection techniques using remotely-sensed data. Int J Remote Sens 10(6):989–1003
65. Snyder D, Kerekes J, Fairweather I, Crabtree R, Shive J, Hager S (2008) Development of a web-based application to evaluate target finding algorithms. In: Proceedings of IEEE IGARSS, vol 2, pp 915–918
66. Song A, Choi J, Han Y, Kim Y (2018) Change detection in hyperspectral images using recurrent 3D fully convolutional networks. Remote Sens 10(1827):1–22
67. Song C, Woodcock CE, Seto KC, Lenney MP, Macomber SA (2001) Classification and change detection using Landsat TM data: when and how to correct atmospheric effects? Remote Sens Environ 75(2):230–244
68. SpecTIR: Featured Instruments—ProSpecTIR VS. https://www.spectir.com/technology
69. Stocker, A., Villeneuve, P.: Generalized chromodynamic detection. In: Proceedings of IEEE IGARSS, pp. 1–4
70. Theiler J (2008) Quantitative comparison of quadratic covariance-based anomalous change detectors. Appl Opt 47(28):12–26
71. Theiler J (2008) Subpixel anomalous change detection in remote sensing imagery. In: Proceedings of IEEE southwest symposium on image analysis and interpretation (SSIAI), pp 165–168
72. Theiler J (2013) Spatio-spectral anomalous change detection in hyperspectral imagery. In: Proceedings of IEEE global conference on signal and information processing

73. Theiler J (2018) Anomaly testing: Chapter 19 in statistical methods for materials science: the data science of microstructure characterization. CRC Press
74. Theiler J, Adler-Golden SM (2008) Detection of ephemeral changes in sequences of images. In: Proceedings of 37th IEEE applied imagery pattern recognition workshop (AIPR)
75. Theiler J, Foy BR, Wohlberg B, Scovel C (2010) Parametric probability distributions for anomalous change detection. In: Proceedings of military sensing symposia (MSS)
76. Theiler J, Perkins S (2006) Proposed framework for anomalous change detection. In: Proceedings of 23rd international conference on machine learning (ICML)
77. Theiler J, Scovel C, Wohlberg B, Foy BR (2010) Elliptically-contoured distributions for anomalous change detection in hyperspectral imagery. IEEE Geosci Remote Sens Lett 7(2):271–275
78. Theiler J, Wohlberg B (2012) Local co-registration adjustment for anomalous change detection. IEEE Trans Geosci Remote Sens 50(8):3107–3116
79. Townshend JRG, Justice CO, Gurney C, McManus J (1992) The impact of misregistration on change detection. IEEE Trans Geosci Remote Sens 30(5):1054–1060
80. USGS: Earth Observing-1 (EO-1)—Archived Aug 1, 2018. https://eo1.usgs.gov
81. USGS: Landsat missions. https://www.usgs.gov/landsat
82. Van der Meer F (2006) The effectiveness of spectral similarity measures for the analysis of hyperspectral imagery. Int J Appl Earth Obs Geoinf 8(1):3–17
83. Virag LA, Colwell JE (1987) An improved procedure for analysis of change in thematic mapper image-pairs. In: Proceedings of international symposium on remote sensing of environment, pp 1101–1110
84. Vongsy K, Karimkhan S, Shaw AK, Wicker D (2007) Change detection for hyperspectral imagery. In: Proceedings of SPIE, vol 6565, 656516
85. Vongsy K, Mendenhall MJ (2016) Integrating spatial & spectral information for change detection in hyperspectral imagery. In: Proceedings of IEEE WHISPERS, pp 1–4
86. Wang Q, Yuan Z, Du Q, Li X (2019) GETNET: A General End-to-End 2-D CNN framework for hyperspectral image change detection. IEEE Trans Geosci Remote Sens 57(1):3–13
87. Warner T (2005) Hyperspherical direction cosine change vector analysis. Int J Remote Sens 5(6):1201–1215
88. Winter ME (1999) N-FINDER: an algorithm for fast autonomous spectral end-member determination in hyperspectral data. In: Proceedings of SPIE, vol 3753, pp 266–276
89. Wohlberg B, Theiler J (2009) Improved change detection with local co-registration adjustments. In: Proceedings of IEEE WHISPERS, pp 1–4
90. Wu C, Du B, Zhang L (2013) A subspace-based change detection method for hyperspectral images. IEEE J Sel Top Appl Earth Obs Remote Sens 6(2):815–830
91. Wu C, Du B, Zhang L (2018) Hyperspectral anomalous change detection based on joint sparse representation. ISPRS J Photogramm Remote Sens 146, 137–150
92. Yang Z, Mueller R (2007) Spatial-spectral cross-correlation for change detection—a case study for citrus coverage change detection. In: Proceedings of ASPRS
93. Yokoya N, Zhu X (2015) Graph regularized coupled spectral unmixing for change detection. In: Proceedings of IEEE WHISPERS
94. Zelinski ME, Henderson J, Smith M (2014) Use of Landsat 5 for change detection at 1998 Indian and Pakistani nuclear test sites. IEEE J Sel Top Appl Earth Obs Remote Sens 7(8):3453–3460
95. Zelinski ME, Henderson JR, Held EL (2018) Image registration and change detection for artifact detection in remote sensing imagery. In: Proceedings of SPIE, vol 10644, pp 1–18
96. Ziemann A, Messinger DW, Basener B (2010) Iterative convex hull volume estimation in hyperspectral imagery for change detection. In: Proceedings of SPIE, vol 7695, p 76951I
97. Ziemann A, Ren CX, Theiler J (2019) Multi-sensor anomalous change detection at scale. In: Proceedings of SPIE, vol 10986, p 1098615
98. Ziemann A, Theiler J (2017) Simplex ACE: a constrained subspace detector. Opt Eng 56(8):1–13
99. Zollweg J, Schlamm A, Gillis DB, Messinger D (2011) Change detection using mean-shift and outlier-distance metrics. In: Proceedings of SPIE, vol 8048, p 804808

Chapter 13
Recent Advances in Hyperspectral Unmixing Using Sparse Techniques and Deep Learning

Shaoquan Zhang, Yuanchao Su, Xiang Xu, Jun Li, Chengzhi Deng and Antonio Plaza

Abstract Spectral unmixing is an important technique for remotely sensed hyperspectral image interpretation that expresses each (possibly mixed) pixel vector as a combination of pure spectral signatures (endmembers) and their fractional abundances. Recently, sparse unmixing and deep learning have emerged as two powerful approaches for spectral unmixing. In this chapter, we focus on two particularly innovative contributions. First, we provide an overview of recent advances in semi-supervised sparse unmixing algorithms, with particular emphasis on techniques that include spatial–contextual information for a better scene interpretation. These algorithms require a spectral library of signatures available a priori to conduct the unmixing. Then, we describe new developments in the use of deep learning for spec-

S. Zhang · C. Deng
Jiangxi Province Key Laboratory of Water Information Cooperative Sensing
and Intelligent Processing, Nanchang Institute of Technology, Nanchang
330099, China
e-mail: zhangshaoquan@nit.edu.cn

C. Deng
e-mail: dengchengzhi@126.com

Y. Su · J. Li (✉)
Guangdong Provincial Key Laboratory of Urbanization and Geo-Simulation,
School of Geography and Planning, Sun Yat-sen University, Guangzhou
510275, China
e-mail: lijun48@mail.sysu.edu.cn

Y. Su
e-mail: suych3@mail2.sysu.edu.cn

X. Xu
Zhongshan Institute, University of Electronic Science and Technology of China,
Zhongshan 528402, China
e-mail: xuxiang8@mail2.sysu.edu.cn

A. Plaza
Hyperspectral Computing Laboratory, Department of Technology of Computers
and Communications, Escuela Politécnica, University of Extremadura,
10003 Cáceres, Spain
e-mail: aplaza@unex.es

© Springer Nature Switzerland AG 2020
S. Prasad and J. Chanussot (eds.), *Hyperspectral Image Analysis*,
Advances in Computer Vision and Pattern Recognition,
https://doi.org/10.1007/978-3-030-38617-7_13

377

tral unmixing purposes, focusing on a new fully unsupervised deep auto-encoder network (DAEN) method. Our experiments with simulated and real hyperspectral datasets demonstrate the competitive advantages of these innovative approaches over some well-established unmixing methods, revealing that these methods are currently at the forefront of hyperspectral unmixing.

13.1 Introduction

13.1.1 Spectral Unmixing

Hyperspectral remote sensing sensors collect spectral information from the Earth's surface using hundreds of narrow and contiguous wavelength bands [1]. It has been widely applied in various fields, such as target detection, material mapping, and material identification [2]. However, due to insufficient spatial resolution and spatial complexity, pixels in remotely sensed hyperspectral images are likely to be formed by a mixture of pure spectral constituents (*endmembers*) rather than a single substance [3]. The existence of mixed pixels complicates the exploitation of hyperspectral images [4]. Spectral unmixing, aimed at estimating the fractional abundance of the pure spectral signatures or endmembers, was proposed to deal with the problem of spectral mixing and effectively identifies the components of the mixed spectra in each pixel [5].

Unmixing algorithms rely on specific mixing models, which can be characterized as either linear or nonlinear [5, 6]. On the one hand, the linear model assumes that the spectral response of a pixel is given by a linear combination of the endmembers present in the pixel. On the other hand, the nonlinear mixture model assumes that the incident radiation interacts with more than one component and is affected by multiple scattering effects [3, 7]. As a result, nonlinear unmixing generally requires prior knowledge about object geometry and the physical properties of the observed objects [8]. The linear mixture model exhibits practical advantages, such as ease of implementation and flexibility in different applications. In this chapter, we will focus exclusively on the linear mixture model.

Under the linear mixture model, a group of unmixing approaches has been proposed [9–13]. Depending on whether a spectral library is available or not, we classify these methods into two categories, i.e., unsupervised and semi-supervised unmixing algorithms. With the wide availability of spectral libraries, sparse unmixing [8], as a semi-supervised approach in which mixed pixels are expressed in the form of combinations of a number of pure spectral signatures from a large spectral library, is able to handle the drawbacks introduced by such *virtual* endmembers and the unavailability of pure pixels.

13.1.2 Sparse Unmixing

The sparse unmixing approach exhibits significant advantages over unsupervised approaches, as it does not need to extract endmembers from the hyperspectral data or estimate the number of endmembers. Another advantage of sparse unmixing is that it provides great potential for accurate estimation of the fractional abundances, as all endmembers are normally represented in the library. However, these algorithms fully rely on the availability of a library *in advance*, and hence their semi-supervised nature.

The success of sparse unmixing relies on the fact that the unmixing solution is sparse, as the number of endmembers used to represent a mixed pixel is generally much smaller than the number of spectral signatures in the library [8]. As a result, new algorithms have been developed to enforce the sparsity on the solution. The sparse unmixing algorithm via variable splitting and augmented Lagrangian (SUnSAL) [8] adopts the ℓ_1 regularizer on the abundance matrix, which aims at introducing sparsity through the spectral domain that a pixel is unlikely to be mixed by a high number of components. The introduction of SUnSAL brought new insights into the concept of sparse unmixing. However, the real degree of sparsity is beyond the reach of the ℓ_1 regularizer due to the imbalance between the number of endmembers in the library and the number of components that generally participate in a mixed pixel. New algorithms have been developed in order to perform a better characterization of the degree of sparsity. Some techniques have focused on the introduction of new orders over the sparse regularizer such as the collaborative SUnSAL (CLSUnSAL) algorithm [14] and the graph-regularized $\ell_{1/2}$-NMF (GLNMF) method [15]. Other algorithms have introduced weighting factors to penalize the nonzero coefficients on the sparse solution [16], such as the reweighted sparse unmixing method [17] and the double reweighted sparse unmixing (DRSU) algorithm [18]. Although these methods obtained promising results, they consider pixels in a hyperspectral data as independent entities, and the spatial–contextual information in the hyperspectral image is generally disregarded. Since hyperspectral images generally follow specifical spatial arrangements by nature, it is important to consider spatial information for their characterization [19].

Following this observation, several algorithms have focused on incorporating spatial correlation into the final solution. For instance, the sparse unmixing via variable splitting augmented Lagrangian and total variation (SUnSAL-TV) [20] represents one of the first attempts to include spatial information in sparse unmixing. It exploits the spatial information via a first-order pixel neighborhood system. Similar to SUnSAL, SUnSAL-TV opened new avenues and brought new insights into the concept of spatial sparse unmixing, which is able to promote piece-wise transitions in the estimated abundances. However, its performance strongly relies on the parameter settings [21]. At the same time, its model complexity results in a heavy computational cost, further limiting its practical application potential. New developments aimed at fully exploiting the spatial correlation among image features (and further imposing sparsity on the abundance matrix) have been mainly developed along two directions. High-order neighborhood information over spatial regularizers has been introduced to reach this goal. For instance, the nonlocal sparse unmixing (NLSU) algorithm [22] can take advantage of high-order structural information. However, the neighborhood of the pixel changes randomly, thus limiting the continuity of spectral information. Another drawback of NLSU is that its model is more complex than that of SUnSAL-TV, which limits its practical application. Spatially weighted factors (aimed at characterizing spatial information through the inclusion of a weight on the sparse regularizer) have also been used to account for the spatial information in sparse unmixing. For example, the local collaborative sparse unmixing (LCSU) uses a spatial weight to impose local collaborativity, thus addressing some of the issues observed in SUnSAL-TV (including oversmoothed boundaries and blurred abundance maps) [23]. With similar complexity as the SUnSAL-TV, the LCSU exhibits similar unmixing performance as the SUnSAL-TV. This indicates that using spatial weights (as compared to spatial regularizers) has good potential in terms of improved unmixing performance and computational complexity. In [24], the spectral–spatial weighted sparse unmixing (S^2WSU) is proposed, which simultaneously exploits the spectral and spatial information contained in hyperspectral images via weighting factors, aiming at enhancing the sparsity of the solution. As a framework, the S^2WSU algorithm with its open structure, it is able to accept multiple types of spectral and spatial weighting factors, thus providing great flexibility for the exploration of different spatial scenarios, such as edge information, nonlocal similarity, homogeneous neighborhood information, etc.

13.1.3 Deep Learning for Spectral Unmixing

With advances in computer technology, learning-based approaches for unmixing have achieved a fast development in the past few years. Joint Bayesian unmixing is a typical example of learning-based approaches, which leads to good abundance estimates due to the incorporation of a full additivity (i.e., sum-to-one) and nonnegativity constraints [25–27]. Approaches based on artificial neural networks (ANNs) have also been developed for the learning of abundance fractions, assuming the prior knowledge of the endmember signatures [28–30]. These approaches exhibit better performance when compared with handcrafted methods, but they assume that endmembers are known in advance and, therefore, need to incorporate endmember extraction algorithms to perform unmixing. More recently, as a common tool for deep learning, auto-encoders have achieved a fast development in unmixing applications. Nonnegative sparse auto-encoder (NNSAE) and denoising auto-encoder were employed to obtain the endmember signatures and abundance fractions simultaneously for unmixing, with advanced denoising and intrinsic self-adaptation capabilities [31–33]. However, their strength lies in the aspect of noise reduction and they exhibit limitations when dealing with outliers. Due to the fact that outliers likely lead to initialization problems, their presence can bring strong interference to the unmixing solutions. In [34], a stacked nonnegative sparse auto-encoder (SNSA) is proposed to address the issue of outliers. For linear mixing model (LSM)-based hyperspectral unmixing, the physical meaning of the model implies the sum-to-one on abundance fractions when every material in a pixel can be identified [3, 33, 35]. However, similar to the NMF-based approaches, SNSA adopts an additivity penalty on the abundance coefficients. The additivity penalty denotes that a penalty coefficient is used for controlling approximation of the sum-to-one. As this is not a hard constraint, the sum-to-one constraint is not necessarily ensured [34].

In [36], the fully unsupervised deep auto-encoder network (DAEN) unmixing method was recently proposed to address the presence of outliers in hyperspectral data. The DAEN has two main steps. In the first step, the spectral features are learned by the stacked auto-encoders (SAEs), aiming at generating good initializations for the network. In the second step, it employs a variational auto-encoder (VAE) to perform unmixing for the estimation of the endmembers and abundances. VAE combines variational inference to perform unsupervised learning and inherit auto-encoder architecture which can be trained with gradient descent [37]. Different from conventional auto-encoders, VAEs include a reparameterization which strictly ensures the abundance sum-to-one constraint during unmixing. Compared with other NMF-based algorithms, the DAEN has three main advantages: (1) with the use of SAEs, it can effectively tackle the problem of outliers and generate a good initialization of the unmixing network; (2) with the adoption of a VAE, it can ensure the nonnegativity and sum-to-one constraints, resulting in the good performance on abundance estimation; and (3) the endmember signatures and abundance fractions are obtained simultaneously. We emphasize the fully unsupervised nature of DAEN as one of its most powerful features.

13.1.4 Contributions of This Chapter

In this chapter, we focus on two types of techniques that are currently at the forefront of spectral unmixing. First, we provide an overview of advances in sparse unmixing algorithms, which can improve over traditional sparse unmixing algorithms by including spatial–contextual information that is crucial for a better scene interpretation. As these algorithms are semi-supervised and dependent on a library, we then describe new developments in the use of deep learning to perform spectral unmixing in fully unsupervised fashion, focusing on the DAEN method. Our experiments with simulated and real hyperspectral datasets demonstrate the competitive advantages of these innovative approaches over some well-established unmixing methods.

The remainder of this paper is organized as follows. The principles of sparse unmixing theory are presented in Sect. 13.2. The DAEN unmixing method is described in detail in Sect. 13.3. Section 13.4 describes several experiments to evaluate sparse unmixing algorithms. Section 13.5 describes several experiments to evaluate the DAEN algorithm. Finally, Sect. 13.6 concludes with some remarks and hints at plausible future research lines.

13.2   Sparse Unmixing Techniques

13.2.1   Sparse Versus Spectral Unmixing

The linear mixture model assumes that the spectral response of a pixel in any given spectral band is a linear combination of all of the endmembers present in the pixel at the respective spectral band. For each pixel, the linear model can be written as follows:

$$
\mathbf{y} = \mathbf{M}\boldsymbol{\alpha} + \mathbf{n}
$$
$$
\text{s.t.:} \quad \alpha_j \geq 0, \ \sum_{j=1}^{q} \alpha_j = 1, \tag{13.1}
$$

where $\mathbf{y}$ is a $d \times 1$ column vector (the measured spectrum of the pixel), d denotes the number of bands. $\mathbf{M}$ is a $d \times q$ matrix containing q pure spectral signatures (endmembers), $\boldsymbol{\alpha}$ is a $q \times 1$ vector containing the fractional abundances of the endmembers, and $\mathbf{n}$ is a $d \times 1$ vector collecting the errors affecting the measurements at each spectral band. The so-called abundance nonnegativity constraint (ANC) ($\alpha_j \geq 0$ for ($j = 1, 2, \ldots, q$)) and the abundance sum-to-one constraint (ASC)($\sum_{j=1}^{q} \alpha_j = 1$).

Sparse unmixing reformulates (13.1) assuming the availability of a library of spectral signatures a priori as follows:

$$
\mathbf{y} = \mathbf{A}\mathbf{h} + \mathbf{n}, \tag{13.2}
$$

where $\mathbf{h} \in \mathbb{R}^{m \times 1}$ is the fractional abundance vector compatible with spectral library $\mathbf{A} \in \mathbb{R}^{d \times m}$ and m is the number of spectral signatures in $\mathbf{A}$.

Assuming that the dataset contains n pixels organized in the matrix $\mathbf{Y} = [\mathbf{y}_1, \ldots, \mathbf{y}_n] \in \mathbb{R}^{d \times n}$ we may write then

$$
\mathbf{Y} = \mathbf{A}\mathbf{H} + \mathbf{N} \ \text{s.t.:} \quad \mathbf{H} \geq 0, \tag{13.3}
$$

where $\mathbf{N} = [\mathbf{n}_1, \ldots, \mathbf{n}_n] \in \mathbb{R}^{d \times n}$ is the error. $\mathbf{H} = [\mathbf{h}_1, \ldots, \mathbf{h}_n] \in \mathbb{R}^{m \times n}$ denotes the abundance maps corresponding to library $\mathbf{A}$ for the observed data $\mathbf{Y}$, and $\mathbf{H} \geq 0$ is the so-called abundance nonnegativity constraint (ANC). It should be noted that we explicitly enforce the ANC constraint without the abundance sum-to-one constraint (ASC), due to some criticisms about the ASC in the literature [8].

As the number of endmembers involved in a mixed pixel is usually very small when compared with the size of the spectral library, the abundance matrix $\mathbf{H}$ is sparse. With these considerations in mind, the unmixing problem can be formulated as an $\ell_2 - \ell_0$ optimization problem,

$$
\min_{\mathbf{H}} \ \frac{1}{2}\|\mathbf{A}\mathbf{H} - \mathbf{Y}\|_F^2 + \lambda\|\mathbf{H}\|_0 \ \text{s.t.:} \quad \mathbf{H} \geq 0, \tag{13.4}
$$

where $|| \cdot ||_F$ is the Frobenius norm and λ is a regularization parameter. Problem (13.4) is nonconvex and difficult to solve [38, 39]. The SUnSAL alternatively uses the $\ell_2 - \ell_1$ norm to replace the $\ell_2 - \ell_0$ norm and solves the unmixing problem as follows [40]:

$$\min_{\mathbf{H}} \quad \frac{1}{2} ||\mathbf{AH} - \mathbf{Y}||_F^2 + \lambda ||\mathbf{H}||_{1,1} \quad \text{s.t.:} \quad \mathbf{H} \geq 0, \tag{13.5}$$

where $||\mathbf{H}||_{1,1} = \sum_{i=1}^{n} ||\mathbf{h}_i||_1$ with $\mathbf{h}_i$ $(i = 1, \ldots, n)$ being the ith column of $\mathbf{H}$. SUnSAL solves the optimization problem in (13.5) efficiently using the ADMM [40]. However, as stated before, the real degree of sparsity is generally beyond the reach of the ℓ_1 regularizer.

13.2.2 Collaborative Regularization

Similar to (13.5), in [14], an $\ell_{2,1}$ mixed norm (called collaborative regularization) was proposed, which globally imposes sparsity among the endmembers in collaborative fashion for all pixels. According to the collaborative sparse unmixing model described in [14], the objective function can be defined as follows:

$$\min_{\mathbf{H}} \frac{1}{2} ||\mathbf{AH} - \mathbf{Y}||_F^2 + \lambda \sum_{k=1}^{m} ||\mathbf{h}^k||_2 \quad \text{s.t. } \mathbf{h} \geq 0, \tag{13.6}$$

where $\mathbf{h}^k$ denotes the k-th line of $\mathbf{H}$ $(k = 1, 2, \ldots, m)$ and $\sum_{k=1}^{m} ||\mathbf{h}^k||_2$ is the so-called $\ell_{2,1}$ mixed norm. Note that the main difference between SUnSAL and CLSUn-SAL is that the former employs pixel-wise independent regressions, while the latter enforces joint sparsity among all the pixels.

13.2.3 Total Variation Regularization

In order to take into account the spatial information of the image, a total variation (TV) regularizer can be integrated with SUnSAL (called SUnSAL-TV) to promote spatial homogeneity among neighboring pixels [20]:

$$\min_{\mathbf{H}} \quad \frac{1}{2} ||\mathbf{AH} - \mathbf{Y}||_F^2 + \lambda ||\mathbf{H}||_{1,1} + \lambda_{TV} TV(\mathbf{H})$$
$$\text{s.t.:} \quad \mathbf{H} \geq 0, \tag{13.7}$$

where $TV(\mathbf{H}) \equiv \sum_{\{k,i\} \in \mathcal{N}} ||\mathbf{h}_k - \mathbf{h}_i||_1$, $\mathcal{N}$ represents the set of (horizontal and vertical) pixel neighbors in the image, and $\mathbf{h}_k$ denotes a series of the neighboring pixels of $\mathbf{h}_i$ in abundance matrix $\mathbf{H}$. SUnSAL-TV shows great potential to exploit the spatial information for sparse unmixing. However, it may lead to oversmoothness and blurred boundaries.

13.2.4 Local Collaborative Regularization

In [23], the proposed LCSU assumes that endmembers tend to appear localized in spatially homogeneous areas instead of distributed over the full image. The proposed approach can also preserve global collaborativity (e.g., in the case that an endmember appears in the whole image), since it generalizes to global collaborativity through local searching:

$$\min_{\mathbf{h}} \frac{1}{2}||\mathbf{AH} - \mathbf{Y}||_F^2 + \lambda \sum_{k=1}^{m} \sum_{i=1}^{n} ||\mathbf{h}_{x \in \mathcal{N}(i)}^k||_2$$
$$\text{s.t. } \mathbf{h} \geq 0, \tag{13.8}$$

where $\mathbf{h}^k$ denotes the kth line of matrix $\mathbf{H}$ ($k = 1, 2, \ldots, m$), $\sum_{k=1}^{m} ||\mathbf{h}^k||_2$ is the so-called $\ell_{2,1}$ mixed norm, $\mathcal{N}(i)$ is the neighborhood of pixel i ($i = 1, 2, \ldots, n$), and λ is a regularization parameter controlling the degree of sparseness. The main difference between the proposed approach and SUnSAL-TV is that LCSU imposes collaborative sparsity among neighboring pixels, while SUnSAL-TV aims at promoting piece-wise smooth transitions in abundance estimations. In other words, SUnSAL-TV enforces that neighboring pixels share similar fractional abundances for the same endmember, while LCSU focuses on imposing local collaborativity among the full set of endmembers, thus addressing problems observed in SUnSAL-TV such as oversmoothed or blurred abundance maps. The main difference between problem (13.8) and problem (13.6) is that LCSU introduces spatial information to promote *local* collaborativity, while CLSUnSAL focuses on *global* collaborativity. In comparison with CLSUnSAL, the proposed LCSU assumes that neighboring pixels share the same support. This is more realistic, as a given endmember is likely to appear localized in a spatially homogeneous region rather than in the whole image.

13.2.5 Double Reweighted Regularization

Inspired by the success of weighted ℓ_1 minimization in sparse signal recovery, the double reweighted sparse unmixing and total variation (DRSU-TV) [41] was proposed to simultaneously exploit the spectral dual sparsity as well as the spatial smoothness of fractional abundances, as follows:

$$\min_{\mathbf{H}} \quad \frac{1}{2}||\mathbf{AH} - \mathbf{Y}||_F^2 + \lambda||(\mathbf{W}_{\text{spe2}}\mathbf{W}_{\text{spe1}}) \odot \mathbf{H}||_{1,1} + \lambda_{\text{TV}}\text{TV}(\mathbf{H}), \quad \text{s.t.:} \quad \mathbf{H} \geq 0,$$
$$\tag{13.9}$$

where the operator $\odot$ denotes the element-wise multiplication of two variables. The first regularizer $\lambda||(\mathbf{W}_{\text{spe2}}\mathbf{W}_{\text{spe1}}) \odot \mathbf{X}||_{1,1}$ introduces a prior with spectral sparsity, where λ is the regularization parameter, $\mathbf{W}_{\text{spe1}} = \{w_{\text{spe1},ki} | k = 1, \ldots, m, i = 1, \ldots, n\} \in \mathbb{R}^{m \times n}$ and $\mathbf{W}_{\text{spe2}} = \text{diag}(w_{\text{spe2},11}, \ldots, w_{\text{spe2},kk}, \ldots, w_{\text{spe2},mm}) \in \mathbb{R}^{m \times m}$, for $k = 1, \ldots, m$, are the dual weights, with $\mathbf{W}_{\mathrm{spe1}}$ being the original weight introduced in [16] aimed at penalizing the nonzero coefficients on the solution and $\mathbf{W}_{\mathrm{spe2}}$ promoting nonzero row vectors. The latter regularizer $\lambda_{\mathrm{TV}} \mathrm{TV}(\mathbf{H})$ exploits the spatial prior with λ_{TV} being the parameter controlling the degree of smoothness. It can be seen that DRSU-TV incorporates a TV-based regularizer to enforce the spatial smoothness of abundances compared to DRSU.

In [41], Problem (13.9) is optimized via ADMM under an iterative scheme. The dual weights $\mathbf{W}_{\mathrm{spe1}}$ and $\mathbf{W}_{\mathrm{spe2}}$ are updated as follows, at iteration $t + 1$:

$$w_{\mathrm{spe1},ki}^{(t+1)} = \frac{1}{h_{ki}^{(t)} + \varepsilon}, \qquad (13.10)$$

where $\varepsilon > 0$ is a small positive value and

$$w_{\mathrm{spe2},kk}^{(t+1)} = \frac{1}{||\mathbf{H}^{(t)}(k, :)||_2 + \varepsilon}, \qquad (13.11)$$

where $\mathbf{H}^{(t)}(k, :)$ is the kth row in the estimated abundance of the tth iteration. Notice that, as shown in (13.10) and (13.11), it is suggested that large weights be used to discourage nonzero entries in the recovered signal, while small weights encourage nonzero entries. DRSU-TV, exploiting the spectral and spatial priors simultaneously under the sparse unmixing model, exhibits good potential in comparison with the ℓ_1- or TV-based methods. However, as an adaptation of the ℓ_1- and TV-based approach, the limitations of DRSU-TV are associated with the use of a regularizer-based spatial prior. That is, the computational complexity is similar to that of SUnSAL-TV. Such high computational complexity constrains the practical applications of DRSU-TV. Furthermore, the unmixing performance of the method is sensitive to the regularization parameter λ_{TV}.

13.2.6 Spectral–Spatial Weighted Regularization

In [24], the $\mathrm{S}^2\mathrm{WSU}$ algorithm is developed, which aims at exploiting the spatial information more efficiently for sparse unmixing purposes. As opposed to the approaches that exploit a regularizer-based spatial prior (which have one additional parameter for the spatial regularizer and often exhibit high complexity), the $\mathrm{S}^2\mathrm{WSU}$ algorithm includes the spatial correlation via a weighting factor, resulting in good computational efficiency and less regularization parameters. Let $\mathbf{W}_{\mathrm{spe}} \in \mathbb{R}^{m \times m}$ be the spectral weighting matrix and $\mathbf{W}_{\mathrm{spa}} \in \mathbb{R}^{m \times n}$ be the spatial one. Following [16], the objective function of the $\mathrm{S}^2\mathrm{WSU}$ is given as follows:

$$\min_{\mathbf{H}} \frac{1}{2}||\mathbf{AH} - \mathbf{Y}||_F^2 + \lambda||(\mathbf{W}_{\mathrm{spe}}\mathbf{W}_{\mathrm{spa}}) \odot \mathbf{H}||_{1,1}, \quad \text{s.t.: } \mathbf{H} \geq 0. \qquad (13.12)$$

For the spectral weighting factor $\mathbf{W}_{spe}$, relying on the success of [14, 18], it adopts row collaborativity to enforce joint sparsity among all the pixels. Similar to the $\mathbf{W}_{spe2}$ in DRSU-TV, the $\mathbf{W}_{spe}$ aims at enhancing the sparsity of the endmembers in the spectral library. In detail, at iteration $t + 1$, it can be updated as

$$\mathbf{W}_{spe}^{(t+1)} = \text{diag}\left[\frac{1}{||\mathbf{H}^{(t)}(1, :)||_2 + \varepsilon}, \ldots, \frac{1}{||\mathbf{H}^{(t)}(m, :)||_2 + \varepsilon}\right]. \quad (13.13)$$

For the spatial weighting factor $\mathbf{W}_{spa}$, let $w_{spa,ki}^{(t+1)}$ be the element of the kth line and ith row in $\mathbf{W}_{spa}$ at iteration $t + 1$, it incorporates the neighboring information as follows:

$$w_{spa,ki}^{(t+1)} = \frac{1}{f_{x\in\mathcal{N}(i)}(h_{kx}^{(t)}) + \varepsilon}, \quad (13.14)$$

where $\mathcal{N}(i)$ denotes the neighboring set for element h_{ki}, and $f(\cdot)$ is a function explicitly exploiting the spatial correlations through the neighborhood system. It uses the neighboring coverage and importance to incorporate the spatial correlation as follows:

$$f(h_{ki}) = \frac{\sum_{x\in\mathcal{N}(i)} \theta_{kx} h_{kx}}{\sum_{x\in\mathcal{N}(i)} \theta_{kx}}, \quad (13.15)$$

where $\mathcal{N}(i)$ corresponds to the neighboring coverage and θ represents the neighborhood importance. It considers the 8-connected (3×3 window) for algorithm design and experiments. With respect to the neighboring importance, for any two entries k and i, we compute it as follows:

$$\theta_{ki} = \frac{1}{im(k, i)}, \quad (13.16)$$

where function $im(\cdot)$ is the important measurement over the two elements h_k and h_i. Let (a, b) and (c, d) be the spatial coordinates of h_k and h_i. The European distance is specifically considered, that is, $\theta_{ki} = 1/\sqrt{(a - c)^2 + (b - d)^2}$.

It should be noted that the optimization problem of S^{2}WSU can be iteratively solved by an outer–inner looping scheme, where the inner loop updates the unmixing coefficients via ADMM and the outer loop updates the spectral and spatial weights [24].

13.3 Deep Learning for Hyperspectral Unmixing

As one of the very few unsupervised approaches available, the deep auto-encoder network (DAEN) unmixing method specifically addresses the presence of outliers in hyperspectral data [36]. In the following subsections, we describe the different processing modules that compose this promising approach for deep hyperspectral unmixing.

13.3.1 NMF-Based Unmixing

Let $\mathbf{Y} \equiv [\mathbf{y}_1, \ldots, \mathbf{y}_n] \in \mathbb{R}^{d \times n}$ be matrix representation of a hyperspectral dataset with n spectral vectors and d spectral bands. Under the linear mixing model, we have [3, 42]

$$\mathbf{Y} = \mathbf{WH} + \mathbf{N} \qquad (13.17)$$
$$\text{s.t.: } \mathbf{H} \geq 0, \mathbf{1}_m^T \mathbf{H} = \mathbf{1}_n^T,$$

where $\mathbf{W} \equiv [\mathbf{w}_1, \ldots, \mathbf{w}_m] \in \mathbb{R}^{d \times m}$ is the mixing matrix containing m endmembers, $\mathbf{w}_i$ denotes the ith endmember, $\mathbf{H} \geq 0$ and $\mathbf{1}_m^T \mathbf{H} = \mathbf{1}_n^T$ are the so-called abundance nonnegativity and sum-to-one constraints, which stem from a physical interpretation of the abundance vectors, and $\mathbf{1}_m = [1, 1, \ldots, 1]^T$ is a column vector of size m (the notation $[\cdot]^T$ stands for vector or matrix transpose). Finally, $\mathbf{N} \in \mathbb{R}^{d \times n}$ is the error matrix that may affect the measurement process (e.g., noise). It should be noted that the symbol naming in this section is not the same as the naming in Sect. 13.2. In addition, we have a detailed description of each symbol.

For a given observation $\mathbf{Y}$, unmixing aims at obtaining the mixing matrix $\mathbf{W}$ and the abundance matrix $\mathbf{H}$. In this work, we tackle the simultaneous estimation of $\mathbf{W}$ and $\mathbf{H}$ by seeking a solution with the following NMF-based optimization:

$$(\mathbf{W}, \mathbf{H}) = \arg\min_{\mathbf{W}, \mathbf{H}} \frac{1}{2} \|\mathbf{Y} - \mathbf{WH}\|_F^2 + \mu \, f_1(\mathbf{W}) + \lambda f_2(\mathbf{H}), \qquad (13.18)$$

where $\| \cdot \|_F^2$ denotes the Frobenius norm, $f_1(\mathbf{W})$ and $f_2(\mathbf{H})$ are two regularizers on the mixing matrix $\mathbf{W}$ and the abundance fractions $\mathbf{H}$, respectively, with μ and λ being the regularization parameters.

13.3.2 Deep Auto-Encoder Network

In this section, the DAEN unmixing method [36] is described (illustrated in Fig. 13.1), where $\mathbf{U}$ and $\mathbf{V}$ are the latent variables (LV) of the reparameterization of the VAE, respectively. As shown in Fig. 13.1, the endmember matrix $\mathbf{W}$ corresponds to the last weight matrix of the decoder in VAE, and the abundance $\mathbf{H}$ is estimated from the hidden layers of VAE, while $\widehat{\mathbf{W}}$ and $\widehat{\mathbf{H}}$ denote the initializations for VAE generated by SAEs, respectively.

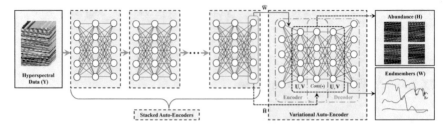

Fig. 13.1 The flowchart of the proposed DAEN, which includes two parts, i.e., stacked auto-encoders (SAEs) and a variational auto-encoder (VAE). The stacked auto-encoders (SAEs) generate the initializations $\widehat{\mathbf{W}}$ and $\widehat{\mathbf{H}}$ for the VAE, while the VAE performs the NMF-based unmixing aiming at obtaining the endmembers $\mathbf{W}$ and abundances $\mathbf{H}$, respectively

13.3.3 Stacked Auto-Encoders for Initialization

Based on the geometry assumption that endmembers are generally located around the vertices of the data simplex, we use a pure pixel-based method to extract a set of candidate pixels as the training set for the SAEs. Specifically, we adopt VCA to obtain a set of k candidates, with $k > m$. As VCA considers random directions in the subspace projection [3, 43], we run it for p times, resulting in q candidates, with $q = p \cdot k$. These q candidates are then grouped into m training sets $\{\mathbf{C}_i\}_{i=1}^{m}$ based on the spectral angle distance (SAD) and clustering, with $\mathbf{C}_i = [\mathbf{c}_1, \ldots, \mathbf{c}_{i_n}] \in \mathbb{R}^{d \times i_n}$ and i_n is the number of samples in $\mathbf{C}_i$. Let $\mathbf{c}_{i_o}$ and $\mathbf{c}_{j_o}$ be the cluster centers of $\mathbf{C}_i$ and $\mathbf{C}_j$, respectively. For any candidate $\mathbf{c}_{i_s}$ in $\mathbf{C}_i$, for $i_s = 1, \ldots, i_n$, we have $\mathrm{SAD}(\mathbf{c}_{i_o}, \mathbf{c}_{i_s}) \leq \mathrm{SAD}(\mathbf{c}_{j_o}, \mathbf{c}_{i_s})$, for any $j = 1, \ldots, m$ and $j \neq i$, where

$$\mathrm{SAD}(\mathbf{c}_{i_o}, \mathbf{c}_{i_s}) = \arccos\left(\frac{[\mathbf{c}_{i_o}, \mathbf{c}_{i_s}]}{\|\mathbf{c}_{i_o}\| \cdot \|\mathbf{c}_{i_s}\|}\right). \tag{13.19}$$

In this work, for p and k, we empirically set $p = 30$ and $k = 3$ m, respectively. By enforcing nonnegativity, the training of SAEs minimizes the reconstruction error as follows:

$$\min \sum_{s=1}^{i_n} \|\mathbf{c}_s - \widehat{\mathbf{w}}_i\|_2^2, \tag{13.20}$$

where $\widehat{\mathbf{w}}_i$ is the reconstructed signature of the ith endmember and $\widehat{\mathbf{W}} = [\widehat{\mathbf{w}}_1, \ldots, \widehat{\mathbf{w}}_m]$ are the reconstructed endmember matrix. Following [44], the reconstructed signature is denoted as

$$\widehat{\mathbf{w}}_i = \mathbf{M}_i f(\mathbf{M}_i^T \mathbf{C}_i), \tag{13.21}$$

where $\mathbf{M}_i$ is the matrix of weights between the input and hidden neurons or those from hidden to output neurons, and $f(\cdot)$ is the activation function [44] given by

$$f(\mathbf{g}_i) = \frac{1}{1 + \exp(-\mathbf{a}_i . * \mathbf{g}_i - \mathbf{b}_i)}, \tag{13.22}$$

where $\mathbf{g}_i = \mathbf{M}_i^T \mathbf{C}_i$, $\mathbf{a}_i$ and $\mathbf{b}_i$ are parameters aimed at controlling the information transmission between neurons, and $.*$ is the dot product, i.e., element-wise operator. Notice that the number of input neurons and output neurons is the same as the hidden neurons, while the number of hidden neurons here is set as the number of bands. Then, we can use a gradient rule to update $\mathbf{a}_i$ and $\mathbf{b}_i$ as follows:

$$\begin{cases} \Delta \mathbf{a}_i = \gamma(1 - (2 + \frac{1}{\tau})f_i + \frac{1}{\tau}f_i^2), \\[2mm] \Delta \mathbf{b}_i = \gamma \frac{1}{\mathbf{b}_i} + \mathbf{g}_i \Delta \mathbf{a}_i, \end{cases} \tag{13.23}$$

where γ and τ are hyper-parameters in the learning process controlling the mean activity level of the desired output distribution. Following the empirical settings in [44], we set $\gamma = 0.0001$ and $\tau = 0.2$. With the aforementioned definition in hand, the learning reduces to the following update rule:

$$\Delta \mathbf{M}_i \Leftarrow \eta \Delta \widehat{\mathbf{w}}_i f_i^T + |\mathbf{M}_i|, \tag{13.24}$$

where $\Delta \widehat{\mathbf{w}}_i$ is the gradient of candidate i for update, $|\mathbf{M}_i|$ enforces the weight matrix to be nonnegative, and η is an adaptive learning rate. In this work, following [44], we set $\eta = \hat{\eta}(\|f_i\|^2 + \epsilon)^{-1}$ with $\hat{\eta} = 0.002$, where $\epsilon = 0.001$ is a small parameter to ensure the positivity of η.

Finally, let $\widehat{\mathbf{w}}_i^t$, $\widehat{\mathbf{w}}_i^{t+1}$ be the reconstructions from the t-th and $(t+1)$-th autoencoders, respectively. The SAEs ends when $\|\widehat{\mathbf{w}}_i^{t+1} - \widehat{\mathbf{w}}_i^t\|_2^2$ converges.

After the endmember matrix $\widehat{\mathbf{W}}$ is reconstructed, based on the linear mixing model (13.17), the abundances $\widehat{\mathbf{H}}$ can be obtained via the fully constrained least square (FCLS) [42]. In the learning of the VAE, $\widehat{\mathbf{W}}$ and $\widehat{\mathbf{H}}$ are used as initializations of $\mathbf{W}$ and $\mathbf{H}$, respectively.

13.3.4   Variational Auto-Encoders for Unmixing

First, let us recall the NMF-based objective function in (13.18), which contains two regularizers on the mixing matrix and abundance matrix, respectively. For the first regularizer $f_1(\mathbf{W})$ on the mixing matrix, following [11], we have

$$f_1(\mathbf{W}) = \text{MinVol}(\mathbf{W}), \tag{13.25}$$

where $\text{MinVol}(\cdot)$ is a function aiming at enclosing all the pixels into the simplex constructed by the endmembers. Specifically, following [11], we set $\text{MinVol}(\mathbf{W}) = \|\det(\mathbf{W})\|$, with $\|\det(\mathbf{W})\|$ being the volume defined by the origin and the columns of $\mathbf{W}$.

With respect to regularizer $f_2(\mathbf{H})$ on the abundance matrix, in order to ensure the nonnegativity and sum-to-one constraints, we employ the variational auto-encoder (VAE) to penalize the solution of $\mathbf{H}$, denoted as

$$f_2(\mathbf{H}) = \text{VAE}(\mathbf{H}), \tag{13.26}$$

where the neurons of all hidden layers are set as the number of endmembers, while the number of inputs and outputs corresponds to the number of pixels.

With these definitions in mind, we obtain the following objective function:

$$
\begin{aligned}
(\mathbf{W}, \mathbf{H}) = \arg\min_{\mathbf{W}, \mathbf{H}} \frac{1}{2} \|\mathbf{Y} - \mathbf{W}\mathbf{H}\|_F^2 \\
+ \mu \, \text{MinVol}(\mathbf{W}) + \lambda \text{VAE}(\mathbf{H}).
\end{aligned}
\tag{13.27}
$$

In the following, we present the VAE-based regularizer in detail. Let $\mathbf{U}$ and $\mathbf{V}$ be the LV, we define $f_2(\mathbf{H})$ as

$$f_2(\mathbf{H}(\mathbf{U}, \mathbf{V})) = \left\| \frac{1}{2n} (\mathbf{1}_{m \times n} + \ln \mathbf{V}^2 - \mathbf{U}^2 - \mathbf{V}^2) \mathbf{1}_n \right\|_2^2, \tag{13.28}$$

where $\mathbf{1}_{m \times n} \in \mathbb{R}^{m \times n}$ with all elements being 1, and vector $\mathbf{1}_n = [1, \ldots, 1]^T \in \mathbb{R}^n$, $\mathbf{U} = \{\mathbf{u}_1, \ldots, \mathbf{u}_n\} \in \mathbb{R}^{m \times n}$, $\mathbf{V} = \{\mathbf{v}_1, \ldots, \mathbf{v}_n\} \in \mathbb{R}^{m \times n}$. The derivation of (13.28) is shown in [36]. Following [37], let $\mathbf{u}_j = [u_{1,j}, \ldots, u_{m,j}]^T \in \mathbb{R}^m$ and $\mathbf{v}_j = [v_{1,j}, \ldots, v_{m,j}]^T \in \mathbb{R}^m$ be the reparameters of LV, we define $h_{i,j} = \text{Cons}(u_{i,j}, v_{i,j})$, where $\text{Cons}(\cdot)$ represents a decay function as follows:

$$
\text{Cons}(u_{i,j}, v_{i,j}) =
\begin{cases}
u_{i,j} + \sigma v_{i,j}, \; 0 < (u_{i,j} + \sigma v_{i,j}) < 1 \\
\\
0, \quad \text{otherwise},
\end{cases}
\tag{13.29}
$$

where σ is a parameter that, as indicated in [25], can be obtained via Monte Carlo (MC) sampling. In order to meet the abundance sum-to-one constraint, we have

$$h_{m,j} = 1 - \sum_{i=1}^{m-1} h_{i,j}. \tag{13.30}$$

The objective function in (13.27) is a combinational problem, which is nonconvex, and therefore it is difficult to solve. In [36], it proposes an iterative scheme to optimize $\mathbf{W}$ and $\mathbf{H}$, respectively, both of which are solved by a gradient descent method. The first-order derivatives of the objective function are computed as follows:

$$\begin{cases} \nabla_{\mathbf{U}}(\mathbf{W}, \mathbf{H}) = d(\mathbf{U}) - \frac{2\lambda}{n}\mathbf{z}(\mathbf{1}_n)^T . * \mathbf{U}, \\[3mm] \nabla_{\mathbf{V}}(\mathbf{W}, \mathbf{H}) = d(\mathbf{V}) + \frac{2\lambda}{n}\mathbf{z}(\mathbf{1}_n)^T . * (\ln \mathbf{V}./\mathbf{V} - \mathbf{V}), \end{cases} \tag{13.31}$$

where ./ is the dot division, $\mathbf{z} = \frac{1}{2n}(\mathbf{1}_{m \times n} + \ln \mathbf{V}^2 - \mathbf{U}^2 - \mathbf{V}^2)\mathbf{1}_n$. $d(\mathbf{U})$ and $d(\mathbf{V})$ are gradients of reconstructed errors, which are

$$\begin{cases} d(\mathbf{U}) = \mathbf{W}^T(\mathbf{WH} - \mathbf{Y})). * \mathbb{C}_{cons}, \\[3mm] d(\mathbf{V}) = \sigma \mathbf{W}^T(\mathbf{WH} - \mathbf{Y}). * \mathbb{C}_{cons}, \end{cases} \tag{13.32}$$

where $\mathbb{C}_{cons}$ is an indicative function, $\mathbb{C}_{cons} = \mathbf{1}_{m \times n}\{0 < (\mathbf{U} + \sigma \mathbf{V}) < 1\}$. For more details, the derivation of (13.31) is given in [36].

Algorithm 1 DAEN for hyperspectral unmixing

Input: dataset $\mathbf{Y}$.
Output: endmembers $\mathbf{W}$, abundances $\mathbf{H}$.
Step 1. /* *SAE for initialization* */
1. **Initialization:** $\mathbf{M}_i$.
2. Set hyper-parameters following [44].
3. Obtain $p \times k$ candidates via VCA[43].
repeat
    4. Update $\{\widehat{\mathbf{w}}_i\}_{i=1}^m$ in (13.21).
    5. Update $\{\mathbf{M}_i\}_{i=1}^m$ in (13.24).
until convergence
6. Compute $\widehat{\mathbf{H}}$ via FCLS[42].
Step 2. /* *VAE for unmixing* */
7. **Initialization:** $\mathbf{U}$ and $\mathbf{V}$.
repeat
    8. Update $\Delta \mathbf{H}$ in (13.33).
    9. Update $\Delta \mathbf{W}$ in (13.34).
until convergence

With respect to the updates of $\mathbf{H}$ and $\mathbf{W}$, we employ the gradient descent method for the solutions as follows:

$$\mathbf{H} \Leftarrow \mathbf{H} + \Delta \mathbf{H},$$

and

$$\mathbf{W} \Leftarrow \mathbf{W} + \Delta \mathbf{W},$$

where $\Delta \mathbf{H}$ and $\Delta \mathbf{W}$ are the gradients for $\mathbf{H}$ and $\mathbf{W}$, respectively. Specifically,

- For $\mathbf{H}$, we have

$$\Delta \mathbf{H} = -\varphi(\nabla_{\mathbf{U}}(\mathbf{W}, \mathbf{H}) + \sigma \nabla_{\mathbf{V}}(\mathbf{W}, \mathbf{H})), \tag{13.33}$$

where φ is the learning rates that can be estimated by the Armijo rule [45].

- For **W**, we obtain Δ**W** via Adadelta [46] as follows:

$$\Delta\mathbf{W} = -\frac{\text{RMS}[\Delta\mathbf{W}]}{\text{RMS}[\nabla\mathbf{W}(\mathbf{W}, \mathbf{H})]}\nabla_{\mathbf{W}}(\mathbf{W}, \mathbf{H}), \qquad (13.34)$$

where RMS[$\cdot$] is the root-mean-square [46]. The first-order derivatives of the objective function (13.27) are calculated as follows:

$$\nabla_{\mathbf{W}}(\mathbf{W}, \mathbf{H}) = (\mathbf{WH} - \mathbf{Y})\mathbf{H}^T + \mu d(\text{MinVol}(\mathbf{W})). \qquad (13.35)$$

where $d(\text{MinVol}(\mathbf{W}))$ is the gradient for the volume function, which can be computed as the one in [47].

Finally, a pseudocode of the proposed DAEN is given in Algorithm 1. As shown in Algorithm 1, DAEN consists of two main parts, a set of SAEs for initialization and one VAE for unmixing. Specifically, in Line 1, $\mathbf{M}_i$ is randomly initialized. In Line 2, the hyper-parameters are set following [44], while in Line 3, the candidate samples used for training are generated via VCA. In Lines 4 and 5, $\{\widehat{\mathbf{w}}_i\}$ and $\{\mathbf{M}_i\}$ are iteratively updated until SAE terminates. In Line 6, it computes the abundance estimation $\widehat{\mathbf{H}}$ via FCLS. In Line 7, the LV variables, **U** and **V**, are randomly initialized. Finally, in Lines 8 and 9, the endmember matrix **W** and the abundance matrix **H** are iteratively updated, respectively.

13.4 Experiments and Analysis: Sparse Unmixing

In this section, we illustrate the unmixing performance of these sparse unmixing methods using simulated hyperspectral datasets. For quantitative analysis, the signal-to-reconstruction error (SRE, measured in dB) is used to evaluate the unmixing accuracy. For comparative purposes, the results obtained by SUnSAL [8], SUnSAL-TV [20], LCSU [23], DRSU [18], DRSU-TV [41], and S^{2}WSU [24] algorithms are reported. Let $\widehat{\mathbf{h}}$ be the estimated abundance, and **h** be the true abundance. The SRE(dB) can be computed as follows:

$$\text{SRE(dB)} = 10 \cdot \log_{10}(E(\|\mathbf{h}\|_2^2)/E(\|\mathbf{h} - \widehat{\mathbf{h}}\|_2^2)), \qquad (13.36)$$

where $E(\cdot)$ denotes the expectation function. Furthermore, we use another indicator, i.e., the probability of success p_s, which is an estimate of the probability that the relative error power be smaller than a certain threshold. It is formally defined as follows: $p_s \equiv P(\|\widehat{\mathbf{h}} - \mathbf{h}\|^2/\|\mathbf{h}\|^2 \leq threshold)$. In our case, the estimation result is considered successfully when $\|\widehat{\mathbf{h}} - \mathbf{h}\|^2/\|\mathbf{h}\|^2 \leq 3.16$ (5 dB). This threshold was demonstrated to be appropriate in [8]. The larger the SRE (dB) or the p_s, the more accurate the unmixing results.

13.4.1 Simulated Datasets

The spectral library that we use in our synthetic image experiments is a dictionary of minerals extracted from the United States Geological Survey (USGS) library.[1] Such library, denoted by $\mathbf{A}$, contains $m = 240$ materials (different mineral types), with spectral signatures with reflectance values consisting of $L = 224$ spectral bands and distributed uniformly in the interval 0.4–2.5 μm. Following the work in [20], simulated data cube is generated with 100×100 pixels and nine spectral signatures (Adularia GDS57 Orthoclase, Jarosite GDS99 K Sy 200C, Jarosite GDS101 Na Sy 200, Anorthite HS349.3B, Calcite WS272, Alunite GDS83 Na63, Howlite GDS155, Corrensite CorWa-1, Fassaite HS118.3B.), which are randomly chosen from the spectral library $\mathbf{A}$. The fractional abundances are piece-wise smooth, i.e., they are smooth with sharp transitions; moreover, they are subject to the ANC and ASC. These data can reveal the spatial features quite well for the different unmixing algorithms. For illustrative purposes, Fig. 13.2 shows the true abundance maps of the endmembers. After generating the data cube, it was contaminated with i.i.d. Gaussian noise, for three levels of the signal-to-noise (SNR) ratio: 30, 40, and 50 dB.

Table 13.1 shows the SRE (dB) and p_s results achieved by the different tested algorithms under different SNR levels. For all the tested algorithms, the input parameters have been carefully tuned for optimal performance. From Table 13.1, we can see that the methods of using double weights (DRSU, DRSU-TV, and S^{2}WSU) have obtained better SRE (dB) results than other algorithms in all cases. Furthermore, the S^{2}WSU achieved better SRE (dB) results than the competitors in all cases, which indicates that the inclusion of a spatial factor in the sparse regularizer can further promote the spatial correlation on the solution and improve the unmixing performance. The p_s obtained by the S^{2}WSU is also much better than those obtained by other algorithms in the case of low SNR values, which reveals that the inclusion of spatial information leads to high robustness. Based on the aforementioned results, we can conclude that the spatial weighted strategy offers the potential to improve sparse unmixing performance.

13.4.2 Real Hyperspectral Data

In this section, we resort to the well-known Airborne Visible Infrared Imaging Spectrometer (AVIRIS) Cuprite dataset for evaluation of the proposed approach, which is a common benchmark for validation of spectral unmixing algorithms. The data are available online in reflectance units.[2] The portion used in experiments corresponds to a 350×350-pixel subset of the scene, with 224 spectral bands in the range 0.4–2.5 μm and nominal spectral resolution of 10 nm. Prior to the analysis, bands 1–2,

---

[1] Available online at http://speclab.cr.usgs.gov/spectral.lib06.

[2] http://aviris.jpl.nasa.gov/html/aviris.freedata.html.

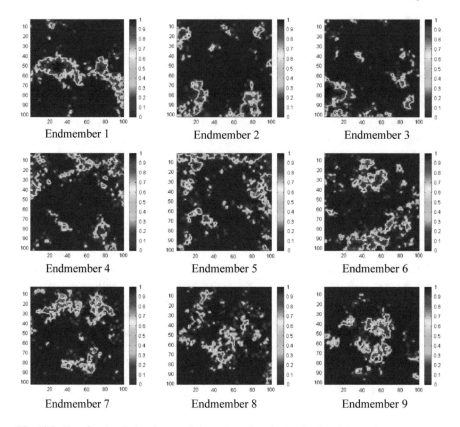

Fig. 13.2 True fractional abundances of the endmembers in the simulated data cube

105–115, 150–170, and 223–224 were removed due to water absorption and low
SNR, leaving a total of 188 spectral bands. The spectral library used in this experiment is the same library **A** used in our simulated experiments and the noisy bands
are also removed from **A**. The classification maps of these materials produced by
Tricorder software[3] are also displayed. Figure 13.3 shows a mineral map produced
in 1995 by USGS, in which the Tricorder 3.3 software product [48] was used to map
different minerals present in the Cuprite mining district. The USGS map serves as
a good indicator for qualitative assessment of the fractional abundance maps produced by the different unmixing algorithms. Note that the publicly available AVIRIS
Cuprite data were collected in 1997 but the Tricorder map was produced in 1995. In
addition, the true abundances of the real hyperspectral data are unavailable. Thus, we
can only make a qualitative analysis of the performances of different sparse unmixing
algorithms by comparing their estimated abundances with the mineral maps.

Figure 13.4 conducts a qualitative comparison between the classification maps
produced by the USGS Tricorder algorithm and the fractional abundances estimated

[3]http://speclab.cr.usgs.gov/PAPER/tetracorder.

Table 13.1 SRE(dB) and p_s scores achieved after applying different unmixing methods to the simulated data Cube 1 (the optimal parameters for which the reported values were achieved are indicated in the parentheses)

Algorithm	SNR = 30 dB		SNR = 40 dB		SNR = 50 dB	
	SRE (dB)	p_s	SRE (dB)	p_s	SRE (dB)	p_s
SUnSAL	8.4373	0.7946	15.1721	0.9886	23.0894	1
	($\lambda = 2e{-}2$)		($\lambda = 5e{-}3$)		($\lambda = 1e{-}3$)	
SUnSAL-TV	11.4304	0.9470	17.7695	0.9998	26.1655	1
	($\lambda = 1e{-}2; \lambda_{TV} = 4e{-}3$)		($\lambda = 5e{-}3; \lambda_{TV} = 1e{-}3$)		($\lambda = 2e{-}3; \lambda_{TV} = 2e{-}4$)	
LCSU	11.4317	0.9463	18.1793	0.9999	26.2194	1
	($\lambda = 3e{-}2$)		($\lambda = 7e{-}3$)		($\lambda = 1e{-}3$)	
DRSU	14.9876	0.9745	29.6861	1	41.1967	1
	($\lambda = 3e{-}3$)		($\lambda = 1e{-}3$)		($\lambda = 6e{-}4$)	
DRSU-TV	18.8630	0.9994	30.9403	1	41.1967	1
	($\lambda = 2e{-}3; \lambda_{TV} = 2e{-}3$)		($\lambda = 2e{-}3; \lambda_{TV} = 4e{-}4$)		($\lambda = 6e{-}4; \lambda_{TV} = 0$)	
S^{2}WSU	20.5709	0.9995	31.9461	1	41.4053	1
	($\lambda = 5e{-}3$)		($\lambda = 3e{-}3$)		($\lambda = 6e{-}4$)	

by SUnSAL, SUnSAL-TV, LCSU, DRSU, DRSU-TV, and S^{2}WSU algorithms for three highly representative minerals in the Cuprite mining district (Alunite, Buddingtonite, and Chalcedony). In this experiment, the regularization parameters used for SUnSAL, LCSU, DRSU, and S^{2}WSU were empirically set to $\lambda = 0.001$, $\lambda = 0.001$, $\lambda = 0.0001$, and $\lambda = 0.002$, respectively, while the parameters for SUnSAL-TV and DRSU-TV were set to $\lambda = 0.001$, $\lambda_{TV} = 0.001$ and $\lambda = 0.002$, $\lambda_{TV} = 0.0001$, respectively. As shown in Fig. 13.4, all the algorithms obtained reasonable unmixing results, with high abundances for the pixels showing the presence of the considered minerals. This indicates that the sparse unmixing algorithms can lead to good interpretation of the considered hyperspectral dataset. However, it can be seen that some of the abundance maps (e.g., Buddingtonite mineral) estimated by SUnSAL and SUnSAL-TV look noisy and the results obtained by SUnSAL-TV are oversmoothed. In addition, DRSU yields abundance maps without good spatial consistency of the minerals of interest (e.g., Chalcedony mineral), and we can also find that the abundances estimated by S^{2}WSU algorithms are generally comparable or higher in the regions classified as respective minerals in comparison to DRSU. Finally, the *sparsity* obtained by SUnSAL, SUnSAL-TV, LCSU, DRSU, DRSU-TV, and S^{2}WSU are 0.0682, 0.0743, 0.0734, 0.0430, 0.0423, and 0.0420, respectively. These small differences lead to the conclusion that the proposed approaches use a smaller number of elements to explain the data, thus obtaining higher sparsity. Therefore, from a qualitatively viewpoint, we can conclude that the S^{2}WSU method exhibits good potential to improve the results obtained by other algorithms in real analysis scenarios.

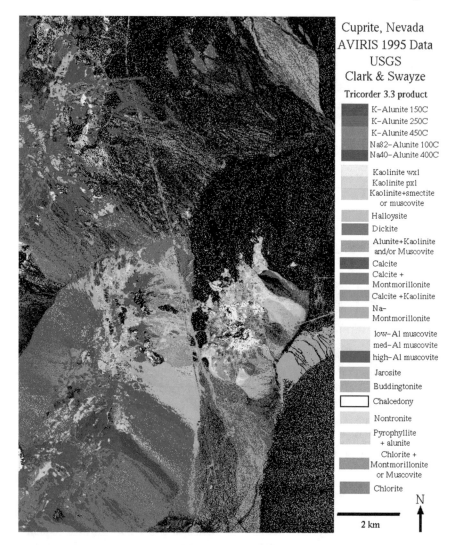

Fig. 13.3 USGS map showing the location of different minerals in the Cuprite mining district in Nevada

13.5 Experiments and Analysis: Deep Learning

In this section, the DAEN approach is applied to two real hyperspectral images: Mangrove [49] and Samson [50] datasets for evaluation. In these experiments, the parameters involved in the considered algorithms follow the settings in the simulated experiments, i.e., we use $\mu = 0.1$ and $\lambda = 0.1$, respectively.

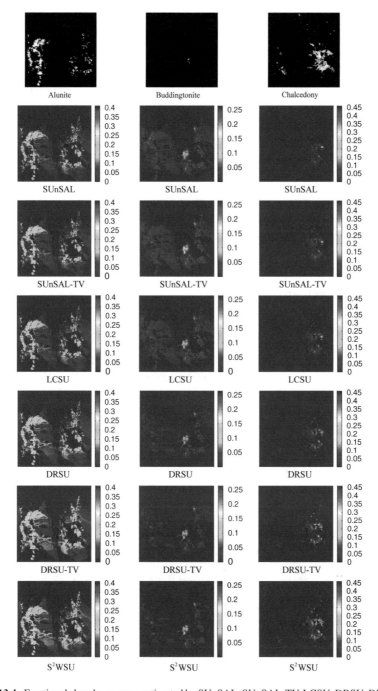

Fig. 13.4 Fractional abundance maps estimated by SUnSAL, SUnSAL-TV, LCSU, DRSU, DRSU-TV, and S^{2}WSU as compared to the classification maps produced by USGS Tricorder software for the considered 350 × 350-pixel subset of the AVIRIS Cuprite scene

We compare the DAEN approach presented in this work with other advanced unmixing algorithms, specifically with the N-FINDR [51], VCA [43], MVC-NMF [47], Bayesian [25], PCOMMEND [52], and SNSA [34] methods.

Three indicators, i.e., SAD, reconstruction error (RE), and root-mean-square error (RMSE) are used to measure the accuracy of the unmixing results, which are defined as follows:

$$\begin{cases} \mathrm{SAD}(\mathbf{w}_i, \widehat{\mathbf{w}}_i) = \arccos\left(\frac{[\mathbf{w}_i, \widehat{\mathbf{w}}_i]}{\|\mathbf{w}_i\| \cdot \|\widehat{\mathbf{w}}_i\|}\right), \\ \mathrm{RE}(\{\mathbf{y}_j\}_{j=1}^n, \{\widehat{\mathbf{y}}_j\}_{j=1}^n) = \frac{1}{n}\sum_{j=1}^n \sqrt{\|\mathbf{y}_j - \widehat{\mathbf{y}}_j\|_2^2}, \\ \mathrm{RMSE}(\widehat{\mathbf{h}}_j, \mathbf{h}_j) = \frac{1}{n}\sum_{i=1}^n \sqrt{\|\mathbf{h}_j - \widehat{\mathbf{h}}_j\|_2^2}, \end{cases} \qquad (13.37)$$

where $\widehat{\mathbf{w}}_i$ and $\mathbf{w}_i$ denote the extracted endmember and the library spectrum, $\widehat{\mathbf{y}}_j$ and $\mathbf{y}_j$ are the reconstruction and original signature of pixel j, and $\widehat{\mathbf{h}}_j$ and $\mathbf{h}_j$ are the corresponding estimated and actual abundance fractions, respectively.

13.5.1 Mangrove Dataset

The Mangrove data is an EO-1 Hyperion (hyperspectral) image which has been obtained from the USGS Earth Resources Observation and Science (EROS) Center through a data acquisition request to the satellite data provider [49], and collected over the Henry Island of the Sunderban Biosphere Reserve of West Bengal, India. After applying atmospheric correction, we have converted the radiance data to reflectance units by using FLAASH model in ENVI software, and the endmembers (pure signatures of mangrove species) have been identified by a ground survey of the study area, including *Avicennia, Bruguiera, Excoecaria, Phoenix*. The Mangrove data, as shown in Fig. 13.5, includes 137×187 pixels and 155 bands, with a spatial resolution of 30 m. For detailed information of the Mangrove data, we refer to [49].

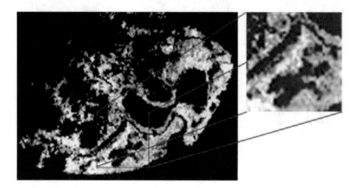

Fig. 13.5 The 45×45 pixel subscene of the Mangrove data used in our experiment

Table 13.2 SADs (in radians) and REs along with their standard deviations obtained by different methods for the Mangrove data from 10 Monte Carlo runs, where the best results are in bold

Mineral	N-FINDR	VCA	MVC-NMF	Bayesian	PCOMMEND	SNSA	DAEN
Avicennia	0.1495 ± 0%	0.9602 ± 4.65%	0.1973 ± 2.18%	0.1001 ± 3.05%	0.0992 ± 2.81%	0.0995 ± 4.26%	**0.0968 ±** 2.58%
Bruguiera	0.8235 ± 0%	0.8824 ± 7.79%	0.9825 ± 7.53%	1.5957 ± 4.16%	0.9564 ± 3.95%	0.1103 ± 3.84%	**0.1025 ±** 4.03%
Excoecaria	0.7361 ± 0%	0.7377 ± 1.50%	0.0904 ± 4.08%	0.0963 ± 2.96%	0.0876 ± 5.73%	0.0880 ± 3.27%	**0.0863 ±** 1.96%
Phoenix	0.1306 ± 0%	1.3063 ± 0.12%	0.9782 ± 3.93%	0.9624 ± 3.93%	1.7065 ± 8.23%	0.0711 ± 4.41%	**0.0706 ±** 3.72%
Mean SAD	0.4599	0.9717	0.5621	0.6886	0.7124	0.0922	**0.0890**
RE	0.0822 ± 0%	0.0980 ± 0.40%	0.0392 ± 2.36%	0.0129 ± 4.57%	0.0162 ± 5.19%	0.0057 ± 0.25%	**0.0050 ±** 0.13%

In our experiment, a subscene with 45×45 pixels of the Mangrove data has been used to further evaluate the proposed DAEN. Following [49], the considered subscene contains four endmembers, i.e., $m = 4$.

Table 13.2 presents the obtained quantitative results from the Mangrove data. It can be seen that the DAEN achieved very promising results for the four considered mangrove spices. However, the other competitors ended up with errors when detecting or estimating the endmembers. This is due to the fact that, according to our observation, the Mangrove scene contains many outliers across the whole image, which brings a lot of difficulties for general unmixing methods. This point was verified by our experiment, in which we detected a total of 17 outliers. For illustrative purposes, Fig. 13.6 scatterplots the unmixing results obtained by the considered methods, in which the detected outliers are also illustrated. From Fig. 13.6, we can observe that the DAEN produced good unmixing results for this dataset, while all the other methods resulted in problems.

Finally, for illustrative purposes, the estimated endmember signatures, along with their ground references, and the corresponding abundance maps obtained by the DAEN are shown in Fig. 13.7. Effective results can be observed from these figures.

In summary, our experiments with this challenging Mangrove dataset demonstrate the effectiveness of the DAEN for real scenarios with outliers, which is a general situation in real problems.

13.5.2 Samson Dataset

In this experiment, we use the Samson dataset which includes 156 bands covering the wavelengths from 0.401 to 0.889 μm, and 95×95 pixels, as shown Fig. 13.8,

Fig. 13.6 Unmixing results for the subscene of the Mangrove data, where the data are projected onto the first two principal components (PCs)

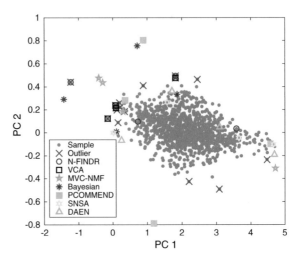

for validation [50]. There are three endmembers including Soil, Tree, and Water in the ground truth image.

Table 13.3 demonstrates the obtained quantitative results for the considered methods. It can be observed that the proposed DAEN obtained the best mean SAD and RMSE. For illustrative purposes, the endmember signatures and the estimated abundances are shown in Fig. 13.9. These figures reveal that the endmembers and abundances, estimated from DAEN, have good matches with regard to the corresponding ones in the ground truth.

13.6   Conclusions and Future Work

Spectral unmixing provides a way to quantitatively analyze sub-pixel components in remotely sensed hyperspectral images [19]. Sparse unmixing has been widely used as a semi-supervised approach that requires the presence of a library of spectral signatures. In this context, spectral–spatial sparse unmixing methods, which aim at collaboratively exploiting spectral and spatial–contextual information, offer a powerful unmixing strategy in case a complete spectral library is available a priori. If no spectral library is available *in advance*, we suggest the fully unsupervised deep auto-encoder network (DAEN) unmixing as a powerful approach that can effectively deal with the presence of outliers in hyperspectral data. Our experimental results reveal that the two aforementioned techniques are currently at the forefront of spectral unmixing. Specifically, we empirically found that the S^2WSU algorithm consistently achieves better unmixing performance than other advanced spectral unmixing algorithms in case a spectral library is available. This implies that the integration of spectral and spatial–contextual information via the considered spectral–spatial weighted strat-

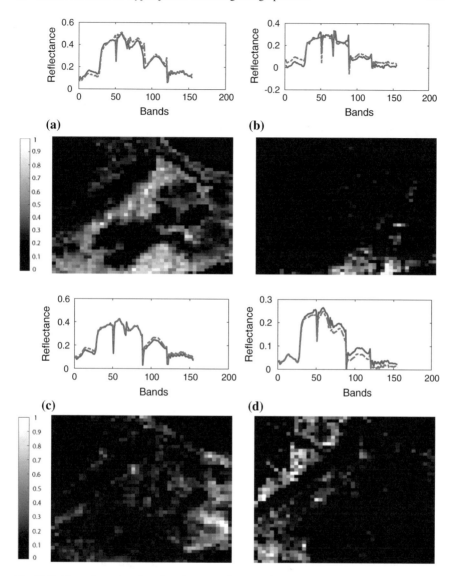

Fig. 13.7 The estimated endmember signatures (in red), along with the ground reference (in blue) and their corresponding abundance maps by the proposed DAEN. **a** Avicennia, **b** Bruguiera, **c** Excoecaria, **d** Phoenix

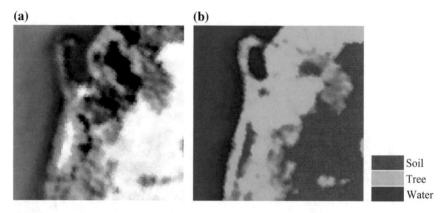

Fig. 13.8 The Samson image (**a**) and its corresponding ground truth (**b**)

Table 13.3 SADs (in radians) and REs along with their standard deviations obtained by different methods for the Samson data from 10 Monte Carlo runs, where the best results are in bold

Mineral	N-FINDR	VCA	MVC-NMF	Bayesian	PCOMMEND	SNSA	DAEN
Soil	0.0713 ± 0%	0.0627 ± 1.85%	**0.0402** ± 3.87%	0.1062 ± 7.25%	0.2849 ± 4.35%	0.0410 ± 5.02%	0.0405 ± 2.76%
Tree	0.0495 ± 0%	0.0501 ± 7.82%	0.0261 ± 3.62%	0.0610 ± 8.34%	0.0505 ± 6.14%	0.0205± 2.89%	**0.0196** ± 3.52%
Water	0.0408 ± 0%	0.0273 ± 3.74%	0.0304 ± 5.29%	0.0364 ± 2.48%	0.0716 ± 4.34%	0.0291 ± 2.59%	**0.0279** ± 3.83%
Mean SAD	0.0539	0.0467	0.0322	0.0679	0.1357	0.0302	**0.0293**
RMSE	0.9572 ± 0%	0.8926 ± 1.35%	0.6430 ± 0.98%	0.7501 ± 1.63%	0.9439 ± 2.35%	0.6143 ± 3.37%	**0.6097** ± 3.62%
RE	0.0129 ± 0%	0.0116 ± 0.19%	0.0075 ± 0.86%	0.0103 ± 0.42%	**0.0057** ± 1.15%	0.0066 ± 0.25%	0.0062 ± 0.85%

egy has great potential in improving unmixing performance. Our experiments also indicate that the fully unsupervised DAEN approach can handle problems with significant outliers more effectively than other popular spectral unmixing approaches. This is an important observation, since the presence of outliers is common in real problems and traditional unmixing algorithms are often misguided by outliers (that can be also understood as endmembers due to their singularity). Our future work will focus on exploring the combination of sparse unmixing and deep learning algorithms to further improve the unmixing performance.

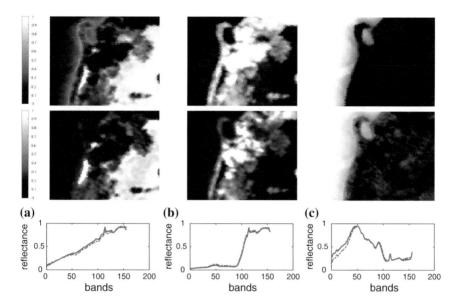

Fig. 13.9 Results obtained by the proposed DAEN on the Samson dataset. Top: Ground truth abundance maps on Samson data. Middle: Estimated abundance maps from the proposed DAEN. Bottom: Estimated endmember signatures (in red) along with their corresponding reference signatures (in blue). **a** Soil, **b** tree, **c** water

References

1. Landgrebe D (2002) Hyperspectral image data analysis. IEEE Signal Process Mag 19(1):17–28
2. Bioucas-Dias JM, Plaza A, Camps-Valls G, Scheunders P, Nasrabadi N, Chanussot J (2013) Hyperspectral remote sensing data analysis and future challenges. IEEE Geosci Remote Sens Mag 1(2):6–36
3. Bioucas-Dias JM, Plaza A, Dobigeon N, Parente M, Du Q, Gader P, Chanussot J (2012) Hyperspectral unmixing overview: geometrical, statistical, and sparse regression-based approaches. IEEE J Sel Topics Appl Earth Observ Remote Sens 5(2):354–379
4. Plaza A, Du Q, Bioucas-Dias JM, Jia X, Kruse FA (2011) Foreword to the special issue on spectral unmixing of remotely sensed data. IEEE Trans Geosci Remote Sens 49(11):4103–4110
5. Keshava N, Mustard JF (2002) Spectral unmixing. IEEE Signal Process Mag 19(1):44–57
6. Dobigeon N, Tourneret JY, Richard C, Bermudez JCM, McLaughlin S, Hero AO (2014) Nonlinear unmixing of hyperspectral images: models and algorithms. IEEE Signal Process Mag 31(1):82–94
7. Ma WK, Bioucas-Dias JM, Chan TH, Gillis N, Gader P, Plaza AJ, Ambikapathi A, Chi CY (2014) A signal processing perspective on hyperspectral unmixing: Insights from remote sensing. IEEE Signal Process Mag 31(1):67–81
8. Iordache M-D, Bioucas-Dias JM, Plaza A (2011) Sparse unmixing of hyperspectral data. IEEE Trans Geosci Remote Sens 49(6):2014–2039 Jun
9. Nascimento JM, Bioucas-Dias JM (2005) Vertex component analysis: a fast algorithm to unmix hyperspectral data. IEEE Trans Geosci Remote Sens 43(4):898–910 Apr
10. Boardman J, Kruse F, Green R (1995) Mapping target signatures via partial unmixing of AVIRIS data. In: Proceedings of the JPL airborne earth science workshop, pp 23–26

11. Li J, Agathos A, Zaharie D, Bioucas-Dias JM, Plaza A, Li X (2015) Minimum volume simplex analysis: a fast algorithm for linear hyperspectral unmixing. IEEE Trans Geosci Remote Sens 53(9):5067–5082

12. Zhang S, Agathos A, Li J (2017) Robust minimum volume simplex analysis for hyperspectral unmixing. IEEE Trans Geosci Remote Sens 55(11):6431–6439

13. Berman M, Kiiveri H, Lagerstrom R, Ernst A, Dunne R, Huntington JF (2004) ICE: a statistical approach to identifying endmembers in hyperspectral images. IEEE Trans Geosci Remote Sens 42(10):2085–2095

14. Iordache M-D, Bioucas-Dias JM, Plaza A (2014) Collaborative sparse regression for hyperspectral unmixing. IEEE Trans Geosci Remote Sens 52(1):341–354

15. Lu X, Wu H, Yuan Y, Yan P, Li X (2013) Manifold regularized sparse NMF for hyperspectral unmixing. IEEE Trans Geosci Remote Sens 51(5):2815–2826

16. Candès EJ, Wakin MB, Boyd SP (2008) Enhancing sparsity by reweighted l_1 minimization. J Fourier Anal Appl 14(5):877–905

17. Zheng CY, Li H, Wang Q, Philip Chen CL (2016) Reweighted sparse regression for hyperspectral unmixing. IEEE Trans Geosci Remote Sens 54(1):479–488

18. Wang R, Li HC, Liao W, Pižurica A (2016) Double reweighted sparse regression for hyperspectral unmixing. In: Proceedings of the IEEE international geoscience remote sensing symposium. pp 6986–6989

19. Shi C, Wang L (2014) Incorporating spatial information in spectral unmixing: a review. Remote Sens Environ 149:70–87

20. Iordache M-D, Bioucas-Dias JM, Plaza A (2012) Total variation spatial regularization for sparse hyperspectral unmixing. IEEE Trans Geosci Remote Sens 50(11):4484–4502

21. Zhao X, Wang F, Huang T, Ng MK, Plemmons RJ (2013) Deblurring and sparse unmixing for hyperspectral images. IEEE Trans Geosci Remote Sens 51(7):4045–4058

22. Zhong Y, Feng R, Zhang L (2014) Non-local sparse unmixing for hyperspectral remote sensing imagery. IEEE J Sel Topics Appl Earth Observ Remote Sens 7(6):1889–1909

23. Zhang S, Li J, Liu K, Deng C, Liu L, Plaza A (2016) Hyperspectral unmixing based on local collaborative sparse regression. IEEE Geosci Remote Sens Lett 13(5):631–635

24. Zhang S, Li J, Li HC, Deng C, Plaza A (2018) Spectral-spatial weighted sparse regression for hyperspectral image unmixing. IEEE Trans Geosci Remote Sens 56(6):3265–3276

25. Dobigeon N, Moussaoui S, Coulon M, Tourneret JY, Hero AO (2009) Joint Bayesian endmember extraction and linear unmixing for hyperspectral imagery. IEEE Trans Signal Process 57(11):4355–4368

26. Dobigeon N, Moussaoui S, Tourneret J-S, Carteret C (2009) Bayesian separation of spectral sources under non-negativity and full additivity constraints. Signal Process 89(12):2657–2669

27. Schmidt F, Schmidt A, Treguier E, Guiheneuf M, Moussaoui S, Dobigeon N (2010) Implementation strategies for hyperspectral unmixing using Bayesian source separation. IEEE Trans Geosci Remote Sens 48(11):4003–4013

28. Plaza J, Plaza A, Perez R, Martinez P (2009) On the use of small training sets for neural network-based characterization of mixed pixels in remotely sensed hyperspectral images. Pattern Recogn 42(11):3032–3045

29. Plaza J, Plaza A (2010) Spectral mixture analysis of hyperspectral scenes using intelligently selected training samples. IEEE Geosci Remote Sens Lett 7(2):371–375

30. Giorgio LA, Frate FD (2011) Pixel unmixing in hyperspectral data by means of neural networks. IEEE Trans Geosci Remote Sens 49(11):4163–4172

31. Guo R, Wang W, Qi H (2015) Hyperspectral image unmixing using autoencoder cascade. In: Proceedings of the 7th workshop hyperspectral image signal processing: evolution in Remote Sensing, pp 1–4

32. Su Y, Marinoni A, Li J, Plaza A, Gamba P (2017) Nonnegative sparse autoencoder for robust endmember extraction from remotely sensed hyperspectral images. In: IEEE international geoscience and remote sensing symposium (IGARSS), pp 205–208

33. Qu Y, Qi H, uDAS: an untied denoising autoencoder with sparsity for spectral unmixing. IEEE Trans Geosci Remote Sens to be published. https://doi.org/10.1109/TGRS.2018.2868690

34. Su Y, Marinoni A, Li J, Plaza J, Gamba P (2018) Stacked nonnegative sparse autoencoders for robust hyperspectral unmixing. IEEE Geosci Remote Sens Lett 15(9):1427–1431
35. Hapke B (1993) Theory of reflectance and emittance spectroscopy. Cambridge Univ. Press, Cambridge, UK
36. Su Y, Li J, Plaza A, Marinoni A, Gamba P, Chakravortty S, DAEN: deep autoencoder networks for hyperspectral unmixing. IEEE Trans Geosci Remote Sens, to be published. https://doi.org/10.1109/TGRS.2018.2890633
37. Kingmaand DP, Welling M (2014) Auto-encoding variational Bayes. In: 2014 the international conference on learning representations (ICLR), pp 1–14
38. Candès EJ, Tao T (2005) Decoding by linear programming. IEEE Trans Inf Theory 51(12):4203–4215
39. Candès EJ, Tao T (2006) Near-optimal signal recovery from random projections: universal encoding strategies. IEEE Trans Inf Theory 52(12):5406–5425
40. Bioucas-Dias JM, Figueiredo M (2010) Alternating direction algorithms for constrained sparse regression: application to hyperspectral unmixing. In: Proceedings of the 2nd workshop hyperspectral image signal processing: evolution in Remote Sensing, pp 1–4
41. Wang R, Li HC, Pizurica A, Li J, Plaza A, Emery WJ (2017) Hyperspectral unmixing using double reweighted sparse regression and total variation. IEEE Geosci Remote Sens Lett 14(7):1146–1150
42. Heinz DC, Chang C-I (2001) Fully constrained least squares linear spectral mixture analysis method for material quantification in hyperspectral imagery. IEEE Trans Geosci Remote Sens 39(3):529–545
43. Nascimento JMP, Dias-Bioucas JM (2005) Vertex component analysis: a fast algorithm to unmix hyperspectral data. IEEE Trans Geosci Remote Sens 43(4):898–910
44. Lemme A, Reinhart RF, Steil JJ (2012) Online learning and generalization of parts-based image representations by non-negative sparse autoencoders. Neural Netw 33:194–203
45. Nocedal J, Wright SJ (2006) Numerical Optimization. Springer Science, New York, NY, Springer, USA, pp 33–36
46. Zeiler MD, ADADELTA: an adaptive learning rate method. Comput Sci. arXiv:1212.5701
47. Miao L, Qi H (2007) Endmember extraction from highly mixed data using minimum volume constrained nonnegative matrix factorization. IEEE Trans Geosci Remote Sens 45(3):765–777
48. Clark R, Swayze G, Livo K, Kokaly R, Sutley S, Dalton J, McDougal R, Gent C (2003) Imaging spectroscopy: earth and planetary remote sensing with the USGS tetracorder and expert systems. J Geophys Res 108(E12):5131–5135 Dec
49. Chakravortty S, Li J, Plaza A (2017) A technique for subpixel analysis of dynamic mangrove ecosystems with time-series hyperspectral image data. IEEE J Sel Topics Appl Earth Observ Remote Sens 11(4):1244–1252
50. Zhu F, Wang Y, Fan B, Xiang S, Meng G, Pan C (2014) Spectral unmixing via data-guided sparsity. IEEE Trans Signal Process 23(12):5412–5427 Dec
51. Winter ME (1999) N-FINDR: an algorithm for fast autonomous spectral endmember determination in hyperspectral data. In: Proceedings of the SPIE image symposium V, vol 3753, pp 1–10
52. Zare A, Gader P, Bchir O, Frigui H (2013) Piecewise convex multiple-model endmember detection and spectral unmixing. IEEE Trans Geosci Remote Sens 51(5):2853–2862

Chapter 14
Hyperspectral–Multispectral Image Fusion Enhancement Based on Deep Learning

Jingxiang Yang, Yong-Qiang Zhao and Jonathan Cheung-Wai Chan

Abstract Hyperspectral image (HSI) contains rich spatial and spectral information, which is beneficial for identifying different materials. HSI has been applied in many fields, including land-cover classification and target detection. However, due to limited photonic energy, there are trade-offs between spatial resolution, bandwidth, swath width, and signal-noise-ratio. One outcome is that the spatial resolution of HSI is often moderate, which may lead to the spectral mixture of different materials in each pixel. In addition, for some Earth Observation applications such as urban mapping and fine mineral exploration, it is required to have a high spatial resolution image. Compared with HSI, multispectral imagery (MSI) often has wider spectral bandwidth and higher spatial resolution. HSI-MSI fusion for HSI resolution enhancement aims at fusing MSI with HSI, and generating HSI of higher resolution is an important technology. Having enormous capacity in feature extraction and representing mapping function, deep learning has shown great potential in HSI resolution enhancement. Two issues exist when we apply deep learning to HSI-MSI fusion: (1) how to jointly extract spectral–spatial deep features from HSI and MSI, (2) and how to fuse the extracted features and then generate high-resolution HSI. In this chapter, we first review the recent advances in HSI resolution enhancement technologies, particularly in HSI-MSI fusion technology, and then present our solution for HSI-MSI fusion based on a two-branch convolutional neural network.

J. Yang (✉)
School of Computer Science and Engineering, Nanjing University of Science and Technology, Nanjing 210094, People's Republic of China
e-mail: yang123jx@mail.nwpu.edu.cn

Y.-Q. Zhao
School of Automation, Northwestern Polytechnical University, Xi'an 710072, People's Republic of China
e-mail: zhaoyq@nwpu.edu.cn

J. C.-W. Chan
Department of Electronics & Informatics, Vrije Universiteit Brussel, 1050 Brussels, Belgium
e-mail: jcheungw@etrovub.be

© Springer Nature Switzerland AG 2020
S. Prasad and J. Chanussot (eds.), *Hyperspectral Image Analysis*,
Advances in Computer Vision and Pattern Recognition,
https://doi.org/10.1007/978-3-030-38617-7_14

407

14.1   Introduction

Hyperspectral image (HSI) is collected in contiguous spectral bands over a certain electromagnetic spectral range per pixel [1, 2]. The rich spectral and spatial information in HSI is beneficial for discriminating different materials in the scene. HSI has been widely applied in many fields, including land-cover classification [3], target detection [4], environment monitoring [5], and agriculture management [6]. However, the photonic energy in the process of hyperspectral imaging is limited. Due to the trade-off between spatial resolution, spectral bandwidth, swath width, and signal-noise-ratio, the spatial resolution of HSI is often moderate [7–9]. Most future spaceborne hyperspectral imaging missions, such as Environmental Mapping and Analysis Program (EnMAP), are with spatial resolution 30 m. The moderate resolution would lead to endmember spectral mixture and consequently degrade the discriminative ability of HSI [10, 11]. Some Earth Observation applications, such as mineral exploration [12], fine urban mapping [13], and sub-pixel target detection [14], require HSI with high spatial resolution. Therefore, enhancing the spatial resolution of HSI is of significance for many applications. Updating imaging hardware, e.g., increasing the optical aperture, could enhance the resolution of HSI, but it is often expensive. Enhancing the resolution of HSI via image processing method is more realistic.

Compared with HSI, the spatial resolution of multispectral (MSI) and panchromatic image is often finer with more information on surface structure in the scene. With the increasing availability of multispectral and panchromatic imaging missions at high resolution (HR), it is possible to collect HR MSI and panchromatic image over the same scene in the similar period as the HSI [7, 8]. HR MSI and panchromatic image could provide complementary information for spatial enhancement of HSI. According to the requirement of auxiliary data, the HSI resolution enhancement technologies can be classified as follows:

1. HSI super-resolution (SR), it can be classified into two categories, single- and multi-frame SR. Single-frame SR generates the HR HSI using only one low resolution (LR) HSI, while multiple-frame SR needs multiple LR HSIs of the same scene with sub-pixel shifts, as shown in Fig. 14.1a;
2. Hyperspectral pan-sharpening, which fuses LR HSI with HR panchromatic image taken over the same scene and generates HR HSI, as shown in Fig. 14.1b;
3. HSI-MSI fusion, which fuses LR HSI with HR MSI taken over the same scene and produces HR HSI, as shown in Fig. 14.1c.

The rest of this chapter will be organized as follows. We first provide an overview of the advances of the above three HSI enhancement technologies in Sect. 14.2. Then, our proposed HSI-MSI fusion method based on deep learning is introduced in Sect. 14.3. Some experimental results on simulated and real space borne data are also presented in this section. In Sect. 14.4, we provide some discussions and conclusions.

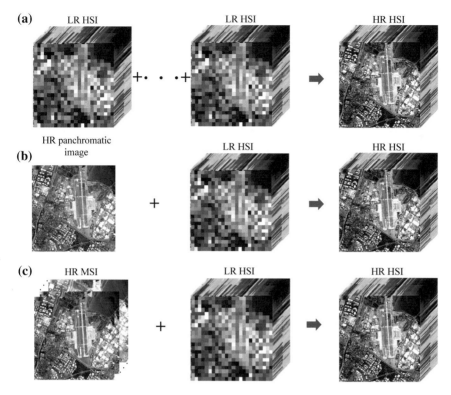

Fig. 14.1 **a** HSI SR, **b** pan-sharpening, **c** HSI-MSI fusion

14.2   Overview of HSI Enhancement

In this section, we will first review the HSI enhancement methods, discuss and analyze their advantages and disadvantages. Then we introduce deep learning and its applications in image resolution enhancement, particularly in HSI resolution enhancement.

14.2.1   HSI Pan-sharpening

Hyperspectral pan-sharpening reconstructs HR HSI by fusing the LR HSI with a HR panchromatic image. Several hyperspectral pan-sharpening methods have been proposed recently, they can be broadly classified into three categories: component substitution methods, multi-resolution analysis methods, and variational methods [7]. Component substitution methods project the LR HSI to transformation space via principal component analysis (PCA) [15] or Gram–Schmidt transformation methods

[16], in which spatial and spectral components are separated. The spatial component of LR HSI is then replaced with the HR panchromatic image. HR HSI can be achieved via inverse transformation. While such methods are simple and fast to be implemented, and they have robustness over the co-registration errors, they suffer from spectral distortion [7]. Multi-resolution analysis based pan-sharpening methods obtain the spatial details of HR panchromatic image via filtering, and then inject the spatial details into LR HSI. Representative methods include smoothing filtered-based intensity modulation (SFIM) [17] and generalized Laplacian pyramid (GLP) [18]. Compared with the component substitution methods, multi-resolution analysis based pan-sharpening methods have lower spectral distortion and are more robust over co-registration errors. Pan-sharpening can also be completed in a variational framework. Because reconstructing HR HSI is an ill-posed problem, prior information is necessary for regularization. Sparsity prior [19], low rank prior [20], and non-local similarity prior [21] could be used for regularization. Convex optimization is often involved in variational methods, which means heavier computation and more computing time.

14.2.2 HSI Super-Resolution

With single-frame SR approach, HR HSI can be super-resolved directly from its LR counterpart without the requirement of auxiliary data, so it is the most flexible in real applications. A basic single-frame HSI SR operation involves interpolating the LR HSI band-by-band (e.g., bilinear and bicubic interpolation). Such methods are simple and fast, but the details and structures in the HR HSI are prone to be blurred. An image can be represented sparsely by atoms in a dictionary [22]. Such sparsity prior can be used to regularize the SR problem. A sparse representation based HSI SR method was proposed in [23]. Other than the sparsity prior, a non-local similarity prior [24] was also exploited as a regularization term. In [25], in order to exploit the self-similarity in spatial and spectral domains, a group sparse representation method was proposed for HSI SR.

Another SR option is multi-frame HSI SR. LR HSIs taken over the same scene with sub-pixel shifts could provide image information in sub-pixel scale for HSI resolution enhancement. Many multi-frame HSI SR methods have been proposed, representative methods are non-uniform interpolation method, projective onto convex sets (POCS) method and maximum a posteriori (MAP) method [26]. In [27], in order to improve the performance of non-uniform interpolation, Chan et al. proposed a SR method based on a thin-plate spline non-rigid transform model and then applied it to multi-angular compact high-resolution imaging spectrometer (CHRIS) data. POCS multi-frame HSI SR methods could be combined with prior information such as amplitude constraint and total variation edge smoothing constraint [28]. The impact of multi-frame CHRIS SR on classification and unmixing applications was also investigated in [29, 30]. In [31], a multi-frame HSI sub-pixel mapping method was proposed in MAP framework with different prior models, such as Laplacian, total variation, and bilateral total variation.

14.2.3 HSI-MSI Fusion

HSI-MSI fusion is another option to enhance the resolution of HSI. HR MSI provides not only HR spatial information, but also spectral information. Compared with pan-sharpening, HSI-MSI fusion methods have high spectral fidelity [9]. HSI-MSI fusion technology gathered immense research interests in remote sensing and image processing communities [32–43]. HR HSI can be reconstructed by combining endmembers of LR HSI and abundance of HR MSI. Based on this idea, several unmixing based fusion methods have been proposed. For example, in [32], HSI and MSI were alternatively unmixed by applying nonnegative matrix factorization in a coupled way. HR HSI was reconstructed with the endmembers and the HR abundance under a linear mixture model. This method was also applied to fuse Hyperion HSI with ASTER MSI, and a HSI with 15 m resolution was produced [33]. Similarly, by exploiting the sparsity prior of endmembers, a fusion method based on sparse matrix factorization was proposed in [34], which was used to fuse MODIS HSI with Landsat 7 ETM + MSI and the resolution of the resulted MODIS HSI is enhanced by 8 times [35]. HR HSI can also be reconstructed with a dictionary. In [36] and [37], a spatial dictionary was learned from HR MSI, HR HSI was then reconstructed via joint sparse coding. In [38] and [39], a spectral dictionary was learned from LR HSI, then it was used to reconstruct HR HSI based on the abundance map of MSI. The HSI-MSI fusion problem could also be solved in a variation framework [40–43]. In [40], a variation model was proposed for HSI-MSI fusion with the sparsity prior to HSI exploited as a regularizer. Other than sparsity regularizer, a vector-total-variation regularizer was used in [41], a low rank constraint [42] and a spectral embedding regularizer [43] were designed for fusion. In [44], a maximum a posteriori fusion method was proposed by exploiting the joint statistics of endmembers under a stochastic mixing model.

14.2.4 Deep Learning Based Image Enhancement

Most of the above HSI resolution enhancement methods suffer from three major drawbacks:

1. They are based on hand-crafted features such as the dictionary [36, 37], which can be regarded as low-level feature with limited representative ability;
2. They rely on prior assumptions, such as the linear spectral mixture assumption in [32] and the sparsity assumption in [36]. Quality degeneration may be caused if these assumptions do not fit the problem;

3. Optimization problems are often involved in the testing stage, making the HSI reconstruction time-consuming.

Recently, deep learning has attracted research interests due to its ability to automatically learn high-level features and its high non-linearity [45, 46], which is of great potential to model the complex nonlinear relationship between LR and HR HSIs in both spatial and spectral domains [47]. Compared with hand-crafted features, the features extracted by deep learning are hierarchical. Both low-level and high-level features can be extracted, which would be more comprehensive and robust for reconstructing HR HSI [48, 49]. In addition, deep learning is data-driven. It does not rely on any assumption or prior knowledge. After the off-line training, only feed-forward computation is needed in the testing stage of deep learning, which would make the HSI reconstruction fast. Below we give a summary on recent efforts in deep learning based image enhancement.

Deep learning in artificial intelligence has become a hot topic and has been successfully applied in many fields such as pattern recognition and computer vision [50, 51]. Deep learning is typically a neural network model with deep architecture. Features are automatically extracted from the data via abstraction by multiple layers, which are represented in a hierarchical fashion from low to high levels. Usually, the features extracted by higher layers are more abstract and discriminative than that of low-layers. In addition, the deep learning model is of high non-linearity. Compared with other machine learning models, it has more capacity in representing the mapping functions.

Among the typical deep learning models, convolutional neural network (CNN) is the most widely used model for single image SR enhancement. Several methods have been proposed [52–62]. The success of CNN in image SR could be summarized in three aspects. Firstly, CNN is built upon 2-D convolution computation, which could exploit the spatial context of an image. Secondly, CNN with deep architecture has high capacity and flexibility to represent the mapping function between LR and HR images [63]. Thirdly, compared with other deep learning models such as SAE [64], due to the weight sharing and local connection scheme, CNN often has fewer connections and is less prone to over-fitting [60].

In 2014, Dong et al. proposed a SR CNN network (SRCNN) [17]. As shown in Fig. 14.2, the CNN architecture for SR is composed of several convolutional layers. The input of the network is the LR image, which is up-scaled to the same size as its

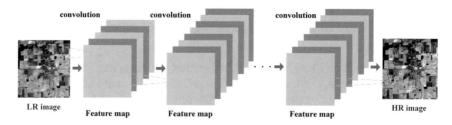

Fig. 14.2 A typical CNN architecture for super-resolution of single image

HR version. In the training stage, the mapping function between the up-scaled LR and HR images can be learned and represented by the CNN. In the testing stage, the HR image is reconstructed from its LR counterpart with the learned mapping function. Inspired by this idea, some other CNN based SR methods have also been proposed. For example, a faster SRCNN (FSRCNN) was proposed by adopting deconvolution layer and a small kernel size in [52]. Kim et al. pointed out that increasing the depth of CNN was helpful for improving the SR performance, a very deep CNN for super-resolution (VDSR) was proposed and trained with residual learning strategy in [53]. Trainable parameters would drastically increase in very deep CNN; to address this issue, a recursive CNN was proposed by sharing the parameters of different layers [54]. Most CNN SR methods employed the high-level features for reconstruction and neglected the low- and mid-level features. In [55] and [56], the authors proposed a residual dense network for SR, in which layers were densely connected to make full use of the hierarchical features. In [57], the authors proposed an end-to-end deep and shallow network (EEDS) composed of a shallow CNN and a deep CNN, which restored the principal components and the high-frequency components of an image, respectively. In order to enhance the sparsity and robustness of the network, the mapping of the wavelet coefficients between the LR and HR images was learned by CNN for the single image SR [58–61]. To address the challenge of super-resolving image by large factors, the authors in [62] proposed progressive deep learning models to upscale image gradually. Similarly, a Laplacian Pyramid SR CNN (LapSRN) was proposed in [65], which could progressively reconstruct high-frequency details of different sub-bands of HR image.

CNN has also been applied for single-frame HSI SR. A deep residual CNN network (DRCNN) with a spectral regularizer was proposed for HSI SR in [66]. In order to exploit the spectral correlation in HSI, the authors in [67] proposed a spectral difference convolutional network (SDCNN) to learn the mapping of spectral differences between the LR and HR HSIs. In [68], the SDCNN was combined with a spatial error correction model to rectify the artifacts produced during the SR. A 3D CNN based HSI SR method was proposed in [69]. The mapping function between the LR and HR HSIs was learned by a 3D CNN, and the spectral–spatial correlation in HSI could be exploited by the 3D convolution.

For pan-sharpening, a HR panchromatic image was stacked with an up-scaled LR MSI to form an input cube, and a pan-sharpening CNN network (PNN) was used to learn the mapping between the input cube and HR MSI [70]. A deep residual PNN (DRPNN) model was proposed to boost PNN by using residual learning in [71]. In [72], in order to preserve image structures, the mapping was learned by a residual network called PanNet in the high-pass filtering domain rather than the image domain. Multi-scale information could be exploited in the mapping learning. In [73], Yuan et al. proposed a multi-scale and multi-depth CNN (MSDCNN) for pan-sharpening, and each layer was constituted by filters with different sizes for the multi-scale features.

CNN has also been applied to HSI-MSI fusion. In [74], the dimensionality-reduced LR HSI and the HR MSI were stacked as a data cube, a 3D CNN was used to learn the mapping between the stacked data cube and HR HSI. In [75], the

LR HSI was enhanced initially via fusing with HR MSI, and a CNN was used to learn the mapping between the initially enhanced HSI and HR HSI.

Despite these progress, several key issues when deep learning is applied to HSI-MSI fusion still need more thorough studies: (1) how to jointly extract spectral–spatial deep features from LR HSI and HR MSI, and how to fuse them and generate HR HSI, (2) how to exploit the spectral correlation in deep learning, which is beneficial for HSI enhancement, (3) how to jointly reconstruct the highly number of hyperspectral bands in deep learning, so as to avoid reconstruction in a band-by-band fashion.

14.3   HSI-MSI Fusion Based on Two-Branch CNN

To address the above issues, in this section, we introduce our solution for HSI-MSI fusion. Specifically, in Sect. 14.3.1, we introduce our proposed spectral–spatial deep feature learning method based on two-branch CNN [3]. Then in Sect. 14.3.2, we describe a HSI-MSI fusion method inspired by the spectral–spatial deep feature learning method in Sect. 14.3.1.

14.3.1   Spectral–Spatial Deep Feature Learning

HSI is rich in spectral and spatial information simultaneously, in order to extract the spectral–spatial deep features from HSI, we proposed a CNN model with two-branch architecture (Two-CNN) in [3], as shown in Fig. 14.3. The model has two branches of CNN, devoting for features in the spectral and spatial domains. The spectral branch takes spectrum s_n of the nth pixel as input, and after l layers of convolutional operation and *max* pooling, we get the output $F_{\text{spe}}^{(l)}(s_n)$ of the spectral branch, which can be regarded as spectral features. It should be noted that input s_n

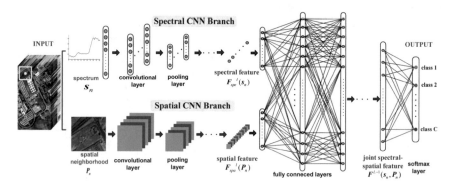

Fig. 14.3 The proposed two-branch CNN for spectral–spatial deep feature learning

is a 1-D signal, so all the convolutional and pooling operations in this branch belong to 1-D computation.

The spatial branch in Two-CNN takes neighboring pixels of the spectrum s_n as the input. HSI has contiguous spectral bands, and noise contaminated bands are often found. The noise can be modeled as additive Gaussian noise with zero mean [76]. In order to fuse spatial information of all bands and suppress the noise, we average all the images over the spectral bands, then spatial neighboring patch $P_n \in \mathbb{R}^{r \times r}$ (r is patch size) of the nth pixel is used as input for the spatial branch, after l convolutional and pooling layers in this branch, the output $F_{spa}^{(l)}(P_n)$ can be regarded as spatial features.

In order to exploit both the spectral and spatial correlation and obtain spectral–spatial joint deep features, we feed simultaneously $F_{spe}^{(l)}(s_n)$ and $F_{spa}^{(l)}(P_n)$ to the fully connected layers. The output of $(l + 1)$th layer is

$$F^{(l+1)}(s_n, P_n) = g\left\{ W^{(l+1)} \cdot [F_{spe}^{(l)}(s_n) \oplus F_{spa}^{(l)}(P_n)] + b^{(l+1)} \right\}, \qquad (14.1)$$

where $W^{(l+1)}$ and $b^{(l+1)}$ are the weight matrix and bias of the fully connected layer, respectively. $\oplus$ means concatenating the spectral features and spatial features. $g(.)$ is activation function. After several fully connected layers, the output of the last fully connected layer can be regarded as final spectral–spatial joint deep features. If the extracted feature is applied to land-cover classification, a softmax regression layer can be used as the final layer to predict the probability distribution of each class:

$$p(n) = \frac{1}{\sum_{k=1}^{C} e^{W_k^{(L)} F^{(L-1)}(s_n, P_n)}} \begin{bmatrix} W_1^{(L)} F^{(L-1)}(s_n, P_n) \\ W_2^{(L)} F^{(L-1)}(s_n, P_n) \\ \vdots \\ e^{W_C^{(L)} F^{(L-1)}(s_n, P_n)} \end{bmatrix}, \qquad (14.2)$$

where $W_k^{(L)}$ ($k = 1, 2, \ldots, C$) is the kth row of weight matrix of Lth softmax regression layer. C is the number of classes. $p(n) \in \mathbb{R}^C$ is a vector with C elements, it is the probability of assigning the nth pixel to each class. $F^{(L-1)}(s_n, P_n)$ is the final output of feature extraction, it conveys both spectral and spatial information, and can be treated as the spectral–spatial joint deep features of the nth pixel.

14.3.2 HSI-MSI Fusion

In the HSI-MSI fusion task, deep features should be extracted jointly from the LR HSI and the HR MSI. We extract spectral features from the LR HSI and spatial features from the HR MSI, then fuse these features and generate the spectrum of HR HSI. Inspired by the deep feature learning and the two-branch CNN model in the previous sub-section, we propose a two-branch CNN based HSI-MSI fusion method (Two-CNN-Fu) [77, 78].

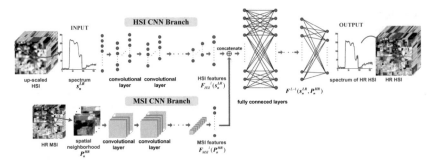

Fig. 14.4 The proposed two-branch CNN architecture for HSI-MSI fusion

The architecture of Two-CNN-Fu is shown in Fig. 14.4. There are two CNN branches in the network devoting features from the LR HSI and the HR MSI. The LR HSI is firstly up-scaled to the same size with the HR MSI, and features are extracted by the two branches from the spectrum of each pixel in the up-scaled HSI and its corresponding spatial neighborhood in the HR MSI. The HSI branch takes spectrum s_n^{LR} of the nth pixel in the up-scaled HSI as input, and after l layers of convolutional operation, we could extract features $F_{HSI}^{(l)}(s_n^{LR})$ from the LR HSI. The input s_n^{LR} is 1-D signal, all of the convolutional operations in this branch reduce to 1-D computation, so all of the convolutional kernels and feature maps per convolutional layer in this branch reduce to 1-D case.

In order to fuse the spatial information in the MSI of the same scene, the corresponding spatial neighboring block $P_n^{HR} \in \mathbb{R}^{r \times r \times b}$ in the MSI (as shown in the red box in Fig. 14.4) of the nth pixel is used as input for the MSI branch, where r is block size (it is fixed to 31×31 in the experiment), b is the number of bands of MSI. All bands in MSI are used for fusion in this branch. After l convolutional layers in this branch, we can extract features $F_{MSI}^{(l)}(P_n^{HR})$ from the MSI.

It is noted that $F_{HSI}^{(l)}(s_n^{LR})$ and $F_{MSI}^{(l)}(P_n^{HR})$ are obtained by vectorizing the feature maps of the HSI and MSI branches, respectively. In order to fuse the information of HSI and MSI, we concatenate the extracted features $F_{HSI}^{(l)}(s_n^{LR})$ and $F_{MSI}^{(l)}(P_n^{HR})$, then simultaneously feed them to the FC layers. FC layers are adopted here because it could fully fuse the information of HSI and MSI. After several FC layers, the output of the last FC layer is the reconstructed spectrum of the expected HR HSI

$$\hat{s}_n^{HR} = W^{(L)} \cdot F^{(L-1)}(s_n^{LR}, P_n^{HR}) + b^{(L)}, \tag{14.3}$$

where $W^{(L)}$ and $b^{(L)}$ are the weight matrix and the bias of the Lth FC layer, respectively. $F^{(L-1)}(s_n^{LR}, P_n^{HR})$ is the feature vector of the $(L-1)$th FC layer. All the convolutional kernels, weight matrices, and bias values in the network are trained in an end-to-end fashion. In the testing stage, we extract the spectrum of the up-scaled HSI from each pixel and its corresponding neighborhood block in the HR MSI, then feed them to the trained network. The output of the network is the spectrum of the expected HR HSI. After putting back the reconstructed spectrum to each pixel, a HR HSI could be obtained.

In the training stage, the LR HSI is firstly up-scaled to the same size as the HR MSI. Although the up-scaled HSI has the same size as the HR HSI, it is still blurring, as shown in Fig. 14.4. The purpose of up-scaling is to match the size of LR HSI to that of HR MSI and HR HSI. This strategy has also been adopted in other deep learning based image super-resolution methods, for example in [47, 53]. The deep learning model is trained to learn the mapping between up-scaled HSI and HR HSI. In the testing stage, we should also firstly up-scale the LR HSI to the size of the HR HSI using the same interpolation algorithm as the training stage, then feed it to the trained deep learning network, and a HR HSI with better quality could be recovered.

There are also some other advantages of the architecture of Two-CNN-Fu. Firstly, the deep learning network extracts features from the spectrum of LR HSI, as such the spectral correlation of HSI could be exploited by deep learning. Secondly, the deep learning network would output the spectrum of the expected HR HSI, so all of the bands of HR HSI could be jointly reconstructed, which is beneficial for reducing the spectral distortion. Thirdly, we propose to extract features from the spectrum of LR HSI. Compared with extracting features from a 3D HSI block, it involves less computation with simpler network complexity.

All the convolutional kernels, weight matrices, and bias values in the network are trained by minimizing the reconstruction error of the HR HSI spectra. *Frobenius* norm is used to measure the reconstruction error in the loss function [47]. The set of training samples is denoted as $\left\{s_n^{\mathrm{LR}}, P_n^{\mathrm{HR}}, s_n^{\mathrm{HR}}\right\}$, $(n = 1, 2, \ldots N)$, the loss function is written as

$$J = \frac{1}{N} \sum_{n=1}^{N} \left\| s_n^{\mathrm{HR}} - \hat{s}_n^{\mathrm{HR}} \right\|_F^2, \tag{14.4}$$

where N is number of training samples. For the nth training sample, s_n^{LR} is the spectrum of LR HSI, P_n^{HR} is the corresponding spatial neighborhood in HR MSI, and s_n^{HR} is the spectrum of HR HSI. $\hat{s}_n^{\mathrm{HR}}$ is the reconstructed spectrum of HR HSI. The loss function is optimized using stochastic gradient descent (SGD) method with standard back-propagation [79].

14.3.3 Experimental Results

In this section, the performance of the proposed fusion algorithm is evaluated on several datasets, including simulated datasets and real spaceborne data.

The network parameters of our deep learning model are given in Table 14.1. The deep learning network needs to be initialized before training. All of the convolutional kernels and weight matrices of the FC layers are initialized from a Gaussian random distribution with standard variance 0.01 and mean 0. The bias values are initialized

Table 14.1 The parameter setting of the network architecture

Number of filters per conv. layer	20 (HSI branch) 30 (MSI branch)
Size of filter per conv. layer	45×1 with stride 1 (HSI branch) 10×10 with stride 1 (MSI branch)
Number of neurons per FC layer	450 (The first two FC layers) Number of HSI bands (Last FC layer)
Number of conv. layers	3 (MSI branch) 3 (HSI branch)
Number of FC layers	3

to be 0. The parameters involved in standard stochastic gradient descent method are learning rate, momentum, and batch size. The learning rate is fixed as 0.0001, momentum is set to 0.9, and the batch size is set to 128. The number of training epochs is set to 200.

We compare our method with other state-of-the-art fusion methods: coupled non-negative matrix factorization (CNMF) method [32], sparse spatial–spectral representation method (SSR) [38], and Bayesian sparse representation method (BayesSR) [39]. The MATLAB codes of these methods are released by the original authors. The parameter settings in the compared methods first follow the suggestions from the original authors; we then empirically tune them to achieve the best performance. The number of endmembers is a key parameter for the CNMF method; it is set to 30 in the experiment. The parameters in the SSR method include the number of dictionary atoms, the number of atoms in each iteration, and the spatial patch size, which are set to 300, 20, and 8×8, respectively. The parameters in the BayesSR method consist of the number of inferencing sparse coding in Gibbs sampling process and the number of iterations of dictionary learning, which are set to 32 and 50,000, respectively. Note that all the compared methods fuse LR HSI with HR MSI. Although there are some deep learning-based HSI super-resolution methods, such as 3D-CNN [69], they only exploited LR HSI. Therefore, they are not used for comparison.

14.3.3.1 Experimental Results on Simulated Datasets

Two simulated datasets are used in the experiment. The first dataset was collected by airborne visible infrared imaging spectrometer (AVIRIS) sensor, which consists of four images captured over Indian Pines, Moffett Field, Cuprite, and Lunar Lake sites with dimensions $753 \times 1923, 614 \times 2207, 781 \times 6955$, and 614×1087, respectively. The spatial resolution is 20 m. The dataset was taken in the range of 400–2500 nm with 224 bands. After discarding the water absorption bands and noisy bands, there are 162 bands remained. The second one is simulated Environmental Mapping and Analysis Program (EnMAP) data, which was acquired by HyMap sensor over Berlin district in August 2009 [80]. The size of this data is 817×220 with spatial resolution 30 m. There are 244 spectral bands in the range of 420–2450 nm.

The above HSI datasets are regarded as HR HSI and reference image, both of LR HSI and HR MSI are simulated from them. The LR HSI is generated from the reference image via spatial Gaussian down-sampling. The HR MSI is obtained by spectral degrading the reference image with the spectral response function of Landsat-7 multispectral imaging sensor as filters. There are six spectral bands of the simulated MSI, which cover 450–520, 520–600, 630–690, 770–900, 1550–1750, and 2090–2350 nm spectral regions. We crop two sub-images with size 256×256 from AVIRIS *Indian pines* and *Moffett Field* dataset and one sub-image with size 256×160 from EnMAP *Berlin* dataset as testing data. 50,000 samples are extracted for training each Two-CNN-Fu model. There is no overlapping between the testing and the training regions.

The fusion performance is evaluated by peak-signal-noise-ratio (PSNR, dB), structural similarity index measurement (SSIM) [81], feature similarity index measurement (FSIM) [82], and spectral angle mean (SAM). We calculate PSNR, SSIM, and FSIM on each band, and then the mean values over all the bands are given. The indices on the three testing datasets are given in Tables 14.2 and 14.3. The best indices values are highlighted in bold.

It can be seen that our proposed Two-CNN-Fu method has competitive performance on the three testing datasets. In Table 14.2, the PSNR, SSIM, and FSIM of our results are higher than those of compared methods, which means that our fusion results are closer to the original HR HSI, with fewer errors. The SSR method is based on spatial–spectral sparse representation; a spectral dictionary is first learned with the sparsity, and then combined with the abundance of MSI to reconstruct the HR HSI. While in the CNMF method, the endmember of LR HSI and the abundance of

Table 14.2 The evaluation indices of different fusion methods on the three testing datasets by a factor of two

Testing data	Index	SSR [38]	BayesSR [39]	CNMF [32]	Two-CNN-Fu
Indian pines	PSNR (dB)	31.5072	33.1647	33.2640	**34.0925**
	SSIM	0.9520	0.9600	0.9650	**0.9714**
	FSIM	0.9666	0.9735	0.9745	**0.9797**
	SAM	3.6186°	3.4376°	3.0024°	**2.6722°**
Moffett Field	PSNR (dB)	28.3483	31.0965	31.4079	**31.7860**
	SSIM	0.9317	0.9499	0.9568	**0.9661**
	FSIM	0.9558	0.9694	0.9734	**0.9788**
	SAM	3.9621°	3.7353°	3.1825°	**2.7293°**
Berlin	PSNR (dB)	30.0746	29.8009	32.2022	**34.8387**
	SSIM	0.9373	0.9272	0.9569	**0.9684**
	FSIM	0.9512	0.9468	0.9705	**0.9776**
	SAM	2.8311°	3.2930°	1.4212°	**1.0709°**

Table 14.3 The evaluation indices of different fusion methods on the three testing datasets by a factor of four

Testing data	Index	SSR [38]	BayesSR [39]	CNMF [32]	Two-CNN-Fu
Indian pines	PSNR(dB)	30.6400	32.9485	32.7838	**33.6713**
	SSIM	0.9516	0.9601	0.9603	**0.9677**
	FSIM	0.9651	0.9730	0.9696	**0.9769**
	SAM	3.7202°	3.5334°	3.1227°	**2.8955°**
Moffett Field	PSNR(dB)	27.3827	29.4564	30.7893	**31.4324**
	SSIM	0.9181	0.9274	0.9509	**0.9621**
	FSIM	0.9477	0.9561	0.9684	**0.9752**
	SAM	4.7584°	4.4500°	3.3972°	**2.8697°**
Berlin	PSNR(dB)	29.7133	29.2131	30.1242	**31.6728**
	SSIM	0.9357	0.9265	0.9464	**0.9531**
	FSIM	0.9516	0.9420	0.9586	**0.9608**
	SAM	2.9062	5.6545	3.8744	**2.2574**

MSI are alternatively estimated in a coupled way, the estimated endmember and the abundance would be more accurate, so CNMF could achieve better performance than SSR. The BayesSR method learns the dictionary in a non-parametric Bayesian sparse coding framework and often performs better than the parametric SSR method. The best performance is achieved by Two-CNN-Fu on the three testing datasets. Two-CNN-Fu extracts hierarchical features, which are more comprehensive and robust than the hand-crafted features in [32, 38, 39]. The performance of Two-CNN-Fu demonstrates the effectiveness and potential of deep learning in the HSI-MSI fusion task. In order to verify the robustness over a higher resolution ratio between LR HSI and HR MSI, we also simulate the LR HSI by a factor of four and then fuse it with MSI. The Two-CNN-Fu also performs better than other methods, as shown in Table 14.3. The PSNR curves over the spectral bands are presented in Fig. 14.5. It can be found that the PSNR values of Two-CNN-Fu are higher than the compared methods in most bands.

It is worth noting that the result of our Two-CNN-Fu method has the lowest spectral distortion among the compared methods in most cases, as shown in Tables

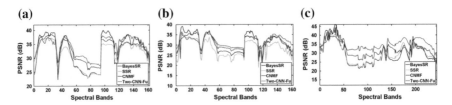

Fig. 14.5 PSNR values of each band of different fusion results by a factor of two, **a** on AVIRIS *Indian pines* data; **b** on AVIRIS *Moffett Field* data; **c** on EnMAP *Berlin* data

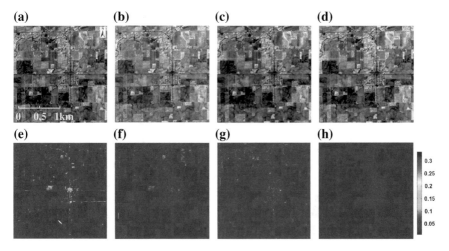

Fig. 14.6 Reconstructed images (band 70) and root mean square error (RMSE) maps of different fusion results by a factor of four. The testing image is cropped from *Indian pines* of AVIRIS data with size 256 × 256. **a** Result of SSR [38], **b** result of BayesSR [39], **c** result of CNMF [32], **d** result of Two-CNN-Fu, **e** RMSE map of SSR [38], **f** RMSE map of BayesSR [39], **g** RMSE map of CNMF [32], **h** RMSE map of Two-CNN-Fu

14.2 and 14.3. Our deep learning network directly learns the mapping between the spectra of LR and HR HSIs. The objective function Eq. (14.4) for training the network aims at minimizing the error of the reconstructed spectra of HR HSI. In addition, instead of reconstructing HSI in a band-by-band way, our deep learning model jointly reconstructs all bands of HSI. These two characteristics are beneficial for reducing the spectral distortion.

We present parts of the reconstructed HSIs in Figs. 14.6, 14.7, and 14.8. In order to visually evaluate the quality of different fusion results, we also give pixel-wise root mean square error (RMSE) maps, which reflect the errors of reconstructed pixels over all the bands. It is clear that the fusion result of our Two-CNN-Fu method has fewer errors than the compared methods. The compared methods rely on hand-crafted features such as the dictionary. Their RMSE maps have materials-related patterns, which may be caused by the errors introduced in dictionary learning or endmember extraction. Our Two-CNN-Fu method reconstructs the HR HSI based on the mapping function between LR and HR HSIs, which is trained by minimizing the error of the reconstructed HR HSI, so the fusion result of Two-CNN-Fu has fewer errors.

14.3.3.2 Experimental Results on Real Data

In order to investigate the applicability of the proposed method, we apply it to real spaceborne HSI-MSI data fusion. The HSI data was collected by Hyperion sensor, which is carried on Earth Observing-1 (EO-1) satellite. This satellite was launched in

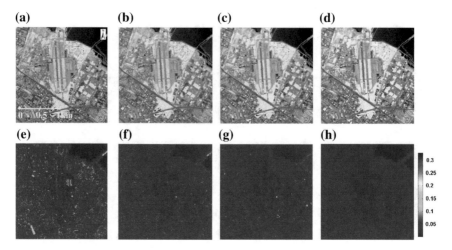

Fig. 14.7 Reconstructed image (band 80) and root mean square error (RMSE) maps of different fusion results by a factor of two. The testing image is cropped from *Moffett Field* of AVIRIS data with size 256 × 256. **a** Result of SSR [38], **b** result of BayesSR [39], **c** result of CNMF [32], **d** result of Two-CNN-Fu, **e** RMSE map of SSR [38], **f** RMSE map of BayesSR [39], **g** RMSE map of CNMF [32], **h** RMSE map of Two-CNN-Fu

November 2000. The MSI data was captured by the Sentinel-2A satellite, launched in June 2015. The spatial resolution of Hyperion HSI is 30 m. There are 242 spectral bands in the spectral range of 400–2500 nm. The Hyperion HSI suffers from noise; after removing the noisy bands and water absorption bands, 83 bands remained. The Sentinel-2A satellite provides MSIs with 13 bands. We select four bands with 10 m spatial resolution for the fusion. The central wavelengths of these four bands are 490 nm, 560 nm, 665 nm, and 842 nm, and their bandwidths are 65 nm, 35 nm, 30 nm, and 115 nm, respectively.

The Hyperion HSI and the Sentinel-2A MSI in this experiment were taken over Lafayette, LA, USA in October and November 2015, respectively. We crop sub-images to 341 × 365 and 1023 × 1095 as study areas from the overlapped region of the Hyperion and Sentinel data, as shown in Fig. 14.9. The remainder of the overlapped region is used for training the Two-CNN-Fu network.

In this experiment, our goal is to fuse the 30 m HSI with the 10 m MSI, and then generate a 10 m HSI, so a Two-CNN-Fu network that could enhance HSI by a factor of three should be trained. In the training stage, we first down-sample the 30 m Hyperion HSI and 10 m Sentinel-2A MSI into 90 m and 30 m, by a factor of three, respectively. Then we train a Two-CNN-Fu network that could fuse the 90 m HSI with the 30 m MSI and reconstruct the original 30 m HSI. This network could enhance HSI by a factor of three. We assume that it could be transferred to the fusion task of 30 m HSI and 10 m MSI. By applying the trained network to the 30 m HSI and the 10 m MSI, an HSI with 10 m resolution could be reconstructed. The network parameters are set according to Table 14.1, except that the number of convolutional

(a) (b) (c) (d)

(e) (f) (g) (h)

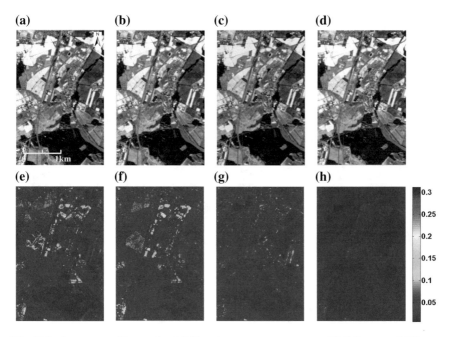

Fig. 14.8 Reconstructed image (band 200) and root mean square error (RMSE) maps of different fusion results by a factor of two. The testing image is cropped from *Berlin* of EnMAP data with size 256 × 160. **a** Result of SSR [38], **b** result of BayesSR [39], **c** result of CNMF [32], **d** result of Two-CNN-Fu, **e** RMSE map of SSR [38], **f** RMSE map of BayesSR [39], **g** RMSE map of CNMF [32], **h** RMSE map of Two-CNN-Fu

layers in the HSI branch is one, because we only use 83 bands of the Hyperion data. In this case, the maximal number of convolutional layers in the HSI branch is one.

The fusion results of different methods are presented in Fig. 14.10. The size of the fusion result is 1023 × 1095. In order to highlight the details of the fusion results, we also zoom in some areas in Figs. 14.11 and 14.12. It is clear that there is some noise in the results of SSR and CNMF, as shown in Fig. 14.11b, d. In Fig. 14.12, we also find that some details in the results of SSR and CNMF are blurred, as indicated in the dashed box. The results of BayesSR and Two-CNN-Fu are sharper and cleaner, and our Two-CNN-Fu method produces the HR HSI with higher spectral fidelity. After compared with the original LR images, it is clear that the spectral distortion of BayesSR is heavier than our Two-CNN-Fu results. The color of the BayesSR results seems to be darker than the original LR image. Spectral distortion would affect the accuracy of many applications such as classification.

It is noted that the temporal difference is about one month between the acquisitions of Hyperion HSI and Sentinel-2A MSI. Some changes may occur during this month, which may be one of the factors that lead to the spectral distortion. Even though nearly all of the fusion results in Figs. 14.11 and 14.12 suffer from the spectral distortion,

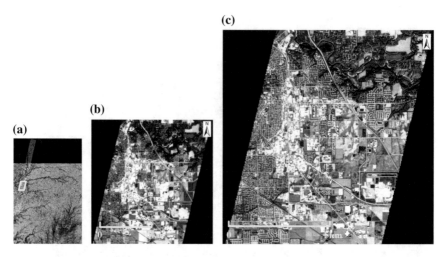

Fig. 14.9 The experiment data taken over Lafayette, **a** illustration of Hyperion and Sentinel-2A data, the red part is Hyperion data, the green part is Sentinel-2A data, the white part is the overlapped region, the yellow line indicates the study area; **b** color composite (bands 31, 21, 14) of Hyperion data in the study area with size 341 × 365; **c** color composite (bands 4, 3, 2) of Sentinel-2A data in the study area with size 1023 × 1095

our Two-CNN-Fu method generates results with less errors, which demonstrates the robustness of the proposed method.

In order to assess the performance quantitatively, we evaluate the fusion results using the no-reference HSI quality assessment method in [83], which would give quality scores for each reconstructed HSI. In this no-reference assessment method, some pristine HSIs are needed as training data to learn the benchmark quality-sensitive features. We use the original LR Hyperion data after discarding the noisy bands as training data. The quality score measures the distance of the reconstructed HSI and the pristine benchmark; a lower score value means better quality. The quality scores of different fusion results are given in Table 14.4. The best index is highlighted in bold. The scores of our fusion result are better or competitive with other compared methods.

Land-cover classification is one of the important applications of HSI. We test the effect of different fusion methods on the land-cover classification. Land-cover groundtruth is obtained from open street map (OSM). According to the OSM data, there are 12 classes of land-covers in the study area. We select parts from each class as ground truth, as shown in Fig. 14.13 and Table 14.5. Two classifiers, support vector machine (SVM) [84] and canonical correlation forests (CCF) [85], are used in the experiment due to their stability and good performance. The SVM classifier is implemented with the LIBSVM toolbox [84], and the radial basis function is used as kernel function of SVM. The regularization parameters in SVM are determined by five-fold cross-validation in the range of $[2^{-10}, 2^{-9}, \ldots, 2^{19}, 2^{20}]$. The parameter involved in the CCF classifier is the number of trees; we set it to 200 in the experiment.

(a)   **(b)**

(c)   **(d)**

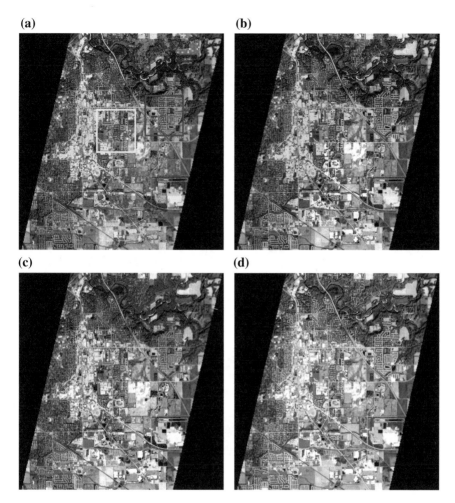

Fig. 14.10 False color composite (bands 45, 21, 14) of different Hyperion–Sentinel fusion results, size of the enhanced image is 1023 × 1095 with 10 m resolution, **a** result of SSR [38]; **b** result of BayesSR [39]; **c** result of CNMF [32]; **d** result of Two-CNN-Fu

Fifty samples of each class are randomly chosen for training; the remainder of the groundtruth is used as testing samples. We repeat the classification experiment 10 times, and then report the mean value and standard variance of overall accuracy in Table 14.6. The best indices are highlighted in bold.

In Table 14.6, it can be found that the classification accuracy of our fusion result is higher than that of the other three fusion methods. Both SVM and CCF classifiers obtain competitive classification accuracy, and the classification results have a similar trend on the two classifiers. As we can observe in Figs. 14.11 and 14.12, the spectral distortion and noise of our fusion method is less than that of other methods, which may explain why our classification accuracy is higher. The classification map

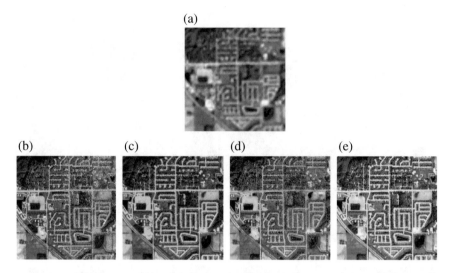

Fig. 14.11 False color composite (bands 45, 21, 14) of the enlarged area in the blue box of Fig. 14.10, size of the area is 200 × 200, **a** the original 30 m Hyperion data; **b** fusion result of SSR [38]; **c** fusion result of BayesSR [39]; **d** fusion result of CNMF [32]; **e** fusion result of Two-CNN-Fu

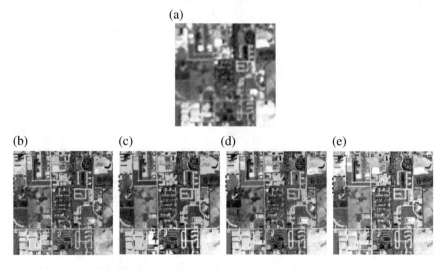

Fig. 14.12 False color composite (bands 45, 21, 14) of the enlarged area in the yellow box of Fig. 14.10, size of the area is 200 × 200, **a** the original 30 m Hyperion data; **b** fusion result of SSR [38]; **c** fusion result of BayesSR [39]; **d** fusion result of CNMF [32]; **e** fusion result of Two-CNN-Fu

Table 14.4 The no-reference quality assessment scores of different results on Hyperion–Sentinel fusion

Methods	SSR [38]	BayesSR [39]	CNMF [32]	Two-CNN-Fu
Scores [83]	22.8317	20.9626	22.8317	**20.2425**

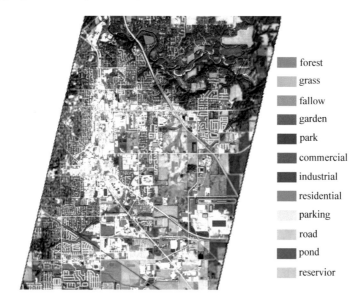

forest
grass
fallow
garden
park
commercial
industrial
residential
parking
road
pond
reservior

Fig. 14.13 The ground truth labeled from each class

Table 14.5 The number of ground truth labeled in the study area

Class name	Training samples	Testing samples
Forest	50	1688
Grass	50	466
Fallow	50	1856
Garden	50	226
Park	50	836
Commercial	50	548
Industrial	50	1618
Residential	50	524
Parking	50	918
Road	50	1053
Pond	50	375
Reservoir	50	397
Total	600	10,505

Table 14.6 The overall accuracy (OA) of different fusion results

Classifier	SSR [38]	BayesSR [39]	CNMF [32]	Two-CNN-Fu
SVM	81.53 ± 1.18%	77.01 ± 0.97%	86.54 ± 0.98%	**89.81 ± 0.86%**
CCF	85.04 ± 0.64%	80.74 ± 0.73%	89.75 ± 1.50%	**94.15 ± 0.47%**

Fig. 14.14 The classification map of Hyperion–Sentinel fusion result by Two-CNN-Fu method, the classifier is CCF

of Two-CNN-Fu fusion results is given in Fig. 14.14, where most of the land-covers can be classified correctly; even some details such as roads and residential areas can be classified well with the fusion enhanced image. Misclassification of some land-covers, such as forests and gardens, may be caused by the similarity in spectra between these two land-covers. The experiment has demonstrated that our proposed deep learning based fusion method has great potential when used with real space-borne HSI-MSI fusion, and the reconstructed HR HSI could result in competitive classification performance.

The implementation code of the Two-CNN-Fu algorithm is available online: https://github.com/polwork/Hyperspectral-and-Multispectral-fusion-via-Two-branch-CNN.

14.4 Conclusions and Discussions

In this chapter, we discussed the advances in HSI resolution enhancement and illustrated our proposed deep learning based fusion method for enhancement. How to jointly extract spectral–spatial deep features from the LR HSI and the HR MSI, and how to fuse them are the key issues when we apply deep learning to HSI-MSI fusion.

We address these issues by designing a two-branch CNN. There are two branches of CNN in our network, extracting features from both the LR HSI and HR MSI. The HSI branch of the Two-CNN-Fu extracts features from the spectrum of each pixel from the LR HSI, its corresponding spatial neighborhood in HR MSI is used as input of the MSI branch. The features of LR HSI and HR MSI are then fed to fully connected layers, where HSI and MSI can be fully fused. The output of the fully connected layers is the spectrum of the expected HR HSI.

Despite the achieved progress, there are still some issues that should be further addressed in the future. Firstly, if we adopt a HSI-MSI fusion approach, the HSI and MSI may not be captured in the exact same period, the illumination condition on the surface may change [86]. It is necessary to enhance the robustness of HSI-MSI fusion algorithm over temporal difference, illumination condition, and co-registration error. Secondly, different types of noise exist in real HSI. For example, in our experiment in Sect. 14.3.3.2, Hyperion HSI suffers from severe stripping noise and deadlines. After discarding these noisy bands, there are only 83 bands remained. How to enhance the robustness of HSI SR algorithm over the noise is another issue for SR with real spaceborne data. Finally, how to assess the impact of HSI enhancement on various applications such as classification also needs more research. In [8], different HSI enhancement methods were applied to several datasets and their impact on land-cover classification is compared. It was found that the HSI enhancement method with high SR performance may not lead to satisfactory classification performance. Tackling classification and SR in a unified model and exploiting useful information from the intermediate classification result of the enhanced HSI may help to improve the SR performance as well as classification performance.

Acknowledgements This work is supported by the National Natural Science Foundation of China (61771391, 61371152), the Science, Technology and Innovation Commission of Shenzhen Municipality (JCYJ20170815162956949), the National Natural Science Foundation of China and South Korean National Research Foundation Joint Funded Cooperation Program (61511140292), the Fundamental Research Funds for the Central Universities (3102015ZY045), the China Scholarship Council for joint Ph.D. students (201506290120), and the Innovation Foundation of Doctor Dissertation of Northwestern Polytechnical University (CX201621).

References

1. Zhao Y, Pan Q, Cheng Y (2011) Imaging spectropolarimetric remote sensing and application. National Defense Industry Press, Beijing
2. Zhao Y, Yi C, Kong SG, Pan Q, Cheng Y (2016) Multi-band polarization imaging and applications. Springer, Berlin Heidelberg
3. Yang J, Zhao YQ, Chan JCW (2017) Learning and transferring deep joint spectral-spatial features for hyperspectral classification. IEEE Trans Geosci Remote Sens 55(8):4729–4742
4. Nasrabadi NM (2014) Hyperspectral target detection: an overview of current and future challenges. IEEE Signal Process Mag 31(1):34–44

5. Jay S, Guillaume M, Minghelli A et al (2017) Hyperspectral remote sensing of shallow waters: considering environmental noise and bottom intra-class variability for modeling and inversion of water reflectance. Remote Sens Environ 200:352–367

6. Liang L, Di L, Zhang L et al (2015) Estimation of crop LAI using hyperspectral vegetation indices and a hybrid inversion method. Remote Sens Environ 165:123–134

7. Loncan L, de Almeida LB, Bioucas-Dias JM, Briottet X, Chanussot J et al (2015) Hyperspectral pansharpening: a review. IEEE Geosci Remote Sens Mag 3(3):27–46

8. Yokoya N, Grohnfeldt C, Chanussot J (2017) Hyperspectral and multispectral data fusion: a comparative review of the recent literature. IEEE Geosci Remote Sens Mag 5(2):29–56

9. Ghamisi P, Yokoya N, Li J, Liao W et al (2017) Advances in hyperspectral image and signal processing: a comprehensive overview of the state of the art. IEEE Geosci Remote Sens Mag 5(4):37–78

10. Bioucas-Dias JM, Plaza A, Dobigeon N et al (2012) Hyperspectral unmixing overview: geometrical, statistical, and sparse regression-based approaches. IEEE J Sel Top Appl Earth Obs Remote Sens 5(2):354–379

11. Dobigeon N, Tourneret JY, Richard C et al (2014) Non-linear unmixing of hyperspectral images: models and algorithms. IEEE Signal Process Mag 31(1):82–94

12. Yokoya N, Chan JC-W, Segl K (2016) Potential of resolution-enhanced hyperspectral data for mineral mapping using simulated EnMAP and Sentinel-2 images. Remote Sens 8(3)

13. Chen F, Wang K, Van de Voorde T, Tang TF (2017) Mapping urban land cover from high spatial resolution hyperspectral data: an approach based on simultaneously unmixing similar pixels with jointly sparse spectral mixture analysis. Remote Sens Environ 196:324–342

14. Zhang L, Zhang L, Tao D et al (2014) Hyperspectral remote sensing image subpixel target detection based on supervised metric learning. IEEE Trans Geosci Remote Sens 52(8):4955–4965

15. Shahdoosti HR, Ghassemian H (2016) Combining the spectral PCA and spatial PCA fusion methods by an optimal filter. Inf Fusion 27:150–160

16. Dalla Mura M, Vivone G, Restaino R et al (2015) Global and local Gram-Schmidt methods for hyperspectral pansharpening. In: IEEE international geoscience and remote sensing symposium, pp 37–40

17. Liu JG (2000) Smoothing filter-based intensity modulation: a spectral preserve image fusion technique for improving spatial details. Int J Remote Sens 21(18):3461–3472

18. Aiazzi B, Alparone L, Baronti et al (2006) MTF-tailored multiscale fusion of high-resolution MS and Pan imagery. Photogramm Eng Remote Sens 72(5):591–596

19. Ghahremani M, Ghassemian H (2016) A compressed-sensing-based pan-sharpening method for spectral distortion reduction. IEEE Trans Geosci Remote Sens 54(4):2194–2206

20. Yang S, Zhang K, Wang M (2017) Learning low-rank decomposition for pan-sharpening with spatial-spectral offsets. IEEE Trans Neural Netw Learn Syst 20(8):3647–3657

21. Garzelli A (2015) Pansharpening of multispectral images based on nonlocal parameter optimization. IEEE Trans Geosci Remote Sens 53(4):2096–2107

22. Peleg T, Elad M (2014) A statistical prediction model based on sparse representations for single image super-resolution. IEEE Trans Image Process 23(6):2569–2582

23. Zhao YQ, Yang J, Chan JC-W (2014) Hyperspectral imagery super-resolution by spatial–spectral joint nonlocal similarity. IEEE J Sel Top Appl Earth Obs Remote Sens 7(6):2671–2679

24. Ren C, He X, Teng Q et al (2016) Single image super-resolution using local geometric duality and non-local similarity. IEEE Trans Image Process 25(5):2168–2183

25. Li J, Yuan Q, Shen H, Meng X, Zhang L (2016) Hyperspectral image super-resolution by spectral mixture analysis and spatial-spectral group sparsity. IEEE Geosci Remote Sens Lett 13(9):1250–1254

26. Zhang H, Yang Z, Zhang L, Shen H (2014) Super-resolution reconstruction for multi-angle remote sensing images considering resolution differences. Remote Sens 6(1):637–657

27. Chan JC-W, Ma J, Kempeneers P, Canters F (2010) Superresolution enhancement of hyperspectral CHRIS/Proba images with a thin-plate spline nonrigid transform model. IEEE Trans Geosci Remote Sens 48(6):2569–2579

28. Ma J, Chan JC-W (2012) Superresolution reconstruction of hyperspectral remote sensing imagery using constrained optimization of POCS. In: IEEE international geoscience and remote sensing symposium, pp 7271–7274
29. Chan JC-W, Ma J, Van de Voorde T, Canters F (2011) Preliminary results of superresolution enhanced angular hyperspectral (CHRIS/Proba) images for land-cover classification. IEEE Geosci Remote Sens Lett 8(6):1011–1015
30. Demarchi L, Chan JC-W, Ma J, Canters F (2012) Mapping impervious surfaces from superesolution enhanced CHRIS/Proba imagery using multiple endmember unmixing. ISPRS J Photogramm Remote Sens 72:99–112
31. Xu X, Zhong Y, Zhang L, Zhang H (2013) Sub-pixel mapping based on a MAP model with multiple shifted hyperspectral imagery. IEEE J Sel Top Appl Earth Obs Remote Sens 6(2):580–593
32. Yokoya N, Yairi T, Iwasaki A (2012) Coupled nonnegative matrix factorization unmixing for hyperspectral and multispectral data fusion. IEEE Trans Geosci Remote Sens 50(2):528–537
33. Yokoya N, Mayumi N, Iwasaki A (2013) Cross-calibration for data fusion of EO-1/Hyperion and Terra/ASTER. IEEE J Sel Top Appl Earth Obs Remote Sens 6(2):419–426
34. Lanaras C, Baltsavias E, Schindler K (2015) Hyperspectral super-resolution by coupled spectral unmixing. In: Proceedings of the IEEE international conference on computer vision, pp 3586–3594
35. Huang B, Song H, Cui H et al (2014) Spatial and spectral image fusion using sparse matrix factorization. IEEE Trans Geosci Remote Sens 52(3):1693–1704
36. Zhu XX, Bamler R (2013) A sparse image fusion algorithm with application to pan-sharpening. IEEE Trans Geosci Remote Sens 51(5):2827–2836
37. Zhu XX, Grohnfeldt C, Bamler R (2016) Exploiting joint sparsity for pansharpening: the J-SparseFI algorithm. IEEE Trans Geosci Remote Sens 54(5):2664–2681
38. Akhtar N, Shafait F, Mian A (2014) Sparse spatio-spectral representation for hyperspectral image super-resolution. In: European conference on computer vision, pp 63–78
39. Akhtar N, Shafait F, Mian A (2015) Bayesian sparse representation for hyperspectral image super resolution. In: Proceedings of the IEEE conference on computer vision and pattern recognition, pp 3631–3640
40. Wei Q, Bioucas-Dias J, Dobigeon N, Tourneret JY (2015) Hyperspectral and multispectral image fusion based on a sparse representation. IEEE Trans Geosci Remote Sens 53(7):3658–3668
41. Simões M, Bioucas-Dias J, Almeida LB, Chanussot J (2015) A convex formulation for hyperspectral image superresolution via subspace-based regularization. IEEE Trans Geosci Remote Sens 53(6):3373–3388
42. Veganzones MA, Simoes M, Licciardi G et al (2016) Hyperspectral super-resolution of locally low rank images from complementary multisource data. IEEE Trans Image Process 25(1):274–288
43. Zhang K, Wang M, Yang S (2017) Multispectral and hyperspectral image fusion based on group spectral embedding and low-rank factorization. IEEE Trans Geosci Remote Sens 55(3):1363–1371
44. Eismann MT, Hardie RC (2005) Hyperspectral resolution enhancement using high-resolution multispectral imagery with arbitrary response functions. IEEE Trans Geosci Remote Sens 43:455–465
45. Zhang L, Zhang L, Du B (2016) Deep learning for remote sensing data: a technical tutorial on the state of the art. IEEE Geosci Remote Sens Mag 4(2):22–40
46. Zhu XX, Tuia D, Mou L et al (2017) Deep learning in remote sensing: a review. IEEE Geosci Remote Sens Mag 5(4):8–36
47. Dong C, Loy CC, He K, Tang X (2016) Image super-resolution using deep convolutional networks. IEEE Trans Pattern Anal Mach Intell 38(2):295–307
48. LeCun Y, Bengio Y, Hinton G (2015) Deep learning. Nature 521(7553):436
49. Hinton GE, Salakhutdinov RR (2006) Reducing the dimensionality of data with neural networks. Science 313(5786):504–507

50. Jordan MI, Mitchell TM (2015) Machine learning: trends, perspectives, and prospects. Science 349(6245):255–260
51. Bengio Y (2009) Learning deep architectures for AI. Found Trends Mach Learn 2(1):1–127
52. Dong C, Loy CC, Tang X (2016) Accelerating the super-resolution convolutional neural network. In: European conference on computer vision, pp 391–407
53. Kim J, Lee JK, Lee KM (2016) Accurate image super-resolution using very deep convolutional networks. In: The IEEE conference on computer vision and pattern recognition, pp 1646–1654
54. Kim J, Lee JK, Lee KM (2016) Deeply-recursive convolutional network for image super-resolution. In: The IEEE conference on computer vision and pattern recognition, pp 1637–1645
55. Tai Y, Yang J, Liu X, Xu C (2017) Memnet: a persistent memory network for image restoration. In: Proceedings of the IEEE conference on computer vision, pp 4539–4547
56. Zhang Y, Tian Y, Kong Y, Zhong B, Fu Y (2018) Residual dense network for image super-resolution. In: The IEEE conference on computer vision and pattern recognition
57. Wang Y, Wang L, Wang H, Li P (2016) End-to-end image super-resolution via deep and shallow convolutional networks. arXiv:1607.07680
58. Huang H, He R, Sun Z, Tan T (2017) Wavelet-SRNet: a wavelet-based CNN for multi-scale face super resolution. In: Proceedings of the IEEE conference on computer vision and pattern recognition, pp 1689–1697
59. Guo T, Mousavi HS, Vu TH, Monga V (2017) Deep wavelet prediction for image super-resolution. In: The IEEE conference on computer vision and pattern recognition workshops, pp 104–113
60. Bae W, Yoo JJ, Ye JC (2017) Beyond deep residual learning for image restoration: persistent homology-guided manifold simplification. In: The IEEE conference on computer vision and pattern recognition workshops, pp 1141–1149
61. Liu P, Zhang H, Zhang K, Lin L, Zuo W (2018) Multi-level wavelet-CNN for image restoration. In: Proceedings of the IEEE conference on computer vision and pattern recognition workshops, pp 773–782
62. Wang Y, Perazzi F, McWilliams B, Sorkine-Hornung A et al (2018) A fully progressive approach to single-image super-resolution. arXiv:1804.02900
63. Zhang K, Zuo W, Chen Y, Meng D, Zhang L (2017) Beyond a Gaussian denoiser: residual learning of deep cnn for image denoising. IEEE Trans Image Process 26(7):3142–3155
64. Cui Z, Chang H, Shan S et al (2014) Deep network cascade for image super-resolution. In: European conference on computer vision, pp 49–64
65. Lai WS, Huang JB, Ahuja N, Yang MH (2017) Deep laplacian pyramid networks for fast and accurate superresolution. IEEE Conf Comput Vis Pattern Recognit 2(3):624–632
66. Wang C, Liu Y, Bai X (2017) Deep residual convolutional neural network for hyperspectral image super-resolution. In: International conference on image and graphics, pp 370–380
67. Li Y, Hu J, Zhao X, Xie W, Li J (2017) Hyperspectral image super-resolution using deep convolutional neural network. Neurocomputing 266:29–41
68. Hu J, Li Y, Xie W (2017) Hyperspectral image super-resolution by spectral difference learning and spatial error correction. IEEE Geosci Remote Sens Lett 14(10):1825–1829
69. Mei S, Yuan X, Ji J, Zhang Y, Wan S, Du Q (2017) Hyperspectral image spatial super-resolution via 3D full convolutional neural network. Remote Sens 9
70. Masi G, Cozzolino D, Verdoliva L, Scarpa G (2016) Pansharpening by convolutional neural networks. Remote Sens 8(7)
71. Wei Y, Yuan Q, Shen H, Zhang L (2017) Boosting the accuracy of multispectral image pansharpening by learning a deep residual network. IEEE Geosci Remote Sens Lett 14(10):1795–1799
72. Yang J, Fu X, Hu Y et al (2017) PanNet: a deep network architecture for pan-sharpening. In: Proceedings of the IEEE conference on computer vision and pattern recognition, pp 5449–5457
73. Yuan Q, Wei Y, Meng X, Shen H, Zhang L (2018) A multiscale and multidepth convolutional neural network for remote sensing imagery pan-sharpening. IEEE J Sel Top Appl Earth Obs Remote Sens 11(3):978–989
74. Palsson F, Sveinsson JR, Ulfarsson MO (2017) Multispectral and hyperspectral image fusion using a 3-D-convolutional neural network. IEEE Geosci Remote Sens Lett 14(5):639–643

75. Dian R, Li S, Guo A, Fang L (2018) Deep hyperspectral image sharpening. IEEE Trans Neural Netw Learn Syst 29(11):5345–5355
76. Zhao YQ, Yang J (2015) Hyperspectral image denoising via sparse representation and low-rank constraint. IEEE Trans Geosci Remote Sens 53(1):296–308
77. Yang J, Zhao YQ, Chan JC-W (2018) Hyperspectral and multispectral image fusion via deep two-branches convolutional neural network. Remote Sens 10(5):800
78. Yang J (2019) Super-resolution enhancement of hyperspectral data through image fusion and deep learning methods. PhD thesis, Vrije Universiteit Brussel
79. Bottou L (2012) Stochastic gradient descent tricks. In: Neural networks: tricks of the trade, pp 421–436
80. Okujeni A, Van Der Linden S, Hostert P (2016) Berlin-urban-gradient dataset 2009—An EnMAP preparatory flight campaign (datasets). Technical Report, GFZ Data Services, Potsdam, Germany
81. Wang Z, Bovik AC, Sheikh HR, Simoncelli EP (2004) Image quality assessment: from error visibility to structural similarity. IEEE Trans Image Process 13(4):600–612
82. Zhang L, Zhang L, Mou X, Zhang D (2011) FSIM: a feature similarity index for image quality assessment. IEEE Trans Image Process 20(8):2378–2386
83. Yang J, Zhao Y, Yi C, Chan JC-W (2017) No-reference hyperspectral image quality assessment via quality-sensitive features learning. Remote Sens 9(4):305
84. Chang CC, Lin CJ (2011) LIBSVM: a library for support vector machines. ACM Trans Intell Syst Technol 2(3):1–27
85. Rainforth T, Wood F (2015) Canonical correlation forests. arXiv:1507.05444
86. Guanter L, Brell M, Chan JC-W et al (2018) Synergies of spaceborne imaging spectroscopy with other remote sensing approaches. Surv Geophys 1–31

Chapter 15
Automatic Target Detection for Sparse Hyperspectral Images

Ahmad W. Bitar, Jean-Philippe Ovarlez, Loong-Fah Cheong and Ali Chehab

Abstract In this work, a novel target detector for hyperspectral imagery is developed. The detector is independent on the unknown covariance matrix, behaves well in large dimensions, distributional free, invariant to atmospheric effects, and does not require a background dictionary to be constructed. Based on a modification of the robust principal component analysis (RPCA), a given hyperspectral image (HSI) is regarded as being made up of the sum of a low-rank background HSI and a sparse target HSI that contains the targets based on a pre-learned target dictionary specified by the user. The sparse component is directly used for the detection, that is, the targets are simply detected at the non-zero entries of the sparse target HSI. Hence, a novel target detector is developed, which is simply a sparse HSI generated automatically from the original HSI, but containing only the targets with the background is suppressed. The detector is evaluated on real experiments, and the results of which demonstrate its effectiveness for hyperspectral target detection especially when the targets are well matched to the surroundings.

A. W. Bitar (✉) · J.-P. Ovarlez
SONDRA Lab/CentraleSupélec, Université Paris-Saclay, 91192
Gif-sur-Yvette, France
e-mail: ahmad.bitar@centralesupelec.fr; ab76@aub.edu.lb

J.-P. Ovarlez
e-mail: jean-philippe.ovarlez@onera.fr; jeanphilippe.ovarlez@centralesupelec.fr

A. W. Bitar · A. Chehab
Department of Electrical and Computer Engineering, American University of Beirut,
Beirut, Lebanon
e-mail: chehab@aub.edu.lb

J.-P. Ovarlez
The French Aerospace Lab (ONERA), Université Paris-Saclay, 91120
Palaiseau, France

L.-F. Cheong
Department of Electrical and Computer Engineering,
National University of Singapore, Singapore, Singapore
e-mail: eleclf@nus.edu.sg

© Springer Nature Switzerland AG 2020
S. Prasad and J. Chanussot (eds.), *Hyperspectral Image Analysis*,
Advances in Computer Vision and Pattern Recognition,
https://doi.org/10.1007/978-3-030-38617-7_15

435

15.1   Introduction

15.1.1   What Is a Hyperspectral Image?

An airborne hyperspectral imaging sensor is capable of simultaneously acquiring the same spatial scene in a contiguous and multiple narrow (0.01–0.02 μm) spectral wavelength (color) bands [1–12]. When all the spectral bands are stacked together, the result is a hyperspectral image (HSI) whose cross-section is a function of the spatial coordinates and its depth is a function of wavelength. Hence, an HSI is a 3-D data cube having two spatial dimensions and one spectral dimension. Thanks to the narrow acquisition, the HSI could have hundreds to thousands of contiguous spectral bands. Having this very high level of spectral detail gives better capability to see the unseen.

Each band of the HSI corresponds to an image of the surface covered by the field of view of the hyperspectral sensor, whereas each pixel in the HSI is a p-dimensional vector, $\mathbf{x} \in \mathbb{R}^p$ (p stands for the total number of spectral bands), consisting of a spectrum characterizing the materials within the pixel. The spectral signature of $\mathbf{x}$ (also known as reflectance spectrum) shows the fraction of incident energy, typically sunlight, that is reflected by a material from the surface of interest as a function of the wavelength of the energy [2, 13].

The HSI usually contains both pure and mixed pixels [14–18]. A pure pixel contains only one single material, whereas a mixed pixel contains multiple materials, with its spectral signature representing the aggregate of all the materials in the corresponding spatial location. The latter situation often arises because hyperspectral images are collected hundreds to thousands of meters away from an object so that the object becomes smaller than the size of a pixel. Other scenarios might involve, for example, a military target hidden under foliage or covered with camouflage material.

15.1.2   Hyperspectral Target Detection: Concept and Challenges

With the rich information afforded by the high spectral dimensionality, hyperspectral imagery has found many applications in various fields, such as astronomy, agriculture [19, 20], mineralogy [21], military [22–24], and in particular, target detection [1, 2, 14, 22, 25–30]. Usually, the detection is built using a binary hypothesis test that chooses between the following competing null and alternative hypothesis: target absent (H_0), that is, the test pixel $\mathbf{x}$ consists only of background, and target present (H_1), where $\mathbf{x}$ may be either fully or partially occupied by the target material.

It is well known that the signal model for hyperspectral test pixels is fundamentally different from the additive model used in radar and communications applications [3, 14]. We can regard each test pixel $\mathbf{x}$ as being made up of $\mathbf{x} = \alpha \mathbf{t} + (1 - \alpha) \mathbf{b}$, where $0 \leq \alpha \leq 1$ designates the target fill-fraction, $\mathbf{t}$ is the spectrum of the target, and $\mathbf{b}$ is the spectrum of the background. This model is known as replacement signal model, and hence, when a target is present in a given HSI, it replaces (that is, removes) an equal part of the background [3]. For notational convenience, sensor noise has been incorporated into the target and background spectra (i.e., the vectors **t** and **b** include noise) [3].

In particular, when $\alpha = 0$, the pixel **x** is fully occupied by the background material (target not present). When $\alpha = 1$, the pixel **x** is fully occupied by the target material and is usually referred to as the full or resolved target pixel. Whereas when $0 < \alpha < 1$, the pixel **x** is partially occupied by the target material and is usually referred to as the subpixel or unresolved target [14].

A prior target information can often be provided to the user. In real-world hyperspectral imagery, this prior information may not be only related to its spatial properties (e.g., size, shape, texture) and which is usually not at our disposal, but to its spectral signature. The latter usually hinges on the nature of the given HSI where the spectra of the targets of interest have been already measured by some laboratories or with some handheld spectrometers.

Different Gaussian-based target detectors (e.g., Matched Filter [31, 32], Normalized Matched Filter [33], and Kelly detector [34]) have been developed. In these classical detectors, the target of interest to detect is known (e.g., its spectral signature is fully provided to the user).

However, the aforementioned detectors present several limitations in real-world hyperspectral imagery. The task of understanding and solving these limitations presents significant challenges for hyperspectral target detection.

- **Challenge one**: One of the major drawbacks of the aforementioned classical target detectors is that they depend on the unknown covariance matrix (of the background surrounding the test pixel) whose entries have to be carefully estimated, especially in large dimensions [35–37], and to ensure success under different environments [13, 27, 38–40]. However, estimating large covariance matrices has been a longstanding important problem in many applications and has attracted increased attention over several decades. When the spectral dimension is considered large compared to the sample size (which is the usual case), the traditional covariance estimators are estimated with a lot of errors unless some covariance regularization methods are considered [35–37]. It implies that the largest or smallest estimated coefficients in the matrix tend to take on extreme values not because this is "the truth", but because they contain an extreme amount of error [35, 36]. This is one of the main reasons why the classical target detectors usually behave poorly in detecting the targets of interest in a given HSI.

  In addition, there is always an explicit assumption (specifically, Gaussian) on the statistical distribution characteristics of the observed data. For instance, most materials are treated as Lambertian because their bidirectional reflectance distribution function characterizations are usually not available, but the actual reflection is likely to have both a diffuse component and a specular component. This latter component would result in gross corruption of the data. In addition, spectra from multiple materials are usually assumed to interact according to a linear mixing model; nonlinear mixing effects are not represented and will contribute to another source of noise.

- **Challenge two**: The classical target detectors that depend on the target to detect **t** use only a single reference spectrum for the target of interest. This may be inadequate since in real-world hyperspectral imagery, various effects that produce variability to the material spectra (e.g., atmospheric conditions, sensor noise, material composition, and scene geometry) are inevitable [41, 42]. For instance, target signatures are typically measured in laboratories or in the field with handheld spectrometers that are at most a few inches from the target surface. Hyperspectral images, however, are collected at huge distances away from the target and have significant atmospheric effects present.

Recent years have witnessed a growing interest in the notion of sparsity as a way to model signals. The basic assumption of this model is that natural signals can be represented as a "sparse" linear combination of atom signals taken from a dictionary. In this regard, two main issues need to be addressed: (1) how to represent a signal in the sparsest way, for a given dictionary? and (2) how to construct an accurate dictionary in order to successfully represent the signal?

Recently, a signal classification technique via sparse representation was developed for the application of face recognition [43]. It is observed that aligned faces of the same object with varying lighting conditions approximately lie in a low-dimensional subspace [44]. Hence, a test face image can be sparsely represented by atom signals from all classes. This representation approach has also been exploited in several other signal classification problems such as iris recognition [45], tumor classification [46], and hyperspectral imagery unmixing [6, 47, 48].

In this context, Chen et al. [49] have been inspired by the work in [43], and developed an approach for sparse representation of hyperspectral test pixels. In particular, each test pixel $\mathbf{x} \in \mathbb{R}^p$ (either target of background) in a given HSI, is assumed to lie in a low-dimensional subspace of the p-dimensional spectral-measurement space. Hence, it can be represented by a very few atom signals taken from the dictionaries, and the recovered sparse representation can be used directly for the detection. For example, if a test pixel $\mathbf{x}$ contains the target (that is, $\mathbf{x} = \alpha\,\mathbf{t} + (1 - \alpha)\,\mathbf{b}$, with $0 < \alpha \leq 1$), it can be sparsely represented by atom signals taken from the target dictionary (denoted as $\mathbf{A}_t$); whereas, if $\mathbf{x}$ is only a background pixel (e.g., $\alpha = 0$), it can be sparsely represented by atom signals taken from the background dictionary (denoted as $\mathbf{A}_b$). Very recently, Zhang et al. [50] have extended the work done by Chen et al. in [49] by combining the idea of binary hypothesis and sparse representation together, obtaining a more complete and realistic sparsity model than in [49]. More precisely, Zhang et al. [50] have assumed that if the test pixel $\mathbf{x}$ belongs to hypothesis H_0 (target absent), it will be modeled by the $\mathbf{A}_b$ only; otherwise, it will be modeled by the union of $\mathbf{A}_b$ and $\mathbf{A}_t$. This in fact yields a competition between the two hypotheses corresponding to the different pixel class label.

These sparse representation methods [49, 50] are independent on the unknown covariance matrix, behave well in large dimensions, distributional free, and invariant to atmospheric effects. More precisely, they can alleviate the spectral variability caused by atmospheric effects, and can also better deal with a greater range of noise phenomena.

- **Challenge three**: The main drawback of these sparse representation methods [49, 50] is the lack of a sufficiently universal dictionary, especially for the background $\mathbf{A}_b$; some form of in-scene adaptation would be desirable. The background dictionary $\mathbf{A}_b$ is usually constructed using an adaptive scheme (a local method) which is based on a dual concentric window centered on the test pixel, with an inner window region (IWR) centered within an outer window region (OWR), and only the pixels in the OWR will constitute the samples for $\mathbf{A}_b$. Clearly, the dimension of IWR is very important and has a strong impact on the target detection performance since it aims to enclose the targets of interests to be detected. It should be set larger than or equal to the size of all the desired targets of interest in the corresponding HSI, so as to exclude the target pixels from erroneously appearing in $\mathbf{A}_b$. However, information about the target size in the image is usually not at our disposal. It is also very unwieldy to set this size parameter when the target could be of irregular shape (e.g., searching for lost plane parts of a missing aircraft). Another tricky situation is when there are multiple targets in close proximity in the image (e.g., military vehicles in long convoy formation). Hence, the construction of $\mathbf{A}_b$ for the sparse representation methods is a very challenging problem since a contamination of it by the target pixels can potentially affect the target detection performance.

15.1.3   Goals and Outline

In this work, we handle all the aforementioned challenges by making very little specific assumptions about the background or target [51, 52]. Based on a modification of the recently developed robust principal component analysis (RPCA) [53], our method decomposes an input HSI into a background HSI (denoted by $\mathbf{L}$) and a sparse target HSI (denoted by $\mathbf{E}$) that contains the targets of interest.

While we do not need to make assumptions about the size, shape, or number of the targets, our method is subject to certain generic constraints that make less specific assumption on the background or the target. These constraints are similar to those used in RPCA [53, 54], including

1. the background is not too heavily cluttered with many different materials with multiple spectra, so that the background signals should span a low-dimensional subspace, a property that can be expressed as the low-rank condition of a suitably formulated matrix [55–61];
2. the total image area of all the target(s) should be small relative to the whole image (i.e., spatially sparse), e.g., several hundred pixels in a million-pixel image, though there is no restriction on a target shape or the proximity between the targets.

Our method also assumes that the target spectra are available to the user and that the atmospheric influence can be accounted for by the target dictionary $\mathbf{A}_t$. This pre-learned target dictionary $\mathbf{A}_t$ is used to cast the general RPCA into a more specific

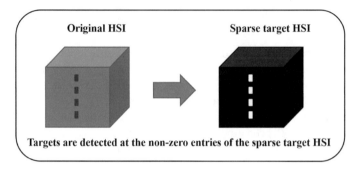

Fig. 15.1 Sparse target HSI: our novel target detector

form, specifically, we further factorize the sparse component **E** from RPCA into the product of **A**$_t$ and a sparse activation matrix **C** [51]. This modification is essential to disambiguate the true targets from the background.

After decomposing a given HSI into the sum of a low-rank HSI and a sparse HSI, the latter will define our detector. That is, the targets are detected at the non-zero entries of the sparse HSI. Hence, a novel target detector is developed, which is simply a sparse HSI generated automatically from the original HSI, but containing only the targets with the background is suppressed (see Fig. 15.1).

> The main advantages of our proposed detector are the following: (1) indepen-
> dent on the unknown covariance matrix; (2) behaves well in large dimensions;
> (3) distributional free; (4) invariant to atmospheric effects via the use of the
> target dictionary **A**$_t$; and (5) does not require a background dictionary to be
> constructed.

This chapter is structured along the following lines. First comes an overview of some related works in Sect. 15.2. In Sect. 15.3, the proposed decomposition model as well as our novel target detector are briefly outlined. Section 15.4 presents real experiments to gauge the effectiveness of the proposed detector for hyperspectral target detection. The chapter ends with a summary of the work and some directions for future work.

15.1.4 Summary of Main Notations

Throughout this chapter, we depict vectors in lowercase boldface letters and matrices in uppercase boldface letters. The notation $(.)^T$ and $\mathrm{Tr}(.)$ stand for the transpose and trace of a matrix, respectively. In addition, $\mathrm{rank}(.)$ is for the rank of a matrix. A variety of norms on matrices will be used. For instance, **M** is a matrix, and $[\mathbf{M}]_{:,j}$ is the jth column. The matrix $l_{2,0}$, $l_{2,1}$ norms are defined by $\|\mathbf{M}\|_{2,0} = \#\left\{ j \; : \; \left\| [\mathbf{M}]_{:,j} \right\|_2 \neq 0 \right\}$, and $\|\mathbf{M}\|_{2,1} = \sum_j \left\| [\mathbf{M}]_{:,j} \right\|_2$, respectively. The Frobenius norm and the nuclear norm (the sum of singular values of a matrix) are denoted by $\|\mathbf{M}\|_F$ and $\|\mathbf{M}\|_* = \mathrm{Tr}\left(\mathbf{M}^T \mathbf{M} \right)^{(1/2)}$, respectively.

15.2   Related Works

Whatever the real application may be, somehow the general RPCA model needs to be subject to further assumptions for successfully distinguishing the true targets from the background. Besides the generic RPCA and our suggested modification discussed in Sect. 15.1.3, there have been other modifications of RPCA. For example, the generalized model of RPCA, named the low-rank representation (LRR) [62], allows the use of a subspace basis as a dictionary or just uses self-representation to obtain the LRR. The major drawback in LRR is that the incorporated dictionary has to be constructed from the background and to be pure from the target samples. This challenge is similar to the aforementioned background dictionary $\mathbf{A}_b$ construction problem. If we use the self-representation form of LRR, the presence of a target in the input image may only bring about a small increase in rank and thus be retained in the background [52].

In the earliest models using a low-rank matrix to represent the background [53, 54, 63], no prior knowledge on the target was considered. In some applications such as Speech enhancement and hyperspectral imagery, we may expect some prior information about the target of interest and which can be provided to the user. Incorporating this information about the target into the separation scheme in the general RPCA model should allow us to potentially improve the target extraction performance. For example, Chen and Ellis [64], and Sun and Qin [65], proposed a Speech enhancement system by exploiting the knowledge about the likely form of the targeted speech. This was accomplished by factorizing the sparse component from RPCA into the product of a dictionary of target speech templates and a sparse activation matrix. The proposed methods in [64] and [65] typically differ on how the fixed target dictionary of speech spectral templates is constructed. Our proposed model in Sect. 15.3 is very related to [64] and [65]. In real-world hyperspectral imagery, the prior target information may not be only related to its spatial properties (e.g., size, shape, and texture) and which is usually not at our disposal, but to its spectral signature. The latter usually hinges on the nature of the given HSI where the spectra of the targets of interest present have been already measured by some laboratories or with some handheld spectrometers.

15.3   Main Contribution

Suppose an HSI of size $h \times w \times p$, where h and w are the height and width of the image scene, respectively, and p is the number of spectral bands. Our proposed modification of RPCA is mainly based on the following steps:

1. Let us consider that the given HSI contains q pixels $\{\mathbf{x}_i\}_{i\in[1,q]}$ of the form:

$$\mathbf{x}_i = \alpha_i \, \mathbf{t}_i + (1 - \alpha_i) \, \mathbf{b}_i, \quad 0 < \alpha_i \leq 1,$$

   where $\mathbf{t}_i$ represents the known target that replaces a fraction α_i of the background $\mathbf{b}_i$ (i.e., at the same spatial location). The remaining $(e - q)$ pixels in the given HSI, with $e = h \times w$, are thus only background ($\alpha = 0$).
2. We assume that all $\{\mathbf{t}_i\}_{i\in[1,q]}$ consist of similar materials, and thus they should be represented by a linear combination of N_t common target samples $\{\mathbf{a}_j^t\}_{j\in[1,N_t]}$, where $\mathbf{a}_j^t \in \mathbb{R}^p$ (the superscript t is for target), but weighted with different set of coefficients $\{\beta_{i,j}\}_{j\in[1,N_t]}$. Thus, each of the q targets is represented as

$$\mathbf{x}_i = \alpha_i \sum_{j=1}^{N_t} \left(\beta_{i,j}\, \mathbf{a}_j^t\right) + (1 - \alpha_i)\, \mathbf{b}_i \quad i \in [1,q].$$

3. We rearrange the given HSI into a two-dimensional matrix $\mathbf{D} \in \mathbb{R}^{e\times p}$, with $e = h \times w$ (by lexicographically ordering the columns). The matrix $\mathbf{D}$ can be decomposed into a low-rank matrix $\mathbf{L}_0$ representing the pure background, a sparse matrix capturing any spatially small signal residing in the known target subspace, and a noise matrix $\mathbf{N}_0$. More precisely, the model is

$$\mathbf{D} = \mathbf{L}_0 + (\mathbf{A}_t\, \mathbf{C}_0)^T + \mathbf{N}_0,$$

   where $(\mathbf{A}_t \mathbf{C}_0)^T$ is the sparse target matrix, ideally with q non-zero rows representing $\alpha_i\{\mathbf{t}_i^T\}_{i\in[1,q]}$, with target dictionary $\mathbf{A}_t \in \mathbb{R}^{p\times N_t}$ having columns representing the target samples $\{\mathbf{a}_j^t\}_{j\in[1,N_t]}$, and a coefficient matrix $\mathbf{C}_0 \in \mathbb{R}^{N_t\times e}$ that should be a sparse column matrix, again ideally containing q non-zero columns each representing $\alpha_i[\beta_{i,1}, \ldots, \beta_{i,N_t}]^T$, $i \in [1,q]$. $\mathbf{N}_0$ is assumed to be independent and identically distributed Gaussian noise with zero mean and unknown standard deviation.
4. After reshaping $\mathbf{L}_0$, $(\mathbf{A}_t\, \mathbf{C}_0)^T$, and $\mathbf{N}_0$ back to a cube of size $h \times w \times p$, we call these entities as "low-rank background HSI", "sparse target HSI", and "noise HSI", respectively.

In order to recover the low-rank matrix $\mathbf{L}_0$ and the sparse target matrix $(\mathbf{A}_t \mathbf{C}_0)^T$, we consider the following minimization problem:

$$\min_{\mathbf{L},\mathbf{C}} \left\{ \tau\, \text{rank}(\mathbf{L}) + \lambda\, \|\mathbf{C}\|_{2,0} + \left\| \mathbf{D} - \mathbf{L} - (\mathbf{A}_t \mathbf{C})^T \right\|_F^2 \right\}, \tag{15.1}$$

where τ controls the rank of $\mathbf{L}$, and λ the sparsity level in $\mathbf{C}$.

15.3.1 Recovering a Low-Rank Background Matrix and a Sparse Target Matrix by Convex Optimization

We relax the rank term and the $\|.\|_{2,0}$ term to their convex proxies. More precisely, we use the nuclear norm $\|\mathbf{L}\|_*$ as a surrogate for the rank$(\mathbf{L})$ term, and the $l_{2,1}$ norm for the $l_{2,0}$ norm.[1]

We now need to solve the following convex minimization problem:

$$\min_{\mathbf{L},\mathbf{C}} \left\{ \tau\, \|\mathbf{L}\|_* + \lambda\, \|\mathbf{C}\|_{2,1} + \left\| \mathbf{D} - \mathbf{L} - (\mathbf{A}_t \mathbf{C})^T \right\|_F^2 \right\}. \tag{15.2}$$

Problem (15.2) is solved via an alternating minimization of two sub-problems. Specifically, at each iteration k,

$$\mathbf{L}^{(k)} = \underset{\mathbf{L}}{\text{argmin}} \left\{ \left\| \mathbf{L} - \left(\mathbf{D} - \left(\mathbf{A}_t\, \mathbf{C}^{(k-1)}\right)^T \right) \right\|_F^2 + \tau\, \|\mathbf{L}\|_* \right\}, \tag{15.3a}$$

$$\mathbf{C}^{(k)} = \underset{\mathbf{C}}{\text{argmin}} \left\{ \left\| \left(\mathbf{D} - \mathbf{L}^{(k)} \right)^T - \mathbf{A}_t\, \mathbf{C} \right\|_F^2 + \lambda\, \|\mathbf{C}\|_{2,1} \right\}. \tag{15.3b}$$

The minimization sub-problems (15.3a), (15.3b) are convex and each can be solved optimally.

---

[1] A natural suggestion could be that the rank of $\mathbf{L}$ usually has a physical meaning (e.g., number of endmembers in background), and thus, why not to minimize the latter two terms in Eq. (15.2) with the constraint that the rank of $\mathbf{L}$ should not be larger than a fixed value d? That is,

$$\min_{\mathbf{L},\mathbf{C}} \left\{ \lambda\, \|\mathbf{C}\|_{2,1} + \left\| \mathbf{D} - \mathbf{L} - (\mathbf{A}_t \mathbf{C})^T \right\|_F^2 \right\}, \quad s.t.\ \text{rank}(\mathbf{L}) \leq d.$$

In our opinion, assuming that the number of endmembers in background is known exactly will be a strong assumption and our work will be less general as a result. One can assume d to be some upper bound, in which case, the suggested formulation is a possible one. However, solving such a problem (with a hard constraint that the rank should not exceed some bound) is in general a NP-hard problem, unless there happens to be some special form in the objective which allows for a tractable solution. Thus, we adopt the soft constraint form with the nuclear norm as a proxy for the rank of $\mathbf{L}$; this is an approximation commonly done in the field and is found to give good solutions in many problems empirically.

Solving sub-problem (15.3a): we solve sub-problem (15.3a) via the singular value thresholding operator [66]. We assume that $\left(\mathbf{D} - \left(\mathbf{A}_t \, \mathbf{C}^{(k-1)} \right)^T \right)$ has a rank equal to r. According to Theorem 2.1 in [66], sub-problem (15.3a) admits the following closed-form solution:

$$
\mathbf{L}^{(k)} = D_{\tau/2} \left(\mathbf{D} - \left(\mathbf{A}_t \, \mathbf{C}^{(k-1)} \right)^T \right)
$$
$$
= \mathbf{U}^{(k)} \, D_{\tau/2} \left(\mathbf{S}^{(k)} \right) \, \mathbf{V}^{(k)T}
$$
$$
= \mathbf{U}^{(k)} \, \text{diag} \left(\left\{ \left(s_t^{(k)} - \tfrac{\tau}{2} \right)_+ \right\} \right) \mathbf{V}^{(k)T}
$$

where $\mathbf{S}^{(k)} = \text{diag} \left(\left\{ s_t^{(k)} \right\}_{1 \le t \le r} \right)$, and $D_{\tau/2}(.)$ is the singular value shrinkage operator. The matrices $\mathbf{U}^{(k)} \in \mathbb{R}^{e \times r}$, $\mathbf{S}^{(k)} \in \mathbb{R}^{r \times r}$, and $\mathbf{V}^{(k)} \in \mathbb{R}^{p \times r}$ are generated by the singular value decomposition (SVD) of $\left(\mathbf{D} - \left(\mathbf{A}_t \, \mathbf{C}^{(k-1)} \right)^T \right)$.

Proof Since the function $\left\{ \left\| \mathbf{L} - \left(\mathbf{D} - \left(\mathbf{A}_t \, \mathbf{C}^{(k-1)} \right)^T \right) \right\|_F^2 + \tau \, \| \mathbf{L} \|_* \right\}$ is strictly convex, it is easy to see that there exists a unique minimizer, and we thus need to prove that it is equal to $D_{\tau/2} \left(\mathbf{D} - \left(\mathbf{A}_t \, \mathbf{C}^{(k-1)} \right)^T \right)$. Note that to understand how the aforementioned closed-form solution has been obtained, we provide in detail the proof steps that have been given in [66].

To do this, let us first find the derivative of sub-problem (15.3a) w.r.t. $\mathbf{L}$ and set it to zero. We obtain

$$
\left(\mathbf{D} - \left(\mathbf{A}_t \, \mathbf{C}^{(k-1)} \right)^T \right) - \hat{\mathbf{L}} = \frac{\tau}{2} \, \partial \left\| \hat{\mathbf{L}} \right\|_* , \tag{15.4}
$$

where $\partial \left\| \hat{\mathbf{L}} \right\|_*$ is the set of subgradients of the nuclear norm. Let $\mathbf{U}_L \, \mathbf{S}_L \, \mathbf{V}_L^T$ to be the SVD of $\mathbf{L}$, it is known [67–69] that

$$
\partial \| \mathbf{L} \|_* = \left\{ \mathbf{U}_L \, \mathbf{V}_L^T + \mathbf{W} \, : \, \mathbf{W} \in \mathbb{R}^{e \times p}, \, \mathbf{U}_L^T \mathbf{W} = \mathbf{0}, \, \mathbf{W} \, \mathbf{V}_L = \mathbf{0}, \, \| \mathbf{W} \|_2 \le 1 \right\} .
$$

Set $\hat{\mathbf{L}} = D_{\tau/2} \left(\mathbf{D} - \left(\mathbf{A}_t \, \mathbf{C}^{(k-1)} \right)^T \right)$ for short. In order to show that $\hat{\mathbf{L}}$ obeys Eq. (15.4), suppose the SVD of $\left(\mathbf{D} - \left(\mathbf{A}_t \, \mathbf{C}^{(k-1)} \right)^T \right)$ is given by

$$
\left(\mathbf{D} - \left(\mathbf{A}_t \, \mathbf{C}^{(k-1)} \right)^T \right) = \mathbf{U}_0 \, \mathbf{S}_0 \, \mathbf{V}_0^T + \mathbf{U}_1 \, \mathbf{S}_1 \, \mathbf{V}_1^T ,
$$

where $\mathbf{U}_0$, $\mathbf{V}_0$ (resp. $\mathbf{U}_1$, $\mathbf{V}_1$) are the singular vectors associated with singular values larger than $\tau/2$ (resp. inferior than or equal to $\tau/2$). With these notations, we have

$$\hat{\mathbf{L}} = D_{\tau/2}\left(\mathbf{U}_0 \, \mathbf{S}_0 \, \mathbf{V}_0^T\right) = \left(\mathbf{U}_0 \left(\mathbf{S}_0 - \frac{\tau}{2}\mathbf{I}\right)\mathbf{V}_0^T\right).$$

Thus, if we return back to Eq. (15.4), we obtain

$$\mathbf{U}_0 \, \mathbf{S}_0 \, \mathbf{V}_0^T + \mathbf{U}_1 \, \mathbf{S}_1 \, \mathbf{V}_1^T - \mathbf{U}_0 \left(\mathbf{S}_0 - \frac{\tau}{2}\mathbf{I}\right)\mathbf{V}_0^T = \frac{\tau}{2}\partial\left\|\hat{\mathbf{L}}\right\|_*,$$

$$\Rightarrow \mathbf{U}_1 \, \mathbf{S}_1 \, \mathbf{V}_1^T + \mathbf{U}_0 \frac{\tau}{2}\mathbf{V}_0^T = \frac{\tau}{2}\partial\left\|\hat{\mathbf{L}}\right\|_*,$$

$$\Rightarrow \left(\mathbf{U}_0 \, \mathbf{V}_0^T + \mathbf{W}\right) = \partial\left\|\hat{\mathbf{L}}\right\|_*,$$

where $\mathbf{W} = \dfrac{2}{\tau}\mathbf{U}_1 \, \mathbf{S}_1 \, \mathbf{V}_1^T$.

By definition, $\mathbf{U}_0^T \, \mathbf{W} = \mathbf{0}$, $\mathbf{W} \, \mathbf{V}_0 = \mathbf{0}$, and we also have $\|\mathbf{W}\|_2 \le 1$.
Hence, $\left(\mathbf{D} - \left(\mathbf{A}_t \, \mathbf{C}^{(k-1)}\right)^T\right) - \hat{\mathbf{L}} = \dfrac{\tau}{2}\partial\left\|\hat{\mathbf{L}}\right\|_*$, which concludes the proof. □

Solving sub-problem (15.3b): (15.3b) can be solved by various methods, among which we adopt the alternating direction method of multipliers (ADMM) [70]. More precisely, we introduce an auxiliary variable $\mathbf{F}$ into sub-problem (15.3b) and recast it into the following form:

$$\left(\mathbf{C}^{(k)}, \mathbf{F}^{(k)}\right) = \underset{\substack{\text{argmin}\\ s.t. \ \mathbf{C}=\mathbf{F}}}{}\left\{\left\|\left(\mathbf{D} - \mathbf{L}^{(k)}\right)^T - \mathbf{A}_t \, \mathbf{C}\right\|_F^2 + \lambda \, \|\mathbf{F}\|_{2,1}\right\}. \tag{15.5}$$

Problem (15.5) is then solved as follows (scaled form of ADMM):

$$\mathbf{C}^{(k)} = \underset{\mathbf{C}}{\text{argmin}}\left\{\left\|\left(\mathbf{D} - \mathbf{L}^{(k)}\right)^T - \mathbf{A}_t \, \mathbf{C}\right\|_F^2 + \frac{\rho^{(k-1)}}{2}\left\|\mathbf{C} - \mathbf{F}^{(k-1)} + \frac{1}{\rho^{(k-1)}}\mathbf{Z}^{(k-1)}\right\|_F^2\right\}, \tag{15.6a}$$

$$\mathbf{F}^{(k)} = \underset{\mathbf{F}}{\text{argmin}}\left\{\lambda \, \|\mathbf{F}\|_{2,1} + \frac{\rho^{(k-1)}}{2}\left\|\mathbf{C}^{(k)} - \mathbf{F} + \frac{1}{\rho^{(k-1)}}\mathbf{Z}^{(k-1)}\right\|_F^2\right\}, \tag{15.6b}$$

$$\mathbf{Z}^{(k)} = \mathbf{Z}^{(k-1)} + \rho^{(k-1)}\left(\mathbf{C}^{(k)} - \mathbf{F}^{(k)}\right), \tag{15.6c}$$

where $\mathbf{Z} \in \mathbb{R}^{N_t \times e}$ is the Lagrangian multiplier matrix, and ρ is a positive scalar.
Solving sub-problem (15.6a):

$$-2\mathbf{A}_t^T\left(\left(\mathbf{D} - \mathbf{L}^{(k)}\right)^T - \mathbf{A}_t \, \mathbf{C}\right) + \rho^{(k-1)}\left(\mathbf{C} - \mathbf{F}^{(k-1)} + \frac{1}{\rho^{(k-1)}}\mathbf{Z}^{(k-1)}\right) = \mathbf{0},$$

$$\Rightarrow \left(2\mathbf{A}_t^T \, \mathbf{A}_t + \rho^{(k-1)}\mathbf{I}\right)\mathbf{C} = \rho^{(k-1)}\mathbf{F}^{(k-1)} - \mathbf{Z}^{(k-1)} + 2\mathbf{A}_t^T\left(\mathbf{D} - \mathbf{L}^{(k)}\right)^T.$$

This implies

$$\boxed{\mathbf{C}^{(k)} = \left(2\mathbf{A}_t^T \, \mathbf{A}_t + \rho^{(k-1)}\mathbf{I}\right)^{-1}\left(\rho^{(k-1)}\mathbf{F}^{(k-1)} - \mathbf{Z}^{(k-1)} + 2\mathbf{A}_t^T\left(\mathbf{D} - \mathbf{L}^{(k)}\right)^T\right)}$$

Solving sub-problem (15.6b):

According to Lemma 3.3 in [71] and Lemma 4.1 in [62], sub-problem (15.6b) admits the following closed-form solution:

$$
[\mathbf{F}]_{:,j}^{(k)} = \max\left(\left\|[\mathbf{C}]_{:,j}^{(k)} + \frac{1}{\rho^{(k-1)}} [\mathbf{Z}]_{:,j}^{(k-1)}\right\|_2 - \frac{\lambda}{\rho^{(k-1)}}, 0\right) \left(\frac{[\mathbf{C}]_{:,j}^{(k)} + \frac{1}{\rho^{(k-1)}} [\mathbf{Z}]_{:,j}^{(k-1)}}{\left\|[\mathbf{C}]_{:,j}^{(k)} + \frac{1}{\rho^{(k-1)}} [\mathbf{Z}]_{:,j}^{(k-1)}\right\|_2}\right)
$$

Proof At the jth column, sub-problem (15.6b) refers to

$$
[\mathbf{F}]_{:,j}^{(k)} = \operatorname*{argmin}_{[\mathbf{F}]_{:,j}} \left\{ \lambda \left\|[\mathbf{F}]_{:,j}\right\|_2 + \frac{\rho^{(k-1)}}{2} \left\|[\mathbf{C}]_{:,j}^{(k)} - [\mathbf{F}]_{:,j} + \frac{1}{\rho^{(k-1)}} [\mathbf{Z}]_{:,j}^{(k-1)}\right\|_2^2 \right\} .
$$

By finding the derivative w.r.t $[\mathbf{F}]_{:,j}$ and setting it to zero, we obtain

$$
- \rho^{(k-1)} \left([\mathbf{C}]_{:,j}^{(k)} - [\mathbf{F}]_{:,j} + \frac{1}{\rho^{(k-1)}} [\mathbf{Z}]_{:,j}^{(k-1)}\right) + \frac{\lambda [\mathbf{F}]_{:,j}}{\left\|[\mathbf{F}]_{:,j}\right\|_2} = \mathbf{0}
$$

$$
\Rightarrow [\mathbf{C}]_{:,j}^{(k)} + \frac{1}{\rho^{(k-1)}} [\mathbf{Z}]_{:,j}^{(k-1)} = [\mathbf{F}]_{:,j} + \frac{\lambda [\mathbf{F}]_{:,j}}{\rho^{(k-1)} \left\|[\mathbf{F}]_{:,j}\right\|_2} . \tag{15.7}
$$

By computing the l_2 norm of (15.7), we obtain

$$
\left\|[\mathbf{C}]_{:,j}^{(k)} + \frac{1}{\rho^{(k-1)}} [\mathbf{Z}]_{:,j}^{(k-1)}\right\|_2 = \left\|[\mathbf{F}]_{:,j}\right\|_2 + \frac{\lambda}{\rho^{(k-1)}} . \tag{15.8}
$$

From Eqs. (15.7) and (15.8), we have

$$
\frac{[\mathbf{C}]_{:,j}^{(k)} + \frac{1}{\rho^{(k-1)}} [\mathbf{Z}]_{:,j}^{(k-1)}}{\left\|[\mathbf{C}]_{:,j}^{(k)} + \frac{1}{\rho^{(k-1)}} [\mathbf{Z}]_{:,j}^{(k-1)}\right\|_2} = \frac{[\mathbf{F}]_{:,j}}{\left\|[\mathbf{F}]_{:,j}\right\|_2} . \tag{15.9}
$$

Consider that

$$
[\mathbf{F}]_{:,j} = \left\|[\mathbf{F}]_{:,j}\right\|_2 \times \frac{[\mathbf{F}]_{:,j}}{\left\|[\mathbf{F}]_{:,j}\right\|_2} . \tag{15.10}
$$

By replacing $\left\|[\mathbf{F}]_{:,j}\right\|_2$ from (15.8) into (15.10), and $\dfrac{[\mathbf{F}]_{:,j}}{\left\|[\mathbf{F}]_{:,j}\right\|_2}$ from (15.9) into (15.10), we conclude the proof. $\square$

15.3.2 Some Initializations and Convergence Criterion

We initialize $\mathbf{L}^{(0)} = \mathbf{0}$, $\mathbf{F}^{(0)} = \mathbf{C}^{(0)} = \mathbf{Z}^{(0)} = \mathbf{0}$, $\rho^{(0)} = 10^{-4}$ and update $\rho^{(k)} = 1.1\,\rho^{(k-1)}$. The criteria for convergence of sub-problem (15.3b) is $\left\| \mathbf{C}^{(k)} - \mathbf{F}^{(k)} \right\|_F^2 \leq 10^{-6}$.

For Problem (15.2), we stop the iteration when the following convergence criterion is satisfied:

$$\frac{\left\| \mathbf{L}^{(k)} - \mathbf{L}^{(k-1)} \right\|_F}{\|\mathbf{D}\|_F} \leq \epsilon \quad \text{and} \quad \frac{\left\| \left(\mathbf{A}_t\, \mathbf{C}^{(k)}\right)^T - \left(\mathbf{A}_t\, \mathbf{C}^{(k-1)}\right)^T \right\|_F}{\|\mathbf{D}\|_F} \leq \epsilon$$

where $\epsilon > 0$ is a precision tolerance parameter. We set $\epsilon = 10^{-4}$.

15.3.3 Our Novel Target Detector: $(\mathbf{A}_t\mathbf{C})^T$

We use $(\mathbf{A}_t\mathbf{C})^T$ directly for the detection.

> Note that for this detector, we require as few false alarms as possible to be deposited in the target image, but we do not need the target fraction to be entirely removed from the background (that is, a very weak target separation can suffice). As long as enough of the target fractions are moved to the target image, such that non-zero support is detected at the corresponding pixel location, it will be adequate for our detection scheme. From this standpoint, we should choose a λ value that is relatively large so that the target image is really sparse with zero or little false alarms, and only the signals that reside in the target subspace specified by $\mathbf{A}_t$ will be deposited there.

15.4 Experiments and Analysis

To obtain the same scene as in Fig. 8 in [72], we have concatenated two sectors labeled as "f970619t01p02_r02_sc03.a.rf" and "f970619t01p02_r02_sc04.a.rfl" from the online Cuprite data [73]. We shall call the resulting HSI as "Cuprite HSI" (see Fig. 15.2). The Cuprite HSI is a mining district area, which is well understood mineralogically [72, 74]. It contains well-exposed zones of advanced argillic alteration, consisting principally of kaolinite, alunite, and hydrothermal silica. It was acquired by the Airborne Visible/Infrared Imaging Spectrometer (AVIRIS) in June 23, 1995 at local noon and under high visibility conditions by a NASAER-2 aircraft flying at

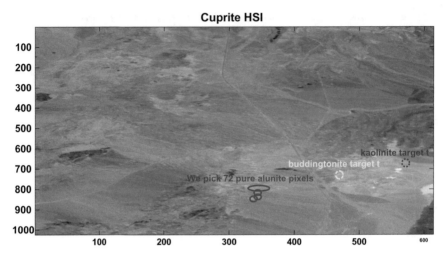

Fig. 15.2 The Cuprite HSI of size $1024 \times 614 \times 186$. We exhibit the mean power in db over the 186 spectral bands

an altitude of 20 km. It is a 1024×614 image and consists of 224 spectral (color) bands in contiguous (of about $0.01 \mu m$) wavelengths ranging exactly from 0.4046 to $2.4573 \mu m$. Prior to some analysis of the Cuprite HSI, the spectral bands 1–4, 104–113, 148–167, and 221–224 are removed due to the water absorption in those bands. As a result, a total of 186 bands are used.[2]

By referring to Fig. 8 in [72], we picked 72 pure alunite pixels from the Cuprite HSI (72 pixels located inside the solid red ellipses in Fig. 15.2) and generate a $100 \times 100 \times 186$ HSI zone formed by these pixels. We shall call this small HSI zone as "Alunite HSI" (see Fig. 15.3), and which will be used for the target evaluations later. We incorporate, in this zone, seven target blocks (each of size 6×3) with $\alpha \in [0.01, 1]$ (all have the same α value), placed in long convoy formation all formed by the same target **t** that we picked from the Cuprite HSI and which will constitute our target of interest to be detected. The target **t** replaces a fraction $\alpha \in [0.01, 1]$ from the background; specifically, the following values of α are considered: 0.01, 0.02, 0.05, 0.1, 0.3, 0.5, 0.8, and 1.

In the experiments, two kinds of target **t** are considered:

1. '**t**' that represents the buddingtonite target,
2. '**t**' that represents the kaolinite target.

More precisely, our detector $(\mathbf{A}_t \mathbf{C})^T$ is evaluated on two target detection scenarios:

- **Evaluation on an easy target (buddingtonite target)**: It has been noted by Gregg et al. [72] that the ammonia in the Tectosilicate mineral type, known as buddingtonite, has a distinct N-H combination absorption at $2.12 \mu m$, a position similar

---

[2]We regret that in our work in [51, 52], we missed to add "221–224" with the other bands that are removed. Adding "221–224" will give exactly a total of 186 bands.

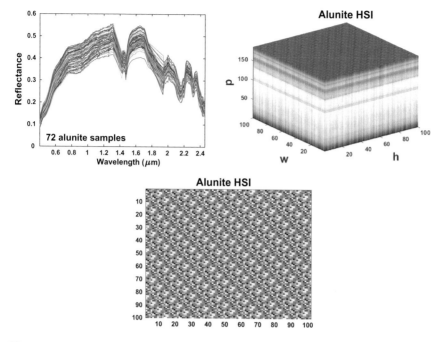

Fig. 15.3 A $100 \times 100 \times 186$ "Alunite HSI" generated by 72 pure alunite samples picked from the Cuprite HSI (72 pixels from the solid red ellipses in Fig. 15.2). For the third image, we exhibit the mean power in db over the 186 spectral bands

to that of the cellulose absorption in dried vegetation, from which it can be distinguished based on its narrower band width and asymmetry. Hence, the buddingtonite mineral can be considered as an "easy target" because it does not look like any other mineral with its distinct 2.12 μm absorption (that is, it is easily recognized based on its unique 2.12 μm absorption band).

In the experiments,[3] we consider the "buddingtonite" pixel at location $(731, 469)$ in the Cuprite HSI (the center of the dash-dotted yellow circle in Fig. 15.2) as the buddingtonite target **t** to be incorporated in the Alunite HSI for $\alpha \in [0.01, 1]$.

- **Evaluation on a challenging target (kaolinite target)[4]**: The paradigm in military applications for hyperspectral imagery seems to center on finding the target but ignoring all the rest. Sometimes, that rest is important, especially if the target is well matched to the surroundings. It has been shown by Gregg et al. [72] that both alunite and kaolinite minerals have overlapping spectral features, and thus, discrimination between these two minerals is a big challenge [72, 75].

In the experiments, we consider the "kaolinite" pixel at location $(672, 572)$ in the

---

[3]The MATLAB code of the proposed detector and experiments is available upon request. Please feel free to contact Ahmad W. Bitar.

[4]We thank Dr. Gregg A. Swayze from the United States Geological Survey (USGS) who has suggested us to evaluate our model (15.2) on the distinction between alunite and kaolinite minerals.

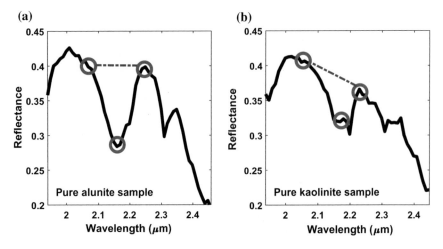

Fig. 15.4 Three-point band depth images for both **a** alunite and **b** kaolinite

Cuprite HSI (the center of the dotted blue circle in Fig. 15.2) as the kaolinite target **t** to be incorporated in the Alunite HSI for $\alpha \in [0.01, 1]$.

Figure 15.4a exhibits a three-point band depth image for our alunite background that shows the locations where an absorption feature, centered near 2.17 μm, is expressed in spectra of surface materials. Figure 15.4b exhibits a three-point band depth image for our kaolinite target that shows the locations where an absorption feature, centered near 2.2 μm, is expressed in spectra of surface materials. As we can observe, there is a subtle difference between the alunite and kaolinite three-point band depth images, showing that the successful spectral distinction between these two minerals is a very challenging task to achieve [75].[5]

15.4.1 Construction of the Target Dictionary $\mathbf{A}_t$

An important problem that requires a very careful attention is the construction of an appropriate dictionary $\mathbf{A}_t$ in order to capture the target well and distinguish it from the background. If $\mathbf{A}_t$ does not well represent the target of interest, our model in (15.2) may fail on discriminating the targets from the background. For example, Fig. 15.5 shows the detection results of our detector $(\mathbf{A}_t \mathbf{C})^T$ when $\mathbf{A}_t$ is constructed from some of the background pixels in the Alunite HSI. We can obviously observe that our detector is not able to capture the targets mainly because of the poor dictionary $\mathbf{A}_t$ constructed.

---

[5]We have been inspired by Fig. 8D-E in [75] to provide a close example of it in this chapter as can be shown in Fig. 15.4.

When $\mathbf{A}_t$ is constructed from background samples

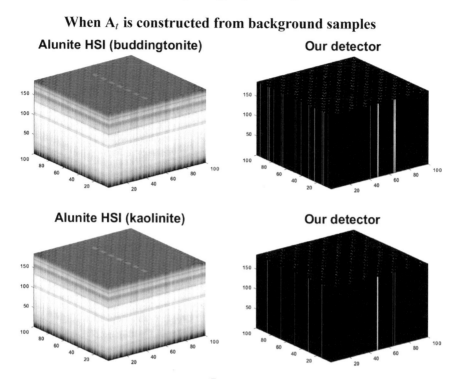

Fig. 15.5 Evaluation of our detector $(\mathbf{A}_t\mathbf{C})^T$ for detecting the buddingtonite and kaolinite target (for $\alpha = 1$) from the Alunite HSI when $\mathbf{A}_t$ is constructed from some background pixels acquired from the Alunite HSI

The target present in the HSI can be highly affected by the atmospheric conditions, sensor noise, material composition, and scene geometry. This may produce huge variations on the target spectra. In view of these real effects, it is very difficult to model the target dictionary $\mathbf{A}_t$ well. But this raises the question on "*how these effects should be dealt with?*".

Some scenarios for modeling the target dictionary have been suggested in the literature. For example, by using physical models and the MODTRAN atmospheric modeling program [76], target spectral signatures can be generated under various atmospheric conditions. For simplicity, we handle this problem in this work by exploiting target samples that are available in some online spectral libraries. More precisely, $\mathbf{A}_t$ can be constructed via the United States Geological Survey (USGS-Reston) spectral library [77]. However, the user can also deal with the Advanced Spaceborne Thermal Emission and Reflection (ASTER) spectral library [78] that includes data from the USGS spectral library, the Johns Hopkins University (JHU) spectral library, and the Jet Propulsion Laboratory (JPL) spectral library.

There are three buddingtonite samples available in the ASTER spectral library and will be considered to construct the dictionary $\mathbf{A}_t$ for the detection of our buddingtonite

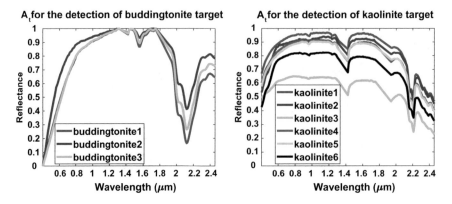

Fig. 15.6 Target dictionaries for the detection of buddingtonite and kaolinite

target (see Fig. 15.6 (first column)); whereas six kaolinite samples are available in the USGS spectral library and will be acquired to construct $\mathbf{A}_t$ for the detection of our kaolinite target (see Fig. 15.6 (second column)).

Note that the Alunite HSI, the buddingtonite target $\mathbf{t}$, the kaolinite target $\mathbf{t}$, and the buddingtonite/kaolinite target samples extracted from the online spectral libraries are all normalized to the values between 0 and 1.

For instance, it is usually difficult to find, for a specific given target, a sufficient number of available samples in the online spectral libraries. Hence, the dictionary $\mathbf{A}_t$ may still be not sufficiently selective and accurate. This is the most reason why problem (15.2) may fail to well capture the targets from the background.

15.4.2 Target Detection Evaluation

We now aim to qualitatively evaluate the target detection performances of our detector $(\mathbf{A}_t \mathbf{C})^T$ on both the buddingtonite and kaolinite target detection scenarios, when $\mathbf{A}_t$ is constructed from target samples available in the online spectral libraries (from Fig. 15.6). As can be seen from Fig. 15.7, our detector is able to detect the buddingtonite targets with no false alarms until $\alpha \leq 0.1$ where a lot of false alarms appear.

For the detection of kaolinite, it was difficult to have a clean detection (without false alarms) even for high values of α. This is to be expected since the kaolinite target is well matched to the alunite background (the kaolinite and alunite have overlapping spectral features), and hence, the discrimination between them is very challenging.

It is interesting to note (results omitted here) that if we consider $\mathbf{A}_t = \mathbf{t}$ (that is, we are searching for the exact signature $\mathbf{t}$ in the Alunite HSI), the buddingtonite and even the kaolinite targets are able to be detected with no false alarms for $0.1 < \alpha \leq 1$. When $\alpha \leq 0.1$, a lot of false alarms appear, but, the detection performances for both the buddingtonite and kaolinite targets remain better than to those in Fig. 15.7.

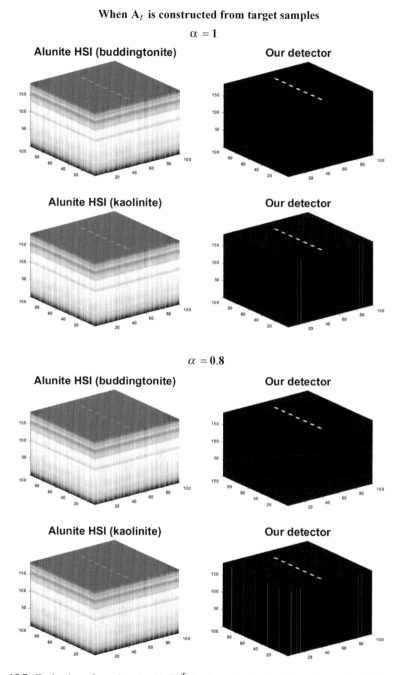

Fig. 15.7 Evaluation of our detector $(\mathbf{A}_t\mathbf{C})^T$ for detecting the buddingtonite and kaolinite target (for $\alpha \in [0.01, \ 1]$) when $\mathbf{A}_t$ is constructed from target samples in the online spectral libraries

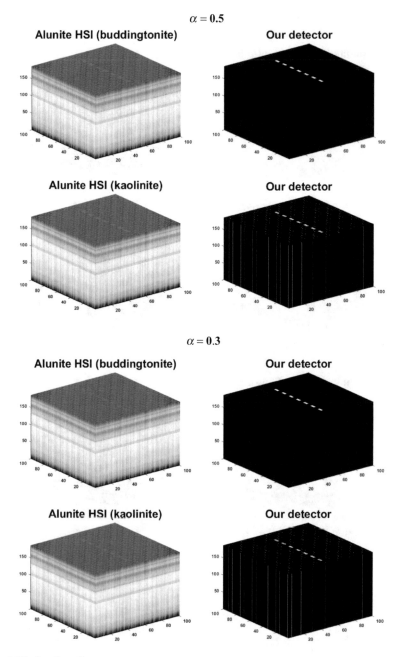

Fig. 15.7 (continued)

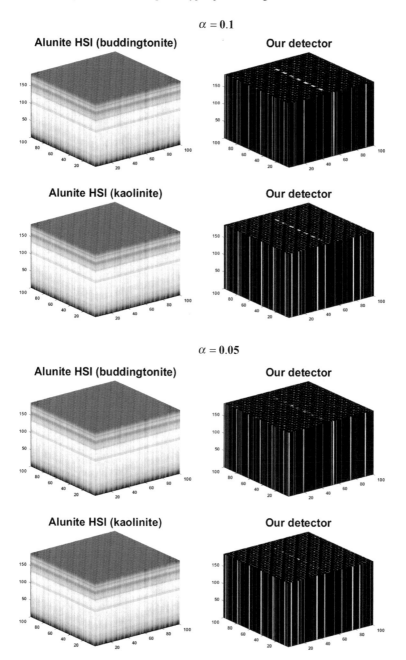

Fig. 15.7 (continued)

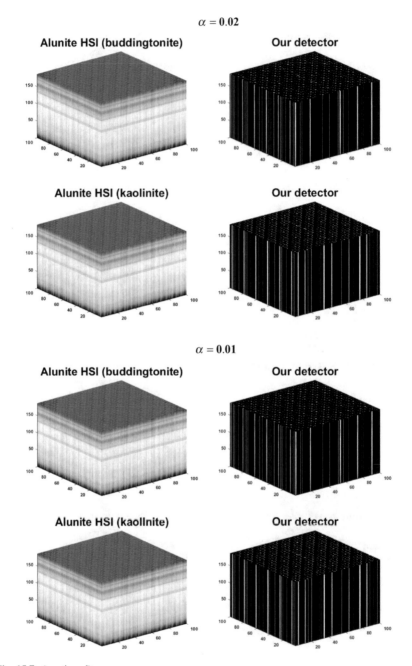

Fig. 15.7 (continued)

15.5 Conclusion and Future Work

In this chapter, the well-known robust principal component analysis (RPCA) is exploited for target detection in hyperspectral imagery. By making assumptions similar to those used in RPCA, a given hyperspectral image (HSI) has been decomposed into the sum of a low-rank background HSI and a sparse target HSI that only contains the targets (with the background is suppressed) [52]. In order to alleviate the inadequacy of RPCA on distinguishing the true targets from the background, we have incorporated into the RPCA imaging, the prior target information that can often be provided to the user. In this regard, we have constructed a pre-learned target dictionary $\mathbf{A}_t$, and thus, the given HSI is decomposed as the sum of a low-rank background HSI $\mathbf{L}$ and a sparse target HSI $(\mathbf{A}_t \mathbf{C})^T$, where $\mathbf{C}$ is a sparse activation matrix.

In this work, the sparse component $(\mathbf{A}_t \mathbf{C})^T$ was only the object of interest, and thus, used directly for the detection. More precisely, the targets are deemed to be present at the non-zero entries of the sparse target HSI. Hence, a novel target detector is developed, which is simply a sparse HSI generated automatically from the original HSI, but containing only the targets of interest with the background is suppressed.

The detector is evaluated on real experiments, and the results of which demonstrate its effectiveness for hyperspectral target detection, especially on detecting targets that have overlapping spectral features with the background.

The l_1 norm regularizer, a continuous and convex surrogate, has been studied extensively in the literature [79, 80] and has been applied successfully to many applications including signal/image processing, biomedical informatics, and computer vision [43, 81–84]. Although the l_1 norm based sparse learning formulations have achieved great success, they have been shown to be suboptimal in many cases [85–87], since the l_1 is still too far away from the ideal l_0 norm. To address this issue, many non-convex regularizers, interpolated between the l_0 norm and the l_1 norm, have been proposed to better approximate the l_0 norm. They include l_q norm ($0 < q < 1$) [88], Smoothly Clipped Absolute Deviation [89], Log-Sum Penalty [90], Minimax Concave Penalty [91], Geman Penalty [92, 93], and Capped-l_1 penalty [86, 87, 94].

In this regard, from problem (15.2), it will be interesting to use other proxies than the $l_{2,1}$ norm, closer to $l_{2,0}$, in order to probably alleviate the $l_{2,1}$ artifact and also the manual selection problem of both τ and λ. But although the non-convex regularizers (penalties) are appealing in sparse learning, it remains a very big challenge to solve the corresponding non-convex optimization problems.

Acknowledgements The authors would greatly thank Dr. Gregg A. Swayze from the United States Geological Survey (USGS) for his time in providing them helpful remarks about the cuprite data and especially on the buddingtonite, alunite, and kaolinite minerals. They would also like to thank the handling editors (Prof. Saurabh Prasad and Prof. Jocelyn Chanussot) and some other anonymous reviewers for the careful reading and helpful remarks/suggestions.

References

1. Shaw G, Manolakis D (2002) Signal processing for hyperspectral image exploitation. IEEE Signal Process Mag 19(1):12–16
2. Manolakis D, Marden D, Shaw G (2003) Hyperspectral image processing for automatic target detection applications. Linc Lab J 14(1):79–116
3. Manolakis DG, Lockwood RB, Cooley TW (2016) Hyperspectral imaging remote sensing: physics, sensors, and algorithms. Cambridge University Press
4. Zhang L, Zhang Q, Du B, Huang X, Tang YY, Tao D (2018) Simultaneous spectral-spatial feature selection and extraction for hyperspectral images. IEEE Trans Cybern 48(1):16–28
5. Zhang L, Zhang Q, Zhang L, Tao D, Huang X, Du B (2015) Ensemble manifold regularized sparse low-rank approximation for multiview feature embedding. Pattern Recognit48(10):3102 – 3112 (Discriminative feature learning from big data for visual recognition)
6. Bioucas-Dias JM, Plaza A, Camps-Valls G, Scheunders P, Nasrabadi N, Chanussot J (2013) Hyperspectral remote sensing data analysis and future challenges. IEEE Geosci Remote Sens Mag 1(2):6–36
7. Plaza A, Benediktsson JA, Boardman JW, Brazile J, Bruzzone L, Camps-Valls G, Chanussot J, Fauvel M, Gamba P, Gualtieri A, Marconcini M, Tilton JC, Trianni (2009) Recent advances in techniques for hyperspectral image processing. Remote Sens Environ 113:S110–S122, (Imaging spectroscopy special issue)
8. Du Q, Zhang L, Zhang B, Tong X, Du P, Chanussot J (2013) Foreword to the special issue on hyperspectral remote sensing: theory, methods, and applications. IEEE J Sel Top Appl Earth Obs Remote Sens 6(2):459–465
9. Gader PD, Chanussot J (2021) Understanding hyperspectral image and signal processing, Wiley
10. Dalla Mura M, Benediktsson JA, Chanussot J, Bruzzone L (2011) The evolution of the morphological profile: from panchromatic to hyperspectral images, Springer, Berlin, Heidelberg, pp 123–146
11. Prasad S, Bruce LM, Chanussot J (2011) Optical remote sensing: advances in signal processing and exploitation techniques. Augmented vision and reality. Springer, Heidelberg
12. Chanussot J, Collet C, Chehdi K (2009) Multivariate image processing, augmented vision and reality. STE Ltd, UK and Wiley, USA
13. Pons JMF (2014) Robust target detection for hyperspectral imaging. PhD thesis, Centrale-Supélec
14. Manolakis D, Lockwood R, Cooley T, Jacobson J (2009) Is there a best hyperspectral detection algorithm? In: Proceeding of SPIE 7334, algorithms and technologies for multispectral, hyperspectral, and ultraspectral imagery XV, 733402, 27 April 2009
15. Villa A, Chanussot J, Benediktsson JA, Jutten C (2011) Unsupervised classification and spectral unmixing for sub-pixel labelling. In: IEEE International geoscience and remote sensing symposium, July, 71–74
16. Yokoya N, Chanussot J, Iwasaki A (2014) Nonlinear unmixing of hyperspectral data using semi-nonnegative matrix factorization. IEEE Trans Geosci Remote Sens 52(2):1430–1437
17. Villa A, Chanussot J, Benediktsson JA, Jutten C (2011) Spectral unmixing for the classification of hyperspectral images at a finer spatial resolution. IEEE J Sel Top Signal Process 5(3):521–533
18. Licciardi GA, Villa A, Khan MM, Chanussot J (2012) Image fusion and spectral unmixing of hyperspectral images for spatial improvement of classification maps. In: IEEE International and geosciencing remote sensing symposium, July, pp 7290–7293
19. Patel NK, Patnaik C, Dutta S, Shekh AM, Dave AJ (2001) Study of crop growth parameters using airborne imaging spectrometer data. Int J Remote Sens 22(12):2401–2411
20. Datt B, McVicar TR, van Niel TG, Jupp DLB, Pearlman JS (2003) Preprocessing EO-1 Hyperion hyperspectral data to support the application of agricultural indexes. IEEE Trans Geosci Remote Sens 41:1246–1259
21. Hörig B, Kühn F, Oschütz F, Lehmann F (2001) HyMap hyperspectral remote sensing to detect hydrocarbons. Int J Remote Sens 22:1413–1422

22. Manolakis D, Shaw G (2002) Detection algorithms for hyperspectral imaging applications. Signal Process Mag IEEE 19(1):29–43
23. Stein DWJ, Beaven SG, Hoff LE, Winter EM, Schaum AP, Stocker AD (2002) Anomaly detection from hyperspectral imagery. IEEE Signal Process Mag 19(1):58–69
24. Eismann MT, Stocker AD, Nasrabadi NM (2009) Automated hyperspectral cueing for civilian search and rescue. Proc IEEE 97(6):1031–1055
25. Manolakis D, Truslow E, Pieper M, Cooley T, Brueggeman M (2014) Detection algorithms in hyperspectral imaging systems: an overview of practical algorithms. IEEE Signal Process Mag 31(1):24–33
26. Frontera-Pons J, Pascal F, Ovarlez J-P (2017) Adaptive nonzero-mean Gaussian detection. IEEE Trans Geosci Remote Sens 55(2):1117–1124
27. Frontera-Pons J, Veganzones MA, Pascal F, Ovarlez J-P (2016) Hyperspectral anomaly detectors using robust estimators. IEEE J Sel Top Appl Earth Obs Remote Sens 9(2):720–731
28. Frontera-Pons J, Ovarlez J-P, Pascal F (2017) Robust ANMF detection in noncentered impulsive background. IEEE Signal Process Lett 24(12):1891–1895
29. Frontera-Pons J, Veganzones MA, Velasco-Forero S, Pascal F, Ovarlez JP, Chanussot J (2014) Robust anomaly detection in hyperspectral imaging. In: IEEE geoscience and remote sensing symposium, July, pp 4604–4607
30. Cavalli RM, Licciardi GA, Chanussot J (2013) Detection of anomalies produced by buried archaeological structures using nonlinear principal component analysis applied to airborne hyperspectral image. IEEE J Sel Top Appl Earth Obs Remote Sens 6(2):659–669
31. Manolakis D, Shaw G, Keshava N (2000) Comparative analysis of hyperspectral adaptive matched filter detectors. In: Proceedings of SPIE 4049, algorithms for multispectral, hyperspectral, and ultraspectral imagery VI, vol 2
32. Nasrabadi NM (2008) Regularized spectral matched filter for target recognition in hyperspectral imagery. IEEE Signal Process Lett 15:317–320
33. Kraut S, Scharf LL (1999) The CFAR adaptive subspace detector is a scale-invariant GLRT. On Signal Process, IEEE Trans 47(9):2538–2541
34. Kelly EJ (1986) An adaptive detection algorithm. IEEE Trans Aerosp Electron Syst 23(1):115–127
35. Ledoit O, Wolf M (2004) A well-conditioned estimator for large-dimensional covariance matrices. J Multivar Anal 88(2):365–411
36. Ledoit O, Wolf M, (2003) Honey, I shrunk the sample covariance matrix. UPF Economics and Business Working Paper, no. 691
37. Bitar AW, Ovarlez J-P, Cheong L-F (2017) Sparsity-based cholesky factorization and its application to hyperspectral anomaly detection. In: IEEE workshop on computational advances in multi-sensor adaptive processing (CAMSAP-17). Curaçao, Dutch Antilles
38. Chen Y, Wiesel A, Hero AO (2010) Robust shrinkage estimation of high-dimensional covariance matrices. In: IEEE Sensor Array and Multichannel Signal Processing Workshop, OCt, 189–192
39. Pascal F, Chitour Y (2014) Shrinkage covariance matrix estimator applied to stap detection. In: IEEE workshop on statistical signal processing (SSP), June, 324–327
40. Pascal F, Chitour Y, Quek Y (2014) Generalized robust shrinkage estimator and its application to stap detection problem. IEEE Trans Signal Process 62(21):5640–5651
41. Healey G, Slater D (1999) Models and methods for automated material identification in hyperspectral imagery acquired under unknown illumination and atmospheric conditions. IEEE Trans Geosci Remote Sens 37(6):2706–2717
42. Thai B, Healey G (2002) Invariant subpixel material detection in hyperspectral imagery. IEEE Trans Geosci Remote Sens 40(3):599–608
43. Wright J, Yang AY, Ganesh A, Sastry SS, Ma Y (2009) Robust face recognition via sparse representation. IEEE Trans Pattern Anal Mach Intell 31(2):210–227
44. Basri R, Jacobs DW (2003) Lambertian reflectance and linear subspaces. IEEE Trans Pattern Anal Mach Intell 25(2):218–233

45. Pillai JK, Patel VM, Chellappa R (2009) Sparsity inspired selection and recognition of iris images. In: 2009 IEEE 3rd international conference on biometrics: theory, applications, and systems, Sept, pp 1–6
46. Hang X, Wu F-X (2009) Sparse representation for classification of tumors using gene expression data. J Biomed Biotechnol 6
47. Guo Z, Wittman T, Osher S (2009) L1 unmixing and its application to hyperspectral image enhancement
48. Bioucas-Dias JM, Plaza A, Dobigeon N, Parente M, Du Q, Gader P, Chanussot J (2012) Hyperspectral unmixing overview: geometrical, statistical, and sparse regression-based approaches. IEEE J Sel Top Appl Earth Obs Remote Sens 5(2):354–379
49. Chen Y, Nasrabadi NM, Tran TD (2011) Sparse representation for target detection in hyperspectral imagery. IEEE J Sel Top Signal Process 5(3):629–640
50. Zhang Y, Du B, Zhang L (2015) A sparse representation-based binary hypothesis model for target detection in hyperspectral images. IEEE Trans Geosci Remote Sens 53(3):1346–1354
51. Bitar AW, Cheong L, Ovarlez J (2018) Target and background separation in hyperspectral imagery for automatic target detection. In: 2018 IEEE international conference on acoustics, speech and signal processing (ICASSP), pp 1598–1602
52. Bitar AW, Cheong L, Ovarlez J (2019) Sparse and low-rank matrix decomposition for automatic target detection in hyperspectral imagery. IEEE Trans Geosci Remote Sens 57(8):5239–5251
53. Candès EJ, Li X, Ma Y, Wright J (2011) Robust principal component analysis? J ACM 58(3):11:1–11:37
54. Wright J, Arvind G, Shankar R, Yigang P, Ma Y (2009) Robust principal component analysis: exact recovery of corrupted low-rank matrices via convex optimization. In: Bengio Y, Schuurmans D, Lafferty JD, Williams CKI, Culotta A (eds) Advances in neural information processing systems, vol 22. Curran Associates, Inc., pp 2080–2088
55. Chen S-Y, Yang S, Kalpakis K, Chang C (2013) Low-rank decomposition-based anomaly detection. In: Proceedings of SPIE 8743 algorithms and technologies for multispectral, hyperspectral, and ultraspectral imagery XIX, vol 8743, pp 1–7
56. Zhang Y, Du B, Zhang L, Wang S (2016) A low-rank and sparse matrix decomposition-based Mahalanobis distance method for hyperspectral anomaly detection. IEEE Trans Geosci Remote Sens 54(3):1376–1389
57. Xu Y, Wu Z, Li J, Plaza A, Wei Z (2016) Anomaly detection in hyperspectral images based on low-rank and sparse representation. IEEE Trans Geosci Remote Sens 54(4):1990–2000
58. Xu Y, Wu Z, Chanussot J, Wei Z (2018) Joint reconstruction and anomaly detection from compressive hyperspectral images using mahalanobis distance-regularized tensor RPCA. IEEE Trans Geosci Remote Sens 56(5):2919–2930
59. Xu Y, Wu Z, Chanussot J, Dalla Mura M, Bertozzi AL, Wei Z (2018) Low-rank decomposition and total variation regularization of hyperspectral video sequences. IEEE Trans Geosci Remote Sens 56(3):1680–1694
60. Bitar AW, Cheong L, Ovarlez J (2017) Simultaneous sparsity-based binary hypothesis model for real hyperspectral target detection. In: 2017 IEEE international conference on acoustics, speech and signal processing (ICASSP), pp 4616–4620
61. Veganzones MA, Simões M, Licciardi G, Yokoya N, Bioucas-Dias JM, Chanussot J (2016) Hyperspectral super-resolution of locally low rank images from complementary multisource data. IEEE Trans Image Process 25(1):274–288
62. Liu G, Lin Z, Yan S, Sun J, Yu Y, Ma Y (2013) Robust recovery of subspace structures by low-rank representation. IEEE Trans Pattern Anal Mach Intell 35(1):171–184
63. Zhou Z, Li X, Wright J, Candés E, Ma Y (2010) Stable principal component pursuit. In: 2010 IEEE International symposium on information theory, pp 1518–1522
64. Chen Z, Ellis DPW (2013) Speech enhancement by sparse, low-rank, and dictionary spectrogram decomposition. In: IEEE Workshop on applications of signal processing to audio and acoustics Oct, 1–4
65. Sun P, Qin J (2016) Low-rank and sparsity analysis applied to speech enhancement via online estimated dictionary. IEEE Signal Process Lett 23(12):1862–1866

66. Cai J-F, Candès EJ, Shen Z (2010) A singular value thresholding algorithm for matrix completion. SIAM J Optim 20(4):1956–1982
67. Candes EJ, Recht B, Exact low-rank matrix completion via convex optimization. In: 2008 46th Annual allerton conference on communication, control, and computing, Sept, pp 806–812
68. Lewis AS (2003) The mathematics of eigenvalue optimization. Math Program 97(1):155–176
69. Watson GA (1992) Characterization of the subdifferential of some matrix norms. Linear Algebr Its Appl 170:33–45
70. Boyd Stephen, Parikh Neal, Chu Eric, Peleato Borja, Eckstein Jonathan (2011) Distributed optimization and statistical learning via the alternating direction method of multipliers. Found Trends Mach Learn 3(1):1–122
71. Yang Junfeng, Yin Wotao, Zhang Yin, Wang Yilun (2009) A fast algorithm for edge-preserving variational multichannel image restoration. SIAM J Imaging Sci 2(2):569–592
72. Swayze GA, Clark RN, Goetz AFH, Livo KE, Breit GN, Kruse FA, Sutley SJ, Snee LW, Lowers HA, Post JL, Stoffregen RE, Ashley RP (2014) Mapping advanced argillic alteration at cuprite, nevada, using imaging spectroscopy. Econ Geol 109(5):1179
73. Airbone Visible/Infrared Imaging Spectrometer. https://aviris.jpl.nasa.gov
74. Swayze GA, Clark RN, Goetz AFH, Chrien TG, Gorelick NS (2003) Effects of spectrometer band pass, sampling, and signal-to-noise ratio on spectral identification using the tetracorder algorithm. J Geophys Res: Planets 108(E9):5105
75. Clark RN, Swayze GA, Livo KE, Kokaly RF, Sutley SJ, Dalton JB, McDougal RR, Gent CA, Imaging spectroscopy: earth and planetary remote sensing with the usgs tetracorder and expert systems. J Geophys Res: Planets 108(E12)
76. Berk A, Bernstein L, Robertson D (1989) MODTRAN: a moderate resolution model for LOW-TRAN 7. Tech. Rep. GL-TR-90-0122, Geophysics Laboratory, Bedford, MA
77. Clark RN, Swayze GA, Gallagher AJ, King TVV, Calvin WM (1993) The U.S. geological survey, digital spectral library: version 1: 0.2 to 3.0 micros. U.S. Geological Survey, Open file report
78. Baldridge AM, Hook SJ, Grove CI, Rivera G (2009) The ASTER Spectral Library Version 2.0. Remote Sens Environ 113:711–715
79. Tibshirani R (1996) Regression shrinkage and selection via the lasso. J R Stat Soc Ser B (Methodol) 58(1):267–288
80. Efron B, Hastie T, Johnstone I, Tibshirani R (2004) Least angle regression. Ann Stat 32(2):407–451
81. Shevade SK, Keerthi SS (2003) A simple and efficient algorithm for gene selection using sparse logistic regression. Bioinformatics 19(17):2246–2253
82. Amir B, Marc T (2009) A fast iterative shrinkage-thresholding algorithm for linear inverse problems. SIAM J Img Sci 2(1):183–202
83. Wright SJ, Nowak RD, Figueiredo MAT (2009) Sparse reconstruction by separable approximation. IEEE Trans Signal Process 57(7):2479–2493
84. Ye Jieping, Liu Jun (2012) Sparse methods for biomedical data. SIGKDD Explor Newsl 14(1):4–15
85. Candès Emmanuel J, Wakin Michael B, Boyd Stephen P (2008) Enhancing sparsity by reweighted l_1 minimization. J Fourier Anal Appl 14(5):877–905
86. Zhang Tong (2010) Analysis of multi-stage convex relaxation for sparse regularization. J Mach Learn Res 11:1081–1107
87. Zhang T (2013) Multi-stage convex relaxation for feature selection Bernoulli 19(5B):2277–2293
88. Foucart S, Lai MJ (2009) Sparsest solutions of underdetermined linear systems via l_q-minimization for $0 < q < 1$. Appl Comput Harmon Anal 26(3):395–407
89. Fan J, Li R (2001) Variable selection via nonconcave penalized likelihood and its oracle properties. J Am Stat Assoc 96(456):1348–1360
90. Candès E, Wakin MB, Boyd SP (2008) Enhancing sparsity by reweighted l1 minimization. J Fourier Anal Appl 14(5):877–905

91. Zhang CH (2010) Nearly unbiased variable selection under minimax concave penalty. Ann Statist 38(2):894–942
92. Geman D, Yang C (1995) Nonlinear image recovery with half-quadratic regularization. IEEE Trans Image Process 4(7)932–946, Jul
93. Trzasko J, Manduca A (2009) Relaxed conditions for sparse signal recovery with general concave priors. IEEE Trans Signal Process 57(11):4347–4354
94. Gong P, Ye J, Zhang C (2012) Multi-stage multi-task feature learning. In: Pereira F, Burges CJC, Bottou L, Weinberger KQ (eds) Advances in neural information processing systems vol 25. Curran Associates, Inc., pp 1988–1996

Index

© Springer Nature Switzerland AG 2020
S. Prasad and J. Chanussot (eds.), *Hyperspectral Image Analysis*,
Advances in Computer Vision and Pattern Recognition,
https://doi.org/10.1007/978-3-030-38617-7